Ecology of Tropical Cities, Volume I

Fabio Angeoletto • Piotr Tryjanowski
Mark D. E. Fellowes
Editors

Ecology of Tropical Cities, Volume I

Natural and Social Sciences Applied to the Conservation of Urban Biodiversity

Springer

Editors
Fabio Angeoletto
Universidade Federal de Rondonópolis
Rondonópolis, Brazil

Piotr Tryjanowski
Institute of Zoology
Poznań University of Life Sciences
Poznań, Poland

Mark D. E. Fellowes
Department of Biological Sciences
Royal Holloway University of London
Egham, Surrey, UK

ISBN 978-3-031-48011-9 ISBN 978-3-031-48012-6 (eBook)
https://doi.org/10.1007/978-3-031-48012-6

This Springer imprint is published by the registered company Springer Nature Switzerland AG
The registered company address is: Gewerbestrasse 11, 6330 Cham, Switzerland

To the Roman Emperor Tiberius Claudius Caesar Augustus Germanicus, the Conqueror of Britannia, scholar and friend of books. He triumphed over those who saw him as an idiot. To Hypatia of Alexandria, Camila Essy and Flora Angeoletto, extraordinary women who I will always follow.

Fabio Angeoletto

My grandparents not only gave me good genes but also instilled in me a curiosity about the world. I am very grateful to them for that.

Piotr Tryjanowski

To Fumio Taguchi and Phil Brennan, ornithologists and educators, who showed the way.

Mark D. E. Fellowes

Foreword

Urban Ecology and Biodiversity in the Tropics: Addressing the Northern Bias and Embracing Global Change

This is an important book that comes at just the right time. The topic—Ecology of Tropical Cities, Volumes I and II—is an urgent one. But the book is not only urgent, it is also exciting: It brings together many voices in more than 40 chapters; it employs a truly comprehensive view of urban ecology; and finally, it incorporates key connections with evolution, social equity, and governance. The book treats biodiversity as an integrated issue that reaches deep into the needs, challenges, and opportunities of diverse tropical societies.

In spite of the relevance of urban ecology to tropical cities, the science has been relatively neglected in the tropical realm. Of course, there is existing work of excellence and impact from tropical colleagues and places, but the majority of ecological work originates in the higher latitudes of the temperate and boreal North. This bias can no longer be tolerated. Indeed, this book powerfully corrects the northern bias in urban ecology. The authors of the chapters are largely residents of tropical countries. The book highlights the work and insights of both more senior researchers and newer generations of tropical experts. It will help expand the global network of urban ecologists, introducing a number of new colleagues to us all. The number of countries represented—29—is also impressive, which is highly appropriate given that something like 125 of the world's countries and territories have—at least in part—a tropical address. Of the inhabited continents, only Europe lacks tropical geography.

Another exciting thing about this book is that it acknowledges the tropics to be a complex crucible of change. Editors Angeoletto, Tryjanowski, and Fellowes have provided a view of tropical change that is not yet widely appreciated, but is quite breathtaking. Of course, global climate change bears down hard on the tropics, as is well known. One outcome of this will be the increasing hazard to tropical cities of climate-driven extremes such as storms, heatwaves, and droughts. But the dimensions and forces of this and other kinds of global change are usefully highlighted in framing the book. Notable is the projection that the extent of the tropics themselves will expand with climate heating. Furthermore, the continuing growth of the Earth's population will be significantly tropical in origin, and importantly, the urban fraction of this rapidly increasing population will be extraordinarily large. These burgeoning tropical cities are spreading into biodiversity hotspots, which is a practical motivation for the book.

Urban places, which include not only cities but increasingly heterogeneous urban-influenced regions, are also experiencing massive changes in their social components. As migration into cities continues apace due to migrants' desires for the opportunities and services available there, or as they flee social and climate disruptions elsewhere, several issues become clear. One is that the capacity for planning and governance of land allocation and infrastructure provision may be limited in many tropical cities. A consequence of the migrations combined with constrained municipal capacities is the importance of under-resourced and often under-represented informal settlements. Economic informality in its many dimensions can be associated with the bundle of social disruptions experienced and adaptations achieved by novel urban populations. All of these social shifts have implications for ecological services and hazards, and for biodiversity.

This cauldron of changes in the tropics is well met by the way the book brings knowledge together. The book uses a consistent social-ecological framing throughout. It sees urban ecology as a comprehensive science, informed and interacting with a wide variety of other sciences and practical professions. It avoids the uniformity of classical conceptions of the urban that are too dependent on industrial, colonial, and Northern models. This last point is especially important because of the diversity of urban dynamics, structures (or fabrics), motivations, opportunities, and needs across the similarly heterogeneous tropical regions. Finally, the cauldron bubbles with the increasingly clear need to confront the growing economic inequities and social exclusions across the world. Equity and justice suffuse the messages and insights of studies summarized in the book.

Readers who delve into this remarkable book will experience the excitement of new insights, new data, and perhaps meet new colleagues from tropical cities. But most importantly they will be invited to engage as scholars or actors in a crucial struggle. As the editors say in Chap. 1: "*The battle for biodiversity will be won or lost in cities*."

Distinguished Senior Scientist
Cary Institute of Ecosystem Studies
Millbrook, New York, NY, USA

Steward T. A. Pickett

Preface

Our Ambition is to Boost the Field of Tropical Urban Ecology

This book (Volumes I and II) is the result of a shared concern we believe strongly echoes with many others. The overlapping of mega-diverse and tropical countries shines light on a surprising pattern: almost all nations accounting for vast biodiversity are developing countries located in the tropics, such as Brazil, Mexico, India, and the Democratic Republic of Congo. These countries share the common feature of fast urbanization within what are incredibly diverse biomes where urban planning and management controls are often limited. This process leads to urban environments where environmental injustices are common. The socioeconomic status of neighborhoods in these cities oftentimes reflects their urban biological diversity; rich zones enjoy green landscapes, whereas the impoverished ones are covered by concrete.

Let's think about Brazil, for example, where most of its 5,570 municipalities are located in two global biodiversity hotspots, the Atlantic Forest and Cerrado biomes. The rapid degradation of these areas emphasizes the need of conservation strategies that go beyond traditional approaches, such as establishing protected areas that include urban spaces in efforts to enhance the conservation of wildlife and flora. It is alarming how many endangered species are found in these urban environments, yet lack protection, despite their visibility.

Thus, we are concerned with tropical cities that urgently need vastly more information, knowledge, and better methodological approaches to preserve their urban biodiversity and vital ecosystem services, which they provide to hundreds of millions of people. Furthermore, reducing the ecological footprint of tropical cities based on better environmental planning and management will be a critical step forward in global biodiversity conservation.

This book's main goal is to boost the visibility of tropical urban ecology. We ought to trigger the interest of young scientists living in tropical nations in this rising discipline, and to provide expert scientists with a broader understanding of the entangled interaction of social, economic, cultural, historical, political, urban planning, biological, evolutionary, and environmental factors influencing tropical urban ecosystems' capacity to support biodiversity. At the same time we are committed to promoting a broader international collaboration capable of avoiding the limitations of *helicopter research*, in order to ensure inclusive and equitable investigations and knowledge exchange moored in the very contexts of the countries involved. We hope to encourage a continuous and productive dialogue between scientists and professionals by sharing such rich knowledge with urban planners, managers, and policy-makers to help developing more diverse, biophilic, and environmentally equitable tropical cities.

Rondonópolis, Brazil — Fabio Angeoletto
Poznań, Poland — Piotr Tryjanowski
Egham, Surrey, UK — Mark D. E. Fellowes

Acknowledgments

This book was projected by more than 200 kind and talented scientists who we, editors, would like to deeply thank. Organizing ideas, texts, and the contributions from such a special team was a herculean task performed under the storm set by a global pandemic. This ambitious project, which is an overview of the ecology in thousands of cities in the tropics, would hardly be successful without the support by João V. Pildervasser (Associate Editor, Life Sciences, Springer Nature), who always showed us a path when we could hardly see one.

We also thank dozens of reviewers whose close reading and recommendations aimed at chapters' improvement were essential. We are deeply honored with the Foreword signed by Professor Steward Pickett, who is one of the most brilliant urban ecologists ever. His work has ultimately boosted studies on cities' ecology and biodiversity at global scale.

Several scientists enlightened us with advises throughout this odyssey and we herein point out our gratitude to Luís Otávio Bau Macedo, Ian MacGregor-Fors, Franco Leandro de Souza, Domingos Sávio Barbosa, Normandes Matos da Silva, Angelo Serpa, Claudio Angelo Correa Gonzaga, Juan Pedro Ruiz Sanz, Alfan A. Rija, Nishant Kumar, Dhanya Bhaskar, and Marc TJ Johnson. Thank you very much Swapnil Kumbhojkar, Reuven Yosef, Abhinav Mehta, and Shrey Rakholia for the cover picture—the Zara leopard, in Jaipur, India. This picture beautifully illustrates the contents in this book.

We also thank Sonia Essy, Antonio Angeoletto (in memoriam), Helena Angeoletto, Franciele Finck, the colleagues at "Programa de Pós-Graduação em Gestão e Tecnologia Ambiental da UFR," Nestor Alexandre Perehouskei (in memoriam), and Bela Serpa (in memoriam). To our fellows at GoodDeal Consultoria Linguistíca. Thanks to directors of "Escola da Floresta Olho da Coruja," Adroaldo Sturmer and Clarisse Hendges Sturmer; to the directors of "Grupo Simpatia de Comunicação," Eloy M. Scheibe and Dirci Ledur, and to journalist Plínio Bortolotti, who have given us precious room in the press to outspread the knowledge we gathered in this book.

Atrocaducapacaustiduplielastifeliferofugahistoriloqualubrimendi multipliorganiperiodi plastipublirapareciprorustisagasimplitenaveloveraviva univora
cidade
city
cité

atrocidade, atrocity, atrocité;
caducidade, caducity, caducité;
capacidade, capacity, capacité;
causticidade, causticity, causticité;
duplicidade, duplicity, duplicité;
elasticidade, elasticity, elasticité;
felicidade, felicity, felicité;
ferocidade, ferocity, ferocité;
fugacidade, fugacity, fugacité;
historicidade, historicity, historicité;
loquacidade, loquacity, loquacité;
lubricidade, lubricity, lubricité;
mendicidade, mendicity, mendicité;
multiplicidade, multiplicity, multiplicité;
organicidade, organicity, organicité;
periodicidade, periodicity, periodicité;
plasticidade, plasticity, plasticité;
publicidade, publicity, publicité;
rapacidade, rapacity, rapacité;
reciprocidade, reciprocity, reciprocité;
rusticidade, rusticity, rusticité;
sagacidade, sagacity, sagacité;
simplicidade, simplicity, simplicité;
tenacidade, tenacity, tenacité;
velocidade, velocity, velocité;
veracidade, veracity, veracité;
vivacidade, vivacity, vivacité;
unicidade, unicity, unicité;
voracidade, verocity, verocité.

Augusto de Campos, Brazilian poet and visual artist

Contents

Contributors

Luis Abdala-Roberts Departamento de Ecología Tropical, Campus de Ciencias Biológicas y Agropecuarias, 17 Universidad Autónoma de Yucatán, Mérida, Yucatán, Mexico

Loyde V. de Abreu-Harbich Faculty of Architecture and Urban Planning, Mackenzie Presbyterian University, São Paulo, Brazil

Larissa Dall Agnol Programa de Pós-Graduação em Gestão e Tecnologia Ambiental, Universidade Federal de Rondonópolis, Rondonópolis, MT, Brazil

Philips Akinwole Biology Department, DePauw University, Greencastle, IN, USA

Fabio Angeoletto Programa de Pós-Graduação em Geografia da Universidade Federal de Rondonópolis, Mato Grosso, Brazil

Programa de Pós-Graduação em Gestão e Tecnologia Ambiental, Universidade Federal de Rondonópolis, Rondonópolis, Brazil

PUMAS, Pessoas Unidas pelo Meio Ambiente, Agricultura & Sociedade, Chapada, Brazil

Diego F. Angulo Interdisciplinary Center for Biological and Human Research, Faculty of Chemical Sciences, Benito Juárez Autonomous University of Oaxaca, Oaxaca, Mexico

Domingos Sávio Barbosa Programa de Pós-Graduação em Gestão e Tecnologia Ambiental, Universidade Federal de Rondonópolis, Rondonópolis, MT, Brazil

Gislaine Elizete Beloto Departamento de Arquitetura e Urbanismo, Universidade Estadual de Maringá, Maringá, PR, Brazil

Dhanya Bhaskar Indian Institute of Forest Management, Bhopal, India

Daniela Biondi Universidade Federal do Paraná, Curitiba, Brazil

Rogério Bobrowski Univeridade Estadual do Centro Oeste, Irati, Brazil

Martha Bonilla-Moheno Red de Ambiente y Sustentabilidad, Instituto de Ecología, A.C. (INECOL), Xalapa, Mexico

Marcos Clair Bovo Universidade Estadual do Paraná, Campo Mourão, Brazil

Aude E. Caizergues Department of Biology, University of Toronto Mississauga, Mississauga, ON, Canada

Rosa Imelda Rojas Caldelas Universidad Autónoma Baja California, Mexicali, Mexico

Diego Carmona Departamento de Ecología Tropical, Campus de Ciencias Biológicas y Agropecuarias, 17 Universidad Autónoma de Yucatán, Mérida, Yucatán, Mexico

Roberta Mendonça De Carvalho University of Pittsburgh, Urban Studies Program and Center for Latin American Studies, Pittsburgh, PA, USA

Silvia Méri Carvalho Universidade Estadual de Ponta Grossa, Ponta Grossa, Paraná, Brazil

Mariana Chávez-Pesqueira Centro de Investigación Científica de Yucatán, Unidad de Recursos Naturales, Mérida, Yucatán, Mexico

Renata Michelon Cocco Universidade Federal de Santa Maria (UFSM), Santa Maria, RS, Brazil

Mayara Henriques Coimbra Departamento de Arquitetura e Urbanismo, Universidade Estadual de Maringá, Maringá, PR, Brazil

Mattheau Comerford Department of Biology, University of Massachusetts Boston, Boston, MA, USA

Ek Del-Val Instituto de Investigaciones en Ecosistemas y Sustentabilidad, Universidad Nacional Autónoma de México, México City, México

Martín Andrés Díaz Universidad Nacional de La Matanza, San Justo, Argentina

Taise Ernestina Prestes Nogueira Duarte SEDUC MT Secretaria de Estado de Educação de Mato Grosso, Rondonópolis, Mato Grosso, Brazil

Marcin K. Dyderski Institute of Dendrology, Polish Academy of Sciences, KórnikPoland

Karen Dyson Dendrolytics, Seattle, WA, USA

Zingara Rocio dos Santos Eurich Universidade Estadual de Ponta Grossa, Ponta Grossa, Brazil

Mark D.E. Fellowes Department of Biological Sciences, Royal Holloway University of London, Egham, Surrey, UK

Antonio Conceição Paranhos Filho Universidade Federal de Mato Grosso do Sul, Campo Grande, MS, Brazil

Christopher Frost BIO5 Institute, University of Arizona, Tucson, AZ, USA

Letícia de Castro Gabriel Universidade Federal de Santa Maria (UFSM), Santa Maria, RS, Brazil

Spyros Gkelis Department of Botany, School of Biology, Aristotle University of Thessaloniki, Thessaloniki, Greece

William Godsoe Department of Pest Management and Conservation Lincoln University, Lincoln, New Zealand

Cláudio Angelo Corrêa Gonzaga Ministério Público do Estado de Mato Grosso, Itiquira, MT, Brazil

Programa de Pós-Graduação em Ciência Ambiental, Universidade de São Paulo, São Paulo, Brazil

César González-Lagos Departamento de Ciencias, Facultad de Artes Liberales, Universidad Adolfo Ibañéz, Santiago, Chile

Centro de Ecología Aplicada y Sustentabilidad (CAPES), Santiago, Chile

Kiyoko M. Gotanda Department of Biological Sciences, Brock University, St. Catharines, ON, Canada

Eliane Guaraldo Universidade Federal de Mato Grosso do Sul, Campo Grande, MS, Brazil

Urvi Gupta Wildlife Institute of India, Dehradun, Uttarakhand, India

Warnell School of Forestry and Natural Resources, University of Georgia, Athens, GAUSA

Gustavo Hallwass Programa de Pós-Graduação em Ecologia Aquática e Pesca, Universidade Federal do Pará (UFPA), Belém, Pará, Brazil

Instituto de Ciência, Tecnologia e Inovação, Programa de Pós-Graduação em Ecologia Aplicada, Universidade Federal de Lavras (UFLA), São Sebastião do Paraíso, Minas Gerais, Brazil

Glen Ray Hood Department of Biological Sciences, Wayne State University, Detroit, MI, USA

Carlos Iñiguez-Armijos Laboratorio de Ecología Tropical y Servicios Ecosistémicos (EcoSs-Lab), Universidad Técnica Particular de Loja, Loja, Ecuador

Marc T. J. Johnson Department of Biology, Centre for Urban Environments University of Toronto Mississauga, Mississauga, ON, Canada

Michael Just Training Lands and Heritage Branch, Engineer Research and Development Center, Vicksburg, MS, USA

Friedrich W. Keppeler Programa de Pós-Graduação em Ecologia Aquática e Pesca, Universidade Federal do Pará (UFPA), Belém, Pará, Brazil

Departamento de Engenharia Civil, Universidade Federal do Rio Grande do Norte (UFRN), Natal, Rio Grande do Norte, Brazil

Laboratório de Ciência Marinha - LABOMAR, Universidade Federal do Ceará (UFC), Fortaleza, Ceará, Brazil

Luan da Silva Klebers Universidade Federal de Santa Maria (UFSM), Santa Maria, RS, Brazil

Nishant Kumar Wildlife Institute of India, Dehradun, Uttarakhand, India

Department of Biology, Edward Grey Institute of Field Ornithology, University of Oxford, Oxford, UK

Mansfield College, University of Oxford, Oxford, UK

School of Human Ecology, Dr B R Ambedkar University Delhi, Delhi, India

THINKPAWS Sustainability Research Foundation, Delhi, India

National Centre for Biological Sciences, Tata Institute of Fundamental Research, Bengaluru, India

Lucila C. Labaki Faculty of Civil Engineering and Architecture and Urban Planning, State University of Campinas, Campinas, Brazil

Arvind Lakshmisha University of Kassel, Witzenhausen, Germany

Azim Premji University, Bengaluru, India

Jalene M. LaMontagne Department of Biology, University of Missouri-St. Louis, St. Louis, MO, USA

Whitney R. Harris World Ecology Center, University of Missouri-St. Louis, St. Louis, MO, USA

Science & Conservation Division, Missouri Botanical Garden, St. Louis, MO, USA

Carlos Lara Departamento de Ecología, Facultad de Ciencias, Universidad Católica de la Santísima Concepción, Concepción, Chile

Alice Rodrigues Lautert Universidade Federal de Santa Maria (UFSM), Santa Maria, RS, Brazil

Adrián Lázaro-Lobo Biodiversity Research Institute (IMIB), CSIC-University of Oviedo-Principality of Asturias, Mieres, Spain

Deleon Leandro Instituto de Biociências (IB), Universidade Federal de Mato Grosso, Cuiabá, Brazil

Eliana Cardoso-Leite Universidade Federal de São Carlos, Sorocaba, Brazil

Ian MacGregor-Fors Ecosystems and Environment Research Programme, Faculty of Biological and Environmental Sciences, University of Helsinki, Lahti, Finland

Harsha Malhotra Wildlife Institute of India, Dehradun, Uttarakhand, India

Homi Bhabha Centre for Science Education, Tata Institute of Fundamental Research, Mumbai, India

Lindon Fonseca Matias Universidade Estadual de Campinas, UNICAMP, Campinas, SP, Brazil

Andreas Matzarakis Faculty of Environment and Natural Resources, University of Freiburg, Freiburg, Germany

Democritus University of Thrace, Komotini, Greece

Amee Mehta Wildlife Institute of India, Dehradun, Uttarakhand, India

Karin Schwabe Meneguetti Departamento de Arquitetura e Urbanismo, Universidade Estadual de Maringá, Maringá, PR, Brazil

Thomas Merritt School of Natural Sciences, Laurentian University, Sudbury, ON, Canada

Camila Leonardo Mioto Programa de Pós-Graduação em Gestão e Tecnologia Ambiental, Universidade Federal de Rondonópolis, Rondonópolis, MT, Brazil

Xoaquín Moreira Misión Biológica de Galicia (MBG), Consejo Superior de Investigaciones Científicas, Pontevedra, Spain

Rodrigo Fernando Moro HCVRN Environment Consultant, Curitiba, PR, Brazil

Rosemeri Segecin Moro Universidade Estadual de Ponta Grossa, Ponta Grossa, Paraná, Brazil

Harini Nagendra Azim Premji University, Bengaluru, India

Sergio S. Negri Graduate Program in Geography, Federal University of Rondonópolis, Rondonópolis, Brazil

Cathiane M. de Oliveira Graduate Program in Geography, Federal University of Rondonópolis, Rondonópolis, Brazil

Victor Parra-Tabla Departamento de Ecología Tropical, Campus de Ciencias Biológicas y Agropecuarias, 17 Universidad Autónoma de Yucatán, Mérida, Yucatán, Mexico

Juraj Paule Botanischer Garten und Botanisches Museum Berlin, Freie Universität Berlin, Berlin, Germany

Nestor Alexandre Perehouskei In Memoriam, Maringá, Brazil

Dhonatan Diego Pessi Programa de Pós-Graduação em Gestão e Tecnologia Ambiental, Universidade Federal de Rondonópolis, Rondonópolis, MT, Brazil

Vera Pfeiffer College of Veterinary Medicine, Washington State University, Washington, DC, USA

Luis Guilherme Aita Pippi Universidade Federal de Santa Maria (UFSM), Santa Maria, RS, Brazil

Tobias Plieninger University of Kassel, Witzenhausen, Germany

University of Göttingen, Göttingen, Germany

Ilga Porth Centre d'étude de la forêt (CEF), Université Laval, Quebec City, QC, Canada

Dulcina de Aquino Hernandez de Oliveira Queiroz Universidade Estadual de Ponta Grossa, Ponta Grossa, PR, Brazil

Renato Leão Rego Departamento de Arquitetura e Urbanismo, Universidade Estadual de Maringá, Maringá, PR, Brazil

Diana Rennison School of Biological Sciences, University of California San Diego, La Jolla, CA, USA

Evandro Retamero Rodrigues Prefeitura Municipal de Ortigueira, Ortigueira, PR, Brazil

José Guilherme Roquette CAEx Ambiental, Ministério Público do Estado de Mato Grosso, Cuiabá, Brazil

César Ángel Peña Salmón UABC-Universidad Autónoma de Baja California, Mexicali, Mexico

Camilo E. Sánchez-Sarria Red de Ambiente y Sustentabilidad, Instituto de Ecología, A.C. (INECOL), Xalapa, Mexico

James Santangelo Department of Integrative Biology, University of California Berkeley, Berkeley, CA, USA

Jeater W. M. C. Santos Graduate Program in Geography, Federal University of Rondonópolis, Rondonópolis, Brazil

Fernando Scherner Pell Frischmann, LondonUK

Adam C. Schneider Department of Biology, University of Wisconsin-La Crosse, La Crosse, WI, USA

Angelo Serpa Universidade Federal da Bahia and Researcher Level 1B, with a Research Productivity Fellowship from the National Council for Scientific and Technological Development, Salvador, Bahia, Brazil

Normandes Matos da Silva Programa de Pós-Graduação em Gestão e Tecnologia Ambiental, Universidade Federal de Rondonópolis, Rondonópolis, Brazil

Renato A. M. Silvano Departamento de Ecologia e Programa de Pós-Graduação em Ecologia—IB, Universidade Federal do Rio Grande do Sul (UFRGS), Porto Alegre, Rio Grande do Sul, Brazil

Paula Sosenski Departamento de Ecología Tropical, Campus de Ciencias Biológicas y Agropecuarias, 17 Universidad Autónoma de Yucatán, Mérida, Yucatán, Mexico

Marco Antonio M. de Souza Graduate Program in Geography, Federal University of Rondonópolis, Rondonópolis, Brazil

Claudio Fabian Szlafzstein Federal University of Pará, Center for Advanced Amazonian Studies (NAEA), Belém, Brazil

Pramila Thapa University of Kassel, Witzenhausen, Germany

Leuphana University of Lueneburg, Lueneburg, Germany

Andreas Thiel University of Kassel, Witzenhausen, Germany

Piotr Tryjanowski Institute of Zoology, Poznan University of Life Sciences, Poznań, Poland

Marcelo C. Vacchiano Ministério Público do Estado de Mato Grosso, Cuiabá, Brazil

Luis-Bernardo Vázquez ECOSUR, San Cristóbal de las, Casas, Mexico

Raquel Weiss Universidade Federal de Santa Maria (UFSM), Santa Maria, RS, Brazil

Kaitlin Stack Whitney Science, Technology & Society Department, Rochester Institute of Technology, Rochester, NY, USA

Part I
Human Factors and Their Influence on Urban Ecosystems

Chapter 1
The Ecology and Biodiversity of Tropical Cities are as Important as they are Unknown

Mark D. E. Fellowes, Piotr Tryjanowski, and Fabio Angeoletto

Abstract Biodiversity reaches its zenith in the tropics. At the same time, tropical cities are growing dramatically and usually with a low capacity for planning and management for biological conservation. A better knowledge of the ecology of these cities is essential to avoid or at least reduce the impacts of rapid urbanization on nature, to conserve biodiversity and to guarantee access to the benefits of nature— their ecosystem services—to citizens. However, a huge scientific bias is evident. While we know that the the biological diversity of the tropics is relatively poorly studied, and the biodiversity and resultating ecology of tropical cities is even less well understood. In this chapter, we will present the book, its two volumes, its objectives and the main themes covered throughout its four parts.

Keywords Urban ecology · Urban biodiversity · Tropical cities · Tropics · Urbanization in the tropics · Tropical urban ecology · Urban biodiversity conservation · Urban planning · Environmental injustice · Urban ecosystem services · Urban green infrastructures

M. D. E. Fellowes
Department of Biological Sciences, Royal Holloway University of London, Egham, Surrey, UK

P. Tryjanowski
Institute of Zoology, Poznan University of Life Sciences, Poznań, Poland

F. Angeoletto (✉)
Programa de Pós-Graduação em Geografia da Universidade Federal de Rondonópolis, Mato Grosso, Brasil

Programa de Pós-Graduação em Gestão e Tecnologia Ambiental, Universidade Federal de Rondonópolis, Rondonópolis, Brasil

PUMAS, Pessoas Unidas pelo Meio Ambiente, Agricultura & Sociedade, Chapada, Brasil
e-mail: angeoletto@ufr.edu.br

F. Angeoletto et al. (eds.), *Ecology of Tropical Cities*,
https://doi.org/10.1007/978-3-031-48012-6_1

1 Homo Urbanus?

We live in an urban world (Angeoletto, 2012; Melchiorri et al., 2018; Goddard et al., 2021). This is a new phenomenon; data collated by Roser and Ritchie (2019) paints a clear picture of change. Fifty years ago as we write, an estimated 1.45 billion people lived in urban areas, and 2.46 billion people lived in non-urban areas. Populations living in both increased, but more rapidly in urban areas, so that in 2007 humanity transitioned from being predominantly rural, to being an urban species. By 2017 an estimated 55% of people lived in urban areas.

This transition has not been equal across the globe (Roser & Ritchie, 2019; United Nations, 2022). Of the larger countries, those in mainland North America south to Mexico (and including in the Caribbean, Cuba), Uruguay, Venezuela, Chile and Argentina in South America, much of Europe (with the exceptions of Ireland, Portugal, Norway and what were then considered the "Eastern Block" countries) including the USSR, Japan, Australia and New Zealand had all passed 50% urban populations by 1960. In 2020, it is largely east Africa, countries in the southern Sahara, south Asia, Laos, Myanmar and Vietnam in south-east Asia, and more widely Papua New Guinea (the least urban country in the world, with 13.4% of its population living in urban areas) in Australasia and Guyana in South America. This contrasts with countries like Singapore and Kuwait (100% urban population), Belgium (98%), Uruguay (95.5%) and Iceland (93.9%).

What is astonishing in some countries is just how quickly this transition has occurred, with countries like Gabon (1960: 17.4% urban; 2020: 90.1% urban), Oman (1960: 16.4%; 2020: 86.3%), Botswana (1960: 3.1%; 2020: 70.1%) and Angola (1960: 10.4%; 2020: 66.9%) showing enormous change in a lifetime. No country has shown a pattern of deurbanization in the same time period. Urbanisation is associated with deep societal change; a move from an agrarian society to a more industrial or service focused society. On average 33.5% of the population of low income (as defined by the UN) countries live in urban areas, contrasting with 81.8% of people in high income countries (Roser & Ritchie, 2019).

2 Urban Ecosystems Are Profoundly Different

Urban ecosystems are inherently artificial, not only in terms of their artificial physical infrastructure, but also in terms of the mix of remaining species, often highly managed, often introduced (Forman, 2014). Urbanization is synonymous with intensive land-use change, altering existing habitats, patterns of biodiversity, local climate, and water and nutrient cycles. Urban ecosystems are exposed to what can be extreme levels of air, water, soil and noise pollution (Forman, 2014). Each has implications for the species that can inhabit the region. The increased temperature of cities affects everything from the timing of bud burst and pollination (Liang et al., 2023) through to the spread of sexually transmitted diseases among ladybirds

(Welch et al., 2001) and changing evolutionary responses to herbivory (Santangelo et al., 2022; Beridze et al., 2023; Leandro et al., 2025, this volume). But perhaps the greatest challenge to biodiversity is through habitat fragmentation. Urban areas are predominantly constructed of impervious materials, and in the centre of many cities the challenge is finding any green space. Many species are unable to adapt to this loss of suitable habitat, as patterns of gene flow, dispersal, and interactions with other species are impaired, so the ratio of generalists and invasive species increases at the cost of fewer local specialists and native species (Rocha & Fellowes, 2020; Santangelo et al., 2022; Caizergues et al., 2024).

As a result, urban ecosystems are relatively homogenous. But they do not have to be. On occasion, urban habitats provide habitats for species that are scarce elsewhere. Moreover, urban ecosystems are managed ecosystems, and can be adapted for both people and nature, should we have the ambition and the resources to do so. Consider Singapore, a country in a city, where all inhabitants live in urban areas. Singapore has put the support of biodiversity at the heart of urban planning, transforming the city from what was a highly polluted region into one described as 'a city in a garden' (National Parks Board, 2019). While Singapore is not perfect, the effort to engage with biodiversity in policy and decision-making, in education, and in protecting nature shows that change is possible.

3 The Benefits of Thinking Like an Ecologist

It is tempting to despair at the degraded nature of urban ecosystems, but there is beauty among the concrete. Urban ecosystems can play a vital role in creating livable, sustainable, and resilient cities.

Urban ecosystems, particularly in the form of urban parks, nature reserves, brownfield sites and in particular domestic gardens and yards, serve as crucial habitats for diverse flora and fauna, contributing to the conservation of local and sometimes even endangered species. Preserving and enhancing urban green spaces can help mitigate biodiversity loss and maintain ecological balance within urban environments (Soanes et al., 2023; Lima et al., 2023; Moroń et al., 2024. It is here that people come into contact with nature, and there is little doubt of the physical and mental health benefits of access to green spaces within urban areas. While challenging to evidence (Lee & Maheswaran, 2011), urban ecosystems provide opportunities for physical activities, stress reduction, and social interactions, enhancing overall well-being and quality of life for urban dwellers (Browning et al., 2022). They also provide a key educational opportunity, as we see a continued loss in awareness of the natural world and the benefits of nature in each new generation. We can only care about what we know and what we value. It is not just individual benefits, as these green areas serve as gathering spaces, fostering social interactions and community engagement. Parks, gardens, and green corridors create opportunities for people to come together, strengthening social bonds and fostering a sense of belonging within urban communities (Browning et al., 2022). These areas also offer educational and recreational opportunities. Natural environments in cities provide spaces for environmental education, outdoor learning, and leisure activities. Society benefits from green space, connecting people with nature and fostering environmental awareness (Naeem et al., 2016).

But there are many practical benefits of a more diverse and green urban ecology. Urban ecosystems provide services such as air purification, water filtration, climate regulation, and waste decomposition. Trees and vegetation in cities mitigate air pollution, reduce the urban heat island effect, and contribute to the overall well-being of urban residents. Urban ecosystems can support urban agriculture initiatives, promoting local food production and reducing the carbon footprint associated with long-distance food transportation (Nader Heydari et al., 2025, Ecology of Tropical Cities, Volume II). Green roofs, community gardens, and urban farming contribute to food security and improve access to fresh and nutritious produce. Furthermore, well-designed and well-maintained urban ecosystems contribute to increased property values and attract investment. Green spaces, parks, and tree-lined streets enhance the aesthetic appeal of neighbourhoods, making them more desirable places to live, work, and invest (Hoffimann et al., 2017). Indeed, urban wealth can indirectly be assessed by ease of access to biodiverse green space (Farkas et al., 2022).

Green infrastructure, such as green roofs, rain gardens, and urban forests, provides cost-effective solutions for stormwater management, flood control, and heat mitigation, and reduces the burden on traditional infrastructure systems. Green infrastructure offers long-term cost savings and enhances the resilience of cities to climate change impacts, which presents a clear and immediate danger to the well-being of many urbanites (Thapa et al., 2025, this volume).

4 Who Studies What, and Where?

As with any predominantly field-based discipline, who studies, where they study and what is studied matters. Urban ecology is a relatively new discipline, but there is little doubt that much of the initial work has been conducted by scholars from the global north, with the discipline dominated by researchers from North America and Europe in particular (Angeoletto, 2012). This in turn has resulted in a focus on particular types of urban ecologies—those in longer developed, wealthier urban areas, where green infrastructure and education are embedded in societal norms.

But this is not where most urbanisation is occurring, with the most rapid population shift largely seen in the global south, where it is not only an increase in the proportion of people living in cities, but also the huge increase in populations seen in the new mega-cities such as Lagos, Manila, Sao Paulo, Mumbai and Dhaka, and perhaps even greater effects seen in the multitude of new mid-sized cities of the world.

In recent years, a glance through the pages of the journal *Urban Ecosystems* will show that there has been a welcome change in both the breadth of locations and the origins of researchers. In the latest issue (Volume 26, Issue 1) as we write, there are field studies from Brazil (2), China, Ghana, India, Iran, Mexico, Nigeria, and Tunisia, in addition to the five papers from the USA, and singletons from Australia, Austria, Canada, Germany and the UK. That points to a welcome increase in interest in urban ecology from regions previously understudied, and they do need to be studied, because tropical urban ecology presents some notable differences from other regions, and results in key challenges to be addressed. While the tide is changing, these systems have not been studied to any sensible degree. *That is the point of this book.*

Moreover, urban ecology research draws upon an increasingly diverse range of disciplines, including ecology, urban planning, geography, sociology, and environmental science. Scholars with interdisciplinary backgrounds contribute to a holistic understanding of urban ecosystems, incorporating both biophysical and social dimensions. Collaborative research involving multiple institutions and disciplines, crossing geographic boundaries (and indeed some studies are now global), has become increasingly common in urban ecology. Nevertheless, efforts should continue to enhance diversity and inclusivity within the field of urban ecology, as urban ecology is a human science in many ways, and by promoting opportunities for underrepresented groups and regions, by establishing networks that encourage long-term and effective collaboration and knowledge exchange, will broaden the perspectives represented in urban ecology research. Such collaborations foster cross-pollination of ideas and methodologies, resulting in more comprehensive and impactful research outcomes.

5 Where Are the Tropics?

We can simply consider the tropics to be all areas located between 23.4° North and South, since these zones receive direct solar radiation at some point during the year (Corlett, 2013). However, a wider set of different environmental and climatic features has also been used to classify regions as tropical, or not (Feeley & Stroud, 2018). Perhaps not surprisingly, such a definition is not universally agreed, since different ecological studies have been conducted based on different definitions of what constitutes a tropical region. Feeley and Stroud (2018) reviewed the eight most common criteria used in ecological and evolutionary studies to define a tropical region, and depending on what criteria were used, Earth's tropical land surface ranged from 15.47% to 44.1%. To some extent these criteria are therefore arbitrary (Feeley & Stroud, 2018). Briefly, there is a substantial number of possible climatic or ecological definitions for tropics, but there is no mechanism available to enable a consensual definition for them (Corlett, 2013). The definition for tropics remains under debate, although there is consensus among environmental scientists that tropical regions are the most biodiverse places on the planet, as well as the habitat of the largest number of endangered species (Clarke et al., 2017). We have taken a simple approach, and use the classic definition of tropical regions, i.e., the region between latitudes 23.4°N and 23.4°S—which corresponds to the tropics of Cancer and Capricorn, respectively. According to this definition, 33.4% of Earth's land surface is tropical (Feeley & Stroud, 2018).

6 Tropical Urban Ecosystems Are Different

At a superficial level, tropical urban ecosystems display distinct features when compared to their counterparts in wealthy countries due to factors such as climate, regional biodiversity, socio-economic conditions, and cultural contexts. Each matters, and each interacts with the other.

Tropical regions are possess extraordinary levels of biodiversity, hosting a wide array of species, many of which are found nowhere else in the world (Cabral et al., 2025, Ecology of Tropical Cities, Volume II). While the urbanization of tropical cities poses a significant threat to this unique biodiversity, necessitating targeted conservation strategies to safeguard native flora and fauna, urbanization also provides an opportunity to consider the benefits of biodiversity and green infrastructure for people and for nature.

While tropical regions are more diverse, they also face considerable physical challenges that differ from urban areas in more temperate zones. Tropical urban ecosystems are exposed to high temperatures, intense rainfall, and greater susceptibility to extreme weather events such as hurricanes and cyclones. Climate change further exacerbates these challenges, requiring adaptation strategies that account for the vulnerability of tropical cities and their ecosystems to changing climatic conditions. Green infrastructure can help mitigate these challenges through the provision of valuable ecosystem services, such as carbon sequestration, water regulation, and air quality improvement (Forman, 2014). However, the relatively rapid urbanization and associated land-use changes in tropical cities (and in many cases lack of planning regulation) undermines the provision of these services. Building resilience in tropical urban ecosystems necessitates a comprehensive understanding of their dynamics and the implementation of appropriate management strategies.

This is challenging, as tropical cities often experience rapid population growth and extensive informal settlements. The informal sector plays a crucial role in the local economy and requires specific approaches to urban planning, infrastructure development, and social service provision. Addressing the challenges of informal settlements is essential for promoting equitable and inclusive urban development. In common with cities around the world, biodiverse and usable urban greenspace is more often associated with wealth than absolute need, and in some regions the lack of planning control renders planning for biodiversity, environment and well-being almost impossible.

The significant socio-economic disparities and pockets of poverty often seen in tropical exacerbates this challenge, and can only be addressed by targeted interventions to ensure inclusive growth, access to basic services, and equitable distribution of resources. Promoting social and economic equity is crucial for sustainable urban development in tropical regions. Furthermore, the people who dwell in some urban areas in tropical zones have moved there from communities with significant traditional knowledge and cultural practices where an understanding of nature is held close. Integrating this knowledge into urban planning processes can enhance sustainable development, foster social cohesion, and preserve the cultural identity of tropical cities. Recognizing and respecting indigenous rights and cultural heritage are integral to sustainable urban governance. Just as we see cities as being a place that can homogenize biodiversity, cities can also homogenize identity. We must appreciate, value and sustain that which is different.

Pushing back against this ambition is poverty and inequality. Tropical urban ecosystems provide opportunities for urban agriculture, forestry and ecotourism, as well as for education and well-being, and of course for the maintenance of biodiversity itself. Harnessing the potential of these sectors while ensuring their sustainability and minimizing negative environmental impacts can contribute to the development of a green economy in tropical urban areas. This relies upon an integrated and participatory planning approach, where diverse stakeholders, including government agencies, local communities, academia, and civil society, are engaged. Context-specific policies and strategies that address the unique challenges and opportunities of tropical cities are needed.

To achieve this we must promote knowledge exchange and capacity building initiatives to bridge the gap between scientific research, policy formulation, and practical implementation. Strengthening local institutions, investing in research and education, and fostering interdisciplinary collaboration can facilitate evidence-based decision-making and enhance the resilience of tropical urban ecosystems. But we must start with knowledge, and this book aims to play a part in meeting that need.

Tropical urban ecosystems exhibit distinct characteristics that differentiate them from urban ecosystems in wealthy countries. Understanding these differences is vital for effective urban planning, management, and sustainable development in tropical cities. By acknowledging the ecological, social, and economic distinctions of tropical urban ecosystems, policymakers, researchers, and practitioners can work towards creating resilient, inclusive, and environmentally sustainable cities in tropical regions. By embracing diversity and interdisciplinary approaches, we can advance our understanding of urban environmental challenges and contribute to more sustainable and resilient cities worldwide.

7 This Book and Its Two Volumes

As noted, we took an open view of what area the tropics (and therefore urban ecology) encompasses. In part this is simply because the tropics have been expanding by approximately 1° per decade since the 1970s due to climate change (Lucas et al., 2014). This expansion process is resulting in the migration of some tropical species to higher latitudes to find suitable climate conditions (Feeley et al., 2011; Feeley & Stroud, 2018). At the same time, areas bordering the tropics face quite similar socioeconomic and urban challenges. Most of the future expansion of urban regions will take place in low-income, mega-diverse tropical countries.

Carmona et al. (2025, this volume, Chap. 11) estimated that tropical cities grow 1.4 faster than temperate cities (2.29 and 1.6 average annual rate of change in tropical and temperate cities considering the period between 1950 and 2020, respectively). Close to 43% of the world's population currently lives in the tropics. In 2050 this region will contribute with 60% of the total urban land coverage around the world, in which more than 50% of the world's population will reside (State of the Tropics, 2020). Almost all tropical cities occur in developing countries (97.5%) and only 15 tropical cities are associated to high income countries. In the year 2100, 71% of the mega cities (population population ≥ ten million) will occur in the tropics (Hoornweg & Pope, 2017; United Nations, 2018; Carmona et al., 2025, this volume).

Cities in tropical developing countries, such as Brazil and India urbanize themselves quite fast; moreover, they are located in megadiverse ecosystems or biomes (or in global biodiversity hotspots, like several cities listed in the current book) and have low urban planning capacity. These cities have failed to conserve urban biological diversity because, among other reasons, they do not actively act in its conservation process by curbing the advance of real estate speculations on urban forest fragments.

Thus, why must urban biodiversity be preserved? Conservation biology has traditionally divided the world into regions of pristine and degraded habitats. However, this mindset has changed: conservation biology must address habitats where humans live in and help explain how we can share these novel anthropogenic habitats with wild species (Rosenzweig, 2001). Indeed, although the rate of protected areas has been increasing worldwide since 1990, the number of threatened species keeps on growing (PNUMA, 2011), and with it, the need to find alternative ways of living with nature.

Tropical cities can act as additional spaces for biological diversity. Urban spaces focused on biodiversity are necessary in cities located in global biodiversity hotspots. Most of the 5570 cities in Brazil—which is a megadiverse country—are located in two global biodiversity hotspots, namely the Cerrado and Atlantic Forest biomes. The uncontrolled devastation of these biomes is another issue corroborating the argument that tropical urban ecosystems should be used as additional spaces for biological conservation purposes (Angeoletto et al., 2022a; Gonzaga et al. 2024).

Human—social, economic, cultural, urban, political and legal—factors can influence biodiversity patterns in different cities. It is quite common for environmental injustices committed in tropical cities to produce upper middle-class neighborhoods with satisfactory tree coverage and poor neighborhoods where trees are either scarce or non-existent. This inequality also means bad news for wildlife, since large low-income neighborhoods with poor vegetation cover provide less support to urban biological diversity (Schell et al., 2020; Lima et al., 2023). It is essential understanding the ecology of tropical cities to enable the conservation of urban biological diversity. Studies about urban ecology should have interdisciplinary nature, i.e., social and natural sciences' theories and methods should be combined to each other in order to gather information about human influence on urban biota.

This combination was one of our main goals in the present book. The editors actively sought to combine ideas deriving from the social and natural sciences. Social (geographers, architects, urban planners, lawyers, journalists and sociologists) and natural scientists (biologists, ecologists, agronomists, forest engineers, geneticists and veterinarians) wrote chapters that, together, paint a picture of the ecology and biodiversity of tropical cities. Our second goal lies on gathering data, knowledge and techniques to provide insights for scientists and urban planners to help inform conservation plans for urban biodiversity (for example, Silva et al., 2025, this volume). Finally, this book gives voice to dozens of researchers from countries poorly represented in ecological journals. In an intense selection process, we have gathered 220 authors from five continents, 29 countries and 117 institutions. They have authored chapters in a remarkable collective work focused on improving our knowledge of tropical urban ecology.

This is the first book (in two volumes) —published in English—exclusively dedicated to the ecology and biodiversity of tropical cities (Angeoletto et al., 2022b). It is divided in two volumes with four different sections. Volume I - **Human factors and their influence on urban ecosystems**: that section comprises 11 chapters focused on introducing and addressing social, economic, cultural, political, architectural and urban factors, as well as their influence on the ecology of tropical cities. **Natural and social sciences applied to urban biodiversity conservation**: that section comprises 10 chapters focused on presenting projects and studies about ecology of tropical, as well as their application in biological diversity planning and conservation processes. In Volume II - **The flora of tropical cities**: nine chapters focused on introducing and addressing aspects of urban plant diversity and ecology. **The fauna of tropical cities**: 15 chapters focused on introducing and addressing aspects of urban animal diversity and ecology.

The battle for biodiversity will be won or lost in cities. Although cities occupy a small part (from 1% to 6%) of Earth's land, they have huge ecological footprints. Architect Giulio Carlo Argan depicted—in a poetic and pessimistic way—the branches launched by cities along the biosphere, when he stated that nature is no longer beyond the city walls. Cities no longer have walls, they expand themselves as desperate cement mazes and form sordid peripheries of shacks. The city is still the city beyond the city, the city of highways and industrially cultivated fields (Argan, 1993).

There are many challenges ahead of us, such as producing biodiversity-friendly tropical cities where all citizens can democratically share all urban ecosystem services, regardless of who they are or where they come from. There is also room for optimism and hope: the present book is a useful tool to help overcome these challenges, reducing urban ecological footprints and developing effective policies focused on the conservation of biodiversity in tropical cities.

References

Angeoletto, F. (2012). *Planeta ciudad: Ecología urbana y planificación de ciudades medias de Brasil (Tesis doctoral)*. Departamento de Ecología, Universidad Autónoma de Madrid, Madrid.

Angeoletto, F., Tryjanowski, P., Santos, J., Martinez-Miranzo, B., Leandro, D., Bohrer, J., et al. (2022a). Will Brazilian city dwellers actively engage in urban conservation? A case study with the charismatic neotropical Blue-and-yellow macaw (Ara ararauna). *Birds, 3*(2), 234–244.

Angeoletto, F., Tryjanowski, P., Carmona, D., & Fellowes, M. (2022b). Científicos de 31 países producen libro sobre ecología y biodiversidad de ciudades tropicales. *Ecosistemas, 31*(1), 2260–2260.

Argan, G. C. (1993). *História da Arte Como História da Cidade*. Editora. Martins Fontes.

Beridze, B., Mahamoud-Issa, M., Zmudzinski, M., Bednarz, P. A., Filutowska, Z. K., Sikora, B., Winiarska, D., Carmona, D., & Angeoletto, F. (2023). Welcome to the (urban) jungle: Comments on a new science, biological evolution in cities. *European Journal of Ecology, 2*, 1–6.

Browning, M. H. E. M., Rigolon, A., McAnirlin, O., & (Violet) Yoon, H. (2022). Where greenspace matters most: A systematic review of urbanicity, greenspace, and physical health. *Landscape and Urban Planning, 217*, 104233. https://doi.org/10.1016/j.landurbplan.2021.104233

Cabral, L., et al. (2025). Give urban biodiversity a chance! A forest fragment harbors high bird-species richness in a tropical city. In F. Angeoletto, P. Tryjanowski, & M. Fellowes (Eds.), *Ecology of tropical cities: Biodiversity, People & Places*. Volume II. Springer Nature.

Caizergues, A. E., Santangelo, J. S., Ness, R. W., Angeoletto, F., Anstett, D. N., Anstett, J., ... & Johnson, M. T. (2024). Does urbanisation lead to parallel demographic shifts across the world in a cosmopolitan plant?. *Molecular Ecology, 33* (7), e17311.

Carmona, D., et al. (2025). Urban biological evolution in tropical cities. In F. Angeoletto, P. Tryjanowski, & M. Fellowes (Eds.), *Ecology of tropical cities: Natural and social sciences applied to the conservation of urban biodiversity*. Springer Nature.

Clarke, D. A., et al. (2017). Does biodiversity-ecosystem function literature neglect tropical ecosystems? *Trends in Ecology & Evolution, 32*, 320–323.

Corlett, R. T. (2013). Where are the subtropics? *Biotropica, 45*, 273–275.

Farkas, J. Z., Kovács, Z., & Csomós, G. (2022). The availability of green spaces for different socio-economic groups in cities: A case study of Budapest, Hungary. *Journal of Maps, 18*(1), 97–105. https://doi.org/10.1080/17445647.2022.2079433

Feeley, K. J., Silman, M. R., Bush, M. B., Farfan, W., Cabrera, K. G., Malhi, Y., Meir, P., Revilla, N. S., Quisiyupanqui, M. N. R., & Saatchi, S. (2011). Upslope migration of Andean trees. *Journal of Biogeography, 38*, 783–791.

Feeley, K. J., & Stroud, J. T. (2018). Where on earth are the "tropics"? *Frontiers of Biogeography, 10*(1–2).

Forman, R. T. T. (2014). *Urban ecology: Science of cities*. Cambridge University Press.

Goddard, M. A., Davies, Z. G., Guenat, S., Ferguson, M. J., Fisher, J. C., Akanni, A., et al. (2021). A global horizon scan of the future impacts of robotics and autonomous systems on urban ecosystems. *Nature Ecology & Evolution, 5*(2), 219–230.

Gonzaga, C. A. C., Roquette, J. G., Angeoletto, F., Silgueiro, V. D. F., Macedo, L. O. B., Valdiones, A. P., & da Silva, N. M. (2024). The Public Prosecutor's Office's experience using Global Forest Watch to monitor and deter deforestation in the Cerrado. *Environmental Conservation*, 1-6.

Heydari, N., et al. (2025). Urban agriculture: Issues and challenges in Iran. In F. Angeoletto, P. Tryjanowski, & M. Fellowes (Eds.), *Ecology of tropical cities: Biodiversity, People & Places*. Volume II. Springer Nature.

Hoffimann, E., Barros, H., & Ribeiro, A. I. (2017). Socioeconomic inequalities in green space quality and accessibility—evidence from a Southern European City. *International Journal of Environmental Research and Public Health, 14*(8), 916. https://doi.org/10.3390/ijerph14080916

Hoornweg, D., & Pope, K. (2017). Population predictions for the world's largest cities in the 21st century. *Environment and Urbanization, 29*, 195–216.

Lucas, C., Timbal, B., & Nguyen, H. (2014). The expanding tropics: A critical assessment of the observational and modeling studies. *Climate Change, 5*, 89–112.

Lee, A. C. K., & Maheswaran, R. (2011). The health benefits of urban green spaces: A review of the evidence. *Journal of Public Health, 33*(2), 212–222. https://doi.org/10.1093/pubmed/fdq068

Leandro, D., Johnson, M., et al. (2025). GLUEing the world: The GLobal urban evolution project and its importance to urban biodiversity conservation. In F. Angeoletto, P. Tryjanowski, & M. Fellowes (Eds.), *Ecology of tropical cities: Natural and social sciences applied to the conservation of urban biodiversity*. Springer Nature.

Liang, H., He, Y.-D., Theodorou, P., & Yang, C.-F. (2023). The effects of urbanization on pollinators and pollination: A meta-analysis. *Ecology Letters, 00*, 1–14. https://doi.org/10.1111/ele.14277

Lima, G. F., Correa Santos, J. W. M., Albertin, R. M., Martínez-Miranzo, B., Souza, F. L., & Angeoletto, F. (2023). Backyards are a way to promote environmental justice and biodiversity conservation in Brazilian cities. *Diversity, 15*(7), 815.

Melchiorri, M., Florczyk, A. J., Freire, S., Schiavina, M., Pesaresi, M., & Kemper, T. (2018). Unveiling 25 years of planetary urbanization with remote sensing: Perspectives from the global human settlement layer. *Remote Sensing, 10*(5), 768. https://doi.org/10.3390/rs10050768

Moroń, D., Beim, M., Gudowska, A., Angeoletto, F., Celary, W., Cwajna, A., ... & Tryjanowski, P. (2024). Evaluating tramway infrastructure on biodiversity and ecosystem services. *Scientific Reports, 14* (1), 9394.

Naeem, S., Chazdon, R., Duffy, J. E., Prager, C., & Worm, B. (2016). Biodiversity and human well-being: An essential link for sustainable development. *Proceedings of the Royal Society B: Biological Sciences, 283*, 20162091. https://doi.org/10.1098/rspb.2016.2091

National Parks Board. (2019). *Conserving biodiversity: Singapore's national biodiversity strategy and action plan*. Available online at https://www.nparks.gov.sg/-/media/nature-conservation-master-plan.ashx

Ritchie, H., & Roser, M. (2019). *Urbanization*. Published online at OurWorldInData.org. Retrieved from: 'https://ourworldindata.org/urbanization'

Rocha, E. A., & Fellowes, M. D. E. (2020). Urbanisation alters ecological interactions: Ant mutualists increase and specialist insect predators decrease on an urban gradient. *Scientific Reports, 10*, 6406.

Rosenzweig, M. L. (2001). Loss of speciation rate will impoverish future diversity. *Proceedings of the National Academy of Sciences, 98*(10), 5404-5410.

Santangelo, J. S., et al. (2022). Global urban environmental change drives adaptation in white clover. *Science, 375*, 1275–1281. https://doi.org/10.1126/science.abk0989

Schell, C. J., Dyson, K., Fuentes, T. L., Des Roches, S., Harris, N. C., Miller, D. S., et al. (2020). The ecological and evolutionary consequences of systemic racism in urban environments. *Science, 369*(6510), eaay4497.

Silva, et al. (2025). A low-cost method to detect invasive species in the urban environment—an example of a drone-based workflow application. In F. Angeoletto, P. Tryjanowski, & M. Fellowes (Eds.), *Ecology of tropical cities: Natural and social sciences applied to the conservation of urban biodiversity*. Springer Nature.

Soanes, K., Taylor, L., Ramalho, C. E., Maller, C., Parris, K., Bush, J., Mata, L., Williams, N. S. G., & Threlfall, C. G. (2023). Conserving urban biodiversity: Current practice, barriers, and enablers. *Conservation Letters, 16*, e12946. https://doi.org/10.1111/conl.12946

State of the Tropics. (2020). *State of the tropics 2020 report*. James Cook University. https://www.researchgate.net/publication/343441325_State_of_the_Tropics_2020_Report

PNUMA (Programa de las Naciones Unidas para el Medio Ambiente). 2011. Anuario PNUMA – Temas Emergentes en Nuestro Medio Ambiente Global. PNUMA. Nairobi, Kenya.

Thapa, P., et al. (2025). Social-ecological dynamics of green and blue infrastructure in the Indian megacity of Bengaluru. In F. Angeoletto, P. Tryjanowski, & M. Fellowes (Eds.), *Ecology of tropical cities: Natural and social sciences applied to the conservation of urban biodiversity*. Springer Nature.

United Nations, Department of Economic and Social Affairs, Population Division. (2018). *World urbanization prospects: The 2018 revision, Online Edition.*

United Nations, Department of Economic and Social Affairs, Population Division. (2022). *World Population Prospects 2022: Data Sources*. (UN DESA/POP/2022/DC/NO. 9).

Welch, V. L., Sloggett, J. J., Webberley, K. M., & Hurst, G. D. D. (2001). Short-range clinal variation in the prevalence of a sexually transmitted fungus associated with urbanisation. *Ecological Entomology, 26*, 547–550. https://doi.org/10.1046/j.1365-2311.2001.00352.x

Chapter 2
Could Cities Be an Object of Study for Ecology?

Martín Andrés Díaz and Fabio Angeoletto

Abstract Ecology is undoubtedly the discipline that has had the greatest expansion in association with environmental problems. It could be thought then that the conservation of biodiversity in cities is within the domain of application of the discipline. For this reason, there are authors who affirm that the same laws and theories that ecology uses to understand and explain natural ecosystems can be applied to urban ecosystems. However, within the discipline, there is a strong debate as to whether ecology has mature laws and theories as other areas of science. Currently, the discussion is centered between those who think that ecology is a science lacking mature laws and theories, and on the other hand, authors who in recent years have defended the opposite position, presenting reconstructed theories and their associated laws that have allowed to answer the discussion about its epistemological status. From this state of affairs we propose to discuss whether urban ecosystems, with the patterns and processes of the organisms that inhabit cities, are within the domain of the laws and theories of ecology. We will also analyze whether the concept of ecosystem that is fundamental in ecology today can be applied to cities or urban ecosystems. Finally, we will discuss whether the problems of biodiversity conservation in cities are ecological problems or whether a more complex interdisciplinary approach is needed to establish measures for their preservation.

Keywords Urban ecology · Ecological theories · Biodiversity conservation · Socio-ecological approach · Ecology in the City · Ecology of the City · Ecology for the City · Urban ecosystems · Urban biodiversity

M. A. Díaz (✉)
Universidad Nacional de La Matanza, San Justo, Argentina
e-mail: martind@unlam.edu.ar

F. Angeoletto
Programa de Pós-Graduação em Geografia da Universidade Federal de Rondonópolis, Mato Grosso, Brasil

Programa de Pós-Graduação em Gestão e Tecnologia Ambiental, Universidade Federal de Rondonópolis, Rondonópolis, Brasil

PUMAS, Pessoas Unidas pelo Meio Ambiente, Agricultura & Sociedade, Chapada, Brasil

F. Angeoletto et al. (eds.), *Ecology of Tropical Cities*,
https://doi.org/10.1007/978-3-031-48012-6_2

1 Introduction

For several decades, scientific and non-governmental organizations around the world have been alarming about the environmental crisis we are going through. According to IPCC-IBBES this situation can be summarized by stating that "Climate change impacts and biodiversity loss are two of the most important challenges and risks for human societies; at the same time climate and biodiversity are intertwined through mechanistic links and feedbacks" (Pörtner et al., 2021). We can conclude that environmental problems represent today the central point of the discussion agenda in all aspects of human activities and that to solve them we seek to make use of available science and technology, both to predict the adverse effects and to adapt to the changes expected in the coming decades.

Among these environmental problems is the impact of anthropic activities on natural ecosystems. Ellis et al. (2021) state that "Multiple studies confirm that ecosystems across most of the terrestrial biosphere, from 75% to 95% of its area, have now been reshaped to some degree by human societies". Are these areas impacted by human activities the object of study of ecology? Niemela (1999) mentions that "Expansion of cities and towns is a significant cause of conversion of land to highly human modified urban landscapes. From an ecological point of view, urbanization can have both favourable and adverse effects on biotic communities." and then concludes "Although urban ecosystems are governed by the same ecological "laws" as rural ecosystems, the relative importance of certain ecological patterns and processes differs between the two types of ecosystems.". Finally, to be more specific, we can ask: Can urban ecosystems be the object of study of ecology?

This idea that environmental problems are associated with ecology has increased with the progressive rise of environmental concern, which for many authors has prompted the development of this discipline by others that are also related to environmental problems. That is why Castle (2001) stated that "The rise in the scientific importance of ecology followed the environmentalism of the 1960s, to the point that ecology is now often understood as environmental thought or action". But according to other authors, ecology has also had a significant impact on environmental issues, so Di Pasquo (2013) mentions that "disciplinary ecology had a demarcating function in delimiting environmental issues". But the relationship between ecology and environmental issues has presented obstacles, so Odum (1977) mentioned that when movements of environmental concern began to grow in the late 1960 some ecologists resented the public use of their discipline.

What are the obstacles between ecology and environmental problems? What role do these obstacles play in the problem of biodiversity conservation in cities? First, we can mention the notion of scale. Di Pasquo et al. (2011) mentioned that "Throughout this journey, we have pointed out that disciplinary ecology and the discourse developed from the environmental crisis have referred to different scales". It has long been discussed and established that in ecology the scale, characterized by grain and extent, determines for a given observed pattern the processes and factors involved in its explanation. The obstacle is established because many authors consider that environmental problems (climate change, loss of species, etc.) occur at broad scales (regional or global) and most of the theories of ecology are applied at smaller scales (local), which means that these would not be in their domain of application.

Another obstacle that can be discussed is whether urban ecosystems can be the object of study of ecology. Let us bear in mind that ecology began in 1866, when Ernst Haeckel characterized "the whole science of the relations of organisms to their environment" as "Oecology" (Haeckel, 1866, II). At that time, the discipline focused on the study of the individual and its adaptations to the environment, but since then and up to the present time, ecology has incorporated new objects of study (populations, communities, ecosystems, etc.). So, there is much to discuss in order to establish whether urban ecosystems are a new object of study of ecology.

Another obstacle is that although Niemela (1999) has an optimistic position and mentions that ecology will be able to use its "laws" to explain the patterns and processes of urban ecosystems, there has been a discussion within the discipline for several decades about the maturity of its laws and theories. The current discussion in which ecology finds itself can be summarized in two main positions. On the one hand, those philosophers and especially ecologists who claim that ecology does not possess mature theories and laws capable of successfully predicting and explaining the patterns and processes of the discipline (Berryman, 2003; Murray, 2001; Peters, 1991). The other position is held by authors who defend ecology, arguing that it possesses theories and laws capable of being applied to the phenomena studied (Díaz, 2019; Lorenzano & Díaz, 2020; Scheiner & Willig, 2008). So, one question remains to be resolved. Does ecology have mature theories and laws to apply to urban ecosystems?

The last obstacle we are going to mention is due to the fact that ecology is prior to environmental problems and then we can ask ourselves some questions: Can the theories of ecology have a central role in the explanation and prediction of the environmental problems of cities? Are the biodiversity conservation problems of cities within the domain of application of the theories of ecology? Are the conservation problems of cities an ecological problem? To answer these questions, we must start from Castle's (2001) assertion that "Throughout its history, ecology has been plagued by scepticism about its scientific status because of its roots in natural history" and consider whether ecology possesses the theoretical immaturity claimed by its critics.

So, in spite of the interrelation between ecology and environmental problems, the role of ecological theories seems to be in doubt. After two centuries of existence we have a large number of undoubtedly established ecological patterns and processes, however, ecology has been discussing for decades which are the appropriate conceptual constructs to explain and predict them (Díaz, 2021; Lorenzano & Díaz, 2020; Scheiner & Willig, 2011). This situation raises doubts about the role that ecology and especially its theories can play in the problem of biodiversity conservation in cities or urban ecology, an environmental problem that affects millions of people in the world.

The general objective of this work is to discuss the current status of ecology theories and what role they may have in explaining and predicting the conservation of biodiversity in cities. To achieve this we will discuss the obstacles mentioned above, especially if this environmental problem and its object of study (urban ecosystems) are within the application domains of ecology theories and also if ecology is the only discipline involved in this problem.

2 First Obstacle: The Philosophical Problems of Ecology

We will start by searching for the answer to one of the questions we mentioned previously. Does ecology have mature theories and laws to apply to urban ecosystems? If ecology does not possess mature theories as many of its critics claim, we would not be able to discuss or resolve the other obstacles raised, for that reason, we begin by resolving this question first. We will then review the general discussion that has taken place in recent decades within ecology about the maturity of its laws and theories.

Ecology has received in last decades, both from within ecologists (Berryman, 2003; Murray, 2001; O'Hara, 2005; Peters, 1991), as well as from philosophers of science who have dealt with the discipline (Ibarra & Larrañaga, 2011; Paslaru, 2009; Raerinne, 2011) several critiques. Peters (1991) mentioned that prediction is the absolute criterion by which scientific theories are judged and he asserts that the theories of ecology lack this property. In the same sense MacIntosh (1980) stated that ecology was not and is not a predictive science. The cause of this problem, according to these authors, is that ecology is a new science and therefore presents a theoretical immaturity that has not yet allowed it to establish general laws and theories based on the cloud of mathematical models established for each of the patterns studied.

One of the problems is that due to the absence of well-established theories there is ambiguity in the conception of theory within ecology. Robert May (1981) mentions that "Ecological theories come in various forms", such as simple mathematical models, complex or even taking the form of verbal models. Scheiner and Willig (2008) state that "The confusion exists because the term theory is used in reference to a variety of types of systems at different levels of specificity". Recently, Marquet et al. (2014) mention that "In ecology, there is currently no consensus regarding the definition, role, and generality of theories, as discussed in several essays." and then that "It is no surprise that ecologists use the label theory to refer to many things".

All these arguments have led to derogatory statements such as that of Lawton (1999) who indicated that "Some science, areas of physics, has important universal laws, and ecology deeply envies this because it does not". Ecology is then relegated to a minor science, envious of those of great scientific success. The conclusions are stark, "On both sides of the debate, there is a widespread belief that ecology is different from physics because it lacks general laws and because it is not a predictive science (and, therefore, is not 'hard')." (Turchin, 2001).

In defense of ecology there are different positions that attempt to argue against the criticisms from different metatheoretical perspectives. There are authors who consider that ecology has theories and laws to explain patterns of organisms, but they use an intuitive or informal characterization of law and theory. An example is the conception of Robert May (1981) who mentioned that theories can be mathematical equations or models (this is the general idea in ecology) or verbal models. These two views are the most common in ecology, in the case of verbal models it is not enough to state that a theory is "a set of statements that explain a broad set of circumstances" (Ford, 2004), which is a similar plan to that in the classical conception of the philosophy of science. A theory cannot be understood as only a mathematical equation (Falguera, 1994), but this view was generalized by the need to make ecology a predictive science.

Another position has been to affirm that there is a nomological equivalence in certain laws of ecology and some of physics, given that they apply to phenomena that are similar in their functioning. Authors who have defended this conception are Berryman (2003), Ginzburg (1986), Ginzburg and Colyvan (2004) and Turchin (2001), among others. They propose an equivalence between Newton's laws and certain equations of population ecology. Ginzburg (1986) mentions when discussing Malthus' Law that "Populations conserve exponential growth unless they do not. The argument that the laws of both disciplines are an identical type of statement is based on the idea that, if those of physics meet the requirements posed by the classical conception to be legaliform, i.e. universality, unlimited scope, no particular objects and universal predicate (Lorenzano, 2007), those of ecology would also meet them.

The last position we are going to mention in the defence of the theories of ecology is perhaps the one that has advanced the most in achieving the theoretical maturity of the discipline. It takes into account the scientific theory conception of semanticism and/or metatheoretical structuralism and they argue that the classical or inherited school cannot provide a solution to the philosophical problems of ecology (Pickett et al., 2010). This position has allowed the informal and formal presentations of the great part of the constituent theories of ecology (Díaz, 2019; Díaz & Lorenzano, 2017; Scheiner & Willig, 2011).

In 2001 Bertram Murray ensures that "The deductive-nomological structure of theories has received much criticism" and that "Beatty (1981, 1987), Lloyd (1988) and Thompson (1989) have offered an alternative through the semantic structure of theories". The author mentions that Thompson (1989) established that "The semantic conception offers a way to formalize theories of physics and evolution that does not involve distortion and that does not take into account the unwanted features of the syntactic conception".

In the same year David Castle (2001) publishes a very important work for theories in ecology, "A semantic view of ecological theories". Like Murray (2001), the author considers "The central dogma of the inherited conception is that the structure of scientific theories can be analyzed only in terms of their linguistic formulation" and that "When applied to individual sciences, the inherited vision has not been successful". Castle (2001) considers the semantic conception as the successor and the vision to adopt in ecology, since he understands that the main theories of ecology are "representative of several features of the semantic conception of theories".

The idea of a theory as a system with a structure and various components was adopted to carry out the informal reconstruction of the components of the main theories of the discipline in the works of Samuel Scheiner and Michael Willig (2008, 2011), seeking to establish a general theory of ecology. In their works they present eleven theories of ecology: ecological niche, foraging, population dynamics, succession and gradients, among others. They clearly establish the propositions and the domain of each of them, but they do not solve the problem of the position of laws. However, these reconstructions constitute the first explanation of the theories of ecology under a common conceptual framework different from the classical conception.

Taking the general proposals of Castle (2001) but within the framework of the structuralist metatheory, Díaz and Lorenzano (2017), Lorenzano and Díaz (2020) reconstructed the theory of population dynamics in ecology and Díaz (2019) the theory of spatial gradients of species richness. These authors identified all the components of these theories, as well as their fundamental laws and the corresponding theoretical networks in which the different equations were included as terminal specializations. All these works determined that ecology has mature theories and elucidated the epistemological status of laws and models (mathematical equations) as representations of special laws that apply to particular cases.

As with the work of Scheiner and Willig (2008, 2011), it was possible to establish that within ecology there are different levels of generality, on the one hand, the theories and their laws and on the other, the models represented by the mathematical equations. After this review, the initial question can be answered in the affirmative. Ecology has mature theories and therefore we will analyze whether these could be applied to the conservation of biodiversity in cities.

3 Second Obstacle: The Scale of Cities

The next obstacle we are going to discuss is related to the spatial scale of the cities. As mentioned by Di Pasquo et al. (2011), despite the relationship that is supposed to exist between ecology and environmental problems, many of the discipline's theories do not apply to the scales at which certain problems occur. We must remember that one of the main contributions to this issue was made by Levin in 1992 in his work "On the problem of pattern and scale in ecology". Over the years both Levin (1992) and other authors have established that "To understand a system, it is important to study it at the appropriate scale and develop models that bridge across scales." (Chavé, 2013). Niemela (1999) stated that "In addition to these, the question of ecological scale needs to be considered when investigating urban landscapes, especially for the attributes of species diversity patterns."

It is important that we discuss the question of scale in order to understand what possible diversity patterns might be involved in cities and also to determine whether ecology has theories that explain and can be applied to them. In addition, it is important to discuss whether the problem of scale mismatch between ecology and environmental problems mentioned by Di Pasquo et al. (2011) also arises for biodiversity conservation in cities. First, let us recall that cities occupy approximately 1% to 6% of the Earth's surface (Alberti et al., 2003) and that for example the size of the world's one hundred and twenty-three largest cities, measured as the surface area occupied by the urban area in km^2, range from 11,600 km^2 of Greater New York to 181 km^2 of Faisalabad. On average, smaller cities have a surface area of approximately 100 km^2.

On the other hand, in ecology we can take for example the work of Field et al. (2008), who mention that the extent used in ecology (distance between the most distant sampling sites) can be divided into <10 km, 10–100 km, 100–1000 km and >1000 km. These same authors point out that the Grain (size of the sampling unit) can be classified into three different levels, small (<10 km^2), medium (10–500 km^2) and large (>500 km^2). There are different classifications for scales in ecology, but in all of them it is observed that cities are generally located among the local or smaller categories of ecology, both in extent and grain.

To answer the question of whether scale is an obstacle for ecology to study the environmental problems of biodiversity conservation in cities, we can start from what Di Pasquo et al. (2011) state, "Throughout this journey, we have pointed out that disciplinary ecology and the discourse developed from the environmental crisis have referred to different scales". These authors pointed out that this was due to the fact that "in particular, the relationship would be expressed in what we will understand as a "perception of global phenomena" within the analyzed discourses". Then the drawback would be determined because environmental phenomena occurred at broad scales and most of the theories of ecology would apply to smaller scales. Diaz (2021) states that in many cases it is assumed that the domain of environmental problems is narrower than what they really are (ecological deviation) and in other cases the theories of ecology are required to explain phenomena that are not included in their domain of application (explanatory deviation).

Now, according to Di Pasquo et al. (2011), ecology had a need to change the scale of its studies to address global environmental problems, which led to the development of new ecological disciplines such as landscape ecology and macroecology. However, cities are at the smaller scales and historically ecology has worked on those scale sizes. Until about the 1980s, ecological studies usually used what some authors called the "anthropocentric" scale (Cueto, 2006). So much affinity can be established between the scale of ecology and cities that Wiens (1989) mentioned that "Ecologists deal with phenomena that are intuitively familiar, and we are therefore more likely to perceive and study such phenomena on anthropocentric scales that accord with our own experiences."

So, we can conclude that the scale of cities is not an obstacle for the theories of ecology, since we can observe that most of the theories developed from its beginnings until the 1980s apply to scales similar to those of cities. Among others, we can mention population dynamics, ecological succession, community ecology, ecosystem ecology, ecological niche theory, competition theory, species distribution and abundance theory, and species richness gradients, among others (Di Pasquo et al, 2011; Kolasa, 2011).

In order to discuss the conservation of biodiversity in cities, it is necessary to establish what ecological processes and patterns can be found at these scales and also what environmental problems may be affecting them (Santangelo et al., 2022; Souza et al., 2023). According to Niemela (1999) "The geographical scale of research is often relatively large, spanning a city and its surroundings. Consequently, urban ecological research often takes place at the landscape scale." This author states that in urban ecosystems or landscapes the processes that determine biodiversity (α and β) are mainly affected by the connection between patches, succession and invasion of species. As we have mentioned, for these scales, patterns and processes ecology has theories that can be applied to the conservation of organisms.

It is important to remember that ecology has established that the scale, especially the grain, determines which patterns, processes and factors are involved (Levin, 1992). We can also mention what was established by O'Brien (2006), who describes that at local scales, biological dynamics determine local heterogeneity where microclimatic, terrain, edaphic and slope factors, among others, intervene. The result of these processes and factors are the patterns of species richness and diversity, population dynamics and niche occupation, among others.

We conclude then that the scale of cities is not an obstacle to apply the knowledge of ecology, that is, its laws and theories. On the contrary, it has been at these scales that ecology has developed its main theories since its beginnings. We will discuss later what is the role of ecology in the problems of biodiversity conservation in cities, more specifically, whether they are ecological problems or involve the work of other scientific disciplines.

4 Third Obstacle: Urban Ecosystems

The next obstacle we are going to discuss in relation to ecology and its application to the conservation of biodiversity in cities is that the discipline is a science developed to be applied to natural ecosystems and therefore urban ecosystems are not among its objects of study. A review of the history of ecology reveals numerous changes in its objects of study over time. Since its beginnings with Haeckel in 1886, ecology has determined patterns, processes and mechanisms associated with different objects of study that the discipline has included in its scope of application (individuals, populations, communities, guilds, ecosystems, etc.) (Cooper, 2003; Jax, 2001).

Haeckel's ecology (1866) is considered to be a continuation of the naturalist tradition (Jax, 2001) and numerous authors refer to this period as the new natural history (Adams, 1917), although it is now referred to as the old ecology (McIntosh, 1980). The aim was to determine the environmental requirements of the individuals of a species, studying the adaptations they had to possess to survive in a specific environment (Cooper, 2003; Mayr, 1995), the object of study at this stage being the individual.

After that stage, what Kolasa (2011) considers to be the first theory developed within ecology, population growth models, was established. At the beginning of the nineteenth century we find the work of Malthus and Verhulst and their population dynamics equations applied to the problem of populations. Population ecology studies the patterns of distribution and abundance of individuals of the same species in space and/or time. The first equations were joined by Lotka's competition and predation equations and Pearl's curves, among others, but all of them had as their object of study the populations of organisms, which was added to the ecology of the individual.

A new object of study appeared in 1877, when Karl August Möbius proposed the term biocenosis to refer to the biotic community, and community ecology with its patterns of distribution in space and time made its appearance. So drastic was the change that Shelford stated at the time that "Ecology is the science of communities" (Simberloff, 1980). Ecology was then transformed into a discipline that contained at least the ecology of the individual, populations and communities and all their associated patterns.

Finally, the most recent history of ecology involves all the changes that occurred in the last half of the twentieth century. Perhaps the most important one adopted at this time, but proposed decades earlier by Tansley (1935), is the establishment of the ecosystem as the object of study in the discipline. Trying to overcome the Clements/Gleason dilemma, Tansley characterises the ecosystem as the complete system (in the sense of physics) involving not only organisms and their relationships but also the physical factors that make up their environment (Tansley, 1935). Thus, after more than a hundred years of history and with the inclusion of diverse objects of study, it can be affirmed as Jax (2001) states that ecology is transformed into "A bush, not a tree" and plant, animal, terrestrial, aquatic, macroecology, landscape, palaeoecology and others appear. This transforms ecology into a dynamic and complex scientific discipline.

Each time the discipline has included a new object of study, the ecologists who practice it establish new methodological approaches, conceptual constructs (laws, theories, rules and/or models) and concepts to explain the patterns, processes and mechanisms involved. This has determined the coexistence of different disciplinary matrices that are recognised by the problems, theories and methodologies they have used, implying different historical moments in the discipline (Cooper, 2003; McIntosh, 1980). We can return to our initial question and discuss now what happens to all these objects of study in ecology and urban systems.

Can cities be an object of study for ecology? According to Pickett and Grove (2009) while the concept of ecosystems was introduced to solve problems of succession theory and ecological communities, the study of urban ecosystems has become a new challenge for ecology over the last 50 years or so, and they state that "However, there is controversy about whether or how well the ecosystem concept is suited to urban habitats" (Pickett & Grove, 2009). To resolve this question they examined Arthur Tansley's original presentation of the ecosystem concept, and explored how the ecological context of the concept has changed, and they concluded that the fundamental concept of the ecosystem is well suited to urban ecological studies.

Is only the ecosystem concept applicable to cities? We can take into account what Savard et al. (2000) mention when they state that "However, biodiversity concepts can easily be applied to the urban ecosystem itself". Within them one can then find the different levels of organisation and recognise hierarchical attributes of the three components of biodiversity: composition, structure and function (Noss, 1990). Ecology can then also be concerned with cities and their populations, communities and associated emergent properties. Thus, depending on the spatial scale of the city involved, patch structure, species richness, population dynamics, habitat structure and fragmentation and other ecological attributes can be studied (Noss, 1990).

Where is the difference between natural and urban ecosystems? According to Scheiner and Willig (2008) ecology is "The study of the spatial and temporal patterns of the distribution and abundance of organisms, including causes and consequences." As we saw above these patterns and processes are present in urban systems, but as Alberti et al. (2003) states "What makes urban regions different from many other ecosystems is that in these regions humans are a dominant component". This situation determines that the environmental factors causing the observed patterns and processes are associated with anthropogenic activities and not with natural processes. For that reason, Alberti et al. (2003) states that in cities "Traffic congestion, air pollution, and urban sprawl emerge from local-scale interactions among variables such as topography, transportation infrastructure, individual mobility patterns, real estate markets, and social preferences."

We can extend this idea by taking Bennett et al. (2009), who mention that the ecological processes that contribute to sustaining biodiversity can be grouped into seven categories (climatic, primary productivity, habitat formation, interaction between organisms, movement of organisms and disturbance regimes). These authors state that these natural processes are affected by anthropogenic threats or risks such as climate change, habitat degradation and loss, alteration and reduction of hydrological flows, addition of nutrients and chemicals to ecosystems, unsustainable use of natural resources and the impact of species invasions.

One of the characteristics of urban systems, then, is that these anthropogenic threats are of greater significance in urban systems than in natural ecosystems. However, this does not imply that the biodiversity properties mentioned by Noss (1990) cannot be studied by ecology. An urban species gradient, unlike a species gradient in a natural ecosystem, will have variables associated with anthropogenic processes as causal factors, but this does not limit the application of ecology to the study of these patterns; on the contrary, it adds new objects of study and complexity to a discipline that has broadened its domain of application since its beginnings when it only studied the ecology of the individual.

It is necessary to stress that urban systems are not an exclusive problem of ecology, as Niemela (1999) stated "To summarize, urban ecology is a diverse field of research that forms a continuum from "pure" ecology in the urban setting to a combination of ecology and social sciences to examine urban systems." But, on the contrary, urban ecosystems can be the object of study of the discipline, as Alberti et al. (2003) state when mentioning that "To understand specific sets of interactions between humans and ecological processes taking place in urbanizing regions, we propose to examine cities as emergent phenomena - phenomena that cannot be explained simply by studying the properties of their individual parts. Cities are complex ecological entities, which have their own internal rules of behaviour, growth and evolution." EVO? EVOLUTION?

Finally, we can conclude that cities as urban ecosystems with their populations and communities are not an obstacle for ecology, since in its complex history the discipline has been incorporating new objects of study that have broadened its domain of application by studying new patterns and processes. As Pickett and Grove (2009) state "It is important to recognize that all ecological concepts have a core definition as well as attendant models that apply that idea to the material world. The core concept of any successful idea tends to be stable, whereas the models evolve to account for new knowledge, new scales of observation, and new theoretical perspectives". So, the concepts and theories of the different branches of ecology through the development of new models can be applied to new objects of study such as in this case cities. How this happens with ecological knowledge will be discussed in the next section.

5 Final Obstacle: Domain of Application of Ecological Theories

In this final section we will try to answer some of the questions we posed at the beginning of our paper: Can theories of ecology play a central role in the explanation and prediction of environmental problems in cities? Are the problems of biodiversity conservation in cities within the scope of application of theories of ecology? Are the conservation problems of cities an ecological problem?

Regarding the first question we can consider a general statement by Castle (2001) in relation to conservation problems:

> Unfortunately, appeals to ecological theories have neither helped to explain past declines and extinctions, nor have they predicted them reliably. In this context, the lessons of conservation biology have discredited ecological theories, and have justifiably caused conservationists to regard ecological theories as irrelevant to their work.

This idea shows that, despite the apparent interrelationship between ecology and environmental problems, the role of ecological theories was sometimes questioned. This situation was due to several drawbacks that we can affirm have been overcome by the discipline. As already mentioned, work in recent years (Díaz, 2019; Díaz & Lorenzano, 2017; Scheiner & Willig, 2011) has shown that ecology has mature theories that are therefore capable of being applied to patterns at different scales and of explaining and predicting their behaviour.

In order to answer the questions posed in relation to the theories of ecology and in the particular case of cities, we must take into account what Díaz (2021) mentions as an obstacle in these situations when he states that "the underlying problem is associated with misconceptions about the domains of application of both elements of this discussion". We can also refer to Niemela (1999) when he mentions that "human ecosystem models" must take into account human impact by identifying social factors that are associated with ecological factors to explain changes and patterns in urban environments. According to Diaz (2021) in these situations two obstacles can be found that must be taken into account in order to be able to apply the theories of ecology correctly to urban ecosystems.

The first issue is that it is often assumed that the domain of environmental problems is narrower than they actually are. Díaz (2021) called this situation ecological drift. Ecological drift is considered to be the idea that environmental problems are exclusively within the domain of ecology. This deviation obviously prevents the proper conceptualisation of environmental problems and the finding or proposing of appropriate management solutions. An important precedent in this sense in urban ecology is according to Pickett et al. (2016) the paradigm called "Ecology in the city". According to these authors "The ecology in cities paradigm employs a biological and bio-ecological approach".

Pickett et al. (2016) note that in this approach ecologists use the theoretical and methodological tools of their discipline to study habitat or ecosystems within cities that are similar to those studied outside of cities. According to these authors "The habitat types chosen for ecological study in cities were often close structural analogs to those outside of cities and included forest patches, parks, cemeteries, meadows, vacant lots, wastelands, streams, and wetlands". This approach has been used, for example, to explain some species conservation problems in cities, such as changes in biodiversity in an urban park in Boston (Drayton & Primack, 1996) or changes in bird abundance in small green spaces in Baltimore (Rega et al., 2015), among others.

However, the idea that a biodiversity conservation problem can be isolated and thought of as solely an ecological problem helps to explain part of it, but not all of it. Jax (2001) raises the consequences of this view when he mentions that "The demands on ecology and ecologists were high and often exceeded the current possibilities of the discipline, even more so as ecology was claimed by some environmentalists to be both scientific and normative. However, ecology cannot provide any values and objectives for environmental conservation, because these are, by definition, beyond the scope of this science." But is this really the situation between ecology and environmental problems?

The solution to this obstacle is achieved by the correct application of the theories and laws of ecology within its scope of application. To avoid ecological drift we must conceptualise the situation properly in order to give the theories of ecology their proper role. Environmental problems are due to the intervention of anthropogenic factors on natural systems or the alteration of ecological processes due to the use of resources by humans that incorporate threats to ecosystems. The whole set of natural and anthropogenic processes, mechanisms and patterns involved in an environmental problem constitute a complex system that cannot be explained by ecological knowledge alone.

For this reason, Pickett et al. (2016) point out that over time urban ecology matured establishing two other new paradigms, the "ecology of" and the "ecology for" that take this situation into account. In both cases, a socio-ecological approach (ecology of) or a transdisciplinary approach (ecology for) is used, incorporating a more complex analysis that allows understanding how anthropic processes affect the ecological processes traditionally studied by the theories of ecology. In the same vein Willig and Scheiner (2011) mention that a new discipline is emerging, socio-ecology, which is shaped by the integration of social, environmental and engineering sciences and has as its domain of application coupled human and natural systems.

These paradigms that involve the integration of ecology with other disciplines cannot encounter another of the obstacles raised by Díaz (2021), which consists of requiring the theories of ecology to explain phenomena that are not included in their domain of application, called by this author as explanatory deviation. This deviation can be characterised as the conception that the domain of ecology theories includes patterns and processes involved in an environmental problem that are not within the scope of ecology. Schneiner and Willig (2011) mention that "Each theory or model applies to a domain" and the domain "is the scope in space, time and phenomenon addressed by a theory or model". This notion must be associated in ecology with the concept of scale, which is central and indispensable in its explicitness as part of theories. The role of the theories of ecology in environmental problems is only elucidated if these two afore mentioned deviations are eliminated.

In this sense Di Pasquo et al. (2011) argue that ecology often works on ecological patterns and processes at small scales, and environmental problems occur at larger scales. However, this is not the case for cities, as mentioned above, since most of them are within the spatial domain of the main theories of ecology. But when analysing an environmental problem such as the conservation of biodiversity in urban ecosystems, part of the patterns and processes involved are not in the domain of the theories of ecology, and it is therefore a mistake to think that only a theory of ecology can explain the problem under study.

So, in relation to the questions posed at the beginning of this chapter, we can answer that if we avoid the ecological and explanatory detour mentioned above and take into account that environmental problems are not entirely ecological, but that a part of them is within the domain of ecology, then ecology can play an important role in environmental problems. Now, from the point of view of ecology, can a theory of ecology incorporate anthropogenic factors into its models and maintain its domain of application and its character as an ecological theory?

To answer this question, we will look at some examples of urban ecology work. One of the most studied problems in urban ecology are the gradients of species richness along urban-rural environments, in these problems we study the variation in species richness, diversity and abundance along transects that pass through these types of environments. In this case, the Gradient Theory of Species Richness as reconstructed by Díaz (2019), Scheiner and Willig (2005, 2011) and Colwell (2011), which explains variation in species richness as a function of environmental factors spatially associated with species richness, can be applied.

The theory's domain of application is to explain the environmental-spatial variation in the species richness of a taxon whose species are present in a set of sites in a given region (Díaz, 2019). With this domain in mind, we can see that numerous works on urban gradients fall into this domain. Leveau (2019) studied the variation in bird species richness at sites located in urban-rural gradients in three cities in the Pampean region of Argentina. The gradient analysed by Leveau (2019) corresponds to the ecological gradients of productivity that Díaz (2019) determined to be part of the theory of gradients of species richness. We can conclude that these types of work fall within the domain of application of gradient theory as do others carried out in different parts of the world.

The theory explains gradients by establishing the environmental factors associated with spatial variation in the species richness of biota of a given taxon at a certain scale of study across a set of sites. Thus, in the case of Leveau (2019) he establishes the relationship between species richness and primary productivity (NDVI) as has been done for numerous gradients at different scales and regions. According to Díaz (2019), from local gradients, such as a forest or a city, to larger ones such as a region or a continent, what varies is the concrete mathematical form that the equation takes, depending on the specific relationship established between environmental factors and species richness, but they are all specialisations of the same theory. But this situation does not change the application of the gradient theory of ecology since, as Willig and Scheiner (2011) stated, ecological relationships can vary as a function of space or time scale and so do the environmental factors involved.

In the same sense, let us think of other situations, those where the environmental factors involved are not of natural origin but anthropogenic, as in the case of Willigalla and Fartmann (2012), who found an association of *Odonata* species richness in urban-rural gradients in Central Europe with the number and type of water bodies present. Again Willig and Scheiner (2011) mention that as we change scales the main environmental factors involved change, but all these cases are included in models of the same theory. In the case of urban ecosystems, the factors that determine the variation in species richness of the taxa studied are associated with anthropogenic processes or threats, but this does not mean that they cannot be included as specialisations of gradient theory.

By establishing in the case of gradient theory, or any other theory of ecology, that it can be applied to the problems of biodiversity conservation in cities, among other things it allows us to determine the environmental factors involved and in this way we can obtain an ecological understanding of the processes involved in the problem and be able to predict the behaviour of the systems. In this way we can answer the questions posed at the beginning and establish that the theories of ecology can play a central role in the explanation and prediction of environmental problems in cities, given that these are within their domain of application. We must remember as we have mentioned that the conservation problems of cities are not exclusively a problem of ecology and that other disciplines are needed to analyse them holistically.

As a field of research that combines the natural sciences, the social sciences and the humanities, urban ecologist recognizes the evolution of urban landscapes as an inherently intertwined social and ecological process (Dooling et al., 2007; Fellowes et al., 2020). For this reason, a plethora of authors highlights the interdisciplinary nature of urban ecology (Grove & Burch, 1997; Collins et al., 2000; Dow, 2000; Grimm et al., 2000; May, 2004; Musacchio & Wu, 2004; Pickett et al., 2004; Felson & Pickett, 2005; Gratani & Varone, 2006; Marco et al., 2008; Lubbe et al., 2010; Kendal et al., 2010; Rumble et al., 2019; Goddard et al., 2021; Angeoletto et al., 2022; Souza et al., 2023). According to Grimm et al. (2008) and Angeoletto et al. (2019), urban ecology is a science that integrates the theories and methodologies of the natural and social sciences to investigate patterns and processes of urban ecological systems. Dow (2000) defines urban ecology as a new interdisciplinary field in which the integration between ecology and social sciences results in one of the most active intersections between biophysical and social processes.

Maybe the greatest obstacle to achieving a greater understanding of the ecology of tropical cities lies in the low frequency wich biologists and natural scientists adopt interdisciplinary approach, i.e., take social science theories and methods to produce knowledge about ecology and urban biodiversity. Conservation biologists and other natural scientists often do not have the knowledge and training to implement studies on the human dimensions of urban biodiversity conservation (Angeoletto & Santos, 2015; Voigt & Kindgston, 2016). They should adopt an interdisciplinary approach and work in partnership with social sciences researchers, in order to overcome that gap.

References

Adams, C. (1917). The new natural history-ecology. *American Museum Journal, 7*, 491–494.

Alberti, M., Marzluff, J. M., Ryan, C., Shulenberger, E., & Zumbrunnen, C. (2003). Integrating humans into ecology: Opportunities and challenges for studying urban ecosystems. *BioScience, 53*(12), 1169–1179.

Angeoletto, F., & Santos, J. W. M. C. (2015). Los biólogos brasileños no habitan en el planeta ciudad: por qué es urgente formar ecólogos urbanos. *Revista Espaço Acadêmico, 14*(165), 74–82.

Angeoletto, F., Leandro, D. D. S., & Fellowes, M. D. (2019). The consequences of Brazil's lack of transport planning is written in the blood of sparrows. *Urban Geography, 40*(8), 1191–1197.

Angeoletto, F., Tryjanowski, P., Santos, J., Martinez-Miranzo, B., Leandro, D., Bohrer, J., et al. (2022). Will Brazilian city dwellers actively engage in urban conservation? A case study with the charismatic neotropical Blue-and-yellow macaw (Ara ararauna). *Birds, 3*(2), 234–244.

Beatty, J. (1981). What's wrong with the received view of evolutionary theory? In P. D. Asquith & T. Nickles (Eds.), *Proceedings of the biennial meeting of the philosophy of science association* (Vol. 1980(2), pp. 397–426). Philosophy of Science Association.

Beatty, J. (1987). On behalf of the semantic view. *Biology and Philosophy, 2*, 17–23.

Bennett, A. F., Haslem, A., Cheal, D. C., Clarke, M. F., Jones, R. N., Koehn, J. D., et al. (2009). Ecological processes: A key element in strategies for nature conservation. *Ecological Management & Restoration, 10*(3), 192–199.

Berryman, A. (2003). On principles, laws and theory in population ecology. *Oikos, 103*(3), 695–701.

Castle, D. (2001). A semantic view of ecological theories. *Dialectica, 55*(1), 51–65.
Chave, J. (2013). The problem of pattern and scale in ecology: What have we learned in 20 years? *Ecology Letters, 16*, 4–16.
Collins, J. P., Kinzig, A., Grimm, N. B., Fagan, W. F., Hope, D., Wu, J., & Borer, E. T. (2000). A new urban ecology: Modeling human communities as integral parts of ecosystems poses special problems for the development and testing of ecological theory. *American Scientist, 88*(5), 416–425.
Colwell, R. (2011). Biogeographical gradient theory. In S. Schneider & M. Willlig (Eds.), *The theory of ecology*. The University Chicago Press.
Cooper, G. (2003). *The science of the struggle for existence. On the foundations of ecology*. Cambridge University Press.
Cueto, V. R. (2006). escalas en ecología: su importancia para el estudio de la selección de hábitat en aves. *Hornero, 21*(1), 1–13.
Di Pasquo, F. (2013). Una historia de la problemática ambiental y de sus efectos sobre la ecología disciplinar. *Scientiæ Studia, 11*(3), 557–581.
Di Pasquo, F., Folguera, G., & Onna, A. (2011). La ecología disciplinar y la intrusión de la problemática ambiental: hacia la "percepción de fenómenos globales". *Observatorio Medioambiental, 14*, 21–39.
Díaz, M. (2019). *Reconstrucción de la Teoría de gradientes espaciales de riqueza de especies (Tesis doctoral)*. Universidad Nacional de Tres de Febrero.
Diaz, M. (2021). El estatus de las teorías de la ecología y su rol en los problemas ambientales. In F. Bernabé, F. di Pascuo, L. Giri, C. Rendon, & J. Sutz (Eds.), *Filosofía e historia de la ciencia y sociedad en Latinoamérica Vol. 1: Medio ambiente y sociedad / Política científica* (pp. 36–51). Associação de Filosofia e História da Ciência do Cone Sul (AFHIC). ISBN: 978-65-86622-01-0.
Díaz, M., & Lorenzano, P. (2017). La teoría de la dinámica de poblaciones. *Scientie Studia, 15*(2), 307–342.
Dooling, S., Graybill, J., & Greve, A. (2007). Response to Young and Wolf: goal attainment in urban ecology research. *Urban Ecosystems, 10*, 339–347.
Dow, K. (2000). Social dimensions of gradients in urban ecosystems. *Urban Ecosystems, 4*, 255–275.
Drayton, B., & Primack, R. B. (1996). Plant species lost in an isolated conservation area in metropolitan Boston from 1894 to 1993. *Conservation Biology, 10*(1), 30–39.
Ellis, E. C., Gauthier, N., Goldewijk, K. K., Bird, R. B., Boivin, N., Díaz, S., et al. (2021). People have shaped most of terrestrial nature for at least 12,000 years. *Proceedings of the National Academy of Sciences, 118*(17).
Falguera, J. L. (1994). Unidad de noción bajo los usos del término modelo en las ciencias matemáticas y factuales. *Contextos, 23*, 221–244.
Fellowes, M. D., Acquaah-Harrison, K., Angeoletto, F., Santos, J. W., da Silva Leandro, D., Rocha, E. A., ... & Thomas, R. L. (2020). Map-a-mole: Greenspace area influences the presence and abundance of the European mole Talpa europaea in urban habitats. *Animals, 10*(6), 1097.
Felson, A. J., & Pickett, S. T. (2005). Designed experiments: New approaches to studying urban ecosystems. *Frontiers in Ecology and the Environment, 3*(10), 549–556.
Field, R., Hawkins, B., Cornell, H., Currie, D., Diniz-Filho, J., Guégan, J., Kaufman, D., Kerr, J., Mittelbach, G., Oberdorff, T., O'Brien, E., & Turner, J. (2008). Spatial species-richnness gradients across scales: A meta-analysis. *Journal of Biogeography, 36*(1), 132–147.
Ford, E. D. (2004). *Scientific method for ecological research*. Cambridge University Press.
Ginzburg, L. (1986). The theory of population dynamics: I. Back to first principles. *Journal of Theoretical Biology, 122*, 385–399.
Ginzburg, L., & Colyvan, M. (2004). *Ecological orbits how planets move and populations grow*. Oxford University Press.
Goddard, M. A., Davies, Z. G., Guenat, S., Ferguson, M. J., Fisher, J. C., Akanni, A., et al. (2021). A global horizon scan of the future impacts of robotics and autonomous systems on urban ecosystems. *Nature Ecology & Evolution, 5*(2), 219–230.

Gratani, L., & Varone, L. (2006). Carbon sequestration by Quercus ilex L. and Quercus pubescens Willd. and their contribution to decreasing air temperature in Rome. *Urban Ecosystems, 9*, 27–37.
Grimm, N. B., Grove, J. G., Pickett, S. T., & Redman, C. L. (2000). Integrated approaches to long-term studies of urban ecological systems: Urban ecological systems present multiple challenges to ecologists—Pervasive human impact and extreme heterogeneity of cities, and the need to integrate social and ecological approaches, concepts, and theory. *BioScience, 50*(7), 571–584.
Grimm, N. B., Faeth, S. H., Golubiewski, N. E., Redman, C. L., Wu, J., Bai, X., & Briggs, J. M. (2008). Global change and the ecology of cities. *Science, 319*(5864), 756–760.
Grove, J. M., & Burch, W. R. (1997). A social ecology approach and applications of urban ecosystem and landscape analyses: A case study of Baltimore, Maryland. *Urban Ecosystems, 1*, 259–275.
Haeckel, E. (1866). *Generelle Morphologie der Organismen*. Verlag Von Georf Reimer.
Ibarra, A., & Larrañaga, J. (2011). De las redes teóricas a las constelaciones de elementos teóricos: Las prácticas científicas en la Ecología de Poblaciones. *Metatheoria, 1*(2), 167–193.
Jax, K. (2001). History of ecology. In *Encyclopedia of Life Science*. Nature Publishing Group.
Kendal, D., Williams, N. S., & Williams, K. J. (2010). Harnessing diversity in gardens through individual decision makers. *Trends in Ecology and Evolution, 25*(4), 201–202.
Kolasa, J. (2011). Theory makes ecology evolve. In *The theory of ecology* (pp. 21–50).
Lawton, J. (1999). Are the general Laws in ecology? *Oikos, 64*(2), 177–192.
Leveau, L. M. (2019). Primary productivity and habitat diversity predict bird species richness and composition along urban-rural gradients of Central Argentina. *Urban Forestry & Urban Greening, 43*, 126349.
Levin, S. A. (1992). The problem of pattern and scale in ecology: The Robert MacArthur award lecture. *Ecology, 73*(6), 1943–1967.
Lloyd, E. A. (1988). *The structure and confirmation of evolutionary theory. Greenwood Press, New York.*
Lorenzano, C. (2007). La estructura ejemplar de la bioquímica. *Revista de Filosofía, 32*(1), 7–31.
Lorenzano, P., & Díaz, M. A. (2020). Laws, models, and theories in biology: A unifying interpretation. In *Life and evolution: Latin American essays on the history and philosophy of biology* (pp. 163–207). Springer.
Lubbe, C. S., Siebert, S. J., & Cilliers, S. S. (2010). Political legacy of South Africa affects the plant diversity patterns of urban domestic gardens along a socio-economic gradient. *Scientific Research and Essays, 5*(19), 2900–2910.
MacIntosh, R. (1980). The background and some current problems of theretical ecology. *Synthese, 43*, 195–255.
Marco, A., Dutoit, T., Deschamps-Cottin, M., Mauffrey, J. F., Vennetier, M., & Bertaudière-Montes, V. (2008). Gardens in urbanizing rural areas reveal an unexpected floral diversity related to housing density. *Comptes Rendus Biologies, 331*(6), 452–465.
Marquet, P. A., Allen, A., & Brown, J. (2014). On theory in ecology. *SFI Working Paper: 2014-06-018*
May, R. (1981). The role of theory in ecology. *American Zoologist, 21*, 903–910.
May, R. (2004). On the role of the humanities in urban ecology: The case of St. Petersburg. *Urban Ecosystems, 7*, 7–15.
Mayr, E. (1995). *Así es la biología*. Ediciones Debate.
Murray, B. (2001). Are ecological and evolutionary theories scientific? *Biological Review, 76*, 255–289.
Musacchio, L. R., & Wu, J. (2004). Collaborative landscape-scale ecological research: Emerging trends in urban and regional ecology. *Urban Ecosystems, 7*, 175–178.
Niemelä, J. (1999). Is there a need for a theory of urban ecology? *Urban Ecosystems, 3*(1), 57–65.
Noss, R. (1990). Indicators for monitoring biodiversity: A hierarchical approach. *Conservation Biology, 4*(4), 355–364.

O'Brien, E. (2006). Biological relativity to water-energy dynamics. *Journal of Biogeography, 33*, 1868–1888.

O'Hara, R. B. (2005). The anarchis's guide to ecological theory. Or, we don't need no stinkin' laws. *Oikos, 110*(2), 390–393.

Odum, E. (1977). The emergence of ecology as a new integrative discipline. *Science, 195*(4284), 1289–1293.

Paslaru, V. (2009). Ecological explanation between manipulation and mechanism description. *Philos Sci, 76*, 821–837.

Peters, R. (1991). *A critique for ecology*. Cambridge Univ. Press.

Pickett, S. T., & Grove, J. M. (2009). Urban ecosystems: What would Tansley do? *Urban Ecosystems, 12*, 1–8.

Pickett, S. T., Cadenasso, M. L., & Grove, J. M. (2004). Resilient cities: Meaning, models, and metaphor for integrating the ecological, socio-economic, and planning realms. *Landscape and Urban Planning, 69*(4), 369–384.

Pickett, S. T., Kolasa, J., & Jones, C. G. (2010). *Ecological understanding: The nature of theory and the theory of nature*. Elsevier.

Pickett, S. T., Cadenasso, M. L., Childers, D. L., Mcdonnell, M. J., & Zhou, W. (2016). Evolution and future of urban ecological science: Ecology in, of, and for the city. *Ecosystem Health and Sustainability, 2*(7), e01229.

Pörtner, H. O., Scholes, R. J., Agard, J., Archer, E., Arneth, A., Bai, X., Barnes, D., Burrows, M., Chan, L., Cheung, W. L., Diamond, S., Donatti, C., Duarte, C., Eisenhauer, N., Foden, W., Gasalla, M. A., Handa, C., Hickler, T., Hoegh-Guldberg, O., Ichii, K., Jacob, U., Insarov, G., Kiessling, W., Leadley, P., Leemans, R., Levin, L., Lim, M., Maharaj, S., Managi, S., Marquet, P. A., McElwee, P., Midgley, G., Oberdorff, T., Obura, D., Osman, E., Pandit, R., Pascual, U., Pires, A. P. F., Popp, A., Reyes-García, V., Sankaran, M., Settele, J., Shin, Y. J., Sintayehu, D. W., Smith, P., Steiner, N., Strassburg, B., Sukumar, R., Trisos, C., Val, A. L., Wu, J., Aldrian, E., Parmesan, C., Pichs-Madruga, R., Roberts, D. C., Rogers, A. D., Díaz, S., Fischer, M., Hashimoto, S., Lavorel, S., Wu, N., & Ngo, H. T. (2021). *IPBES-IPCC co-sponsored workshop report on biodiversity and climate change*. IPBES and IPCC. https://doi.org/10.5281/zenodo.4782538

Raerinne, J. (2011). Causal and mechanistic explanations in ecology. *Acta Biotheoretica, 59*(3–4), 251–271.

Rega, C. C., Nilon, C. H., & Warren, P. S. (2015). Avian abundance patterns in relation to the distribution of small urban greenspaces. *Journal of Urban Planning and Development, 141*(3), A4015002.

Rumble, H., Angeoletto, F., Connop, S., Goddard, M. A., & Nash, C. (2019). Understanding and applying ecological principles in cities. In *Planning cities with nature: Theories, strategies and methods* (pp. 217–234).

Santangelo, J. S., Ness, R. W., Cohan, B., Fitzpatrick, C. R., Innes, S. G., Koch, S., et al. (2022). Global urban environmental change drives adaptation in white clover. *Science, 375*(6586), 1275–1281.

Savard, J. P. L., Clergeau, P., & Mennechez, G. (2000). Biodiversity concepts and urban ecosystems. *Landscape and Urban Planning, 48*(3–4), 131–142.

Scheiner, S., & Willig, M. (2005). Developing unified theories in ecology as exemplified with diversity gradients. *The American Naturalist, 166*(4):458–469.

Scheiner, S., & Willig, M. (2008). A general theory of ecology. *Theoretical Ecology, 1*, 21–28.

Scheiner, S., & Willig, M. (2011). *The theory of ecology*. The University Chicago.

Simberloff, D. (1980). A succession of paradigm in ecology: Essentialism to materialism and probabilism. *Synthese, 43*, 3–39.

Souza, F. L., Fenoglio, M. S., & Angeoletto, F. (2023). To be a brazilian city dweller, sometimes we must learn to say enough!. *Sustainability, 15*(4), 3699.

Tansley, A. (1935). The use and abuse of vegetational concepts and terms. *Ecology, 16*(3), 284–309.

Thompson, P. (1989). *The structure of biological theories*. State University of New York Press.

Turchin, P. (2001). Does population ecology have general laws? *Oikos, 94*, 17–26.

Voigt, C. C., & Kingston, T. (2016). *Bats in the Anthropocene: Conservation of bats in a changing world* (p. 606). Springer Nature.

Wiens, J. (1989). Spatial scaling in ecology. *Functional Ecology, 3*, 385–397.

Willigalla, C., & Fartmann, T. (2012). Patterns in the diversity of dragonflies (Odonata) in cities across Central Europe. *European Journal of Entomology, 109*(2), 235.

Chapter 3
Green Forest, Gray Cities in the Amazon

Roberta Mendonça De Carvalho and Claudio Fabian Szlafzstein

Abstract This chapter characterizes and analyzes urban trees in Brazilian Amazonian cities and the relationship with geographic and socioeconomic demographics, considering the links between the triad urbanization, environmental conditions, and economic performance. Furthermore, it proposes actions to address the challenges and promote environmental justice. Despite the critical role of urban green cover for environmental management and urban livelihood, the intra-city deficit of green areas is frequently disregarded in the Brazilian Amazon Rainforest, leading to a trend of cities within the green forests turning gray. The region's constant changes away from its green roots are deeply linked with human occupation and urbanization processes, manifested through an urbanization rate of nearly 80%. Inadequate planning and lack of infrastructure results in environmental, social, and spatial inequalities within Brazil and the Amazon Region. Amazonian cities' urban tree cover and GINI index are considerably lower than national averages, indicating socioeconomic and environmental inequalities. Economic discrepancies are likewise reflected in lower degrees of urban infrastructure and regional GDP. This chapter identifies the cities' primary revenue sources to help explain the relatively poor financial performance of Amazonian cities and their comparatively lower percentage of urban trees and infrastructure in correlation. In municipal budgetary design, urban infrastructure and green cover are designated non-mandatory expenses and rely on cities' financial resource surpluses. The lack of investment also contributes for Amazonian cities transforming into vegetation deserts deprived of green spaces. *Grayer* cities suggest that inequalities reflect adaptation to urbanization and wealth distribution. It seems that urban green areas are not among public management priorities, reducing the provision of vital ecosystem services and weakening the necessary foundations for urban sustainability. Considering the current stage of

R. M. De Carvalho (✉)
University of Pittsburgh, Urban Studies Program and Center for Latin American Studies, Pittsburgh, PA, USA
e-mail: rmdecarvalho@pitt.edu

C. F. Szlafzstein
Federal University of Pará, Center for Advanced Amazonian Studies (NAEA), Belém, Brazil

F. Angeoletto et al. (eds.), *Ecology of Tropical Cities*,
https://doi.org/10.1007/978-3-031-48012-6_3

tropical urbanization and a near scenario of minor to no government intervention, the patterns point to a *grayer* future for the Amazon and its cities.

Keywords Urbanization · Amazon · Urban Trees · Urban Deforestation · Sustainable Development

1 Introduction

The rapid global urbanization and increasing climate crises have pushed cities to innovate in environmental management strategies to enhance urban living conditions. One crucial aspect in this regard is the role of urban green cover, including street trees, which has been widely recognized for its ability to enhance the provision of ecosystem services. These services encompass mitigating high temperatures (Chen et al., 2014; Park et al., 2012), improving air quality (Janhäll, 2015; Leung et al., 2011; Elmqvist et al., 2003; Setälä et al., 2013), and sequestering greenhouse gases associated with climate change (Velasco et al., 2016). Developed cities around the world are implementing programs to safeguard and create new green spaces such as parks, tree corridors, urban farms, and vertical gardens (Kabisch & Haase, 2014; Willsher, 2021; City of Paris Let's Vegetate the City, 2021). However, in developing countries, particularly in tropical regions, there are limited examples of conserving and improving urban green spaces, which are crucial components for promoting urban livelihood and well-being (Nath et al., 2018). Within the Brazilian Rainforest, a region experiencing significant urban growth as new urban centers arise and others expand, the intra-city deficit of green space is largely disregarded. Consequently, Amazon's urban areas suffer from the lack and loss of green spaces. In an area known for its forest, it is expected that cities would also exhibit a green landscape. Unfortunately, the current situation often reveals low green coverage and the loss of urban forests suggesting that the cities in yet green forests are turning gray (De Carvalho & Szlafsztein, 2018).

Efforts to conserve and restore forests are closely linked to regional sustainability, biodiversity (Antonelli et al., 2018), environmental degradation, deforestation (Barona et al., 2010; Sonter et al., 2017), climate change (Cook & Vizy, 2008; Soares-Filho et al., 2006), and the well-being of local and indigenous communities (Carneiro da Cunha & de Almeida, 2000). However, these endeavors rarely focus on urban areas or address the impacts of rapid urbanization as drivers of intra-city green loss (Ferreira & Vieira, 2018; Pereira et al., 2016; Rabello & Rodrigues, 2013). In a forest frontier, cities seem to be excluded from the right to be green.

In the last decades, an intense urbanization process in the Brazilian Amazon has transformed the forest's profile, justifying its new label as an "urban forest" (Becker, 2013). The Brazilian Amazon's fast-paced urbanization, reaching over 70%, has been guided by both intra- and inter-regional migration, strongly encouraged by government policies aiming at a regional local economic development that also failed to account for urban environmental planning. Current outcomes of unplanned and austere urban expansion include, but are not limited to, regional poverty and low urban quality of life, often at the expense of the exploitation of natural resources and, consequently, the annihilation of natural environments within and outside urban boundaries (Eloy & Brondizio, 2015; Padoch et al., 2008).

Furthermore, the assessment of urban environmental conditions often tends to restrict foci to aspects such as sanitation, environmental management, solid waste, and air quality (do Nascimento & de Mauro, 2006; Pereira et al., 2020; Tostes & de Carvalho Ferreira, 2017). Unfortunately, vital aspects of urban environmental sustainability, such as urban green cover, are seldom addressed in research or local public policy (Vieira & Panagopoulos, 2020; Amaral et al., 2012), indicating an outdated understanding of the components necessary to promote urban sustainability.

The absence, loss, and restricted access to urban green spaces are frequently linked to inadequate socioeconomic dynamics and significant levels of inequality within cities (Davis et al., 2012; Ernstson, 2013). This pervasive lack of urban greenery undermines the quality of urban environments and hinders the provision of essential ecosystem services in Amazonian cities. In light of these concerns, the primary objective of this chapter is to shed light on the significance of addressing the urbanization process in the Brazilian Amazon. It aims to characterize and analyze the current state of urban trees in Amazonian state capital cities, exploring their relationship with geographic and socioeconomic demographics, and examining the intricate interconnections among urbanization, environmental conditions, and economic performance.

Moreover, this chapter critically evaluates the design of public policies that contribute to lower economic performance and constraints on municipal budgets, thereby exacerbating spatial and economic inequalities, evidencing social and environmental injustice, and placing these cities at the lower end of the economic spectrum within the country. This chapter also proposes a set of public policies and actions to tackle improving and creating urban green areas, and promoting environmental justice in Amazonian cities. The emergence of gray cities in the Amazon region clearly indicates an unsustainable trajectory, reflecting the complex drivers and outcomes of urbanization in the area, ultimately leading the forest and its cities into a *grayer* future.

2 Amazonia: Changes in Forests, Biodiversity, and Cities

The Amazon River Basin, known for its rich biodiversity and expansive rainforest, is as one of the world's most diverse ecosystems. This region harbors approximately 30% of the planet's species of flora and fauna, many of which are endemic. The basin plays a crucial role in maintaining global climate equilibrium, offering invaluable ecosystem services (Laurance et al., 2018; Fontana et al., 2000). This multinational region spans nine countries, 65% falling within Brazilian territory.

In Brazil, the region encompasses various administrative divisions, including the North Region and the Brazilian Legal Amazon (BLA), which also comprises municipalities in the states of Mato Grosso, Tocantins, and Maranhão. For this study, the focus will be on six Brazilian states that are part of the North Region and which territories are fully inserted in the Amazon Rainforest biome. Together, these states cover an area of 3.5 million square kilometers, equivalent to 45% of Brazil's total land area, and are home to 8% of the country's population (IBGE, 2018). In this study, this area will be referred to as the Amazon Region (Fig. 3.1) or simply the Amazon.

Economic activities in the Brazilian Amazon region predominantly rely on unsustainable practices such as the extraction of natural mineral deposits (dos Santos, 2002; Diele-Viegas et al., 2020), hydropower energy production (Latrubesse et al., 2017), and the expansion of agricultural frontiers (Maranhão et al., 2019). Agricultural practices, in particular, drive deforestation through forest clearance for cultivation and logging, contributing significantly to the alarming rates of deforestation and forest fires (Brando et al., 2020; Silva Junior et al., 2021). Despite these challenges, the region still maintains a noteworthy 80% of primary forest cover, primarily due to conservation and preservation efforts within protected areas and indigenous territories (Hansen et al., 2020). In comparison, other Brazilian states have an average primary forest cover of approximately 41% (MapBiomas, 2020) (Fig. 3.2).

Recent territorial occupation fostered by public policies since the 1950s has triggered constant socio and environmental pressure and change (Davidson et al., 2012; Laurance et al., 2001; Phillips et al., 2008; Timpe & Kaplan, 2017). Such changes are associated with modifications in climate conditions (Marengo, 2007; Malhi

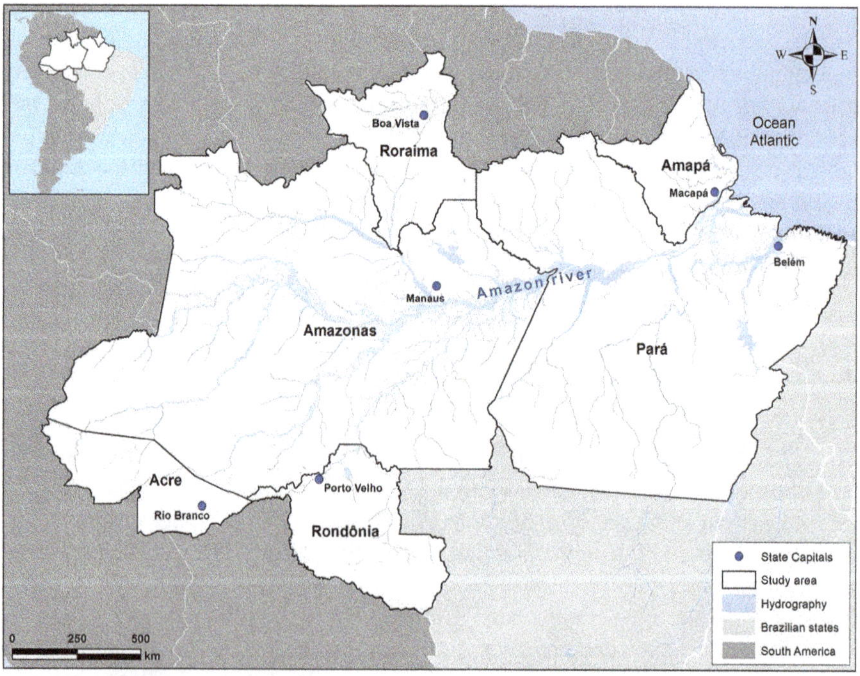

Fig. 3.1 Amazon Region Study Area

Fig. 3.2 Forest Cover in Brazil (MapBiomas, 2020)

et al., 2008), deforestation and environmental degradation (Castello et al., 2013), expansion of mining activities of natural resources (Finer et al., 2008; Oliveira et al., 2013), alteration of water flow dynamics (Shiogama et al., 2011; Tundisi et al., 2014), and human occupation and urbanization processes (Trindade, 2013).

The Brazilian Amazon has experienced significant socio-environmental pressure and change as a result of recent territorial occupation, driven mainly by public policies implemented since the 1950s (Davidson et al., 2012; Laurance et al., 2001; Phillips et al., 2008; Timpe & Kaplan, 2017). These changes have been associated with alterations in climate conditions (Marengo, 2007; Malhi et al., 2008), deforestation and environmental degradation (Castello et al., 2013), expansion of mining activities for natural resources (Finer et al., 2008; Oliveira et al., 2013), changes in water flow dynamics (Shiogama et al., 2011; Tundisi et al., 2014), and urbanization processes (Trindade, 2013) and anthropization.

Despite the importance of preserving the standing forest and the critical roles played by natural environments, urban areas within the forest biome have become significant components of the Amazon's regional landscape. The process of urbanization in the region can be traced back to the European colonial period in the sixteenth century when villages and cities were initially established along the high banks due to the need for communication and trade, and the challenges of clearing dense inland vegetation (Costa & Brondízio, 2011; Costa et al., 2019; Schor & de Oliveira, 2014). However, in the second half of the twentieth century, urbanization patterns shifted as inland cities of varying sizes emerged near newly constructed highways to support agricultural colonization, large infrastructure and energy production projects, and mineral resource exploitation. This recent wave of urbanization in the Amazon is often characterized as the "Boomtown" model, driven by intensive extraction of natural resources and attracting a new urban population seeking economic opportunities (Dufour & Piperata, 2004; Walker & Homma, 1996; Simmons et al., 2002; Perz, 2002; Miranda Rocha, 2011; Isserman et al., 2016; Thypin-Bermeo & Godfrey, 2012). The population growth, primarily due to intra-regional migration, has resulted in ongoing territorial reconfiguration. One of the visible impacts of this influx of people is the loss of natural environments and deforestation both within and outside cities, as the urban fabric expands by infilling existing towns and cities or creating of new ones.

The creation of new political municipalities is a clear indication of population increase, as legal recognition makes clustering of people a prerequisite for new political territorial units. In Brazil, the 1988 federal constitution introduced new requirements for establishing territorial divisions, leading to a significant recent surge in the creation of new municipalities (Brandt, 2010; de Tomio Limas, 2002). Since 1988, 243 new municipalities have been added to the Brazilian Legal Amazon (BLA), accounting for 30% of the total 807 municipalities in 2020 (IBGE, 2012a) (Fig. 3.3).

The population increase in the Amazon region has created a reinforcing loop, with the new municipalities and the urban population escalating from 37% of the total population in 1970 to 76% (IBGE, 2012a). Moreover, 43% of the urban population resides in primate capitals (IBGE, 2018), highlighting the centralization and

Fig. 3.3 New Municipalities in the Brazilian Legal Amazon (1950–1999)

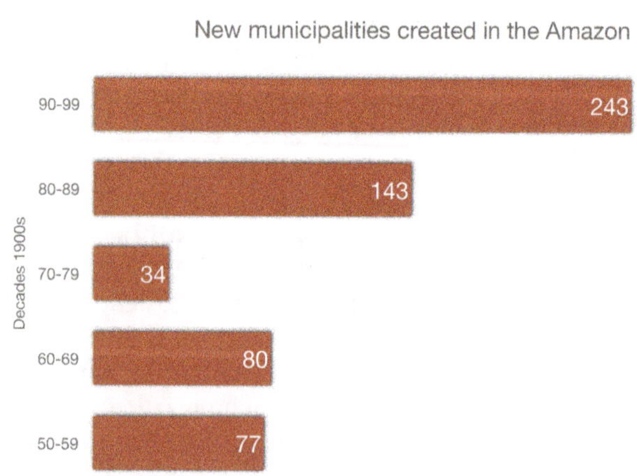

unequal distribution of resources within. Although new cities are being established, the prevalence of urban primacy indicates that the available limited resources are concentrated in state capitals, all of which are primate cities. This rapid urbanization, driven by a significant population increase and characterized by multiple growth spurts, is often associated with inadequate urban planning. Even when planning instruments exist, they often lack monitoring, and implementation falls behind schedule (Barbieri et al., 2009).

The young age of some Amazonian cities might contribute explain the accentuated differences in economic performance even within the region. The new wave of towns further compounds the challenges of promoting sustainable urbanization and developing adequate urban infrastructure, as they lack the initial infrastructure present in other regions of Brazil (Browder & Godfrey, 1997; de Oliveira et al., 2011; Perz, 1999). For example, a city founded in the 1990s, for example, has had less time to mature and implement the necessary infrastructure compared to a city established in the 1600s, such as Belém, the capital of Pará, founded in 1616. The absence of a pre-existing urban structure in newly created cities, many of which surged in a few years, with the concentration of people and the clearance of land, coupled with rapid urbanization and insufficient planning driven by spatial, social, and economic expansion, has led to unexpected outcomes and significant environmental, social, and spatial inequalities at the expenses of environmental justice. While time may have helped older primary capitals like Manaus and Belém address deficiencies in basic infrastructure, these deficiencies tend to be more pronounced in much younger cities.

3 Brazil and the Amazon Region: A Portrait of Inequalities

As the eighth largest economy in the world, Brazil's economic production is primarily based on agriculture and mining extraction and processing, with most of these commodities being exported to meet global demand (CNI, 2020). Public budget management operates at three levels of government: municipal, state, and federal. Resources are mainly allocated to mandatory expenses such as health, education, wages, pensions, and debts (Neduziak & Correia, 2017), leaving limited capital for environmental investments, especially at the municipal level. The budget distribution pyramid reflects the lower values of transfers to cities at its base (IBGE, 2012a). However, disparities and inequalities extend beyond budgetary allocations.

Socioeconomic inequalities are prevalent throughout the country and accentuated within the Amazon region. An inequality map measuring well-being and livelihood across forty indicators including education, health, social services, human rights, and environment, places five out of six Amazonian state capitals in the lowest positions among all Brazilian capitals (Cidades Sustentáveis, 2020). Brazil's GINI coefficient of 53 (on a scale of 1–100) is the highest in South America and among the highest globally (World Bank Gini index (World Bank estimate), 2021). This disparity in wealth distribution creates starkly contrasting worlds. The GINI coefficient of all but one state is worse than the national average. Socioeconomic inequalities become even more pronounced when focusing on capital cities, with a GINI coefficient of 61 in Pará's capital, compared to the national average of 51, exposing a significant portion of the population to the adverse impacts of imbalanced resource access (Cidades Sustentáveis, 2020) (Fig. 3.4). While some states exhibit higher inequality in their capital cities and others in their overall averages, the overall GINI values indicate a more unequal scenario in the Amazon compared to the rest of Brazil.

The contrast between the lush green forests and the gray urban landscapes adds another layer of inequality, highlighting the complex interplay between resource distribution and socioeconomic dynamics. The disparities in urbanization, GDP, and urban green cover contribute to this multifaceted landscape of inequality. Consequently, spatial inequalities in urban areas are reflected in the paradox of lower green cover, connecting socioeconomic processes with environmental and social injustice.

Addressing these inequalities and promoting more sustainable urban and non-urban development in the Brazilian Amazon requires concerted efforts at all levels of government and society. It necessitates equitable distribution of resources and comprehensive planning, monitoring, and implementation of environmental and social policies. Only by promoting inclusive and sustainable urbanization, fostering economic diversification, and ensuring access to essential services and infrastructure, the Amazon region could aim towards a more balanced and equitable future.

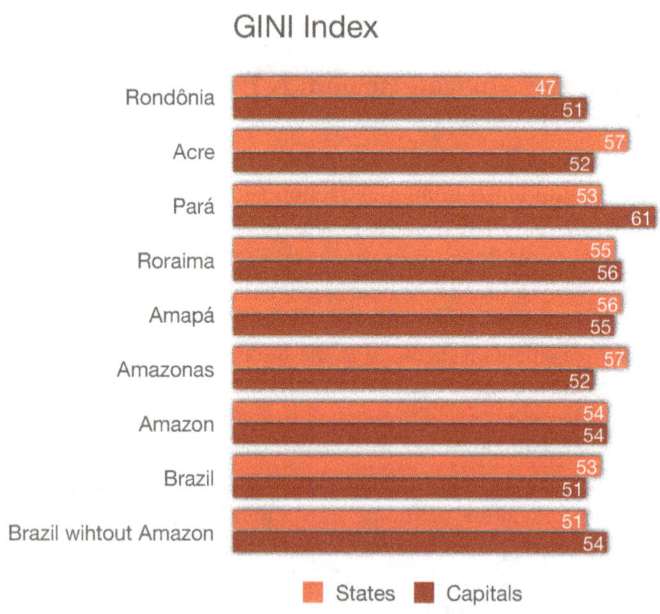

Fig. 3.4 Average of GINI Index in Brazil, Amazonian States, and Amazon Region. Scale 1–100 (Cidades Sustentáveis, 2020)

4 Urbanization, GDP, and Urban Forest Nexus

Urban infrastructure and economic inequalities are closely intertwined with spatial and environmental disparities, particularly in the availability and access to urban forests and green areas. Research highlights that cities with lower percentages of urban forests often experience a deficit in access to green spaces, contributing to reduced urban quality and livelihood (Wu & Kim, 2021; He et al., 2020). Unequal access to green areas has detrimental effects, especially in tropical climate cities characterized by year-round high temperatures and humidity, exacerbating challenges related to essential services such as sewage, water, and electricity (de Oliveira et al., 2011). Thus, the presence or absence of green areas significantly impacts the urban environment‘s quality, particularly in tropical cities.

For the Brazilian government, cities' urbanization levels are measured by infrastructure elements on urban roads, such as public lighting, maintenance holes, sidewalks, and pavement (IBGE, 2012a). Such characteristics and factors are closely associated with historical and cultural processes, different managerial styles and priorities, and the population's social and economic conditions (Brito et al., 2018; Santos, 2010). Heterogeneity within Brazilian cities has its origins in such processes.

The Brazilian government measures the level of urbanization (*grau de urbanização*) in cities based on infrastructure elements present on urban roads, such as public lighting, maintenance holes, sidewalks, and pavement (IBGE, 2012a). These characteristics and factors are influenced by historical and cultural processes, varying public government managerial styles and priorities, and the social and economic conditions of the population (Santos, 2010; Brito et al., 2018). The heterogeneity observed in Brazilian cities can be traced back to these underlying processes.

When examining urbanization level metrics, Brazilian cities generally demonstrate low urbanization, often below 16%. Only a few exceptions, such as the capital city Brasília, and select municipalities Rio de Janeiro, including the former national capital, Rio de Janeiro city itself, surpass the threshold of 40% urbanization.

However, disparities become even more pronounced when comparing cities in the Amazonian states to the national average. Urbanization levels according to IBGE's methodology, corresponds to the presence of urban infrastructure, and in the Amazon are consistently lower than the national average. Even the state with the highest level of urban infrastructure in the Amazon, Roraima, falls below the national average (IBGE, 2012a) (Fig. 3.5). Even considering the national low levels of urbanization at 11%, if the Amazon region is removed from the count, the national average improves to 16%. The discrepancies in basic urban infrastructure in the Amazon region are even lower. Moreover, all Amazonian but one states fall below 8%, with two out of the six Amazonian states exhibiting a mere 2%.

The significant disparity between the Amazon region and the rest of Brazil is also evident when considering their respective GDP figures. Economic performance outside the Amazon region demonstrates a concentration of large-scale and high-technology agriculture, efficient transportation networks, vertical industrial production, and a stronger presence of third-sector economic activities like tourism, education, and other services. In contrast, the Amazon region lags behind due to its

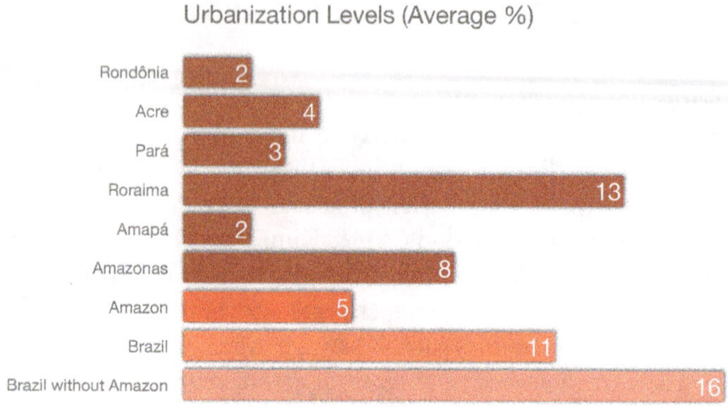

Fig. 3.5 Urbanization level municipal average in Brazil, Amazonian States, and Amazon Region (IBGE 2010)

unequal development conditions, which are closely linked to the lack of adequate infrastructure, resulting in worse economic performance.

The sizable discrepancy dividing the Amazon from the rest of Brazil is also reflected in GDP. Higher economic performances outside the region evidences a concentration of large-scale and high-technology agriculture, transportation networks, vertical industrial production, and a more substantial presence of third-sector economic activities such as tourism, education, and other services. Once more, unequal development conditions place the Amazon at the end of the spectrum, linking lack of infrastructure to poorer economic performance.

The differences in GDP reflect the logical connections between socioeconomic, urban, and environmental inequalities. In 2018, the Amazon region contributed merely 5% to the national GDP. Excluding Rondônia once more, the average GDP per capita of the Amazon states significantly falls below the national average. Even when excluding the Amazon region from the calculation, the average GDP per capita remains lower than the Brazilian performance. The state of Amazonas, despite being the largest state in terms of land area, home to an industrial pole, and the only free trade zone in Brazil, exhibits the lowest GDP, with differences exceeding 100% if compared to the national average without the Amazon region. These disparities in economic performance are magnified when analyzing the Brazilian average without including the Amazon region (IBGE, 2019) (Fig. 3.6), aligning with the performance of level of urbanization.

The unequal geographical distribution of GDP, GDP per capita, and urbanization level in the Amazon region aligns with the low percentage of urban tree cover observed in its cities, highlighting the connection between economic performance and urban environmental (in)justice. While basic urban infrastructure remains a challenge in Amazonian cities, developing of a healthy urban green infrastructure is even further behind. The scarcity of urban represented by urban tree cover significantly affects the prerequisites for a sustainable and thriving city, calling for broader discussions on sustainability beyond the forest.

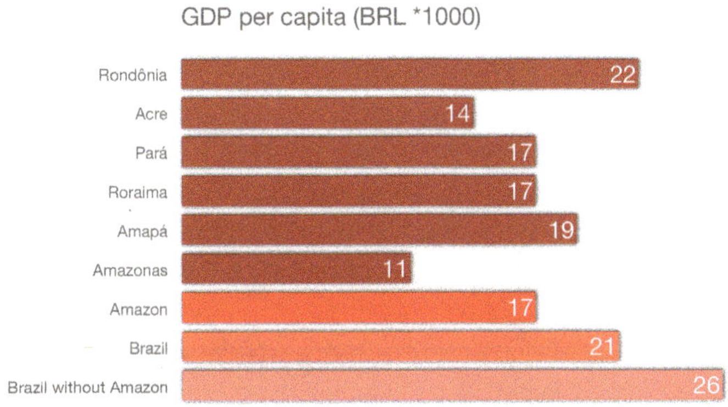

Fig. 3.6 GDP per capita in Brazil, Amazonian States, and Amazon Region in 2018 (IBGE, 2019)

One would expect cities within forested areas to have a higher percentage of urban tree cover, considering their proximity to rich natural environments abundant with vegetation. However, this expectation does not hold true for Amazonian cities. Often, urban areas in the Amazon display sparse green cover, characterized by treeless roads and a lack of public parks and squares.

The Brazilian government measures the level of urban tree cover (*arborização em vias públicas*) based on the percentage of trees along public roads (IBGE, 2012a). While vegetation cover percentages in Amazonian cities are generally lower than the national average, once more they fall significantly below the national average when excluding the Amazon region. The disparity is substantial, with the national average without the Amazon being almost three times higher than the least green state, Acre. None of the Amazonian states come close to reaching the national average. Surprisingly, the urban tree cover average in the Amazon region is only 35%, far below the average of 66% observed elsewhere. Ironically, the largest rainforest in the world contributes to lowering national urban green rates from 66% to 50%. Even the state with the highest percentage of urban tree cover, Amapá, has a rate of only 42%. Conversely, states outside the Amazon, despite having lost most of their native forests, maintain a higher level of urban tree cover, resulting in greener cities overall (Fig. 3.7).

In most cases, the Brazilian national average without the Amazon nearly reaches or surpasses 100% of urban tree cover. Even the smallest difference, found in Amapá, places the national average without the Amazon at a staggering 57% higher level. The average of 35% for urban tree cover in Amazonian states points to cities with lacking greenery. Interestingly, cities situated in historically degraded areas in the last 500 years, such as the Mata Atlântica Forest, exhibit higher percentages of green cover (Freitas et al., 2010). While cities in São Paulo and Mato Grosso boast municipal average values surpassing 80% of urban tree cover, rates in Amazonian cities remain notably lower, peaking at a maximum of only 42%.

These differences are also evident among the capital cities of the six Amazonian states. On average, the proportion of urban tree cover in these capitals is lower

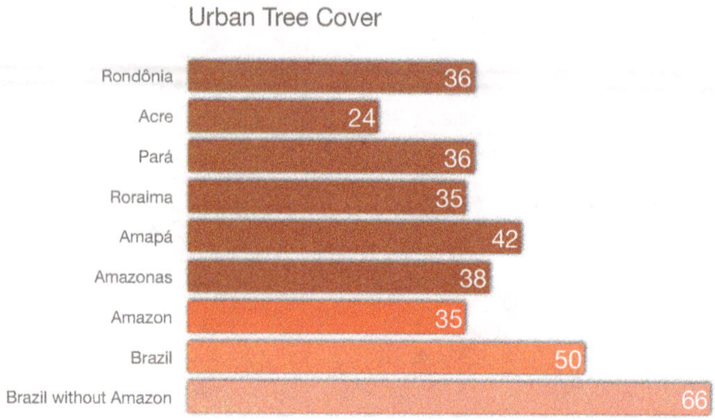

Fig. 3.7 Urban Tree Cover percentage along public roads in Brazil, Amazonian States, and Amazon Region (IBGE, 2012a)

compared to the rest of Brazil. Macapá (AP) and Boa Vista (RR) stand out with the highest rates of urban tree cover, while Rio Branco (AC) lags behind with the lowest coverage (IBGE, 2012a). It is not surprising that Macapá has the highest average among the states, while Acre has the lowest.

To analyze the nexus of urbanization, GDP, and urban tree cover, a comparative scale ranging from the best (1) to the worst (6) conditions was employed, establishing a numerical relationship between these factors (Table 3.1). The same range values were applied across variables to foster the same ground for comparison. The average values at the municipal level reveal that Amapá and Roraima exhibit the most favorable overall conditions, indicating a correlation among the urbanization, GDP and urban green cover. Amapá exhibits the highest proportion of urban trees and the second-best GDP per capita, but its performance in terms of urbanization levels is comparatively lower. In contrast, Roraima excels in urbanization levels but ranks second in urban tree cover. Amazonas showcases the highest GDP and urban tree cover rate. Inversely, Acre demonstrates a correlation between the conditions, but it has the poorest urban tree cover along public roads, which can be attributed to limited resources and challenges related to urbanization.

In addition to historical, social, cultural, and political factors contributing to lower urbanization and urban green cover in Amazonian cities, the responsibility for improving urbanization level and urban green cover falls exclusively on municipal governance. Therefore, the availability of financial resources dramatically influences the ability of cities to address these challenges.

Municipal financial resources are closely tied to their debt capacity, determined by their access to credit from the national treasury, revenues from municipal taxes, and mandatory transfers from the federal government and states. The municipal budget is divided into mandatory and non-mandatory payment categories. Mandatory expenses encompass items such as civil servant salaries, social security, education, and healthcare, while costs associated with urbanization and urban tree cover falls under non-mandatory expenses.

Table 3.1 The numerical relationship between GDP per capita, Urbanization, and Urban Tree Cover in the Amazon states (IBGE, 2012a)

	Municipal Factors				
State	GDP	Urbanization	Urban Tree	Total	Ranking
Acre	4	3	5	12	4
Amazonas	5	2	2	9	2
Amapá	2	5	1	8	1
Pará	3	4	3	10	3
Rondônia	1	5	3	9	2
Roraima	3	1	4	8	1

Table 3.2 Urban Tree Cover Average (%) of municipalities with the best debt payment conditions in the Amazon region (IBGE, 2012a)

Municipality	State	Urban Tree
Apui	Amazonas	47
Guajará		81
Silves		61
Cumaru do Norte	Pará	79
Santarém		43
Boa Vista	Roraima	48

Access to credit from the national treasury is evaluated based on three indicators: indebtedness, current savings, and the liquidity index (Brasil, 2017). The liquidity index categorizes municipalities into three descending levels of payment capacity. In the Amazon region, 17 municipalities were ranked at the highest level of payment capacity in 2021. Among these, one is in Acre six are in Amazonas, eight are in Pará, and two are in Roraima. Notably, six of these municipalities surpass the average urban tree cover percentage of 35% in the Amazon region (Brasil, 2021a) (Table 3.2). This scenario reinforces the evidence of a strong connection between urban green cover percentages and the availability of public financial resources at the municipal level.

In Brazil, municipal tax revenue accounts for only 10% of the total municipal resources. The main municipal tax, Urban Territorial Property (IPTU), faces significant challenges that hinder overall tax collection capacity in cities nationwide. Properties subject to taxation must meet specific criteria, including being located in areas with paved streets, access to water supply, sewage systems, public lighting, and elementary schools or health units within a 3-kilometer radius. The limited availability of urban infrastructure and the consequently low level of urbanization in the Amazon region undermines the collection capacity of the IPTU framework. Moreover, tax collection is further difficulted due to the absence of a comprehensive real estate registry and the implementation of social policies that grant exemptions to low-income families (de Carvalho Júnior, 2009; Del Fiori et al., 2020). These requirements face even more extreme challenges in the Amazon due to the combination of high poverty rates and recent urbanization trends.

Furthermore, the intricate chain of fund transfers between cities, states, and the national government adds complexity to the financial landscape. Federal transfers are made directly to states and municipalities, with states also transferring resources to municipalities, which becomes the primary source of income for many municipalities. The mechanisms of monetary transfers increase municipal financial resources, but the values fluctuate annually and depend on Brazil's overall tax collection capacity. The current flow of resources indicates a high level of municipal dependency on state and federal funds, particularly pronounced in the Amazon region. The limited resources available to Amazonian states and municipalities directly correspond to the low levels of municipal urbanization and urban tree cover, highlighting the connection between the lack or loss of urban greenery and economic factors.

In 2020, Amazonian states received only 16% of total federal transfers, positioning them among the seven least benefited out of the 28 federal units. This discrepancy gains relevance when considering that these values represent more than 20% of the total budgetary resources of these states. Furthermore, direct federal allocations to Amazonian municipalities accounted for an even lower value of 10% of total transfers in 2020. Municipalities in Roraima, Acre, and Amapá were among the least benefited from direct federal transfers (Brasil, 2021b; IBGE, 2012b). The Amazonian municipalities with the highest urbanization rates are those that receive the highest transfers from the federal government within their respective states, such as the state capitals Macapá (Amapá), Rio Branco (Acre), and Porto Velho (Rondônia).

In most municipalities, the limited financial resources available through transfers are associated with poor urban tree cover. However, there are some small cities outliers identified with high percentages of urban tree cover, such as Normandia (Roraima), Ministro Andreazza (Rondônia), and Brasileia (Acre). These cases, in a different way often result from unplanned urbanization that lacks proper urban infrastructure and, as a consequence, maintains a relic natural forest cover.

5 Final Considerations

The inadequate urban tree cover in Amazonian municipalities reflects a deficiency of green spaces within these cities, whether along streets or in urban parks, creating "vegetation deserts" despite being surrounded by trees. A greater number of "gray" cities in the Amazon region, distinct from other parts of Brazil, where cities with higher proportions of green space coexist with less forested territories, can be attributed to the inequalities arising from the urbanization process and wealth distribution. Within the already unequal country of Brazil, the Amazon region faces even more significant disparities.

While global attention has been focused on preserving and conserving the Amazon Rainforest, which still maintains extensive forested areas compared to other regions of Brazil despite deforestation processes, the urban centers within the Amazon region reveal a lack of trees that extends beyond the boundaries of the tropical forests.

In a national comparison, the Amazonian states consistently exhibit the lowest rates of municipal urban tree cover. These disparities in urban tree cover are also associated with variations in urbanization and GDP across Brazil, underscoring the correlation between lower revenue levels and reduced urban tree cover in the Amazon region. The region's low socioeconomic development, coupled with a history of disorderly urbanization and inadequate adherence to legal regulations on urban and territorial planning, compels decision-makers to allocate scarce resources towards what are considered higher-priority needs. Unfortunately, urban greenery is often overlooked in the list of priorities.

Contrary to what was expected for the Amazon region, the lack of green areas reduces the provision of countless vital ecosystem services for the environmental quality of urban livelihood, weakening the necessary foundations for urban sustainability, particularly in the context of climate change scenarios that urge for mitigation measures across the globe. At the current stage of the tropical urbanization process and considering a near scenario of little government intervention, the patterns point to a less green and more gray future for Amazonian cities.

To address this critical situation, several planned and autonomous strategies and measures are proposed, particularly in the state capital cities. These initiatives aim to counteract the prevailing trends and foster a more sustainable and environmentally conscious urban environment, therefore also promoting environmental justice. By promoting environmental justice, we can strive towards a more equitable and sustainable Amazon region, where all residents have the opportunity to thrive in a healthy and green urban environment.

6 Proposed Actions

A. Consolidate and expand environmental and urban policies aimed at creating municipal parks, with a particular emphasis on those situated along the banks of canals and small rivers in Amazonian cities. These linear parks play a crucial role in enhancing the green cover of the cities. Additionally, they serve as transversal strategies for urban reorganization, particularly benefiting vulnerable populations, reducing flooding risks, and improving the overall quality of life.
B. Strengthen municipal afforestation programs where existing initiatives are available or implement new programs where needed. These programs should encompass preventive maintenance for existing trees and the replacement of damaged ones. It is also important to consider expanding the number of employees involved and investing in their training. Municipal budgets should allocate an appropriate share of resources specifically for these purposes.
C. Given the absence or delayed government actions, there is a need to encourage autonomous measures by the population to promote afforestation. One approach is to inspire the reuse of discarded elements found in city streets for this purpose. For instance, worn-out tires can be repurposed as containers for accumulating soil on impermeable sidewalks. Implementing community gardens can

contribute to increasing the green cover in the city. Despite the challenges posed by heavy rainfall and high temperatures, community gardens serve as urban vegetable gardens, improving the food security of the most disadvantaged populations.

D. Develop communication strategies aimed at disseminating the importance of urban green cover for environmental conservation and climate change mitigation to society, governments, and policy funders. Campaigns should emphasize the significance of maintaining and increasing rates of urban vegetation cover for the well-being of city residents and nearby areas, considering the invaluable ecosystem services provided by vegetation. Raising awareness about these benefits will foster support and commitment towards green initiatives.

References

Amaral, D., Vieira, I., Salomão, R., Almeida, S., & Jardim, M. (2012). The status of conservation of urban forests in eastern Amazonia. *Brazilian Journal of Biology, 72*, 257–265. https://doi.org/10.1590/s1519-69842012000200005

Antonelli, A., Zizka, A., Carvalho, F. A., Scharn, R., Bacon, C. D., Silvestro, D., & Condamine, F. L. (2018). Amazonia is the primary source of Neotropical biodiversity. *Proceedings of the National Academy of Sciences, 115*, 6034–6039. https://doi.org/10.1073/pnas.1713819115

Barbieri, A. F., Monte-Mór, R. L. M., & Bilsborrow, R. E. (2009). Towns in the jungle: Exploring linkages between rural-urban mobility, urbanization and development in the Amazon. *Urban Population-Environment Dynamics in the Developing World: Case Studies and Lessons Learned*, 247–279.

Barona, E., Ramankutty, N., Hyman, G., & Coomes, O. T. (2010). The role of pasture and soybean in deforestation of the Brazilian Amazon. *Environmental Research Letters, 5*. https://doi.org/10.1088/1748-9326/5/2/024002

Becker, B. (2013). *A urbe amazônida: A floresta e a cidade*; ISBN 8576172968.

Brando, P., Macedo, M., Silvério, D., Rattis, L., Paolucci, L., Alencar, A., Coe, M., & Amorim, C. (2020). Amazon wildfires: Scenes from a foreseeable disaster. *Flora, 268*, 151609.

Brandt, C. T. (2010). A criação de municípios após a Constituição de 1988 O impacto sobre a repartição do FPM e a Emenda. *Revista Brasileira de Ciências Sociais, 47*, 59–76.

Brasil. (2017). *Portaria Nº 501, de 23 de novembro de 2017*; Ministério da Fazenda. Diário Oficial da União.

Brasil. (2021a). *Capacidade de Pagamento (CAPAG)*. Estados e Municípios.

Brasil. (2021b). *Transferências a estados e municípios*.

Brito, F., Horta, C. J. G., & Amaral, E. (2018). *A urbanização recente no Brasil e as aglomerações metropolitanas*. https://doi.org/10.31219/osf.io/84b92

Browder, J. O., & Godfrey, B. J. (1997). *Rainforest cities: Urbanization, development, and globalization of the Brazilian Amazon*. Columbia University Press. ISBN 0231106556.

Carneiro da Cunha, M., & de Almeida, W. B. (2000). Indigenous People, Traditional People, and Conservation in the Amazon. *Daedalus, 129*, 315.

Castello, L., Mcgrath, D. G., Hess, L. L., Coe, M. T., Lefebvre, P. A., Petry, P., Macedo, M. N., Renó, V. F., & Arantes, C. C. (2013). The vulnerability of Amazon freshwater ecosystems. *Conservation Letters, 6*, 217–229. https://doi.org/10.1111/conl.12008

Chen, D., Wang, X., Thatcher, M., Barnett, G., Kachenko, A., & Prince, R. (2014). Urban vegetation for reducing heat related mortality. *Environmental Pollution, 192*, 275–284.

Cidades Sustentáveis. (2020). *Mapa da Desigualdade Entre as Capitais Brasileiras 2020*.

City of Paris Let's vegetate the city. Available online: https://www.paris.fr/pages/vegetalisons-la-ville-2459 (accessed on Apr 28, 2021)

CNI (Ed.). (2020). *Brazilian Economy 2019-2020*. Confederação Nacioinal da Indústria, Ed.; CNI.

Cook, K. H., & Vizy, E. K. (2008). Effects of twenty-first-century climate change on the Amazon rain forest. *Journal of Climate, 21*, 542–560. https://doi.org/10.1175/2007JCLI1838.1

Costa, S., & Brondízio, E. (2011). Cities Along the Floodplain of the Brazilian Amazon: Characteristics and Trends. In M. Pinedo-Vasquez, M. L. Ruffino, C. Padoch, & E. S. Brondízio (Eds.), *The Amazon Várzea: The Decade Past and the Decade Ahead* (pp. 83–97). Springer Netherlands. ISBN 978-94-007-0146-5.

Costa, S., do Carmo, M. B. S., & Barja, P. R. (2019). The urban hierarchy at the delta of the Amazon River and the importance of small cities. *Urbe, 11*, 1–17. https://doi.org/10.1590/2175-3369.011.001.AO09

Davidson, E. A., de Araújo, A. C., Artaxo, P., Balch, J. K., Brown, I. F., Bustamante, M. M., Coe, M. T., DeFries, R. S., Keller, M., Longo, M., et al. (2012). The Amazon basin in transition. *Nature, 481*, 321–328. https://doi.org/10.1038/nature10717

Davis, A. Y., Belaire, J. A., Farfan, M. A., Milz, D., Sweeney, E. R., Loss, S. R., & Minor, E. S. (2012). Green infrastructure and bird diversity across an urban socioeconomic gradient. *Ecosphere, 3*, art105. https://doi.org/10.1890/es12-00126.1

de Carvalho Júnior, P. H. B. (2009). *Aspectos distributivos do IPTU e do patrimônio imobiliário das famílias brasileiras*.

De Carvalho, R. M., & Szlafsztein, C. F. (2018). Urban vegetation loss and ecosystem services: The influence on climate regulation and noise and air pollution. *Environmental Pollution, 245*, 844–852. https://doi.org/10.1016/j.envpol.2018.10.114

de Oliveira, F. L., Biasotto, R., Santos Jr, O. A., & Montandon, D. T. (2011). *Os Planos Diretores Municipais Pós-Estatuto da Cidade: balanço crítico e perspectivas*.

de Tomio Limas, F. R. (2002). A Criação de Municípios Após a Constituição de 1988. *Revista Brasileira de Ciências Sociais, 17*.

Del Fiori, D., Costa, R. R., de Souza Pio, J. L., Neves, S. F., & Kuster, E. S. S. (2020). Análise do imposto predial e territorial urbano (IPTU): aproveitamento, arrecadação e desigualdade nos municípios brasileiros. *Revista Econômica do Nordeste, 51*, 25–45.

Diele-Viegas, L. M., Pereira, E. J. D. A. L., & Rocha, C. F. D. (2020). The new Brazilian gold rush: Is Amazonia at risk? *Forest Policy and Economics, 119*, 102270.

do Nascimento, J. A. S., & de Mauro, C. A. (2006). A sustentabilidade ambiental urbana a partir de uma perspectiva espacial: o caso das cidades da Amazônia brasileira. *Mercat Geogr da UFC, 5*, 113–121.

dos Santos, B. A. (2002). Recursos minerais da Amazônia. *Estudos Avançados, 16*, 123–152. https://doi.org/10.1590/s0103-40142002000200009

Dufour, D. L., & Piperata, B. A. (2004). Rural-to-urban migration in Latin America: An update and thoughts on the model. *American Journal of Human Biology, 16*, 395–404. https://doi.org/10.1002/ajhb.20043

Elmqvist, T., Folke, C., Nyström, M., Peterson, G., Bengtsson, J., Walker, B., & Norberg, J. (2003). Response diversity, ecosystem change, and resilience. *Frontiers in Ecology and the Environment, 1*, 488–494. https://doi.org/10.1890/1540-9295(2003)001[0488:rdecar]2.0.co;2

Eloy, L., & Brondizio, E. S. (2015). New perspectives on mobility, urbanization and resource management in riverine Amazonia. *Bulletin of Latin American Research, 34*, 3–18.

Ernstson, H. (2013). The social production of ecosystem services: A framework for studying environmental justice and ecological complexity in urbanized landscapes. *Landscape and Urban Planning, 109*, 7–17. https://doi.org/10.1016/j.landurbplan.2012.10.005

Ferreira, A. E. D. M., & Vieira, I. C. G. (2018). Sustentabilidade urbana na região metropolitana de Santarém, Pará, Brasil nos anos 2000 e 2010. *Economía, sociedad y territorio, xviii*, 763–795. https://doi.org/10.22136/est20181238

Finer, M., Jenkins, C. N., Pimm, S. L., Keane, B., & Ross, C. (2008). Oil and gas projects in the Western Amazon: Threats to wilderness, biodiversity, and indigenous peoples. *PLoS One, 3*. https://doi.org/10.1371/journal.pone.0002932

Fontana, J. D., Tanner, R. D., de Farias, G. S., Rodrigues, W. A., de Jesus, M. A., & Fo, S. A. (2000). Amazon: Fauna, flora, and potential (bio) processing applications. *Economic Botony*, 543–550.

Freitas, S. R., Hawbaker, T. J., & Metzger, J. P. (2010). Effects of roads, topography, and land use on forest cover dynamics in the Brazilian Atlantic Forest. *Forest Ecology and Management, 259*, 410–417. https://doi.org/10.1016/j.foreco.2009.10.036

He, S., Wu, Y., & Wang, L. (2020). Characterizing horizontal and vertical perspectives of spatial equity for various urban green spaces: A case study of Wuhan, China. *Frontiers in Public Health, 8*, 1–11. https://doi.org/10.3389/fpubh.2020.00010

IBGE. (2012a). *Censo Demográfico*. Instituto Brasileiro de Geografia e Estatística.

IBGE. (2012b). Censo Demográfico 2010: Características urbanísticas do entorno dos domicílios. *Censo Demográfico 2010, 41*, 1–81.

IBGE. (2018). *Estimativas da população residente no Brasil e unidades da Federação com data de referência em 1 de julho de 2018*. Instituto Brasileiro de Geografia e Estatística.

IBGE. (2019). *Produto Interno Bruto dos Estados*; Vol. 57.

Isserman, A. M., Merrifield, J. D., Geography, S. E., & Jan, N. (2016). Quasi-experimental control group methods for regional analysis: An application to an energy boomtown and growth pole theory stable. *Economic Geography, 63*, 3–19.

Janhäll, S. (2015). Review on urban vegetation and particle air pollution – Deposition and dispersion. *Atmospheric Environment, 105*, 130–137. https://doi.org/10.1016/j.atmosenv.2015.01.052

Kabisch, N., & Haase, D. (2014). Green justice or just green? Provision of urban green spaces in Berlin, Germany. *Landscape and Urban Planning, 122*, 129–139. https://doi.org/10.1016/j.landurbplan.2013.11.016

Latrubesse, E. M., Arima, E. Y., Dunne, T., Park, E., Baker, V. R., D'Horta, F. M., Wight, C., Wittmann, F., Zuanon, J., Baker, P. A., et al. (2017). Damming the rivers of the Amazon basin. *Nature, 546*, 363–369. https://doi.org/10.1038/nature22333

Laurance, W. F., Cochrane, M. A., Bergen, S., Fearnside, P. M., Delamônica, P., Barber, C., D'Angelo, S., & Fernandes, T. (2001). Environment. The future of the Brazilian Amazon. *Science, 291*, 438–439.

Laurance, W. F., Camargo, J. L. C., Fearnside, P. M., Lovejoy, T. E., Williamson, G. B., Mesquita, R. C. G., Meyer, C. F. J., Bobrowiec, P. E. D., & Laurance, S. G. W. (2018). An Amazonian rainforest and its fragments as a laboratory of global change. *Biological Reviews, 93*, 223–247. https://doi.org/10.1111/brv.12343

Leung, D. Y. C., Tsui, J. K. Y., Chen, F., Yip, W.-K., Vrijmoed, L. L. P., & Liu, C.-H. (2011). Effects of urban vegetation on urban air quality. *Landscape Research, 36*, 173–188.

Malhi, Y., Roberts, J. T., Betts, R. A., Killeen, T. J., Li, W., & Nobre, C. A. (2008). Climate change, deforestation, and the fate of the Amazon. *Science (80-), 319*, 169–172. https://doi.org/10.1126/science.1146961

MapBiomas. (2020). MapBiomas general "handbook" – Algorithm Theoretical Basis Document (ATBD) – Collection 5. *Mapbiomas, 1*, 48.

Maranhão, R. L. A., de Carvalho Júnior, O. A., Hermuche, P. M., Gomes, R. A. T., McManus Pimentel, C. M., & Guimarães, R. F. (2019). The Spatiotemporal Dynamics of Soybean and Cattle Production in Brazil. *Sustainability, 11*, 2150. https://doi.org/10.3390/su11072150

Marengo, J. A. (2007). *Mudanças climáticas globais e seus efeitos sobre a biodiversidade: caracterização do clima atual e definição das alterações climáticas para o território brasileiro ao longo do século XXI*; ISBN 8577380386.

Miranda Rocha, G. (2011). A Redistribuição Espacial da População na Área de Influência da Usina Hidrelétrica Tucuruí (PA).

Nath, T. K., Han, S. S. Z., & Lechner, A. M. (2018). Urban green space and well-being in Kuala Lumpur, Malaysia. *Urban for Urban Green, 36*, 34–41.

Neduziak, L. C. R., & Correia, F. M. (2017). The allocation of government spending and economic growth: A panel data study of Brazilian states. *Revista de Administração Pública, 51*, 616–632. https://doi.org/10.1590/0034-7612155177

Oliveira, L. J. C., Costa, M. H., Soares-Filho, B. S., & Coe, M. T. (2013). Large-scale expansion of agriculture in Amazonia may be a no-win scenario. *Environmental Research Letters, 8*. https://doi.org/10.1088/1748-9326/8/2/024021

Padoch, C., et al. (2008). Urban forest and rural cities: Multi-sited households, consumption patterns, and forest resources in Amazonia. *Ecology and Society, 13*, 2. https://doi.org/10.1108/00346650810863037

Park, M., Hagishima, A., Tanimoto, J., & Narita, K. (2012). Effect of urban vegetation on outdoor thermal environment: Field measurement at a scale model site. *Building and Environment, 56*, 38–46. https://doi.org/10.1016/j.buildenv.2012.02.015

Pereira, F. D. S., Vieira, I. C. G., da Silva Pereira, F., & Célia Guimarães Vieira, I. (2016). Expansão urbana da Região Metropolitana de Belém sob a ótica de um sistema de índices de sustentabilidade. *Revista Ambiente & Água, 11*, 731–744. https://doi.org/10.4136/1980-993X

Pereira, M. M., de Amorim Lima, G. V. B., Crispim, D. L., & Fernandes, L. L. (2020). Análise de sustentabilidade no município de Parauapebas–Amazônia, Pará, Brasil. *Research, Society and Development, 9*, e53932343.

Perz, S. G. (1999). The quality of urban environments in the Brazilian Amazon. *Social Indicators Research*, 181–212.

Perz, S. G. (2002). Population growth and net migration in the Brazilian Legal Amazon, 1970–1996. In C. Wood & R. Porro (Eds.), *Deforestation and land use in the Amazon* (pp. 107–129). University Press of Florida, Gainsville.

Phillips, O. L., Lewis, S. L., Baker, T. R., Chao, K. J., & Higuchi, N. (2008). The changing Amazon forest. *Philosophical Transactions of the Royal Society B: Biological Sciences, 363*, 1819–1827. https://doi.org/10.1098/rstb.2007.0033

Rabello, R. P., & Rodrigues, Z. A. L. (2013). Planejamento e sustentabilidade urbana: ações de proteção dos Igarapés de Manaus. *Rev Meio Ambient e Sustentabilidade, 3*, 80–101.

Santos, A. M. S. P. (2010). Urbanização Brasilieira. *Estud Urbanos, 12*, 103.

Schor, T., & de Oliveira, J. A. (2014). Cities, Rivers and Urban network in the Brazilian Amazon. *Brazilian Geographical Journal: Geosciences and Humanities Research Medium, 5*, 258–276.

Setälä, H., Viippola, V., Rantalainen, A. L., Pennanen, A., & Yli-Pelkonen, V. (2013). Does urban vegetation mitigate air pollution in northern conditions? *Environmental Pollution, 183*, 104–112. https://doi.org/10.1016/j.envpol.2012.11.010

Shiogama, H., Emori, S., Hanasaki, N., Abe, M., Masutomi, Y., Takahashi, K., & Nozawa, T. (2011). Observational constraints indicate risk of drying in the Amazon basin. *Nature Communications*. https://doi.org/10.1038/ncomms1252

Silva Junior, C. H. L., Pessôa, A. C. M., Carvalho, N. S., Reis, J. B. C., Anderson, L. O., & Aragão, L. E. O. C. (2021). The Brazilian Amazon deforestation rate in 2020 is the greatest of the decade. *Nature Ecology and Evolution, 5*, 144–145. https://doi.org/10.1038/s41559-020-01368-x

Simmons, C., Perz, S., Pedlowsky, M., & Silva, L. G. T. (2002). The changing dynamics of land conflict in the Brazilian amazon: The rural-urban complex and its environmental implications. *Urban Ecosystems, 6*, 99–121.

Soares-Filho, B. S., Nepstad, D. C., Curran, L. M., Cerqueira, G. C., Garcia, R. A., Ramos, C. A., Voll, E., McDonald, A., Lefebvre, P., & Schlesinger, P. (2006). Modelling conservation in the Amazon basin. *Nature, 440*, 520–523. https://doi.org/10.1038/nature04389

Sonter, L. J., Herrera, D., Barrett, D. J., Galford, G. L., Moran, C. J., & Soares-Filho, B. S. (2017) *Mining drives extensive deforestation in the Brazilian Amazon*. https://doi.org/10.1038/s41467-017-00557-w

Thypin-Bermeo, S., & Godfrey, B. J. (2012). Envisioning Amazonian frontiers: Place-making in a Brazilian boomtown. *Journal of Cultural Geography, 29*, 215–238. https://doi.org/10.1080/08873631.2012.684948

Timpe, K., & Kaplan, D. (2017). The changing hydrology of a dammed Amazon. *Science Advances, 3*, 1–14. https://doi.org/10.1126/sciadv.1700611

Tostes, J. A., & de Carvalho Ferreira, J. F. (2017). Avaliação da Sustentabilidade na Amazônia: A mesorregião Norte do Amapá. *Revista Brasileira de Gestão e Desenvolvimento Regional, 13*.

Trindade, S.-C. (2013). Uma Floresta Urbanizada? Legado e Desdobramentos de uma Teoria sobre o Significado da Cidade e do Urbano na Amazônia. *Espaço Aberto, 3*, 89–108.

Tundisi, J. G., Goldemberg, J., Matsumura-Tundisi, T., & Saraiva, A. C. F. (2014). How many more dams in the Amazon? *Energy Policy, 74*, 703–708. https://doi.org/10.1016/j.enpol.2014.07.013

Velasco, E., Roth, M., Norford, L., & Molina, L. T. (2016). Does urban vegetation enhance carbon sequestration? *Landscape and Urban Planning, 148*, 99–107. https://doi.org/10.1016/j.landurbplan.2015.12.003

Vieira, T. A., & Panagopoulos, T. (2020). Urban forestry in Brazilian Amazonia. *Sustainability, 12*, 3235. https://doi.org/10.3390/SU12083235

Walker, R., & Homma, A. K. O. (1996). Land use and land cover dynamics in the Brazilian Amazon: An overview. *Ecological Economics, 18*, 67–80. https://doi.org/10.1016/0921-8009(96)00033-X

Willsher, K. (2021). Paris agrees to turn Champs-Élysées into "extraordinary garden". *Guard.*

World Bank Gini index (World Bank estimate). Available online: https://data.worldbank.org/indicator/SI.POV.GINI?locations=BR&most_recent_value_desc=false (accessed on May 18, 2021).

Wu, L., & Kim, S. K. (2021). Exploring the equality of accessing urban green spaces: A comparative study of 341 Chinese cities. *Ecological Indicators, 121*, 107080. https://doi.org/10.1016/j.ecolind.2020.107080

Chapter 4
Growing Cities and Shrinking Fish? Potential Urbanization Effects on Fish and Fisheries in Tropical Rivers

Renato A. M. Silvano, Friedrich W. Keppeler, and Gustavo Hallwass

Abstract Urbanization may have ecological and economic effects on small-scale fisheries in tropical rivers. However, such effects remain poorly known. Our goal is to evaluate the potential effects of urban development of major cities on small-scale freshwater fisheries in the Brazilian Amazon, through literature review and by using some results of our research in Amazonian rivers. We focused on some study cases around large cities in different Amazonian regions, including Santarém (Eastern Amazon), and Manaus (Central Amazon). We discuss the positive and negative effects of the proximity to urban centers. Negative effects may include overfishing near large urban centers and its ecological consequences, such as a decrease on fish size of both individual species and the size structure of the fish community, changes on fish diversity, and declines on fish catches of small-scale fisheries. Other negative effects include anthropic impacts of deforestation and pollution, such as eutrophication, solid waste (micro-plastics) and heavy metals (mercury). Positive effects may include improved market opportunities for fishers, in the form of increased demand,

R. A. M. Silvano (✉)
Departamento de Ecologia e Programa de Pós-Graduação em Ecologia—IB, Universidade Federal do Rio Grande do Sul (UFRGS), Porto Alegre, Rio Grande do Sul, Brasil
e-mail: renato.silvano@ufrgs.br

F. W. Keppeler
Programa de Pós-Graduação em Ecologia Aquática e Pesca, Universidade Federal do Pará (UFPA), Belém, Pará, Brasil

Departamento de Engenharia Civil, Universidade Federal do Rio Grande do Norte (UFRN), Natal, Rio Grande do Norte, Brasil

Laboratório de Ciência Marinha - LABOMAR, Universidade Federal do Ceará (UFC), Fortaleza, Ceará, Brasil

G. Hallwass
Programa de Pós-Graduação em Ecologia Aquática e Pesca, Universidade Federal do Pará (UFPA), Belém, Pará, Brasil

Instituto de Ciência, Tecnologia e Inovação, Programa de Pós-Graduação em Ecologia Aplicada, Universidade Federal de Lavras (UFLA), São Sebastião do Paraíso, Minas Gerais, Brasil

F. Angeoletto et al. (eds.), *Ecology of Tropical Cities*,
https://doi.org/10.1007/978-3-031-48012-6_4

besides higher proximity to consumers, which may shorten the market chain and make it more profitable. Another positive effect would be the engagement of fishers in tourism and other related activities. Finally, we suggested hypotheses to be tested and new avenues for future research.

Keywords Human ecology · Freshwater conservation · Food security · Small-scale fisheries · Fish ecology · Aquatic pollution · Tourism · Fishing pressure

1 Introduction

Most tropical cities are in underdeveloped countries (Sachs, 2001), where population growth is fast with values more than three times higher than in developed countries (Cohen, 2004). Cities' growth in tropical regions is further increased due to high levels of rural-urban migration (Brueckner & Lall, 2015). By 2050, it is expected that more than 2.7 billion people will move from rural areas to urban centers in developing countries (World Bank, 2013). Increasing population size in tropical cities implies in higher demand for natural resources, especially in the surrounding areas. Indeed, anthropogenic impacts, including deforestation (DeFries et al., 2010), pollution (Lamsal et al., 2013), and defaunation (Fa et al., 1995; Wilkie & Carpenter, 1999), often increase with population density and near urban centers.

Fish provide 17% of all animal protein consumed in the world, thus playing an important role in food security (The State of World Fisheries and Aquaculture, 2020). Of this total of consumed fish, 12.7% are wild fish caught in freshwater environments, which covers only 1.2% of the planet's surface (Funge-Smith & Bennett, 2019). Fish are rich in proteins and other essential nutrients (vitamins and minerals) and omega-3 fatty acids, contributing to people's nutritional security and health (Hicks et al., 2019; Kawarazuka & Béné, 2010). The food and nutritional security provided by fish is especially important for low-income populations from developing countries, where over 90% of inland fish catches are directed for local human consumption (Fluet-Chouinard et al., 2018; Funge-Smith & Bennett, 2019). Inland fisheries provide full daily animal protein for 119–158 million people from low-income countries of the world (Fluet-Chouinard et al., 2018; Funge-Smith & Bennett, 2019). Indeed, fish is often the only high-quality protein-rich food resource accessible to poor people, especially from Africa, Asia, and South America (Béné et al., 2009; Kawarazuka & Béné, 2010; McIntyre et al., 2016; Fluet-Chouinard et al., 2018). Furthermore, inland fisheries also provide employment and income for riverine communities, contributing to the alleviation of poverty worldwide (Béné et al., 2009). A recent estimate indicates that more than 25 million people work on fisheries direct or indirectly, generating around U$D 24–43 billion annually (Funge-Smith & Bennett, 2019).

In Brazil, the estimated value of inland fisheries is U$D 828 million, mainly based on Amazonian fisheries (Funge-Smith & Bennett, 2019). The Amazon is the largest drainage basin with the greatest diversity of freshwater fish in the world (Dagosta & De Pinna, 2019; Jézéquel et al., 2020). Amazonian populations strongly rely on fish resources, having one of the highest fish consumption rates in the world and contributing to income generation and promotion of the local economy (Isaac & Almeida, 2011; Hallwass et al., 2013; Isaac et al., 2015; Hallwass & Silvano, 2016; Begossi et al., 2019). Although Amazonian fisheries are highly diffuse with fish landings occurring in thousands of small ports and many smaller fishing communities (Hallwass et al., 2011), larger Amazonian cities concentrate most of the region's human population, thus creating a high demand for fish exploited by small-scale fisheries.

In this chapter, we provide an overview of multiple potential negative and positive effects of urban centers on fish and small-scale fisheries in highly diverse Amazonian rivers.

2 Amazon Inland Fisheries: An Overview

Amazon covers approximately seven million km^2, responsible for more than 10% of the world's discharge of freshwater into the oceans (Goulding et al., 2003; Sioli, 2012). This vast and productive expanse of water is composed of different ecosystems, such as rivers, lakes, and floodplains that are connected by flood pulses associated with seasonal variation (Junk et al., 1989, 2011; Sioli, 2012). Fisheries in the Amazon basin are mostly small-scale, employing canoes and small boats with or without small engines, low capacity of storage and autonomy for trips, and small crew (usually one to three fishers) landing in small ports or scattered fishing communities (Hallwass et al., 2011; Hallwass et al., 2020a, b; Keppeler et al., 2020). The largest boats ($\geq$ 11 m) with higher storage capacity ($\geq$ 3 t), autonomy, and employing large amounts of fishing gears usually land their catches in the large ports from the largest Amazonian cities, such as Manaus, Belém, Santarém, and Porto Velho (Bayley & Petrere Jr., 1989; Batista & Petrere Jr, 2003, 2007; Isaac et al., 2008; Doria et al., 2012). Amazonian fisheries exploit the available environments according to rivers' water level and employing a variety of fishing gears (Hallwass et al., 2011, 2020a, b), although mainly centered on gillnets (Hallwass & Silvano, 2016; Hallwass et al., 2023; Silvano et al., 2017a; Keppeler et al., 2020). The decision regarding which habitats to exploit and gears to use is a daily choice that fishers make based on seasonal variations and fish species targeted (Hallwass et al., 2013; Keppeler et al., 2020). Although many fish species (usually from 30 to 40 species) are regularly exploited in Amazonian small-scale fisheries, a few species (around 10), usually accounts for more than 50% of total biomass landed (Hallwass & Silvano, 2016; Maccord et al., 2007). Therefore, Amazonian fisheries are multispecies but also selective (Hallwass & Silvano, 2016). This selectiveness observed in small-scale fisheries in the Amazon and the Brazilian coast reflects market demands and consumer preferences on fish taste, handling time, which is directly related with the fishbone density, and influences of food taboos (Begossi et al., 2004, 2012).

Populations of several commercial fish species have been overexploited in the Amazon, including the pirarucu (*Arapaimas gigas*), which is one of the largest freshwater fish in the world, among other commercial fish, such as the tambaqui (*Colossoma macropomum*) and large catfishes (Smith, 1985; Petrere et al., 2004; Isaac et al., 2008; Garcia et al., 2009; Hallwass et al., 2020a, b). The growth of the Amazon population increased the demand for pirarucu, thus, from the end of the 1950s onwards, this species has declined in abundance and increased in price (Smith, 1985). More recently, co-management systems have contributed to recover overexploited populations of pirarucu and to allow sustainable fisheries of this species, mostly in more productive white-water rivers (Campos-Silva & Peres, 2016; Castello et al., 2009). Although the pirarucu was historically consumed by both Pre-Columbian indigenous and colonizers, nowadays its meat is consumed mainly in large urban centers by those who can afford it (Campos-Silva & Peres, 2016; Prestes-Carneiro et al., 2016; Veríssimo, 1895).

3 Urbanization and Fisheries in the Brazilian Amazon

3.1 Growing Cities and Shrinking Fish?

Although fisheries are a complex economical, recreational, and subsistence activity, which is affected by many factors at multiple scales (Frawley et al., 2021; Keppeler et al., 2020), it is expected that those fish populations closer to tropical urban centers would suffer stronger fishing pressure due to stronger demand and higher accessibility (Brewer et al., 2012; Maire et al., 2016). A decline in the abundance, body size, and diversity of fish species towards tropical cities has been found in both marine (Brewer et al., 2009) and freshwater ecosystems (Keppeler et al., 2018; Tregidgo et al., 2017). Other studies have indicated that accessibility or proximity to fishing villages may be inversely related to the abundance of tropical coastal and freshwater fishes (Silvano et al., 2014; Silvano et al., 2017b).

Most fishing is size-selective, as fishers usually target large individuals, due to their higher economic value (Pellowe & Leslie, 2020). Archaeological and contemporaneous examples provide evidence that fishing leads to gradual shifts in body size distribution, as large individuals are disproportionally removed from populations and communities become dominated by smaller species (Pauly et al., 2002). Therefore, variables associated with fish body size are important biological indicators of overfishing worldwide (Pauly et al., 2002; Welcomme, 1999). Most fish species with large body sizes are sensitive to fishing pressure due to their relatively low demographic resilience, which results from a series of features, such as late maturation, low abundance, slow growth, and low reproductive effort (Winemiller, 2005).

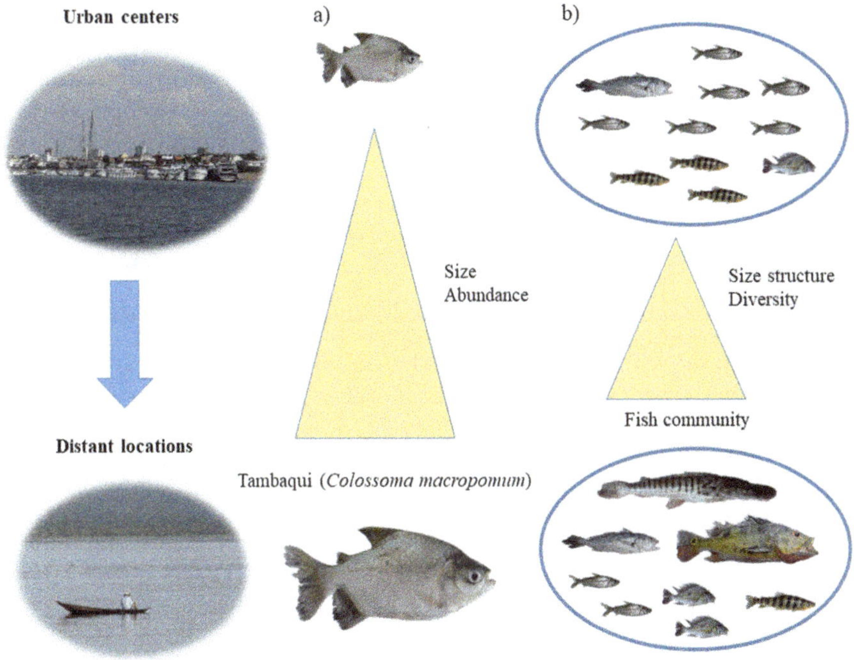

Fig. 4.1 Representation of some of the main documented negative effects of proximity to cities or urban centers on fish and fisheries in the Brazilian Amazon: (**a**) decrease in size and catches of the commercial fish tambaqui (Tregidgo et al., 2017); (**b**) changes in size structure and diversity of fish communities, which tend to be dominated by smaller fish and have fewer species closer to cities (Keppeler et al., 2018). (The photos representing the urban center and distant location are respectively from the city of Santarém and near the village of Santa Cruz (approximately 150 km upstream from the city), in the Tapajós River. © Renato Silvano)

The preferential harvest of large individuals may also cause evolutionary changes as this directional selective pressure benefits individuals with fast development and slow growth, reducing body size at maturity and maximum body sizes at the population level (Fenberg & Roy, 2008). Ultimately, the loss of large individuals may lead to lower productivity and fishery collapse, as large females contribute disproportionally to population replenishment (Barneche et al., 2018).

Changes in fish body size and abundance have been associated with the proximity to tropical cities in the Brazilian Amazon. For example, interviews with fishers indicate a 50% reduction in body size and catch per unit of effort (CPUE) of the fish tambaqui (*Colossoma macropomum*) in sites closer to the city of Manaus, compared to more distant sites along the Purus River (Tregidgo et al., 2017). Tambaqui (Fig. 4.1a) is an iconic frugivorous fish species native to the Orinoco and Amazon basins that can reach up to 1.1 m (total length) and about 40 kg in weight (IGTA 2001). Manaus is the seventh largest city in Brazil with more than 2.2 million people and is in the center of the Amazon basin (https://cidades.ibge.gov.br/brasil/am/manaus/panorama). The tambaqui body size gradient found by Tregidgo et al.

(2017) extends to 1500 km from Manaus and is mainly explained by higher fishing pressure near the city to support the local market. Tregidgo et al. (2017) study agree with the evidence of declines in tambaqui harvest near Manaus from the 80 s (Petrere Jr, 1986) when human population size was only half of what it is today.

At the community scale, the study by Keppeler et al. (2018) shows that fish communities from Amazonian lakes tend to be increasingly dominated by small fish species as they are located closer to urban centers or cities (Fig. 4.1b). Again, fishing pressure from small to medium cities, such as Baião (48,459 inhabitants) in the Tocantins River and Santarém (306,480 inhabitants) in the Tapajós River, may be the main driver of these observed changes in community size structure. However, the Negro River basin, which has more pristine conditions, shows fish communities dominated by larger fishes when compared to other river basins, such as the Tapajós and Tocantins, which are more degraded by deforestation and dams, respectively (Keppeler et al., 2018). Therefore, other factors, such as forest cover and the presence of dams, could also have affected the size patterns observed in the studied fish communities (Keppeler et al., 2018). Furthermore, a study analyzing several factors influencing fish catches by fishers in the Negro and Tapajós rivers, based on records of 5771 fish landings, indicates that the catch per unit of effort of fish landings was positively related to the distance to the urban center, so the aforementioned effects observed on fish could also influence fish catches (Keppeler et al., 2020). A recent study modelling fishing effects on trophic interactions in the Lower Tapajós River predicts than an increase on fishing pressure along the next 30 years (from 2013 to 2042) would cause sharp declines in populations of some large commercial fishes, such as the pirarucu and the large catfish dourada (*Brachyplatystoma rousseauxii*) (Capitani et al., 2021). Therefore, fisheries management planning and related policies should consider urban development in large Amazonian rivers.

A study based on standardized fish samples in 16 sites near fishing communities in the Tapajós and Negro rivers (8 sites in each river) indicates potential negative effects of proximity to larger cities (Santarém in Tapajos and Manaus in Negro) on fish diversity, as those sites closer to cities tend to show lower species richness and greater species homogenization (Nitschke, 2018). Based on the same set of fish sampling data in these two rivers (Silvano, 2020), we observed a positive relationship between distance to the main city (Santarém) and total biomass of fish sampled (e.g. abundance of fish) in the Tapajós River (Fig. 4.2a, $r^2 = 0.75$, d.f. = 7, $p < 0.01$), but we did not observe such relationship between distance to the city (Manaus) and sampled fish biomass in the Negro River (Fig. 4.2b, $r^2 = 0.02$, d.f. = 7, $p = 0.76$).

We argue that the observed negative effects of the proximity to the city on both the diversity and biomass of fish sampled in the Tapajós river could be caused by a stronger fishing pressure of larger commercial boats (Fig. 4.3) coming from the Santarém city. We propose two possible explanations for the observed difference between the two rivers regarding the effects of city's proximity on fish. First, the commercial fishing fleet from Manaus exploits mainly the most productive whitewater rivers, such as the Purus, Solimões, Madeira and Juruá (Batista & Petrere Jr., 2003, 2007; Cardoso et al., 2004), and hence the commercial fishing pressure is relatively lower in the Negro River. Furthermore, all sampling sites in the Negro

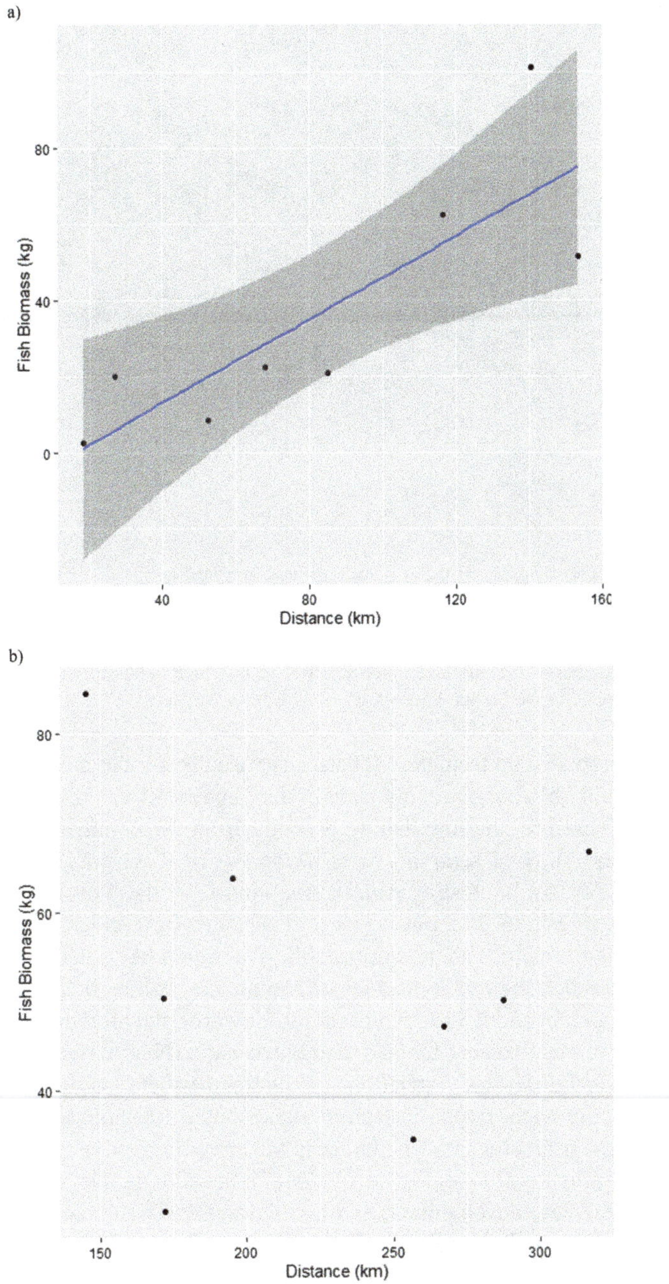

Fig. 4.2 Relationship between distance to the urban center and total biomass of fish sampled in 8 sampling sites located near fishing communities in (**a**) Tapajós River (Santarém as the urban center) and (**b**) Negro River (Manaus as urban center). The results of the regression analyses are in the text; the regression of data for the Tapajós River was made with log-transformed data, to better fit a linear relationship

Fig. 4.3 A large boat with smaller canoes attached to it, which is possibly a commercial fishing boat, photographed in the Lower Amazon River. © Renato Silvano

River were more distant from the Manaus city (usually located more than 200 km of distance, Fig. 4.2b) than sampling sites in the Tapajós River (18 to about 150 km; Fig. 4.2a). Therefore, the low fishing pressure from the commercial fleet and the reduced accessibility of sampling sites in the Negro River could dilute the influences of the city. An interesting pattern emerging from these observations consists of a possible threshold of about 200 km of distance from cities, above which the fishing effects from cities are less noticeable. This could be related to potential constraints imposed on the travel distance of commercial fishing boats, by operational costs related to fuel and ice storage capacity. However, the aforementioned patterns have been observed in clear (Tapajós) and black water (Negro) rivers, which are less productive and hence less targeted by commercial fishers, compared to the more productive white-water rivers. Therefore, the distance threshold related to potential fishing impacts from cities may be larger in the more productive white-water rivers, such as the Purus River (Tregidgo et al., 2017). Second, the Negro River has more protected areas, less development, and less deforestation compared to the Tapajós River, which may at least partially buffer the negative effects of fishing pressure from the city in the Negro River, as observed in other study (Keppeler et al., 2018).

The above-mentioned potential effects of the Amazonian urban centers on fish and small-scale fisheries can have some implications and interactions with Protected Areas in freshwater ecosystems in the Brazilian Amazon, including those Protected Areas that include local people who can use natural resources, such as the Extractive

Reserves (Lopes et al., 2011). Previous studies have shown a lower fishing pressure in extractive reserves, compared to sites located outside these Protected Areas in the Tapajós and Negro river (Hallwass et al., 2020a, b; Keppeler et al., 2020). Furthermore, there is evidence that small-scale fisheries in Protected Areas, such as Extractive Reserves and National Forests, show a higher catch per unit of effort and the presence of larger and more valuable commercial fish species, compared to sites outside these Protected Areas in the Tapajós River (Keppeler et al., 2017; Hallwass et al., 2020a, b). Therefore, the negative effects from urban centers, such as increased fishing pressure, could either be alleviated by or undermine the positive effects of Protected Areas in Amazonian rivers. Furthermore, these potential influences of urban centers (cities) on fish and fisheries could also inform the planning and design of Protected Areas in the Brazilian Amazon. For example, a more remote location distant from urban centers seems to contribute to the maintenance of higher fish abundance and increased fish catches by fishers in extractive reserves and national forests in the rivers Tapajós (Keppeler et al., 2017) and Tocantins (Silvano et al., 2014), besides facilitating enforcement of fishing regulations in the Negro River (Hallwass et al., 2020a, b). These effects of reduced accessibility can be more pronounced when Protected Areas include tributaries of the main rivers, which could thus be more easily surveilled to prevent the incursion of outside fishers, as observed in the Tocantins River (Silvano et al., 2014). On the other hand, Protected Areas could be established in regions closer to urban centers to mitigate the negative effects of increased demand for fish and pressure from commercial fishing on exploited species and fish communities (Keppeler et al., 2018; Tregidgo et al., 2017). We thus suggest that proximity to urban centers could be considered as an additional feature to establish aquatic Protected Areas in the Brazilian Amazon, in addition to other criteria, such as vulnerability to climate change (Frederico et al., 2021) or development projects, such as dams (Latrubesse et al., 2017; Winemiller et al., 2016).

3.2 *Other Potential Negative Effects of Urban Centers*

Besides the aforementioned effects related to increased fishing pressure, urban centers may have other negative effects on fish and fisheries. The Amazonian cities or urban centers are usually associated with numerous infrastructure projects, such as road development (Vilela et al., 2020), housing, and agriculture. These side effects of urban development can increase overall deforestation rates, including clearance of floodplain forests. Recent studies have shown that deforestation in floodplains can reduce fish abundance and diversity in Amazonian rivers (Arantes et al., 2018; Capitani et al., 2021; Lobón-Cerviá et al., 2015) as many fish species rely on the surrounding forests for protection, reproduction, and food resources (e.g., fruits and terrestrial invertebrates) during high water periods (Goulding, 1980; Goulding et al., 1988). Furthermore, deforestation has been linked to reduced fishing yield in Amazonian fisheries, especially in the Lower Amazon River (Castello et al., 2018), but also in the Lower Tapajós River, according to ecological models (Capitani et al., 2021). Therefore, deforestation driven by urban encroachment on riparian and floodplain forests can have detrimental effects on fish and fisheries, which can exacerbate other negative effects from cities, such as increased demand for fish.

Urban development can be also a source of multiple forms of pollution affecting the surrounding aquatic environment, including eutrophication, solid waste (plastics and micro-plastics, among others), and industrial contaminants (Andrade et al., 2019; Giarrizzo et al., 2019; Azevedo-Santos et al., 2021; Angeoletto et al., 2016) (Fig. 4.4). Nevertheless, the potential effects of these sources of urban pollution on Amazonian fish and fisheries remain poorly studied, possibly due to two reasons. First, the large water flow of most Amazonian rivers may dilute pollutants coming from cities, especially during the flood season. The aquatic pollution from cities is thus usually considered to be of less concern than other major impacts or drivers affecting Amazonian fish, such as overfishing (Castello et al., 2013) and habitat alteration through dams (Winemiller et al., 2016) or deforestation (Arantes et al., 2018). Nevertheless, aquatic pollution and eutrophication may indeed occur and be largely unnoticed in Amazonian rivers (e.g., Fig. 4.5). Second, the overwhelming diversity of Amazonian fish species may difficult the evaluation of fish responses to multiple pollution sources, as distinct species may respond in different ways. For example, many Amazonian fishes had developed adaptations to withstand low levels of oxygen in the water (hypoxia) to survive in shallow lakes during the dry season (Val et al., 1998). This adaptation may allow these fish to cope with higher eutrophication levels in waters surrounding cities, although it may not protect against algal toxins and industrial contaminants. One particular pollution source that has gained increased attention in the literature consists in micro-plastics, which can be ingested by fish and thus enter the aquatic food chain, as has been observed in coastal (Lima et al., 2014; Possatto et al., 2011) and freshwater (Azevedo-Santos et al., 2021) ecosystems. A recent study shows ingestion of these micro-plastics by fish in the Xingu River, in the Brazilian Amazon (Andrade et al., 2019), thus evidencing that this form of pollution can be widespread in Amazonian rivers, as observed for freshwater ecosystems in general (Azevedo-Santos et al., 2021). Plastic pollution has received far less attention in freshwater ecosystems, compared to marine ones, but this kind of pollution is common in inland waters, especially near urban areas, affecting fish and other organisms, from plants to mammals (Azevedo-Santos et al., 2021). Although it would be expected that micro-plastic ingestion by freshwater fish would be more accentuated near major cities as a side effect of unplanned urbanization, to our knowledge no published studies have analyzed the effects of micro-plastics near cities, or along a distance gradient from urban centers in the Brazilian Amazon.

The mercury (Hg) released naturally or anthropogenic (from mining, deforestation, burning, among other sources) undergo methylation in the aquatic environment, which transforms it into methylmercury (Crespo-Lopez et al., 2021). This methylmercury (MeHg) is an organic form of mercury, which can transpose any cellular barrier, being thus absorbed by the tissues of several organisms (Crespo-Lopez et al., 2021). The methyl-mercury contamination is one of the most pervasive

a)

b)

Fig. 4.4 Examples of pollution observed in the city of Santarém, in the Tapajós River, Brazilian Amazon: (**a**) solid residues near the river margin during the low water season; (**b**) an unknown, bright blue runoff of liquid waste alongside the street, which ultimately drained into the river. © Renato Silvano

Fig. 4.5 Bright green coloration of the waters in the Tapajós River, during the low water season in 2016, which could be related to a local algal bloom (R.A.M. Silvano and G. Hallwass personal observation) of unknown causes. © Renato Silvano

and concerning kinds of environmental pollution affecting fish and aquatic ecosystems in the Brazilian Amazon (Basta et al., 2023; Crespo-Lopez et al., 2021; Nevado et al., 2010). This metal can bio-magnify along the food chain, thus affecting not only primary consumers but also predators, including humans (Bastos et al., 2015; Lino et al., 2019; Azevedo et al., 2020). Indeed, some populations of riverine and Indigenous people in the Brazilian Amazon, such as those from Madeira and Tapajós rivers, show high levels of mercury concentration, possibly due to fish consumption (Nevado et al., 2010; Hacon et al., 2014; Faial et al., 2015; Hacon et al. 2020). The main source of mercury contamination in Amazonian rivers is gold mining, mostly conducted by illegal mining activities, such as those in the upper reaches of the Tapajós River (Malm, 1998; Nevado et al., 2010). Albeit this mining activity occurs mostly in more remote locations and is thus not related to urban centers, there could be some amount of naturally occurring mercury in the environment, such as in the soil and waters of Amazonian rivers (Crespo-Lopez et al., 2021). Therefore, urban development could accelerate processes of deforestation and soil erosion, which could release this naturally occurring mercury to aquatic ecosystems (Crespo-Lopez et al., 2021). Furthermore, dam construction incentive the urbanization of the region where plants are installed and the reservoirs of these dams affect the cycling of mercury in the Amazon environment, thus intensifying the methylation process and, consequently, increasing the contamination rates of fish and people (Arrifano et al., 2018). Notwithstanding the accumulated knowledge on mercury contamination from gold mining in the Amazon (Malm, 1998; Nevado et al., 2010; Roulet et al.,

1999), the potential of increased mercury contamination of fish, and hence people, driven by urban development has not been studied in detail.

3.3 *Potential Positive Effects of Cities*

The proximity to urban centers may bring some positive effects to small-scale fisheries, for example in the form of better market opportunities or income diversification through tourism and other activities. A limitation faced by most small-scale fishers consists in the reduced market demand for fish in more distant fishing communities, which makes these fishers dependent on middlemen to be able to sell the fish caught. This dependence on middlemen usually increases the length of market chains and reduces profits to fishers, who may not be able to overcome poverty status. For example, a study in a Brazilian Amazon estuary shows that even considering that crab populations have not been reduced and that crab gathering is ecologically sustainable, socioeconomic indicators of crab gatherers had worsened, which may be partially attributed to their low income, as gatherers need to sell crabs to middlemen (Glaser & Diele, 2004). In such a context of impoverished fishers needing a connection to markets to sell their product, the proximity to urban centers may offer more diversified market opportunities, which in some cases may include selling fish to tourists or tourism-related buyers, such as hotels or restaurants. This tourism-driven market may even include creative solutions by fishers to diversify the products' offer and be able to include more fish species in the market. For example, two fishing communities in the Tapajós River (Alter do Chão and Ponta de Pedras), which are closer to the city of Santarém, regularly catch smaller fish from the Hemiodontidae family (locally named charuto, Fig. 4.6), which they process and sell to tourists as appetizers in the river's beaches (Hallwass et al., 2020b). This can be a good way to increase revenues from the market of an abundant, and otherwise undervalued, group of fishes. Indeed, the community of Ponta de Pedras, which is the closest one to the urban center, annually organizes the Charuto festival and harvest much higher biomass of charutos compared to all other communities in the Tapajós River, whereas these fishes are not caught by fishers from the two communities more distant from the city (Silvano et al., 2020).

The Protected Areas that include local people, such as the Extractive Reserves, may also benefit from extra income from tourism coming from the nearby urban center, such as the extractive reserve of Tapajós-Arapiuns in the Tapajós River (Medeiros et al., 2021). However, promising as it may be, a tourism-related economy can bring some caveats for local communities, such as the exclusion from the tourism market or limited access to natural resources, such as fish, as observed in the Brazilian coast (Lopes et al., 2015; Outeiro et al., 2019). Moreover, if not well planned, an uncontrolled increase in tourism can also exacerbate the aforementioned negative effects from urban centers on fish and fisheries, such as overharvesting and increased pollution.

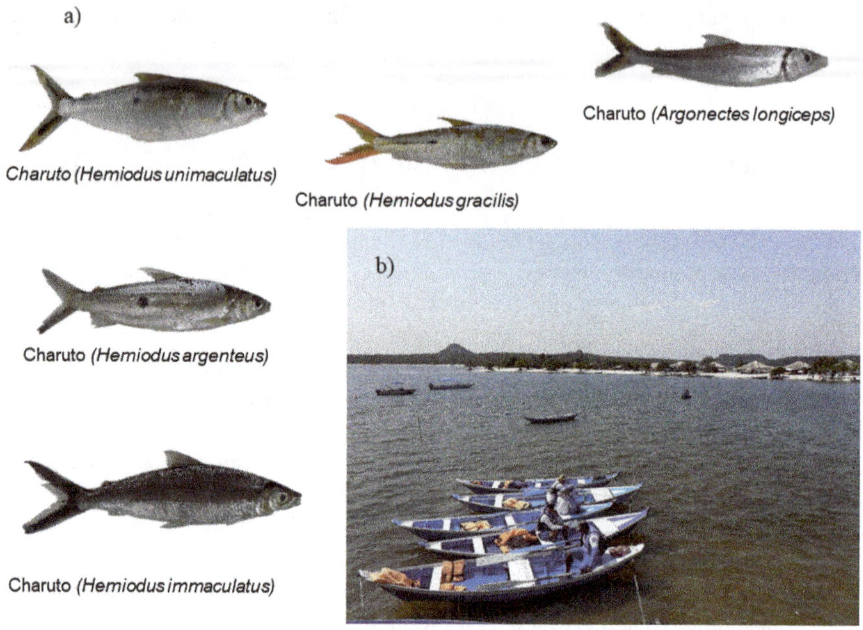

Fig. 4.6 Potential positive effects of the proximity of cities on small-scale fisheries related to tourism in the Tapajós River, Brazilian Amazon: (**a**) the group of fish species locally named charuto (Hemiodontidae), which are regularly sold to tourists; (**b**) boats wait to take tourists to a popular beach in the village of Alter do Chão, approximately 27 km from the city of Santarém. © Renato Silvano. (Fish photos modified from Silvano et al. (2020))

Other economic alternative related to tourism and that could include local riverine people consists of catch and release sport-fishing (Bower et al., 2020; Freire et al., 2016). Although much of this activity usually occur in more remote locations that would have more pristine fish stocks, the urban centers, especially those with larger airports to receive regular passenger flights, can facilitate the access of recreational fishers. Indeed, a well-designed and structured sport fishing operation could generate employment, income, and promote tourism. For example, an estimate for the Negro River calculates an annual revenue of over US$ 22 million from sport fishing in 2011 (Thomé-Souza et al., 2014). Sport-fishing is growing worldwide, with the potential to increase income and economic opportunities (Bower et al., 2020). Moreover, sport-fishing tends to increase in the Amazon because anglers from various countries have identified Amazon as a prime destination for this activity, due to a good pool of target species (Thomé-Souza et al., 2014). Catch-and-release fishing has also been applied in many ecosystems as a strategy to increase fish stocks and catch rates and restore the natural body size distribution of the population and community (Arlinghaus et al., 2007; Sass et al., 2018). Although the results are promising, catch and release may increase the mortality rates of target species and lead to evolutionary changes in fish populations, such as on metabolic rates (Hessenauer et al., 2015). Therefore, this activity needs to be monitored and further studied in Amazonian rivers to ensure its sustainability (Barroco et al., 2017; Freire et al., 2016; Thomé-Souza et al., 2014). Notwithstanding its potential to contribute to livelihoods and economic development of Amazonian fishing communities, the development of sport fishing needs to consider community engagement, sharing of economic benefits and strategies to ensure a sustainable growth and to limit the negative impacts of this activity (Bower et al., 2020).

4 Conclusions

In this chapter, we provided an overview of the potential influences of urban centers (main cities) on fish ecology and small-scale fisheries in large rivers in the Brazilian Amazon. There are still few studies addressing these potential interactions between cities, fish, and fisheries in the Brazilian Amazon and possibly in tropical rivers elsewhere. There are pieces of evidence that proximity to urban centers, possibly through an intensified fishing pressure, has reduced the body size of exploited fish, besides altering the body size distribution and diversity of fish communities, which tend to be dominated by smaller fish and fewer species when closer to cities in the Brazilian Amazon. We commented on other, albeit less documented, potential negative effects of cities on fish, through deforestation and water pollution, as well as some positive effects of cities on small-scale fisheries due to increased and diversified market options. Among the knowledge gaps that could be addressed by future studies, we can propose some research questions. How planning, design, and management of current and future Protected Areas could be adjusted to account for the potential effects of cities on fish and fisheries? To which extent the cities increase or influence aquatic pollution, especially by micro-plastics or methyl-mercury? How to balance the increased opportunities and the potential impacts of proximity to cities on local small-scale fisheries? Besides these, we hope that our chapter could inspire researchers from distinct and interdisciplinary areas to formulate other future promising research questions. An improved understanding of the effects of cities on aquatic ecosystems, fish, and artisanal fisheries will help to amend current problems and avoid future ones, as a way forward to sustainable and ecologically sound urban planning, which would benefit the rich fish diversity and the livelihoods of riverine people in the Amazonian rivers.

Acknowledgments We thank to fishers in the Brazilian Amazon for their invaluable cooperation with our research. Part of the results and publications mentioned in this chapter were funded by the Coordenação de Aperfeicoamento de Pessoal de Nível Superior (CAPES), Brazil, Programa Nacional de Cooperação Acadêmica (Procad)/Ação Novas Fronteiras (NF; 883/2010) and by USAID/National Academy of Sciences of the US (PEER Cycle 4, Grant number: AID-OAA-A-11). One of the authors (R.A.M.S.) gives thanks for research grants from the Conselho Nacional de Desenvolvimento Científico e Tecnológico (CNPq, grant 303393/2019-0).

References

Andrade, M. C., Winemiller, K. O., Barbosa, P. S., Fortunati, A., Chelazzi, D., Cincinelli, A., & Giarrizzo, T. (2019). First account of plastic pollution impacting freshwater fishes in the Amazon: Ingestion of plastic debris by piranhas and other serrasalmids with diverse feeding habits. *Environmental Pollution, 244*, 766–773. https://doi.org/10.1016/j.envpol.2018.10.088

Angeoletto, F., & Santos, J. W. M. C. (2016). Agonizantes: los ríos y arroyos en el contexto ecológico urbano. *Revista Espaço Acadêmico, 16*(183), 104–112.

Arantes, C. C., Winemiller, K. O., Petrere, M., Castello, L., Hess, L. L., & Freitas, C. E. C. (2018). Relationships between forest cover and fish diversity in the Amazon River floodplain. Edited by Robert Arlinghaus. *Journal of Applied Ecology, 55*, 386–395. https://doi.org/10.1111/1365-2664.12967

Arlinghaus, R., Cooke, S. J., Lyman, J., Policansky, D., Schwab, A., Suski, C., Sutton, S. G., & Thorstad, E. B. (2007). Understanding the complexity of catch-and-release in recreational fishing: An integrative synthesis of global knowledge from historical, ethical, social, and biological perspectives. *Reviews in Fisheries Science, 15*, 75–167.

Arrifano, G. P. F., Rodríguez, R. C., Martín-Doimeadios, M. J.-M., Ramírez-Mateos, V., da Silva, N. F. S., Souza-Monteiro, J. R., Augusto-Oliveira, M., et al. (2018). Large-scale projects in the amazon and human exposure to mercury: The case-study of the Tucuruí dam. *Ecotoxicology and Environmental Safety, 147*, 299–305. https://doi.org/10.1016/j.ecoenv.2017.08.048

Azevedo, L. S., Pestana, I. A., da Costa Nery, A. F., Bastos, W. R., & Souza, C. M. M. (2020). Mercury concentration in six fish guilds from a floodplain lake in western Amazonia: Interaction between seasonality and feeding habits. *Ecological Indicators, 111*, 106056. https://doi.org/10.1016/j.ecolind.2019.106056

Azevedo-Santos, V. M., Brito, M. F. G., Manoel, P. S., Perroca, J. F., Rodrigues-Filho, J. L., Paschoal, L. R. P., Gonçalves, G. R. L., et al. (2021). Plastic pollution: A focus on freshwater biodiversity. *Ambio, 50*, 1313–1324. https://doi.org/10.1007/s13280-020-01496-5

Barneche, D. R., Ross Robertson, D., White, C. R., & Marshall, D. J. (2018). Fish reproductive-energy output increases disproportionately with body size. *Science, 360*, 642–645. https://doi.org/10.1126/science.aao6868

Barroco, L. S. A., Freitas, C. E. C., & Lima, Á. C. (2017). Estimation of peacock bass (Cichla spp.) mortality rate during catch-release fishing employing different post-capture procedures. *Brazilian Journal of Biology, 78*, 195–201. https://doi.org/10.1590/1519-6984.18915

Basta, P. C., de Vasconcellos, A. C. S., Hallwass, G., Yokota, D., Pinto, D. O. R., Aguiar, D. S., Souza, C. C., Oliveirada-Costa, M. (2023). Risk Assessment of Mercury-Contaminated Fish Consumption in the Brazilian Amazon: An Ecological Study. *Toxics, 11, 800.* https://doi.org/10.3390/toxics11090800

Bastos, W. R., Dórea, J. G., Bernardi, J. V. E., Lauthartte, L. C., Mussy, M. H., Lacerda, L. D., & Malm, O. (2015). Mercury in fish of the Madeira river (temporal and spatial assessment), Brazilian Amazon. *Environmental Research, 140*, 191–197. https://doi.org/10.1016/j.envres.2015.03.029

Batista, V. S., & Petrere, M., Jr. (2003). Characterization of the commercial fish production landed at Manaus, Amazonas State, Brazil. *Acta Amazonica, 33*, 53–66.

Batista, V. D. S., & Petrere, M., Jr. (2007). Spatial and temporal distribution of fishing resources exploited by the Manaus fishing fleet, Amazonas, Brazil. *Brazilian Journal of Biology, 67*, 651–656.

Bayley, P. B., & Petrere Jr, M. (1989). *Amazon fisheries: Assessment methods, current status and management options*. Canadian special publication of fisheries and aquatic sciences/ Publication speciale canadienne des sciences halieutiques et aquatiques.

Begossi, A., Hanazaki, N., & Ramos, R. M. (2004). Food chain and the reasons for fish food taboos among Amazonian and Atlantic Forest fishers (Brazil). *Ecological Applications, 14*. Wiley Online Library, 1334–1343.

Begossi, A., Salivonchyk, S. V., Hanazaki, N., Martins, I. M., & Bueloni, F. (2012). Fishers (Paraty, RJ) and fish manipulation time: A variable associated to the choice for consumption and sale. *Brazilian Journal of Biology, 72*, 973–975.

Begossi, A., Salivonchyk, S. V., Hallwass, G., Hanazaki, N., Lopes, P. F. M., Silvano, R. A. M., Dumaresq, D., & Pittock, J. (2019). Fish consumption on the Amazon: A review of biodiversity, hydropower and food security issues. *Brazilian Journal of Biology, 79*, 345–357.

Béné, C., Steel, E., Luadia, B. K., & Gordon, A. (2009). Fish as the "bank in the water"–evidence from chronic-poor communities in Congo. *Food Policy, 34*, 108–118.

Bower, S. D., Aas, Ø., Robert Arlinghaus, T., Beard, D., Cowx, I. G., Danylchuk, A. J., Freire, K. M. F., Potts, W. M., Sutton, S. G., & Cooke, S. J. (2020). Knowledge gaps and management priorities for recreational fisheries in the developing world. *Reviews in Fisheries Science & Aquaculture, 28*, 518–535. https://doi.org/10.1080/23308249.2020.1770689

Brewer, T. D., Cinner, J. E., Green, A., & Pandolfi, J. M. (2009). Thresholds and multiple scale interaction of environment, resource use, and market proximity on reef fishery resources in the Solomon Islands. *Biological Conservation, 142*, 1797–1807. https://doi.org/10.1016/j.biocon.2009.03.021

Brewer, T. D., Cinner, J. E., Fisher, R., Green, A., & Wilson, S. K. (2012). Market access, population density, and socioeconomic development explain diversity and functional group biomass of coral reef fish assemblages. *Global Environmental Change, 22*, 399–406. https://doi.org/10.1016/j.gloenvcha.2012.01.006

Brueckner, J. K., & Lall, S. V. (2015). Cities in developing countries. In *Handbook of regional and urban economics* (Vol. 5, pp. 1399–1455). Elsevier. https://doi.org/10.1016/B978-0-444-59531-7.00021-1

Campos-Silva, J. V., & Peres, C. A. (2016). Community-based management induces rapid recovery of a high-value tropical freshwater fishery. *Scientific Reports, 6*, 34745. https://doi.org/10.1038/srep34745

Capitani, L., Angelini, R., Keppeler, F. W., Hallwass, G., & Silvano, R. A. M. (2021). Food web modeling indicates the potential impacts of increasing deforestation and fishing pressure in the Tapajós River, Brazilian Amazon. *Regional Environmental Change, 21*, 42. https://doi.org/10.1007/s10113-021-01777-z

Cardoso, R. S., da Silva, V., Batista, C. H., Júnior, F., & Martins, W. R. (2004). Aspectos econômicos e operacionais das viagens da frota pesqueira de Manaus, Amazônia Central. *Acta Amazonica, 34*, 301–307.

Castello, L., Viana, J. P., Watkins, G., Pinedo-Vasquez, M., & Luzadis, V. A. (2009). Lessons from integrating fishers of arapaima in small-scale fisheries management at the mamirauá reserve, amazon. *Environmental Management, 43*, 197–209. Scopus. https://doi.org/10.1007/s00267-008-9220-5

Castello, L., McGrath, D. G., Hess, L. L., Coe, M. T., Lefebvre, P. A., Petry, P., Macedo, M. N., Renó, V. F., & Arantes, C. C. (2013). The vulnerability of Amazon freshwater ecosystems: Vulnerability of Amazon freshwater ecosystems. *Conservation Letters, 6*, 217–229. https://doi.org/10.1111/conl.12008

Castello, L., Hess, L. L., Thapa, R., McGrath, D. G., Arantes, C. C., Renó, V. F., & Isaac, V. J. (2018). Fishery yields vary with land cover on the Amazon River floodplain. *Fish and Fisheries, 19*, 431–440. https://doi.org/10.1111/faf.12261

Cohen, B. (2004). Urban growth in developing countries: A review of current trends and a caution regarding existing forecasts. *World Development, 32*, 23–51. https://doi.org/10.1016/j.worlddev.2003.04.008

Crespo-Lopez, M. E., Augusto-Oliveira, M., Lopes-Araújo, A., Santos-Sacramento, L., Takeda, P. Y., de Matos Macchi, B., do Nascimento, J. L. M., Maia, C. S. F., Lima, R. R., & Arrifano, G. P. (2021). Mercury: What can we learn from the Amazon? *Environment International, 146*, 106223. https://doi.org/10.1016/j.envint.2020.106223

Dagosta, F. C. P., & De Pinna, M. (2019). The fishes of the Amazon: Distribution and biogeographical patterns, with a comprehensive list of species. *Bulletin of the American Museum of Natural History, 2019*, 1–163.

Hacon, S., S., Oliveira-da-Costa, M., de Souza Gama, C., Ferreira, R., Basta, P. C., Schramm, A., & Yokota, D. (2020). Mercury exposure through fish consumption in traditional communities in the Brazilian northern Amazon. *International Journal of Environmental Research and Public Health, 17*, 5269. https://doi.org/10.3390/ijerph17155269

DeFries, R. S., Rudel, T., Uriarte, M., & Hansen, M. (2010). Deforestation driven by urban population growth and agricultural trade in the twenty-first century. *Nature Geoscience, 3*, 178–181. https://doi.org/10.1038/ngeo756

Doria, C. R., da Costa, M., Ruffino, L., Hijazi, N. C., & da Cruz, R. L. (2012). A pesca comercial na bacia do rio Madeira no estado de Rondônia, Amazônia brasileira. *Acta Amazonica, 42*, 29–40.

Fa, J. E., Juste, J., Del Val, J. P., & Castroviejo, J. (1995). Impact of market hunting on mammal species in Equatorial Guinea. *Conservation Biology, 9*, 1107–1115.

Faial, K., Deus, R., Deus, S., Neves, R., Jesus, I., Santos, E., Alves, C. N., & Brasil, D. (2015). Mercury levels assessment in hair of riverside inhabitants of the Tapajós River, Pará state, Amazon, Brazil: Fish consumption as a possible route of exposure. *Journal of Trace Elements in Medicine and Biology, 30*, 66–76. https://doi.org/10.1016/j.jtemb.2014.10.009

Fenberg, P. B., & Roy, K. (2008). Ecological and evolutionary consequences of size-selective harvesting: How much do we know? *Molecular Ecology, 17*, 209–220. https://doi.org/10.1111/j.1365-294X.2007.03522.x

Fluet-Chouinard, E., Funge-Smith, S., & McIntyre, P. B. (2018). Global hidden harvest of freshwater fish revealed by household surveys. *Proceedings of the National Academy of Sciences of The United States of America*. Scopus, *115*, 7623–7628. https://doi.org/10.1073/pnas.1721097115

Frawley, T. H., Blondin, H. E., White, T. D., Carlson, R. R., Villalon, B., & Crowder, L. B. (2021). Fishers as foragers: Individual variation among small-scale fishing vessels as revealed by novel tracking technology. *Fisheries Research, 238*, 105896. https://doi.org/10.1016/j.fishres.2021.105896

Frederico, R. G., Dias, M. S., Jézéquel, C., Tedesco, P. A., Hugueny, B., Zuanon, J., Torrente-Vilara, G., et al. (2021). The representativeness of protected areas for Amazonian fish diversity under climate change. *Aquatic Conservation: Marine and Freshwater Ecosystems, aqc.3528*, 1158–1166. https://doi.org/10.1002/aqc.3528

Freire, K. M. F., Tubino, R. A., Monteiro-Neto, C., Andrade-Tubino, M. F., Belruss, C. G., Tomás, A. R. G., Tutui, S. L. S., et al. (2016). Brazilian recreational fisheries: Current status, challenges and future direction. *Fisheries Management and Ecology, 23*, 276–290. https://doi.org/10.1111/fme.12171

Funge-Smith, S., & Bennett, A. (2019). A fresh look at inland fisheries and their role in food security and livelihoods. *Fish and Fisheries, 20*, 1176–1195. Scopus. https://doi.org/10.1111/faf.12403

Garcia, A., Tello, S., Vargas, G., & Duponchelle, F. (2009). Patterns of commercial fish landings in the Loreto region (Peruvian Amazon) between 1984 and 2006. *Fish Physiology and Biochemistry, 35*, 53–67. https://doi.org/10.1007/s10695-008-9212-7

Giarrizzo, T., Andrade, M. C., Schmid, K., Winemiller, K. O., Ferreira, M., Pegado, T., Chelazzi, D., Cincinelli, A., & Fearnside, P. M. (2019). Amazonia: The new frontier for plastic pollution. *Frontiers in Ecology and the Environment, 17*, 309–310. https://doi.org/10.1002/fee.2071

Glaser, M., & Diele, K. (2004). Asymmetric outcomes: Assessing central aspects of the biological, economic and social sustainability of a mangrove crab fishery, Ucides cordatus (Ocypodidae), in North Brazil. *Ecological Economics, 49*, 361–373. https://doi.org/10.1016/j.ecolecon.2004.01.017

Goulding, M. (1980). *The fishes and the forest: Explorations in Amazonian natural history*. Univ of California Press.

Goulding, M., Carvalho, M. L., & Ferreira, E. J. G. (1988). *Rio Negro*. Rich life in poor water.

Goulding, Michael, Ronaldo Barthem, and Efrem Ferreira. 2003. The Smithsonian atlas of the Amazon.

Hacon, S., Dórea, J., Fonseca, M., Oliveira, B., Mourão, D., Ruiz, C., Gonçalves, R., Mariani, C., & Bastos, W. (2014). The influence of changes in lifestyle and mercury exposure in riverine populations of the Madeira River (Amazon Basin) near a hydroelectric project. *International Journal of Environmental Research and Public Health, 11*, 2437–2455. https://doi.org/10.3390/ijerph110302437

Hallwass, G., & Silvano, R. A. M. (2016). Patterns of selectiveness in the Amazonian freshwater fisheries: Implications for management. *Journal of Environmental Planning and Management, 59*, 1537–1559.

Hallwass, G., Lopes, P. F., Juras, A. A., & Silvano, R. A. M. (2011). Fishing effort and catch composition of urban market and rural villages in Brazilian Amazon. *Environmental Management, 47*, 188–200.

Hallwass, G., Lopes, P. F. M., Juras, A. A., & Silvano, R. A. M. (2013). Behavioral and environmental influences on fishing rewards and the outcomes of alternative management scenarios for large tropical rivers. *Journal of Environmental Management, 128*, 274–282. https://doi.org/10.1016/j.jenvman.2013.05.037

Hallwass, G., Schiavetti, A., & Silvano, R. A. M. (2020a). Fishers' knowledge indicates temporal changes in composition and abundance of fishing resources in Amazon protected areas. *Animal Conservation, 23*, 36–47. Scopus. https://doi.org/10.1111/acv.12504

Hallwass, G., da Silva, L. H. T., Nagl, P., Clauzet, M., & Begossi, A. (2020b). Small-scale fisheries, livelihoods, and food security of riverine people. In R. A. M. Silvano (Ed.), *Fish and fisheries in the Brazilian Amazon* (pp. 23–39). Springer International Publishing. https://doi.org/10.1007/978-3-030-49146-8_3

Hallwass, G., Keppeler, F. W., Tomazoni-Silva, L. H., Alves, I. A., Isaac, V. J., Almeida, M. C., & Silvano, R. A. M. (2023). 'Disentangling' the advantages from gillnets in freshwater small-scale fisheries in the Brazilian Amazon. *Reviews in Fish Biology and Fisheries*. https://doi.org/10.1007/s11160-023-09771-w

Hessenauer, J.-M., Vokoun, J. C., Suski, C. D., Davis, J., Jacobs, R., & O'Donnell, E. (2015). Differences in the metabolic rates of exploited and unexploited fish populations: A signature of recreational fisheries induced evolution? Edited by Athanassios C. Tsikliras. *PLOS One, 10, e0128336*. https://doi.org/10.1371/journal.pone.0128336

Hicks, C. C., Cohen, P. J., Graham, N. A. J., Nash, K. L., Allison, E. H., D'Lima, C., Mills, D. J., et al. (2019). Harnessing global fisheries to tackle micronutrient deficiencies. *Nature, 574*, 95–98. https://doi.org/10.1038/s41586-019-1592-6

Isaac, V. J., Da Silva, C. O., & Ruffino, M. L. (2008). The artisanal fishery fleet of the lower Amazon. *Fisheries Management and Ecology, 15*, 179–187.

Isaac, V. J., & de Almeida, M. C. (2011). *ElConsumo de pescado en la amazonía brasileña*. FAO.

Isaac, V. J., Almeida, M. C., Giarrizzo, T., Deus, C. P., Vale, R., Klein, G., & Begossi, A. (2015). Food consumption as an indicator of the conservation of natural resources in riverine communities of the Brazilian Amazon. *Anais da Academia Brasileira de Ciências, 87*, 2229–2242. https://doi.org/10.1590/0001-3765201520140250

Jézéquel, C., Tedesco, P. A., Bigorne, R., Maldonado-Ocampo, J. A., Ortega, H., Hidalgo, M., Martens, K., et al. (2020). A database of freshwater fish species of the Amazon Basin. *Scientific Data, 7*, 96. https://doi.org/10.1038/s41597-020-0436-4

Junk, W. J., Bayley, P. B., & Sparks, R. E. (1989). The flood pulse concept in river-floodplain systems. *Canadian Special Publication of Fisheries and Aquatic Sciences, 106*, 110–127.

Junk, W. J., Piedade, M. T. F., Schöngart, J., Mario Cohn-Haft, J., Adeney, M., & Wittmann, F. (2011). A classification of major naturally-occurring Amazonian lowland wetlands. *Wetlands, 31*, 623–640.

Kawarazuka, N., & Béné, C. (2010). Linking small-scale fisheries and aquaculture to household nutritional security: An overview. *Food Security, 2*, 343–357.

Keppeler, F. W., Hallwass, G., & Silvano, R. A. M. (2017). Influence of protected areas on fish assemblages and fisheries in a large tropical river. *ORYX, 51*, 268–279. Scopus. https://doi.org/10.1017/S0030605316000247

Keppeler, F. W., Castro, A., de Souza, G., Hallwass, A. B., Carvalho, M., de Almeida, V., Isaac, J., & Silvano, R. A. M. (2018). Ecological influences of human population size and distance to urban centres on fish communities in tropical lakes. *Aquatic Conservation: Marine and Freshwater Ecosystems, 28*, 1030–1043.

Keppeler, F. W., Hallwass, G., Santos, F., da Silva, L. H. T., & Silvano, R. A. M. (2020). What makes a good catch? Effects of variables from individual to regional scales on tropical small-scale fisheries. *Fisheries Research, 229*, 105571. https://doi.org/10.1016/j.fishres.2020.105571

Lamsal, L. N., Martin, R. V., Parrish, D. D., & Krotkov, N. A. (2013). Scaling relationship for NO $_2$ pollution and urban population size: A satellite perspective. *Environmental Science & Technology, 47*, 7855–7861. https://doi.org/10.1021/es400744g

Latrubesse, E. M., Arima, E. Y., Dunne, T., Park, E., Baker, V. R., d'Horta, F. M., Wight, C., Wittmann, F., Zuanon, J., & Baker, P. A. (2017). Damming the rivers of the Amazon basin. *Nature, 546*, 363–369.

Lima, A. R. A., Costa, M. F., & Barletta, M. (2014). Distribution patterns of microplastics within the plankton of a tropical estuary. *Environmental Research, 132*, 146–155. https://doi.org/10.1016/j.envres.2014.03.031

Lino, A. S., Kasper, D., Guida, Y. S., Thomaz, J. R., & Malm, O. (2019). Total and methyl mercury distribution in water, sediment, plankton and fish along the Tapajós River basin in the Brazilian Amazon. *Chemosphere, 235*, 690–700. https://doi.org/10.1016/j.chemosphere.2019.06.212

Lobón-Cerviá, J., Hess, L. L., Melack, J. M., & Araujo-Lima, C. A. R. M. (2015). The importance of forest cover for fish richness and abundance on the Amazon floodplain. *Hydrobiologia, 750*, 245–255. https://doi.org/10.1007/s10750-014-2040-0

Lopes, P. F. M., Silvano, R. A. M., & Begossi, A. (2011). Extractive and sustainable development reserves in Brazil: Resilient alternatives to fisheries? *Journal of Environmental Planning and Management, 54*, 421–443. https://doi.org/10.1080/09640568.2010.508687

Lopes, P. F. M., Pacheco, S., Clauzet, M., Silvano, R. A. M., & Begossi, A. (2015). Fisheries, tourism, and marine protected areas: Conflicting or synergistic interactions? *Ecosystem Services, 16*, 333–340. https://doi.org/10.1016/j.ecoser.2014.12.003

Maccord, P. F. L., Silvano, R. A. M., Ramires, M. S., Clauzet, M., & Begossi, A. (2007). Dynamics of artisanal fisheries in two Brazilian Amazonian reserves: Implications to co-management. *Hydrobiologia*. Scopus, *583*, 365–376. https://doi.org/10.1007/s10750-006-0486-4

Maire, E., Cinner, J., Velez, L., Huchery, C., Mora, C., Dagata, S., Vigliola, L., Wantiez, L., Kulbicki, M., & Mouillot, D. (2016). How accessible are coral reefs to people? A global assessment based on travel time. Edited by Marcel Holyoak. *Ecology Letters, 19*, 351–360. https://doi.org/10.1111/ele.12577

Malm, O. (1998). Gold mining as a source of mercury exposure in the Brazilian Amazon. *Environmental Research, 77*, 73–78.

McIntyre, P. B., Reidy Liermann, C. A., & Revenga, C. (2016). Linking freshwater fishery management to global food security and biodiversity conservation. *Proceedings of the National Academy of Sciences, 113*, 12880–12885. https://doi.org/10.1073/pnas.1521540113

Medeiros, H. M. N., de Moura Guerreiro, Q. L., Vieira, T. A., da Silva, S. M. S., da Silva Aço Renda, A. I., & Oliveira-Junior, J. M. B. (2021). Alternative tourism and environmental impacts: Perception of residents of an extractive Reserve in the Brazilian Amazonia. *Sustainability, 13*, 2076. https://doi.org/10.3390/su13042076

Nevado, J. J. B., Rodríguez Martín-Doimeadios, R. C., Guzmán Bernardo, F. J., Jimenez Moreno, M., Herculano, A. M., Do Nascimento, J. L. M., & Crespo-López, M. E. (2010). Mercury in the Tapajós River basin, Brazilian Amazon: A review. *Environment International, 36*, 593–608.

Nitschke, P. P. (2018). Influências da pressão pesqueira e áreas de proteção ambiental na comunidade de peixes em dois rios tropicais. MSc. Dissertation, Ecology, Universidade Federal do Rio Grande do Sul (UFRGS).

Outeiro, L., Rodrigues, J. G., Damásio, L. M. A., & Lopes, P. F. M. (2019). Is it just about the money? A spatial-economic approach to assess ecosystem service tradeoffs in a marine protected area in Brazil. *Ecosystem Services, 38*. Scopus. https://doi.org/10.1016/j.ecoser.2019.100959

Pauly, D., Christensen, V., Guénette, S., Pitcher, T. J., Rashid Sumaila, U., Walters, C. J., Watson, R., & Zeller, D. (2002). Towards sustainability in world fisheries. *Nature, 418*, 689–695. https://doi.org/10.1038/nature01017

Pellowe, K. E., & Leslie, H. M. (2020). Size-selective fishing leads to trade-offs between fishery productivity and reproductive capacity. *Ecosphere, 11*. https://doi.org/10.1002/ecs2.3071

Petrere, M., Jr. (1986). Amazon fisheries I-variations in the relative abundance of Tambaqui (Colossoma macropomum CUVIER, 1818) based on catch and effort data of the gill-net fisheries. *Amazoniana: Limnologia et Oecologia Regionalis Systematis Fluminis Amazonas, 9*, 527–547.

Petrere, M., Barthem, R. B., Córdoba, E. A., & Gómez, B. C. (2004). Review of the large catfish fisheries in the upper Amazon and the stock depletion of piraíba (Brachyplatystoma filamentosumLichtenstein). *Reviews in Fish Biology and Fisheries, 14*, 403–414.

Possatto, F. E., Barletta, M., Costa, M. F., Ivar do Sul, J. A., & Dantas, D. V. (2011). Plastic debris ingestion by marine catfish: An unexpected fisheries impact. *Marine Pollution Bulletin, 62*, 1098–1102. https://doi.org/10.1016/j.marpolbul.2011.01.036

Prestes-Carneiro, G., Béarez, P., Bailon, S., Py-Daniel, A. R., & Neves, E. G. (2016). Subsistence fishery at Hatahara (750–1230 CE), a pre-Columbian central Amazonian village. *Journal of Archaeological Science: Reports, 8*, 454–462. https://doi.org/10.1016/j.jasrep.2015.10.033

Roulet, M., Lucotte, M., Farella, N., Serique, G., Coelho, H., Passos, C. J. S., Da Silva, E. D. J., De Andrade, P. S., Mergler, D., & Guimarães, J.-R. D. (1999). Effects of recent human colonization on the presence of mercury in Amazonian ecosystems. *Water, Air, and Soil Pollution, 112*, 297–313.

Sachs, J. (2001). *Tropical underdevelopment. w8119*. National Bureau of Economic Research. https://doi.org/10.3386/w8119

Sass, G. G., Gaeta, J. W., Allen, M. S., Suski, C. D., & Shaw, S. L. (2018). Effects of catch-and-release angling on a largemouth bass (Micropterus salmoides) population in a north temperate lake, 2001–2005. *Fisheries Research, 204*, 95–102.

Silvano, R. A. M. (Ed.). (2020). *Fish and fisheries in the Brazilian Amazon: People, ecology and conservation in black and clear water Rivers*. Springer International Publishing. https://doi.org/10.1007/978-3-030-49146-8

Silvano, R. A. M., Hallwass, G., Lopes, P. F., Ribeiro, A. R., Lima, R. P., Hasenack, H., Juras, A. A., & Begossi, A. (2014). Co-management and spatial features contribute to secure fish abundance and fishing yields in tropical floodplain lakes. *Ecosystems*. Scopus, *17*, 271–285. https://doi.org/10.1007/s10021-013-9722-8

Silvano, R. A. M., Hallwass, G., Juras, A. A., & Lopes, P. F. M. (2017a). Assessment of efficiency and impacts of gillnets on fish conservation in a tropical freshwater fishery. *Aquatic Conservation: Marine and Freshwater Ecosystems, 27*, 521–533. Scopus. https://doi.org/10.1002/aqc.2687

Silvano, R. A. M., Nora, V., Andreoli, T. B., Lopes, P. F. M., & Begossi, A. (2017b). The 'ghost of past fishing': Small-scale fisheries and conservation of threatened groupers in subtropical islands. *Marine Policy, 75*, 125–132. https://doi.org/10.1016/j.marpol.2016.10.002

Silvano, R. A. M., Nitschke, P. P., Vieira, K. C., Nagl, P., Martínez, A. T. R., Chuctaya, J. A., Dutra, M. C. F., Keppeler, F. W., Rowedder, A. R. P., & Andrade, M. C. (2020). Atlas of fish of Tapajós and negro Rivers I: Characiformes. In R. A. M. Silvano (Ed.), *Fish and fisheries in the Brazilian Amazon* (pp. 41–196). Springer International Publishing. https://doi.org/10.1007/978-3-030-49146-8_4

Sioli, H. (2012). *The Amazon: Limnology and landscape ecology of a mighty tropical river and its basin* (Vol. 56). Springer Science & Business Media.

Smith, N. J. H. (1985). The impact of cultural and ecological change on Amazonian fisheries. *Biological Conservation, 32*, 355–373. https://doi.org/10.1016/0006-3207(85)90023-0

The State of World Fisheries and Aquaculture. (2020). FAO. https://doi.org/10.4060/ca9229en

Thomé-Souza, M. J. F., Maceina, M. J., Forsberg, B. R., Marshall, B. G., & Carvalho, Á. L. (2014). Peacock bass mortality associated with catch-and-release sport fishing in the Negro River, Amazonas state, Brazil. *Acta Amazonica, 44*, 527–532.

Tregidgo, D. J., Barlow, J., Pompeu, P. S., de Almeida Rocha, M., & Parry., L. (2017). Rainforest metropolis casts 1,000-km defaunation shadow. *Proceedings of the National Academy of Sciences, 114*, 8655–8659.

Val, A. L., Silva, M. N. P., & Almeida-Val, V. M. F. (1998). Hypoxia adaptation in fish of the Amazon: A never-ending task. *South African Journal of Zoology, 33*, 107–114. https://doi.org/10.1080/02541858.1998.11448459

Veríssimo, J. (1895). *A pesca na Amazônia* (Vol. 3). Livraria classica de Alves.

Vilela, T., Harb, A. M., Bruner, A., da Silva Arruda, V. L., Ribeiro, V., Alencar, A. A. C., Grandez, A. J. E., Rojas, A., Laina, A., & Botero, R. (2020). A better Amazon road network for people and the environment. *Proceedings of the National Academy of Sciences, 117*, 7095–7102. https://doi.org/10.1073/pnas.1910853117

Welcomme, R. L. (1999). A review of a model for qualitative evaluation of exploitation levels in multi-species fisheries. *Fisheries Management and Ecology, 6*, 1–19. https://doi.org/10.1046/j.1365-2400.1999.00137.x

Wilkie, D. S., & Carpenter, J. F. (1999). Bushmeat hunting in The Congo Basin: An assessment of impacts and options for mitigation. *Biodiversity and Conservation, 8*, 927–955.

Winemiller, K. O. (2005). Life history strategies, population regulation, and implications for fisheries management. *Canadian Journal of Fisheries and Aquatic Sciences, 62*, 872–885.

Winemiller, K. O., McIntyre, P. B., Castello, L., Fluet-Chouinard, E., Giarrizzo, T., Nam, S., Baird, I. G., et al. (2016). Balancing hydropower and biodiversity in the Amazon, Congo, and Mekong. *Science, 351*, 128–129. https://doi.org/10.1126/science.aac7082

World Bank. (2013). *Planning, connecting, and financing cities now: Priorities for city leaders*. The World Bank.

Chapter 5
Latin America: Where Urbanization and Poverty Peak Together with Biodiversity?

Camilo E. Sánchez-Sarria, Martha Bonilla-Moheno, Luis-Bernardo Vázquez, and Ian MacGregor-Fors

Abstract Latin America is a region that houses many of the global biodiversity hotspots. Latin American countries are also well known for their social inequality and poverty. Simultaneously, most countries within the region concentrate their population within urban centers. Actually, among the five most populated global megacities, two are Latin American. In this chapter, we describe the region's history of urbanization and socio-economy, examine the urban–biodiversity relationships that prevail in cities, and dissect the relationship that exists between the countries' urbanization, their poverty levels, and their biodiversity (considering bird diversity as a proxy). For practical matters, we focused on major Latin American countries (>3 million people) for which official data are available. Our results show that Latin America is a biodiverse and socio-economically complex region, but it is also heterogeneous. Some countries showed to be tightly linked to one or two of the assessed variables, while others lied in intermediate values. For instance, some countries positively associated with urbanization but not with poverty or biodiversity (e.g., Chile, Uruguay), others related with high urbanization and poverty values (e.g., Puerto Rico, El Salvador), others with high poverty (e.g., Guatemala, Honduras), or high biodiversity (e.g., Peru, Brazil, Colombia, Ecuador, Bolivia). Thus, our results do not only highlight the peculiarities of each country but the heterogeneous reality of these relationships across the region.

Keywords Urban ecology · Sustainability · Bird richness · Global South · Neotropic · Socioeconomic

---

C. E. Sánchez-Sarria · M. Bonilla-Moheno
Red de Ambiente y Sustentabilidad, Instituto de Ecología, A.C. (INECOL), Xalapa, Mexico

L.-B. Vázquez
ECOSUR, San Cristóbal de las Casas, Mexico

I. MacGregor-Fors (✉)
Ecosystems and Environment Research Programme, Faculty of Biological and Environmental Sciences, University of Helsinki, Lahti, Finland
e-mail: ian.macgregor@helsinki.fi

F. Angeoletto et al. (eds.), *Ecology of Tropical Cities*,
https://doi.org/10.1007/978-3-031-48012-6_5

1 Introduction

Latin America has an entangled relationship between its manifold socioeconomic problems and vast biodiversity. In terms of population, the region harbors eight percent of the global population (United Nations, 2019). For many decades, Latin America experienced one of the highest population growth rates across the globe (United Nations, 2019). From 1950 to present day, the number of people tripled, from 200 million to approximately 650 million. According to the United Nations (2019), this number will peak around the year 2050, reaching at least 750 million people. One consequence of this rapid population growth is the high levels of urbanization, resulting in Latin America becoming the second most urbanized region of the world, second only to Northern North America (United Nations, 2019).

Despite the diversity and complexity of the local historical contexts, all countries of the region experienced a transition in their populations' lifestyle from scattered agricultural settlements to densely populated cities and towns, where most of the population now lives (CEPALSTAT, 2020). In fact, in 1950, the median percentage of the population that was considered urban was 36%. This trend drastically changed by 2018, when it was above 73%. By 2050, the urban proportion of the region's population is expected to reach 90% (United Nations, 2019). At present, only six megacities (>10 million inhabitants) in Latin America contain at least 18% of the urban population of the region (United Nations, 2019). Two extreme examples of this pattern are the capitals of the most populated countries in the region, Mexico City (Mexico) and Sao Paulo (Brazil), which have over 20 million people each (UN-ECLAC, 2012). Since the last century, these two metropolises have been the largest cities on the continent and now rank among the 5 most populous cities in the world. However, in recent years the population growth rate of both cities has declined due to migration to mid-size cities, which are also attracting non-urban ('rural') populations and growing at a fast rate (UN-ECLAC, 2012). These population trends are influencing the expansion of other urban areas as well as the number of localities that are now considered large urban centers (Pauchard & Barbosa, 2013). Interestingly, around 63 cities in the region harbor between one and five million people each, representing an important proportion of urbanites (United Nations, 2019).

In hand with its important population, Latin America is an example of inequity and poverty (Martinez-Alier, 1991; Pauchard & Barbosa, 2013; Angeoletto et al., 2019). The economic gap is so large that the top 20% of richest people in the region earn almost 20 times the income of the poorest 20%, with more than 25% of the urban population living in low-income areas (ONU-Habitat, 2012). In the 1960s, poverty was related to a disparity in the relationship between low economic growth and high population growth (Huber, 2002). But in subsequent decades, the population dynamics changed causing new challenges to emerge (e.g., age structure, aging) and new solutions needed (Filgueira & Filgueira, 2002). Between 2002 and 2014, the region experienced significant progress in reducing poverty and social inequality while increasing social protections (e.g., education, health; CEPAL, 2019).

During this period, the rate of poverty fell from 45% to 28%, and extreme poverty decreased from 12% to 8% (CEPAL, 2019; CEPALSTAT, 2020). Nonetheless, over the last 6 years, there have been setbacks in the region; in 2018, ~185 million people lived in poverty, of these almost 40% were in extreme poverty (CEPAL, 2019). Future projections point that for 2030, the rate of poverty will be 25%, and extreme poverty will be 9% (CEPAL, 2019).

In the regional context, the main drivers of poverty are linked to (i) increased migration, (ii) acute and profound demographic transformations, (iii) a notable precariousness of employment, (iv) the increasingly frequent climatic events, and (v) natural disasters (Huber, 2002; Satterthwaite, 2003). The last two factors play a key role in maintaining and increasing poverty as they have higher impacts on low-income areas (Grau & Aide, 2008; Romero-Lankao & Gnatz, 2013; Coates & Nygren, 2020). For example, the Caribbean floods of 2004 were less devastating in the Dominican Republic and Puerto Rico than in Haiti, as their highlands were reforested by the shift from an extractive to an industrial economy, which subsequently reduced agricultural activities and countrywide poverty levels (Grau & Aide, 2008). Similarly, low-income urban spaces are particularly vulnerable to landslides and floods as they are commonly unplanned developments driven by the high and rapid migration of people to the periphery of cities (Romero-Lankao & Gnatz, 2013; Coates & Nygren, 2020). The expansion of hazardous urban areas is particularly worrisome in Latin America, where many cities are located on or near biodiversity hotpots, posing a challenging scenario for the region's sustainability, the health of the people, and the survival of the vast native biodiversity.

Certainly, Latin America is considered the most biodiverse region in the world as it contains refuges for six biodiversity hotspots like the Cerrado's savannas in Brazil, the Andes Mountains which cross from Venezuela to Argentina, and the Mesoamerican Forests in Central America, the third largest hotspot on Earth (Myers et al., 2000; Mittermeier et al., 2011). For example, the Andes are thought to be the richest hotspot on the planet as they harbor more than 45,000 plant and more than 3,000 animal species of which more than 20,000 and 1,500 are endemic, respectively (Myers et al., 2000). Therefore, several Latin American countries (e.g., Brazil, Colombia, Mexico, Ecuador, Peru) are characterized as some of the world's most biodiverse countries. Regrettably, the extraordinary diversity of Latin America is under a continuous threat due to habitat transformation and land-use change, mainly by large-scale cattle ranching and the expansion of agricultural frontier (Miles et al., 2006; Aide et al., 2013; Mustin et al., 2017). Nevertheless, the effect of urbanization on Latin American biodiversity is often overshadowed and underestimated despite it being known as a species-threatening agent in the region (Martinez-Alier, 1991; Pauchard & Barbosa, 2013).

Indeed, Maxwell et al. (2016) in their assessment of the global "big killers" of biodiversity classified urban development (e.g., housing, tourism, recreation, industrial activities) as the third main driver for species loss, behind over-exploitation of the natural resources and agricultural activities, the most important causes of biodiversity loss. Remarkably, the first two "big killers" are the result of the 'metabolism' of cities (Kennedy et al., 2011), process in which a huge amount of resources (e.g., food, energy, materials, services, goods) are brought to and processed in cities every day. Sometimes, in any given city, the resources needed to supply the soaring demand can exceed 200 times the size of its urban sprawl (Wigginton et al., 2016). As mentioned above, this issue is particularly problematic in Latin America where the majority of the population is urban and the countries' economies are mostly based on unsustainable practices of natural resource overexploitation (Rieckmann et al., 2011; Pauchard & Barbosa, 2013). In addition, urban ecology studies in the region barely began in the 1970s, and, despite their increase in the last decade, the discipline is still consolidating and remains underrepresented (MacGregor-Fors & Escobar-Ibáñez, 2017).

For instance, two recent reviews of urban ecology (MacGregor-Fors & Escobar-Ibáñez, 2017; Marzluff, 2017) pinpoint that despite Latin American researchers assessing the ecological patterns in cities, the mechanistic processes underlying these patterns are not well studied. Nevertheless, urban bird ecology is one of the topics most studied. Although several ecological patterns remain similar when contrasted to North America and Europe, other important aspects are different (Ortega-Álvarez & MacGregor-Fors, 2011; MacGregor-Fors & Escobar-Ibáñez, 2017). In Latin American cityscapes, it is more likely to find higher species richness, even up to 341 species within city limits (González-García et al., 2016), and an important representation of insectivores and frugivores birds, including large species such as toucans and macaws, which largely differs from the avifaunas of upland temperate regions (Ortega-Álvarez & MacGregor-Fors, 2011; MacGregor-Fors & Escobar-Ibáñez, 2017). Such differences highlight the need to evaluate how species and populations respond to this rapidly urbanizing region, especially when Latin American cities are expanding on biodiversity hotspots and the pattern and mechanisms of urban ecology may not be entirely comparable to well-studied regions.

In a nutshell, Latin America harbors a unique and complex convergence of high levels of urbanization, socioeconomic inequity, and vast biodiversity (Pauchard & Barbosa, 2013; MacGregor-Fors & Escobar-Ibáñez, 2017). The latter is a result of a historically wide array of abiotic and biotic conditions (Brown, 2014). High levels of urbanization and inequity (both concentrated in the most populated cities of each country) are mainly the result of the accelerated and unplanned migration, settling, and thriving of human populations in cities over the last few decades. Regrettably, the number of landscapes transformed by unplanned urban sprawls and commonly dominated by precarious life quality and massive socioeconomic differences is increasing in the region. These complex urban landscapes threaten human health, biodiversity, the possibility of creating a sustainable region, and, of course, the survivorship of humankind in the future.

The aim of this study was to describe the historical framework and the current state of Latin America from an urban lens through its population trends, economic disparities, and its vast biodiversity. Later and more specifically, we dissect the relationship between urbanization, poverty, and biodiversity (using bird richness as a proxy) for the region as well as for each country. Here, we only focus on the most populous countries (more than three million people) and for which official and comparable socioeconomic data are available. At the end of the chapter, we highlight the peculiarities of each country, explain how these three variables are entangled in a heterogenic network, pose potential future scenarios together with their consequences, and finally, how this information could be merged with public politics on urban development to enhance human health and create cities friendly to biodiversity.

2 Methods

To assess the relationship between urbanization, poverty, and biodiversity in Latin America, we gathered data for the most populated countries of the region (21 countries, >3,000,000 inhab.). From various sources data was retrieved for each country's area, total population, percentage of the population living in urban areas, in poverty and in extreme poverty, the percentage of the urban population in poverty, and, as a proxy for biodiversity, the total number of bird species recorded in its territorial limits. For the socioeconomic statistics of each country, we used data published online by the Economic Commission for Latin America and the Caribbean program (CEPALSTAT, 2020) and the Population Dynamics Division of the United Nations (2019). To retrieve bird richness per country, we used Avibase (2020), an online database information system that compiles and organizes the taxonomic information of all birds in the world. Despite compiling information for Haiti and Cuba, as they have more than three million inhabitants, we excluded them in the analysis as we failed to retrieve updated socioeconomic data.

Although we recognize that the data acquired by each country internally are more precise, we considered that United Nations' socioeconomics statistics are more appropriate for our comparative analysis because it is the most updated available data among countries of the region, which reduces biases in the results and interpretations. Due to the political status of Puerto Rico as a commonwealth of the United States, the socioeconomic statistics were only available in the United States Census Bureau (2020). For both cases, we followed the 'poverty' and 'extreme poverty' concepts placed by the World Bank (2017) in the International Poverty Line, which defines a person as 'poor' when their economic income is less than 1.9 adjusted international dollars per day and is not sufficient to buy a basic basket of goods and services, both food and non-food. A person is defined as 'extremely poor' when their income is less than 1.25 international dollars per day and when they are normally residing in households whose collective income is not sufficient to buy a basic food basket. Furthermore, we approach country biodiversity using avian richness because birds are one of the most studied taxonomic groups worldwide and for which most of their species and their distribution are very well known (Jetz et al., 2012).

We used a non-metric multidimensional scaling (NMDS) analysis to summarize country urbanization level, poverty, and biodiversity (Legendre & Gallagher, 2001; Legendre & Legendre, 2012). Because our aim was to understand the relationship between urbanization, poverty, and biodiversity in Latin America, the NMDS is a suitable analysis to assess similarities and differences among countries (Clarke, 1993; Legendre & Legendre, 2012). The NMDS is an unconstrained ordination method that displays the study objects in a n-dimension plot, allowing us to graphically fit each studied dimension, visually grouping the countries according to their associations with the variables (Legendre & Gallagher, 2001; Oksanen et al., 2019). We ultimately did not include 'urban poverty' and 'extreme poverty' in our analysis for two mains reasons: (1) the absence of data for several countries and (2) their tight correlation with country poverty (urban poverty: $r = 0.972$, $p < 0.001$; extreme poverty: $r = 0.917$, $p < 0.001$). The NMDS was run using the *vegan* (Oksanen et al., 2019) package in R (R Core Team, 2020).

3 Results

As expected, country area and population were highly correlated ($r = 0.927$, $p < 0.001$). Indeed, Brazil was the largest (8,514,880 km^2) as well as the most populated (205,960,000 inhab.) country in our study, while Puerto Rico was the smallest (13,800 km^2) and least populated (3,200,000 inhab.) (Fig. 5.1). Uruguay showed to be the most urbanized country (94.2%) and the one with the lowest percentage of poverty (2.9%). Haiti represented the least urbanized (47.5%) and least biodiverse country (272 spp.). Honduras had the highest percentage of poverty (55.7%). Finally, Colombia has the highest richness in birds (2020) in the region (Table 5.1, Fig. 5.1). From a socioeconomic perspective, the mean percentage of the population that is urban in Latin America was 72.5 (± SD 14.3%), while 28.1 (± SD 14.3%) lived in poverty.

The NMDS (stress: 0.02; Fig. 5.2) depicts the important heterogeneity in the urbanity-poverty-biodiversity relationships among Latin American countries. The graphical representation of the fist two axis of the NMDS shows that some countries are tightly linked to one or two of the assessed variables, while others lie in intermediate values. For instance, some countries positively associated with urbanization but not with poverty or biodiversity (e.g., Chile, Uruguay), others related with high urbanization and poverty values (e.g., El Salvador, Puerto Rico), others with high poverty (e.g., Guatemala, Honduras), or high avian richness (e.g., Peru, Brazil, Colombia, Ecuador, Bolivia). Among the countries that did not show a clear association with the three assessed variables, but were rather average, included Paraguay, Argentina, Costa Rica) (Fig. 5.2).

4 Discussion

Latin America is quite a unique region. First, the region is very large (20 million km^2), encompassing more than 20 countries with a great diversity of histories, cultures, and languages. Certainly, Spanish and Portuguese practices and religion

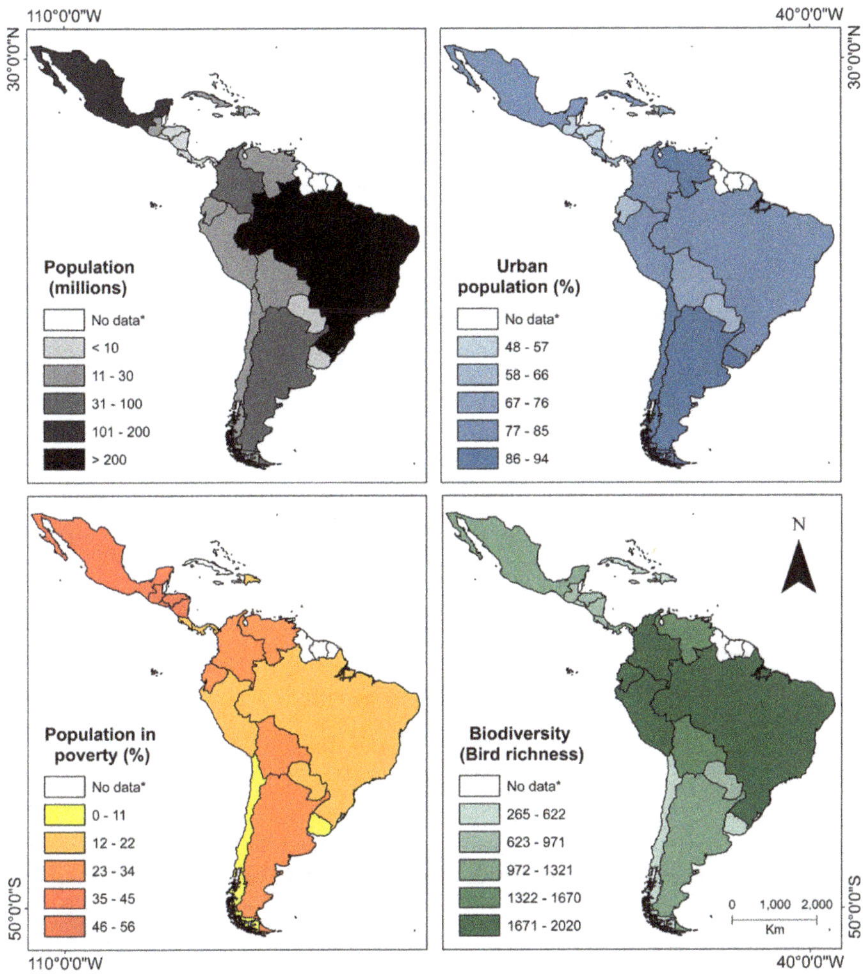

Fig. 5.1 Country population, urban population, population in poverty, and biodiversity (bird species richness) for the 21 most populated countries (>3,000,000 inhabitants) in Latin America

highly influenced the region, but they also occur alongside hundreds—if not thousands—of native indigenous communities and their worldviews. Second and remarkably, the region is home to several of the most biodiverse ecosystems in the world, with high rates of endemism. Lastly and woefully, it is a region with increasing rates of poverty and socioeconomic inequality in a highly urban reality, which is densely concentrated and if not unplanned, following a highly disorganized and economically-driven planning (Benítez et al., 2012). Given these traits, here we explored how—in the most populated countries—urbanization, poverty, and biodiversity are related to each other, for the region as a whole and for each country. In sum and as expected, we found that Latin America is a very diverse and heterogeneous region, with a strong variation of countries and eco-socioeconomic

Table 5.1 Percentage of the population that is urbanized, percentage of the population in poverty, and biodiversity values for the 21 most populated countries of Latin America. The countries are sorted alphabetically. Blank spaces show the absence of updated data for these countries

Country	Urban population (%)	Poverty (%)	Bird richness
Argentina	91.0	24.4	1081
Bolivia	66.8	33.2	1476
Brazil	84.3	19.4	1935
Chile	88.1	10.7	561
Colombia	77.7	29.9	2020
Costa Rica	71.6	16.1	944
Cuba	76.6		392
Dominican Republic	73.7	22	320
Ecuador	62.7	24.2	1727
El Salvador	65.2	34.5	597
Guatemala	52.0	50.5	786
Haiti	47.5		272
Honduras	50.8	55.7	800
Mexico	76.6	41.5	1178
Nicaragua	56.8	46.3	789
Panama	65.2	14.5	1025
Paraguay	63.4	19.5	734
Peru	76.9	16.8	1929
Puerto Rico	93.6	43.1	393
Uruguay	94.2	2.9	508
Venezuela	88.8	28.3	1461

relationships, reason why not all the region is "where urbanization and poverty meet biodiversity" (sensu Escobar-Ibáñez & MacGregor-Fors, 2017) in a similar fashion.

Initially colonized by contemporary Western powers, most countries of Latin America, Africa, and Asia have received very diverse names: "third world", "underdeveloped", "developing", "emerging", and most recently "Global South" (Bolay, 2020). Despite the disparaging implications of these terms, they also describe the socioeconomic transition that countries in these regions are experiencing today. Even so, this transition does not seem to be optimistic when compared to the rest of the world, as inequalities within countries are on the rise both at the country-level and at the regional-level (Bolay et al., 2005; Kanbur & Sumner, 2012). Countries from the "Global South" are following the same patterns of population growth and migration that occurred in the USA and Europe in the past (Dufour & Piperata, 2004; Rodríguez-Vignoli & Rowe, 2018). In this scenario, human settlements are expected to keep growing as they continuously receive waves of non-urban migrant populations (UN-ECLAC, 2012; Bolay, 2020).

Particularly, in Latin America, the migration of human populations to cities can be primarily explained by two factors: (1) the impossibility of small-scale farmers to compete with big enterprises and agricultural international trades, and (2) the cultural and economic 'attractiveness' that an urban life has for the majority of the

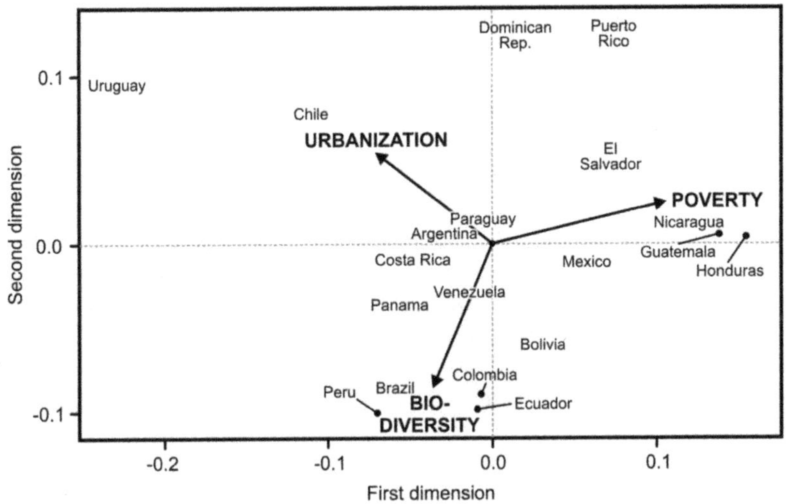

Fig. 5.2 Non-metric multidimensional scaling (NMDS) ordination of 21 Latin American countries by their urban population (urbanization), poverty, and bird richness (biodiversity). Country names are displayed in the figure. The arrows show a positive gradient for each dimension and following the direction of the vector indicates a strong association with it. We excluded Haiti and Cuba in the ordination due to the absence of poverty data

young people as a result of globalization (Dufour & Piperata, 2004; Grau & Aide, 2008; Rodríguez-Vignoli & Rowe, 2018). Nevertheless, other important factors have stimulated the rates of migration such as armed conflicts in Colombia or large-scale natural catastrophes in Honduras (Dufour & Piperata, 2004; Grau & Aide, 2008). Rapid migration of farmers to urban areas has led to the abandonment of agricultural areas, the creation of a skewed distribution of land property (Grau & Aide, 2008; Bolay, 2020) in which a few landowners have acquired huge amounts of land, increasing socioeconomic inequity.

The largest cities are the main economic drivers across the region. Most of the technological, industrial, governmental, and financial powers of countries are concentrated in these cities. Moreover, most urban ecology research has focused on larger cities (>1 million inhab.), while medium and small-sized human agglomerations are rarely studied (MacGregor-Fors & Schondube, 2012; MacGregor-Fors & Escobar-Ibáñez, 2017; Bolay, 2020). This fact should not be overlooked, as urban settings of less than 500,000 inhabitants have the highest population growth rates and are home to nearly 50% of the world's urban population (Bolay & Rabinovich, 2004; UN-ECLAC, 2012; UN, 2019). Yet, the problems faced by small and intermediate cities are not less serious or easier to solve in comparison to bigger cities. Normally, these urban settlements have insufficient budgets to deal with their challenges, which makes them highly dependent on the national governments. Given that these cities are rarely known outside provincial or national borders, they must manage their challenges without a consistent and structured outside support (Bolay, 2020). For example, creating and attracting a skilled and talented staff of workers is very difficult and is one reason why many intermediate and small urban settings do not have reliable and helpful urban planning (Bolay & Rabinovich, 2004; Bolay, 2020).

Cities are undoubtedly a reflection of the history, culture, and idiosyncrasy of humans. Most of the cities, towns, and villages tend to reflect the socioeconomic state of the country as a whole, as well as the sentiment of the general public. Urbanization can seem perfect, coherent, and organized in countries with a better life quality or, on the contrary, dirty, disordered, and chaotic, which is often the case of the majority of Latin American cities, or at least important sectors within them. One notable reason for this is that both individuals and public authorities have been unable to handle and solve problems with the speed and ease with which they arise (Bolay, 2020; Coates & Nygren, 2020). Furthermore, Latin American countries and their urban settlements often face different problems or similar problems of greater magnitude compared to countries in the Global North. Theories and analysis made in developed counties are rooted in a historical, cultural, temporal, social, and economic context that is not necessarily applicable to Latin American reality if we fail to understand and integrate the knowledge and needs of all the different actors who actively participate in the construction of cities and their identity (Bolay, 2018, 2020).

In conclusion, Latin American cities are the living proof of regional heterogeneity in terms of urbanity, poverty and biodiversity. In only one urban settlement, you might find unmanaged and managed greenspaces, artificial and natural bluespaces, densely and sparsely urbanized zones, and high levels of biodiversity both in high-income and low-income areas (MacGregor-Fors & Escobar-Ibáñez, 2017). If the conservation of biodiversity, as well as increases of the well-being of urbanites, is a desired future in Latin America, governments, urban planners, and managers face an increasing number of challenges. The inclusion of well-informed urban ecologists into the decision-making table is not only correct, but necessary (Bolay, 2020). In this sense, Latin American cities are interesting laboratories to study and understand the repercussion of excessive and rapid urbanization on tropical biodiversity. Despite this opportunity, urban ecologists in Latin America face two major challenges: (1) to carry out these kinds of studies in areas where socioeconomic inequality has produced daily public order problems, even in zones where it is already almost impossible to work safely and (2) to achieve the integration of the scientific knowledge on public politics, and decision makers, particularly in public administrations with a lack of interest in nature-based socioeconomic solutions.

References

Aide, T. M., Clark, M. L., Grau, H. R., López-Carr, D., Levy, M. A., Redo, D., Bonilla-Moheno, M., Riner, G., Andrade-Núñez, M. J., & Muñiz, M. (2013). Deforestation and Reforestation of Latin America and the Caribbean (2001–2010). *Biotropica, 45*, 262–271. https://doi.org/10.1111/j.1744-7429.2012.00908.x

Angeoletto, F., Fellowes, M. D., Essi, L., Santos, J. W., Johann, J. M., da Silva Leandro, D., & Mendonca, N. M. (2019b). Ecología urbana y planificación: una convergéncia ineludible. *Revista Electrônica em Gestão, Educação e Tecnologia Ambiental, 23*(1), 1–8.

Avibase. (2020). *Avibase – The world bird database*. https://avibase.bsc-eoc.org/checklist.jsp?lang=EN. Accessed 18 Oct 2020.

Benítez, G., Pérez-Vázquez, A., Nava-Tablada, M., Equihua, M., & Álvarez-Palacios, J. L. (2012). Urban expansion and the environmental effects of informal settlements on the outskirts of Xalapa city, Veracruz, Mexico. *Environment and Urbanization, 24*(1), 149–166. https://doi.org/10.1177/0956247812437520

Bolay, J.-C. (2018). Planning the intermediate city, or how to do better with little: The case of the city of Nueve de Julio, Argentina. *Current Urban Studies, 6*(3), 366–400. https://doi.org/10.4236/cus.2018.63020

Bolay, J.-C. (2020). *Urban planning against poverty: How to think and do better cities in the Global South*. Springer International Publishing.

Bolay, J.-C., & Rabinovich, A. (2004). Intermediate cities in Latin America risk and opportunities of coherent urban development. *Cities, 21*(5), 407–421. https://doi.org/10.1016/j.cities.2004.07.007

Bolay, J.-C., Pedrazzini, Y., Rabinovich, A., Catenazzi, A., & García Pleyán, C. (2005). Urban environment, spatial fragmentation and social segregation in Latin America: Where does innovation lie? *Habitat International, 29*(4), 627–645. https://doi.org/10.1016/j.habitatint.2004.05.003

Brown, J. H. (2014). Why are there so many species in the tropics? *Journal of Biogeography, 41*(1), 8–22. https://doi.org/10.1111/jbi.12228

CEPAL. (2019). *Panorama social de América Latina*. Comisión Económica para América Latina y el Caribe.

CEPALSTAT. (2020). *Databases and statistical publications*. Economic Commission for Latin American and the Caribbean. https://estadisticas.cepal.org/cepalstat/PerfilesNacionales.html?idioma=English. Accessed 18 Oct 2020.

Clarke, K. R. (1993). Non-parametric multivariate analyses of changes in community structure. *Australian Journal of Ecology, 18*(1), 117–143. https://doi.org/10.1111/j.1442-9993.1993.tb00438.x

Coates, R., & Nygren, A. (2020). Urban floods, clientelism, and the political ecology of the state in Latin America. *Annals of the American Association of Geographers, 110*(5), 1301–1317. https://doi.org/10.1080/24694452.2019.1701977

Dufour, D. L., & Piperata, B. A. (2004). Rural-to-urban migration in Latin America: An update and thoughts on the model. *American Journal of Human Biology, 16*(4), 395–404. https://doi.org/10.1002/ajhb.20043

Escobar-Ibáñez, J. F., & MacGregor-Fors, I. (2017). What's new? An updated review of avian ecology in urban Latin America. In I. MacGregor-Fors & J. F. Escobar-Ibáñez (Eds.), *Avian Ecology in Latin American Cityscapes* (pp. 11–31). Springer International Publishing.

Filgueira, C. H., & Filgueira, F. (2002). Models of welfare and model of capitalism: Limits of transferability. In E. Huber (Ed.), *Models of capitalism: Lessons for Latin America* (pp. 127–157). The Pennsylvania State University Press.

González-García, F., Straub, R., Lobato García, J. A., & MacGregor-Fors, I. (2016). Nuevos registros y notas adicionales comentadas sobre la avifauna de la ciudad de Xalapa, Veracruz, México. *Acta Zoologica Mexicana, 32*(3), 253–269. https://doi.org/10.21829/azm.2016.323960

Grau, H., & Aide, M. (2008). Globalization and land-use transitions in Latin America. *Ecology and Society, 13*(2), 16. https://doi.org/10.5751/ES-02559-130216

Huber, E. (Ed.). (2002). *Models of capitalism: Lessons for Latin America*. The Pennsylvania State University Press.

Jetz, W., Thomas, G. H., Joy, J. B., Hartmann, K., & Mooers, A. O. (2012). The global diversity of birds in space and time. *Nature, 491*(7424), 444–448. https://doi.org/10.1038/nature11631

Kanbur, R., & Sumner, A. (2012). Poor countries or poor people? Development assistance and the new geography of global poverty. *Journal of International Development, 24*(6), 686–695. https://doi.org/10.1002/jid.2861

Kennedy, C., Pincetl, S., & Bunje, P. (2011). The study of urban metabolism and its applications to urban planning and design. *Environmental Pollution, 159*(8), 1965–1973. https://doi.org/10.1016/j.envpol.2010.10.022

Legendre, P., & Gallagher, E. D. (2001). Ecologically meaningful transformations for ordination of species data. *Oecologia, 129*(2), 271–280. https://doi.org/10.1007/s004420100716

Legendre, P., & Legendre, L. (2012). *Numerical ecology* (3rd ed.). Elsevier.

MacGregor-Fors, I., & Escobar-Ibáñez, J. F. (Eds.). (2017). *Avian ecology in Latin American cityscapes*. Springer International Publishing.

MacGregor-Fors, I., & Schondube, J. E. (2012). Urbanizing the wild: shifts in bird communities associated to small human settlements. *Revista Mexicana de Biodiversidad, 83*(2). https://doi.org/10.22201/ib.20078706e.2012.2.982

Martinez-Alier, J. (1991). Ecology and the poor: A neglected dimension of Latin American history. *Journal of Latin American Studies, 23*(3), 621–639.

Marzluff, J. M. (2017). A decadal review of urban ornithology and a prospectus for the future. *Ibis, 159*(1), 1–13. https://doi.org/10.1111/ibi.12430

Maxwell, S. L., Fuller, R. A., Brooks, T. M., & Watson, J. E. M. (2016). Biodiversity: The ravages of guns, nets and bulldozers. *Nature News, 536*(7615), 143. https://doi.org/10.1038/536143a

Miles, L., Newton, A. C., DeFries, R. S., Ravilious, C., May, I., Blyth, S., Kapos, V., & Gordon, J. E. (2006). A global overview of the conservation status of tropical dry forests. *Journal of Biogeography, 33*(3), 491–505. https://doi.org/10.1111/j.1365-2699.2005.01424.x

Mittermeier, R. A., Turner, W. R., Larsen, F. W., Brooks, T. M., & Gascon, C. (2011). Global biodiversity conservation: The critical role of hotspots. In F. E. Zachos & J. C. Habel (Eds.), *Biodiversity hotspots: Distribution and protection of conservation priority areas* (pp. 3–22). Springer-Verlag.

Mustin, K., Carvalho, W. D., Hilário, R. R., Costa-Neto, S. V., Silva, C., Vasconcelos, I. M., Castro, I. J., Eilers, V., Kauano, É. E., Mendes-Junior, R. N. G., Funi, C., Fearnside, P. M., Silva, J. M. C., Euler, A. M. C., & Toledo, J. J. (2017). Biodiversity, threats and conservation challenges in the Cerrado of Amapá, an Amazonian savanna. *Nature Conser, 22*, 107–127. https://doi.org/10.3897/natureconservation.22.13823

Myers, N., Mittermeier, R. A., Mittermeier, C. G., da Fonseca, G. A. B., & Kent, J. (2000). Biodiversity hotspots for conservation priorities. *Nature, 403*(6772), 853–858. https://doi.org/10.1038/35002501

Oksanen, J., Blanchet, F. G., Friendly, M., Kindt, R., Legendre, P., McGlinn, D., Minchin, P. R., O'Hara, R. B., Simpson, G. L., Solymos, P., Stevens, M. H. H., Szoecs, E., & Wagner, E. (2019) *Vegan: Community ecology package*.

ONU-Habitat. (2012). *Estado de las ciudades de América Latina y el Caribe 2012: Rumbo a una nueva transición urbana*. Programa de las Naciones Unidas para los Asentamientos Humanos.

Ortega-Álvarez, R., & MacGregor-Fors, I. (2011). Dusting-off the file: A review of knowledge on urban ornithology in Latin America. *Landscape Urban Plan, 101*(1), 1–10. https://doi.org/10.1016/j.landurbplan.2010.12.020

Pauchard, A., & Barbosa, O. (2013). Regional assessment of Latin America: Rapid urban development and social economic inequity threaten biodiversity hotspots. In T. Elmqvist, M. Fragkias, J. Goodness, B. Güneralp, P. J. Marcotullio, R. I. McDonald, S. Parnell, M. Schewenius, M. Sendstad, K. C. Seto, & C. Wilkinson (Eds.), *Urbanization, biodiversity and ecosystem services: Challenges and opportunities* (pp. 589–608). Springer Netherlands.

R Core Team. (2020). *R: A language and environment for statistical computing*. R Foundation for Statistical Computing.

Rieckmann, M., Adomßent, M., Härdtle, W., & Aguirre, P. (2011). Sustainable development and conservation of biodiversity hotspots in Latin America: The case of Ecuador. In F. E. Zachos &

J. C. Habel (Eds.), *Biodiversity hotspots: Distribution and protection of conservation priority areas* (pp. 435–452). Springer-Verlag.

Rodríguez-Vignoli, J., & Rowe, F. (2018). How is internal migration reshaping metropolitan populations in Latin America? A new method and new evidence. *Population Studies, 72*(2), 253–273. https://doi.org/10.1080/00324728.2017.1416155

Romero-Lankao, P., & Gnatz, D. M. (2013). Exploring urban transformations in Latin America. *Current Opinion in Environmental Sustainability, 5*(3), 358–367. https://doi.org/10.1016/j.cosust.2013.07.008

Satterthwaite, D. (2003). The links between poverty and the environment in urban areas of Africa, Asia, and Latin America. *The Annals of the American Academy of Political and Social Science, 590*(1), 73–92. https://doi.org/10.1177/0002716203257095

UN-ECLAC. (2012). *Population, territory and sustainable development. Proceedings of the Expert Group Meeting on Population, Territory and Sustainable Development*. United Nations Economic Commission for Latin America and the Caribbean.

United Nations. (2019). *World urbanization prospects: The 2018 revision. Department of Economic and Social Affairs, Population Dynamics Division*. United Nations.

United States Census Bureau. (2020). *Explore census data*. https://data.census.gov/cedsci/ Accessed 18 Oct 2020.

Wigginton, N. S., Fahrenkamp-Uppenbrink, J., Wible, B., & Malakoff, D. (2016). Cities are the future. *Science, 352*(6288), 904–905. https://doi.org/10.1126/science.352.6288.904

World Bank. (2017). *Monitoring global poverty: Report of the commission on global poverty*. The World Bank.

Chapter 6
Vale Encantado Park: Popular Mobilization for a Conservation Area in a Metropolitan Environment

Angelo Serpa

Abstract Vale Encantado is an area of Atlantic Forest in Salvador, Bahia, which encompasses approximately 100 hectares and is located in the neighborhood of Patamares. It is home to more than 200 tree species and 100 bird species, as well as dozens of species of mammals, reptiles, and amphibians. In 2016, it became an ecological park listed in the Urban Development Master Plan. However, on December 10, 2019, the City Council approved bill n° 299/19 which changed the boundaries of the Park, alleging the outdated cartography was used for its delimitation, which also had not taken into account the status of the land and included private and institutional properties. Since the passage of bill n° 299, a group of local citizens have mobilized against changing the boundaries of the Vale Encantado. They argue that in addition to being a refuge for fauna, its preservation is crucial for the regional microclimate and the environmental health of the entire city. In January 2020, scientific entities recommended the creation of the Vale Encantado Wildlife Refuge. Thousands have signed an online petition supporting the initiative. The mobilization (especially of the middle class) for the creation of this conservation area in Salvador shows the contradictions in a city that is part of the Cities4Forests network, but at the same time it allows the destruction of its vegetation by real estate interests and for transportation such as the BRT (Bus Rapid Transit) and the Metro. The movement for the creation of the Park faces resistance from builders and developers, but has gained support from the population and achieved some victories.

Keywords Vale Encantado · Atlantic forest · Ecological Park · Wildlife refuge · Popular mobilization · Urban development master plan · SOS Vale Encantado collective · Metropolitan environment · Real estate speculation · Salvador-Bahia

---

A. Serpa (✉)
Universidade Federal da Bahia and Researcher Level 1B, with a Research Productivity Fellowship from the National Council for Scientific and Technological Development, Salvador, Bahia, Brazil
e-mail: angserpa@ufba.br

F. Angeoletto et al. (eds.), *Ecology of Tropical Cities*,
https://doi.org/10.1007/978-3-031-48012-6_6

1 The Context

In 2018, the Vale Encantado SOS Collective in partnership with the Mãos da Terra Institute (IMATERRA) prepared, with funds from the Public Ministry of Bahia, technical studies to support the creation of the Vale Encantado Full Protection Conservation Unit (CU). These studies sought to fulfill Bill N° 9069/2016, which instituted the Urban Development Master Plan of the Municipality of Salvador (PDDU 2016) and determined the need to conduct specific research and surveys for the effective implementation of the Municipal Conservation Unit (CU) (SALVADOR, 2016a). For the Collective, the aforementioned CU, named the Vale Encantado Wildlife Refuge, is "an important milestone for the public policy of urban biodiversity conservation in the municipality of Salvador and ensures countless benefits for the quality of life of its population" (IMATERRA; SOS Vale Encantado, 2018, s/p) (Fig. 6.1).

These studies describe the characteristics of the physical and biotic environments, in addition to prioritizing areas for biodiversity conservation in the CU. This work is the basis of the first section of this chapter, in which we will summarize its results to contextualize the mobilization of the Collective and the city's residents for the creation of the Vale Encantado Conservation Unit. The area includes remnants of Atlantic Forest in the surroundings of Avenida Luís Viana Filho (Paralela Avenue), containing a mosaic of ecosystems composed of ombrophilous forest, restinga, and wetlands, which shelter rare animal species endemic to the Atlantic Forest and threatened with extinction. This area is under intense real estate pressure for urban

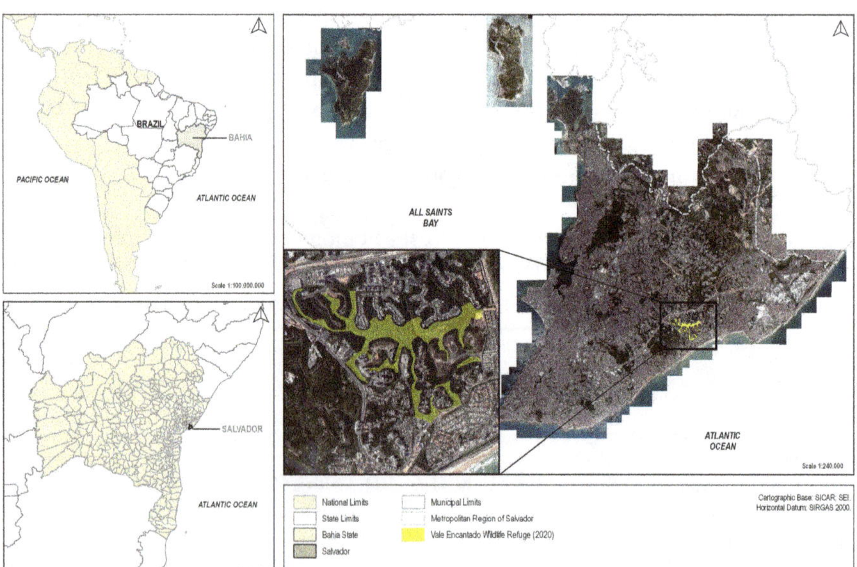

Fig. 6.1 Location map of Vale Encantado in Salvador-Bahia. (Compiled by: Jacileda Santos)

expansion by the middle- to high-income population in Salvador (IMATERRA; SOS Vale Encantado, 2018).

The studies noted the presence of 196 plant species, distributed in 71 families. The ten most common Angiosperm families in number of species were Fabaceae (20 species), Myrtaceae (9), Melastomataceae (9), Poaceae (8), Arecaceae (7), Malvaceae (7), Moraceae (6), Araceae (6), Rubiaceae (6), and Cyperaceae (5). Together, they represent 42% of the total flora identified. The remaining 29 plant families were represented by only one specie, which included 40.8% of the families observed. No endangered species were found in the area, although this result is not conclusive and points to the need for further research (IMATERRA; SOS Vale Encantado, 2018).

Some species with restricted distribution were found in the CU, including mucerengue (*Vochysia lucida*), and piaçava (*Attalea funifera*) and pindoba (*Attalea burretiana*) palms. More than 90% of the species surveyed are native and only 8% are exotic, which indicates the resilience of the vegetation, even in the mostly urbanized context. In general, they are ombrophilous forest species in their different stages of succession, but are mostly in the intermediate regeneration stage. Forested areas in a more advanced regeneration stage of had a similar composition to those in the intermediate stage, but were more diverse and/or larger, with a canopy height exceeding 12 meters (IMATERRA; SOS Vale Encantado, 2018).

The macrofauna cataloged in the 125 hectares of the area subjected to research and surveys included 25 species of anuras, 43 reptiles, 154 birds, and 40 mammals:

> Of the species recorded, 63 occur in only one physiognomy, of which 46 are unique to the Atlantic Forest, 16 to open anthropized areas, and 18 to wetlands (ponds and marshes). Most occupy more than one habitat, with a large predominance of species sharing the ombrophilous forest with other phytophysiognomies (70.6%, n = 185). Of these, 69 are of special interest for conservation, including endangered (n = 4), endemic (n = 17) (...), Atlantic Forest dependent (n = 44), bioindicators (n = 39), sensitive to environmental disturbances (n = 3), some falling into more than one category (IMATERRA; SOS Vale Encantado, 2018, s/p).

Despite the great wealth of macrofauna present in the Vale Encantado, local fauna populations can suffer various impacts, including the risks of being run over, as well as forest fires, deforestation, hunting, and pollution in the area. The consequent loss of species and habitat fragmentation affect the size of populations, reducing genetic variability and diversity. In addition, nesting, dispersal, and migration areas are altered, which increase the possibility of edge effects, causing risks of extinctions and invasion of exotic species (IMATERRA; SOS Vale Encantado, 2018) (Fig. 6.2).

Furthermore, the studies sought to classify the polygonal areas of the Vale Encantado (71.0 ha) and its surroundings (135.6 ha), considering the degree of relevance for protecting the communities of flora and fauna:

> Areas classified according to prioritization categories for biodiversity conservation were numbered in subareas (1 to 8). Of these, three located in the northern portion occupied most of the study area (53.2%) and presented the highest degree of biological importance, "Extremely High". The other subareas predominantly fell into the categories "Very High" (21.4%) and "High" (17.0%) (...). The results of the eight subareas describe the target

Fig. 6.2 Detail of the Vale Encantado area. (Photo by: Jacileda Santos)

elements of flora and fauna to justify the classification, the anthropic pressures, and a set of recommendations for protection, to assist decision-making related to the planning and implementation of actions involving the creation of the Vale Encantado Conservation Unit and the promotion of the sustainable use of the surrounding natural resources (IMATERRA; SOS Vale Encantado, 2018, s/p).

2 The Patamares Neighborhood and the Real Estate Interests at Stake

Patamares is a planned residential neighborhood and, since its inception, was designed to integrate preexisting buildings and green spaces. Built on the site of the former Jaguaripe and Biribeira farms, the neighborhood is the result of urban expansion in a northerly direction. This began with the consolidation of road infrastructure in the 1970s, which allowed greater accessibility to this area of the city. The place was gradually urbanized in the following decades with the construction of several gated residential communities (Santos et al., 2010).

In this context, the Alphaville Salvador subdivision emerged, which is a pioneering enterprise for the higher income population, who were attracted by the presence of Atlantic Forest fragments nearby, which is an incontestable factor of higher real estate values for the area. In the following years, Alphaville Salvador became a magnet to attract new residential real estate developments as well as commercial and service establishments (shopping malls, schools, universities, hotels, supermarkets, etc.) for the middle to high income population,[1] such as Condomínio Le Park, Salvador Ville Condomínio Clube, Wall Street Manhattan Square, Shopping Paralela (Araújo, 2014) and more recently high-rise apartment buildings, such as Greenville.

The appreciation of real estate properties in the area explains the opposition and conflicts over the implementation of the Conservation Unit, including the proposal to create an expressway that would connect the Boca do Rio and Piatã neighborhoods. This would mutilate not only the Vale Encantado, but also the neighboring Pituaçu Park and led to questions about the inclusion of private areas in the polygonal area of the new CU. The expressway would cross Pituaçu Park with a six-lane suspension bridge and divide the Vale Encantado Park, bringing irreversible impact to the last two large forested areas in the Paralela region (http://www.gamba.org.br/noticias/via-do-atlantico-prefeitura-e-governo-do-estado-estao-esperando-para-ver. Accessed on 24/04/2021). The construction of Via do Atlântico did not occur—although some people still defend its implementation, including real estate developers and technicians from public agencies involved with the issue. Nevertheless, the inclusion or not of private areas in the Vale Encantado polygonal continued to generate questions and debates (Fig. 6.3).

On 12/10/2019, the Salvador City Council approved a Municipal Executive Bill (n. 299/19), which instituted the "gated community allotment" and changed the Land Use and Occupancy Ordinance (LOUOS) (SALVADOR, 2016b). In the context of PLE 299, Amendment 3 was approved, which also provided for the alteration of the Park boundaries contained in the System of Areas with Environmental and Cultural Value. A justification for approval was the alleged use of outdated cartography to delimit the Park, which also did not take into account the land ownership situation, with the undue inclusion of private and institutional properties. It was also alleged that parks and CUs are formed by public lands and not by private land. However, the justification was a misunderstanding, because, in section III of chapter I of the LOUOS, the possibility of donation of private areas of environmental value to the Municipality is opened "for the purpose of establishing the CU in the region of the Vale Encantado Ecological Park":

> The text of PL n° 299, which amends, adds articles and tables to Law n° 9148/2016 of the Land Use and Occupancy Ordinance (LOUOS), was rejected by city councilmembers Aladilce Souza (PCdoB) and Marta Rodrigues (PT), in addition to councilmembers Edvaldo

[1] "According to data prepared by the Urban Development Company of the State of Bahia (Bahia, 2016), the neighborhood of Patamares has a resident population of 6156, 53.2% female and 46.7% male. The demographic density is 13.3% inhabitants per hectare. Of the total, 51.8% of the residents declared themselves white, 38% brown, and 8.9% black. The predominant age group is 20 to 49 years old (53.3%). The educational level of head of household is 11 to 14 years of schooling (47.8%). The average income per wage bracket is predominantly between 10 and 20 minimum wages (25.1%). Permanent private houses are the predominant category of houses, and there are no improvised or subnormal private houses. Of the total number of residences, 99.3% have water supply, 81.9% have sewage, and 99.7% have garbage collection. Patamares has 45.3% vegetation cover. The Human Development Index (HDI) is 0.952, considered very high" (IMATERRA; SOS Vale Encantado, 2018, s/p).

Fig. 6.3 Towers of the Alphaville Salvador high-rise apartments. (Photo by: Jacileda Santos)

> Brito (PSD), Helio Ferreira (PCdoB), Marcos Mendes (PSOL), Silvio Humberto (PSB), and Suíca (PT). Councilmember Cezar Leite (PSDB) defended the approval of the original project. (...) When declaring their votes, councilmembers Aladilce and Marta judged unconstitutional the amendments related to the Vale Encantado Park and the Mané Dendê Basin, as well as the alteration of the polygonal zones of two Special Zones of Social Interest (ZEIS) in Ilha dos Frades. Still, according to them, the inclusions alter instruments of the Urban Development Master Plan (PDDU) without technical studies and public hearings (https://www.cms.ba.gov.br/noticias/camara-aprova-condominio-de-lotes-reurb-e-nossos-talentos Accessed on 04/24/2021).

Nevertheless, the technical staff of the Municipal Department for the Sustainable City and Innovation (SECIS) prevailed. The understanding was that the inclusion of private land in the Vale Encantado polygonal should only occur through a formal request from the owners, provided that they showed interest in the objectives of the CU. In other words, the SECIS technicians understood that the creation of the Vale Encantado CU should only include public areas. A technical opinion sent to the Mayor in February 2020, in the draft of the creation decree, suggested that the CU should be delimited with a total area of 692,939.00 m^2. Private areas may be included after the CU is effective, through specific Terms of Agreement and Commitment, to be signed between the Salvador City Government and the interested landowners.

Since the approval of the Urban Development Master Plan in 2016, delimiting the polygonal of the Vale Encantado CU has been a central issue, because its boundaries were defined without conducting technical studies that actually supported the creation of the CU. This justified the specific studies conducted by the Vale Encantado SOS Collective, whose results were synthetically presented in the first section of this chapter. The residents involved in the Collective wanted private areas to be included in the polygon, which was also the wish of some condominiums in

Fig. 6.4 Panoramic view of the Vale Encantado area. (Photo by: Jacileda Santos)

the neighborhood. Therefore, the CU will be proposed as a Wildlife Refuge and not as another category of CU with full protection.[2]

In fact, the polygonal contained in the 2016 Master Plan included private land, generating conflict and opposition to the effective implementation of the CU. In January 2020, the National Institute of Science and Technology in Interdisciplinary and Transdisciplinary Studies, which included 47 laboratories from 31 Brazilian universities, sent a letter to the City Government recommending the creation of the Vale Encantado Wildlife Refuge and highlighting the ecological importance of the area. Thousands of citizens in Salvador have signed an online petition supporting the initiative. Nevertheless, in March 2020, a new delimitation of the Vale Encantado CU was approved without the private land previously contained in the polygon. It included only what the Municipal Secretary of Urban Development considered public land, which was private land donated by the developers of the parcels approved by the City Government in previous years (according to the Land subdivision law of 1976, updated in 2017) (Fig. 6.4).

The implementation of the Via do Atlântico, although no longer defended in public discussions, still guides the actions of some real estate developers and land owners who do not want the creation of a Conservation Unit or a Buffer Zone for the future CU. This is a strategy to revive the possible construction of an expressway, which could be definitively made impossible by an effective Vale Encantado CU. These agents would prefer the creation of an Urban Park, which would allow less problematic disappropriation of public areas in the future, with the allegation that the road would be of public utility, to the detriment of areas in the Vale Encantado polygonal, which could be sacrificed for the implementation of the road, considered unnecessary and with high impact by experts.

[2] In reality, Vale Encantado became an environmental park through a specific municipal decree, number 18,042, on December 6, 2007, an integral conservation unit made up of the lagoons of the Patamares, Greenville and Alphaville subdivisions, located on RA-10 – Itapuã; however, in the 2008 Master Plan, in article 242, it was reclassified as an urban park. The 2016 Master Plan has its polygon defined and, in article 279 (in the guidelines for urban parks), it is recommended, for the Vale Encantado Ecological Park, the preparation of studies aimed at classifying the area into one of the conservation unit categories integral part of the SNUC - National System of Conservation Units. It was from there that the SOS Vale Encantado movement emerged, with a strong presence on social media.

3 SOS VALE ENCANTADO and the Fight for the Vale Encantado

SOS Vale Encantado is an environmental collective of people interested in preserving this area of the city and creating a CU as a wildlife refuge. On its Facebook page, created in June 2018, the Collective presents videos, photos, and posts about its activities in Salvador, from mobilization campaigns to calls for participation in protests and public hearings, as in a post from October 15, 2019:

> An area of Atlantic Forest in Salvador is a candidate to become a WILDLIFE REFUGE. The Vale Encantado encompasses 1 million m^2 of forest that is highly relevant for biodiversity conservation.
>
> In recent years, the Vale Encantado has transformed the Patamares region into a vibrant area for visitors; researchers; and local, national, and international journalists. The environmental collective SOS Vale Encantado and the NGO IMATERRA will defend the project on October 15th, at 2 pm, at the Biology Institute of the Federal University of Bahia (UFBA), with decisive support from the Public Ministry and the interest of the Sustainable City Department (SECIS). SOS Vale Encantado – which brings together dozens of activists and thousands of supporters – and the NGO IMATERRA will present to political, scientific, and social leaders the technical studies that prove the importance of protecting and restoring the Atlantic Forest of Vale Encantado and other regions of Salvador, which is one of the Brazilian capitals with the greatest need to protect green areas.
>
> LET'S CREATE TOGETHER THE FIRST WILDLIFE REFUGE IN SALVADOR? (https://www.facebook.com/events/543709339709434/. Accessed on 27/04/2021).

This was not the only public event held in the long story of the creation of the Vale Encantado CU in Salvador. Several months later, on the morning of March 11, 2020, for example, a public consultation was held regarding the creation of the Vale Encantado Wildlife Refuge (REVIS) in the same External Auditorium at the Biology Institute of the Federal University of Bahia (UFBA), with the presence of the Municipal Sustainability Secretary, SECIS technicians, and a representative of the partner collectives at the opening session. In the minutes of the public inquiry, 176 participants were recorded as attending the event.

Several questions and suggestions resulted from this Inquiry, including the possibility of shared management with a Public Interest Civil Society Organization; greater communication about the CU with the surrounding communities; economic inclusion of low-income communities around the CU in activities compatible with its objectives; implementation of environmental education activities to educate people about the importance of the area; restricting the licensing of businesses in the CU area to socio-environmental actions; increasing the length of the trails; proposals for ecological corridors; compliance with the management plan by the owners of private areas that will be incorporated into the REVIS; preservation of existing parks and creation of new neighborhood parks; and social participation in decision-making about CUs.

The advancement and implementation of the proposals are conditioned on the effective implementation of the CU and the establishment of its Management Council. For example, the possibility of co-management with a Public Interest Civil

Society Organization of the Vale Encantado REVIS depends on an established Management Council of the future CU, which will also be responsible for overseeing the performance of the Public Interest Civil Society Organization and recommending contract termination in case of irregularities. A broader participation of the surrounding population in decision-making about the CU is also conditioned on the existence of a Management Council that can implement a more effective participation plan, including school communities and other stakeholders.

The development of a Management Plan is a prerequisite for the expansion of environmental education activities and programs that already take place in the Vale Encantado Conservation Area. The approval of the plan will allow activities of this nature to be conducted by other interested groups and entities, considering the capacity to receive visitors and minimize possible negative impacts to guarantee the effective preservation of the area. The technical studies of the SOS Vale Encantado and IMATERRA stated that "the Maximum Physical Capacity to be adopted for all trails in the Vale Encantado will be that of the Orelha Preta Gambá trail, with a maximum of 73 people during 4 hours of operation. It is important to emphasize that the maximum number is not a goal to be reached, but a safe parameter" (IMATERRA; SOS Vale Encantado, 2018, s/p). Therefore, the maximum number of daily visitors was defined based on the capacity of the smallest of the six existing trails in the CU (the Gambá de Orelha Preta trail), which also provides the only access to the other trails.

In terms of visitation, currently six trails can be visited by groups in the Vale Encantado CU, with distances ranging from 700 to 1250 meters, with a duration of up to 1:20 hours. The expansion of existing trails or the creation of new ones is also conditioned by daily visitor capacity, which cannot be exceeded without compromising the ecological quality of the REVIS.

According to the technical studies that the calculation of the effective capacity of the CU was based on, conducted by IMATERRA and the Collective SOS Vale Encantado:

> Currently, Vale Encantado is visited by scheduled groups, with the main focus on contemplation and environmental awareness. The group formed by volunteers who are responsible for publicizing and participating in events, welcoming, guiding on the trails, and mobilizing new volunteers is called Hedgehogs. The Hedgehogs include residents from the region and other parts of the city as well as students and professionals in biology, architecture and urbanism, social communication, environmental engineering, tourism, and other backgrounds and focuses, who take turns conducting the activities. Current access to the Vale Encantado is through the Colina "C" gated community in Patamares. The visitor reception procedure, as well as the route, is similar for all activities, except in specific cases such as bird watching and scientific research (IMATERRA; SOS Vale Encantado, 2018, s/p).

"The participant is welcomed at the kiosk of the gated community that has a covered structure, chairs, and toilets, where information is provided about the actions of the SOS Vale Encantado movement, the local ecosystem, the relationship with other forest fragments in the city and instructions for good practices on the trail" (IMATERRA; SOS Vale Encantado, 2018, s/p). In fact, until the effective implementation of the Vale Encantado CU is enacted, the precariousness of the existing infrastructure limits the activities and the number of daily visits. Nevertheless, activities such as bird watching have already become a "tradition" and serve as an introduction for the future CU, gaining new supporters and sympathizers for the effectiveness of the Vale Encantado Wildlife Refuge in the Patamares neighborhood.

With improvements in the infrastructure, expansion of monitor training programs, and more publicity for this type of activity, the CU has enormous potential to attract new visitors/participants. On the other hand, the expansion of bird watching makes it necessary to carry out new studies and research on their impact on bird behavior in the areas of the CU where this activity occurs.

4 In Conclusion: Preservation Versus Valorization of Urban Land – The Contradictions

Salvador is part, together with São Paulo, Belo Horizonte, and Campinas, of the Cities4Forests initiative, which includes 45 cities from six continents. Together, these cities are home to 164.9 million inhabitants. As part of this initiative, cities commit themselves, through their governments, to restoring and conserving their forests, while also seeking to make their inhabitants aware of the benefits of afforestation and the presence of forested areas in the urban context. The initiative in the form of a program was launched at the World Climate Action Summit in San Francisco (USA) in September 2018:

> By joining the program, cities benefit from technical assistance to measure forest cover and prioritize where to plant trees for the maximum benefit of the population, as well as advice on seeking funding to protect watersheds or restore degraded areas (https://cities4forests.com/. Accessed on 28/04/2021).

Even with the integration of programs like this, contradictions remain in Salvador between the preservation/conservation of its forested areas and the strong real estate interests that often threaten the preservation and conservation policies of green areas in the city. Although the available data are not very current, there is strong evidence of continued suppression of the Atlantic Forest in the last decade, such as around Avenida Paralela, with the implementation of real estate projects that have great environmental impact. A survey began in 2010 and presented in 2013, by the Public Ministry of Bahia, for example, shows that, excluding fully protected Conservation Units, Salvador then had 22,504,495.11 m^2 of Atlantic Forest in the initial stage of regeneration, 29,503,103.86 m^2 in the medium stage and 484,354.14 m^2 in the advanced stage, a total of 52,491,953.12 m^2, or 16.9% of the total area of the municipality (Fig. 6.5).

The suppression of vegetation is also a result of the implementation of new urban transit projects, such as the BRT (Bus Rapid Transit) and the Metro. The implementation of the BRT, which has been criticized by environmental movements precisely because of the elimination of hundreds of adult trees and the visual impact of the viaducts, continues with the completion of Sect. 1 and the start of work on Sects. 2

Fig. 6.5 Real estate pressure on the Vale Encantado. (Photo by: Jacileda Santos)

and 3. The construction of Metro Line 2 in the central strip of Paralela Avenue (already completed and in operation) has also eliminated thousands of adult trees, while attracting more real estate development to this area of the city.

All this indicates that beyond the creation of a protected area such as the Vale Encantado Wildlife Refuge, the inclusion of this protected area in the urban/metropolitan fabric beyond its polygonal limits is needed. Santos (2020), for example, proposed the delimitation of Buffer Zones for both the Vale Encantado area and the neighboring Pituaçu Metropolitan Park:

> About 500 meters northeast of Pituaçu Park is the area of Vale Encantado (...) – indicated as a Natural Park by PDDU2016, but with characteristics for the implementation of a Wildlife Refuge (REVIS). This category of CU aims to protect environments that ensure the existence or reproduction of species of local flora and resident or migratory fauna, and that allows remaining green areas belonging to subdivisions in the region to be included, if their owners so wish (BRASIL, 2000). However, its privileged location and all the existing qualified infrastructure make it a focus for the real estate market. Thus, more than providing an area with less impact of urban expansion on the Conservation Unit, the establishment of a Buffer Zone in these spaces will serve to guarantee the preservation of its existence and the limits established by the legal instruments for its establishment. In addition, this element greatly contributes to the integration of the CU with the urban fabric, breaking the paradigm to become something exceptional in relation to what symbolizes the metropolis and the urban (Santos, 2020, p. 182–183).

Criteria for the delimitation of Buffer Zones were not foreseen in the National System of Conservation Units (SNUC); therefore, Santos (2020) stated that each case should be analyzed individually. The author establishes these criteria to simulate a delimitation of the Buffer Zones for both Pituaçu and Vale Encantado, concluding that, beyond the possible delimitation, actions and practices that actually consider the preservation and conservation of Nature in urbanized areas should be encouraged, since "discussing Buffer Zones for Conservation Units located in urban and metropolitan areas is a great challenge, due to the very nature of the territory in question" (Santos, 2020, p. 191).

Like the Buffer Zones, the creation of Ecological Corridors is not a precondition for the creation/definitive delimitation of the Vale Encantado CU; however, it is among the suggestions discussed in the public inquiry held in March 2020, mentioned in Sect. 3 of this chapter. This theme will be the subject of analysis in the Municipal Plan for the Conservation and Recovery of the Atlantic Forest, which has been under preparation since 2019. Technical studies conducted by SOS Vale Encantado in partnership with IMATERRA, to analyze the biological quality of the areas that encompass the CU and its buffer zone, suggest "a corridor consisting of fragments that allow connectivity between existing and planned Conservation Units" in Salvador, "and patches predominantly in medium stages of regeneration," emphasizing that "the conservation of Vale Encantado and other remnants of the Atlantic Forest in the city will be enhanced through the implementation of ecological corridors, which will facilitate the maintenance of ecosystem services necessary for the survival of species" (IMATERRA; SOS Vale Encantado, 2018, s/p).

Waiting for the signature of the city Mayor since the end of 2020, the decree to implement the Vale Encantado REVIS has not yet left paper. Even in this context, there is encouraging news, such as the obtainment by Vale Encantado Park of the title of Outpost of the Atlantic Forest Biosphere Reserve (RBMA), during the 30th Meeting of the National Council (CN-RBMA) held on November 22 and 23, 2021, during the celebrations of 30 years of RBMA and 50 years of the MAB/UNESCO Program—launched in 1971. The Man and the Biosphere Program (MAB) is a result of the "Biosphere Conference" held by UNESCO in Paris in September 1968:

> The ***Atlantic Forest Biosphere Reserve – RBMA***, whose area was recognized by ***UNESCO*** in seven successive phases between 1991 and 2019, was the first unit of the ***World Network of Biosphere Reserves*** declared in Brazil. It is the largest ***Biosphere Reserve*** on the planet, with 89,687,000 hectares, of which 9,000,000 ha are core areas, 38,508,000 ha are buffer zones, and 41,400,000 ha are transition zones, of which approximately 73,238,000 ha are terrestrial areas and 16,449,000 ha are marine areas, in the 17 Brazilian states where the ***Atlantic Forest Biome*** occurs naturally (https://rbma.org.br/n/a-rbma/quem-somos/, emphasis part of the site itself. Accessed on 26/12/2022).

Nevertheless, in Salvador, the bodies involved in the creation of the Vale Encantado REVIS still question whether the Conservation Unit should be implemented by a municipal decree, as it could overlap with the Land Use Planning Law (LOUOS). However, the argument does not seem to hold up, as the Barra Marine Park was created by decree as recently as 2019 and is currently being implemented. It seems that what is really at stake are the opposition and conflicts of interest exposed in this chapter, on whose resolution depends on the effective implementation of the Vale Encantado CU.

Acknowledgments I would like to thank Jacileda Santos for providing the images that illustrate this chapter.

References

Araújo, C. F. de (2014). *Paralela XXI: a expansão urbana no vetor da Avenida Luis Viana Filho Salvador/BA–1968 a 2013*. 2014. 140 f. Dissertação (Mestrado) – Universidade Católica do Salvador. Superintendência de Pesquisa e Pós-Graduação. Mestrado em Planejamento Territorial e Desenvolvimento Social. UCSal, 2014.

IMATERRA; SOS Vale Encantado. (2018). *Estudos Técnicos para Subsidiar a Criação da Unidade de Conservação de Proteção Integral do Vale Encantado*. Disponível em: https://www.imaterra.org/sos-vale-encantado Acesso em: 30/04/2021.

SALVADOR. (2016a). *Lei n° 9.069, de 30 de junho de 2016*. Dispõe sobre o Plano Diretor de Desenvolvimento Urbano do Município de Salvador—PDDU 2016 e dá outras providências.

SALVADOR. (2016b). *Lei n° 9.148 de 8 de setembro de 2016*. Dispõe sobre o Ordenamento do Uso e da Ocupação do Solo do Município de Salvador e dá outras providências.

Santos, J. C. (2020). Zonas de Amortecimento em Unidades de Conservação urbanas. Dois casos em Salvador-Ba-Brasil. *GeoTextos, 16*(2), 173–196, dezembro de 2020.

Santos, E. S., Pinho, J. A. G., Moraes, L. R. S., & Fischer, T. (2010). *O Caminho das Águas em Salvador: Bacias Hidrográficas, Bairros e Fontes*. CIAGS/UFBA/SEMA.

Chapter 7
People, Animals and Waste Form Systemic Links Within Tropical Cities: Case of an Urban Raptor in Delhi, India

Nishant Kumar, Urvi Gupta, Amee Mehta, and Harsha Malhotra

Abstract About 60% of humans now live in cities, placing the 'urban' in a central and defining role for human-animal relationships. Although cities of the Global South markedly differ from their western counterparts, poor research within the tropics precludes assessments for their urban ecological distinctions. To illustrate this case, we discuss how the pace of development and geography of cultural beliefs/practices in Delhi form ties with the ecological responses of animals thriving on human detritus. We studied the population, behaviour and movement ecology of a facultative scavenger, the Black Kite, at 32 sampling units (1 km^2 each) that represent the urban gradient in the megacity. Our multivariate model for logistic distinction between nesting (1) and randomly chosen (0) trees was based on respective comparison of nest-level (micro) and landscape level (macro) covariates. It revealed

N. Kumar (✉)
Wildlife Institute of India, Dehradun, Uttarakhand, India

Department of Biology, Edward Grey Institute of Field Ornithology, University of Oxford, Oxford, UK

Mansfield College, University of Oxford, Oxford, UK

School of Human Ecology, Dr B R Ambedkar University Delhi, Delhi, India

THINKPAWS Sustainability Research Foundation, Delhi, India

National Centre for Biological Sciences, Tata Institute of Fundamental Research, Bengaluru, India
e-mail: nishant.kumar@biology.ox.ac.uk

U. Gupta
Wildlife Institute of India, Dehradun, Uttarakhand, India

Warnell School of Forestry and Natural Resources, University of Georgia, Athens, GA, USA

A. Mehta
Wildlife Institute of India, Dehradun, Uttarakhand, India

H. Malhotra
Wildlife Institute of India, Dehradun, Uttarakhand, India

Homi Bhabha Centre for Science Education, Tata Institute of Fundamental Research, Mumbai, India

F. Angeoletto et al. (eds.), *Ecology of Tropical Cities*,
https://doi.org/10.1007/978-3-031-48012-6_7

how food-affording human activities and practices impacted habitat selection in kites. For nesting, breeding pairs took preference for those parkland/woodlots that were near densely populated areas in the city, thus, enabling easy access to predictable anthropogenic food subsidies. Notably, variations in breeding densities along the urban gradient arise from the availability of nesting sites and their proximity to anthropogenic food sources. We further explored if urban-infrastructure and human factors that mediated access to food-subsidies could also explain patterns in behavioural repertoires. The previous set of variables that elucidated habitat choice by opportunistic kite breeders that exploited human subsidies also explained ordinal patterns we identified in their offspring defense behaviour, concerning the conflicting choice of settlement near human habitations. Despite kites inflicting head injuries during the nesting period, our findings revealed a general empathetic and religiously tolerant attitude toward animals among Delhi's citizens. Furthermore, GPS-tracking of kite movements revealed that South Asian cities are wintering destinations for massive flocks of Black-eared Kites that opportunistically consume human offal from landfills. In uncovering the complex geography of coexistence, we reveal the intricate links between politics, history, socio-economics, and urban planning across various spatio-temporal scales. These factors shape distinct habitat mosaics for urban animals, offering valuable insights into the ecology of tropical cities and their role in conserving urban biodiversity and wildlife management.

Keywords Tropical megacities · Food-subsidies · Ritual feeding · Urbanisation · Human-animal interface · Human-animal conflict · Ethno-zoology · Urban ecology · Synurbic · Biotic homogenisation · Ecosystem service · Social-ecological systems

1 Introduction

Over the past centuries, the formation and expansion of cities have brought about a transformative shift in human civilization, fundamentally altering human interactions with non-human entities within shared spaces (Anonymous, 2016; UNO, 2019). Surprisingly, ecologists have been slow to document the far-reaching impacts of human dominance, primarily driven by infrastructural and lifestyle changes that reshape urban ecosystems. This inertia is evident in the emerging field of urban ecology, which is still in its infancy, and even more so in the Global South where architectural heterogeneity, building density, vegetation, economic activities, and cultural diversity create habitat mosaics for native and invasive species (Rademacher & Sivaramakrishnan, 2013; Angeoletto et al., 2022). Astonishingly, despite projections that nearly 98% of future urbanization will occur in developing countries' cities, there remains a scarcity of comprehensive studies conducted in tropical regions (Marzluff, 2017; UNO, 2019; Angeoletto et al., 2019; Santangelo et al., 2022). As these urban centers continue to expand, encompassing both land and rural populations, it is imperative to address the consequences of urbanization, including the rapid loss of species and biotic homogenization (Grimm et al., 2008). Despite the recurring concerns about impacts of urban sprawl and the subsequent accumulation of human waste, researchers have yet to capture the unique ecological distinctions found in tropical cities, where urbanization remains an ongoing process.

Due to the recent infrastructural pace in tropical cities, urban disparities have led to the formation of highly variable settlement conglomerates in the Global South (Rademacher & Sivaramakrishnan, 2013; Watson, 2013). These urban mosaics are characterized by immigrant communities, shaping both the social-cultural fabric and regional ecological domains. With more than 60% of the global population now residing in cities (UNO, 2019), the significance of studying the impacts of biotic, abiotic, and human factors on biotic communities within and around urban areas has become increasingly important for several reasons. ***Firstly,*** animals exhibit varying abilities to adapt and exploit foraging opportunities in an increasingly urbanized world. ***Secondly***, the commensalism of urban animals in contemporary cities is intricately linked to their demographic responses to urban and human influences. In certain cases, long-term coexistence with humans confers select species with the status of being integral components of the urban social-cultural fabric, often mediating people's connection with nature (e.g. Cox & Gaston, 2016; Fuller et al., 2013; Nilon, 2011). ***Third***, shifting food habits and rising income levels in tropical countries offer significant opportunities for facultative scavengers that rely on carrion and human waste. Many of these facultative or obligate scavengers provide essential ecosystem and sanitary services (e.g., Margalida & Colomer, 2012; Gangoso et al., 2013; Moleón et al., 2014, p. 20; Inger et al., 2016; Richard et al., 2023). ***Fourth***, based on the mutual urban responses in shared spaces, a few species that can withstand human proximity often become ecologically dominant and potentially limit other species, contributing to faunal homogenisation (Carey et al., 2012; McKinney, 2002; Shochat et al., 2010). ***Finally***, focused studies on the highly variable geography of opportunistic species may provide insights into the future of synurbic species (e.g., Angeoletto et al., 2022; Marzluff et al., 2001; Parker & Nilon, 2012).

Current literature primarily explains spatial patterns of anthropogenic impacts on biological populations and communities, overlooking individual life histories in urban ecosystems. This limits our understanding of behavioural mechanisms that mediate human-animal coexistence with 'synurbic' species that have lived within human settlements for centuries (Francis & Chadwick, 2012). These species often show reduced fear of humans or neophilia and may directly cue upon practices that provide food subsidies. Despite the interest in understanding the behavioural bottlenecks and conflicting decisions faced by urban exploiters for wildlife management, the development of mutual human-animal tolerance, as observed in traditional societies, remains an understudied topic. Conflicting decisions arise for animals benefiting from anthropogenic resources, as people act as providers of food subsidies while simultaneously posing potential predation threats due to their proximity to breeding sites such as nests or dens. Additionally, the rise of cities along avifauna migratory flyways introduces novel selection pressures, presenting both new anthropogenic opportunities and threats from overhead transmission wires, wind energy farms, buildings, and artificial lights at night (Drewitt & Langston, 2008; Horton et al., 2019). Considering that tropical landscapes are still undergoing development, studying individual life histories in long-lived organisms provides a quasi-experimental framework to understand how life history traits, such as residency versus migration, are influenced by urban environments (Strohbach et al., 2014). Such research will have significant management implications, offering opportunities to plan "green" and "smart" infrastructure in the Global South that supports the long-term coexistence of wildlife and humans (Hunold & Mazuchowski, 2020; Soanes et al., 2019).

In this chapter, we examine the intricate interplay between infrastructure, human activities, and landscape factors in shaping the ecology, behaviours, and movements of Black Kites (*Milvus migrans*) in Delhi. Drawing from our long-term research encompassing interconnected investigations on habitat selection, behaviours, and life history of kites, we explore the future of the systemic relationships between people, animals, and waste. Our analysis delves into the distinct nature of organism-resource ties in cities compared to relatively natural ecosystems, which are often influenced by selection pressures arising from novel, built environments. Additionally, the chapter provides a comprehensive account of our methodology, shedding light on how these novel selection pressures impact life history strategies in tropical cities like Delhi, ultimately offering insights for promoting sustainable human-animal coexistence.

2 Methodology

Study Species and Area

The study focuses on the Black Kite (*Milvus migrans*), a medium-sized raptor belonging to the *Milvinae* subfamily, which exhibits varying levels of tolerance for human presence, ranging from active avoidance in Europe to synurbic behaviour in South Asia. Currently recognized as the most abundant and successful bird of prey globally (Ferguson-Lees & Christie, 2001), the Black Kite is a generalist and opportunistic feeder capable of achieving high population densities in areas with abundant food resources (Malhotra, 2007; Sergio et al., 2005). It demonstrates remarkable adaptability, occupying habitats that span from natural to fully urbanized landscapes (Ferguson-Lees & Christie, 2001; Ortlieb, 1998). This species' opportunistic nature and ability to exploit human-modified habitats contribute to its generally favourable conservation status (Hagemeijer & Blair, 1997; Sergio, 2003; Thiollay et al., 2004).

Since 2012, we have been conducting research on kites in Delhi, a sprawling megacity with a population exceeding 29 million and an area of 1500 km^2 (Census India, 2011; UNO, 2019). Delhi experiences a semi-arid climate with an average annual precipitation of 640 mm, primarily concentrated during the monsoon season in July and August. The temperature in the region ranges from a minimum mean value of 8.2 °C in winter to a maximum mean value of 39.6 °C in summer. The vegetation in and around Delhi falls under the classification of 'northern tropical thorn forest' (Champion & Seth, 1968).

Due to the polycentric and heterogeneous nature of Delhi, establishing a linear urban-rural gradient presented challenges. Furthermore, limited access to private properties restricted the possibility of designing a single, large continuous study area. Considering the high population densities of kites in our study area, we devised a network comprising 32 sampling plots, each measuring approximately 1 km^2. Employing a stratified-random design, these plots were distributed throughout the city to encompass a range of habitat conditions along the urban gradient. This sampling strategy allowed us to capture the diverse patterns of resource dispersion and accessibility for kites in different habitats, effectively representing the potential nesting habitats of this facultative urban scavenger (see Kumar et al., 2014, 2018a for detailed sampling information).

Four key aspects of Delhi shape the ecology of kites. Firstly, the city is characterized by inadequate solid waste management, resulting in abundant food sources for kites in the form of carrion or refuse. Various facultative scavengers, including kites, crows, egrets, dogs, rats, macaques, pigeons, etc., have access to organic refuse, and it has been estimated that kites alone remove approximately 4000 tonnes of organic waste from Delhi's streets annually (Kumar et al., 2019a, b). Despite its relatively small size, the National Capital ranks among the top regions in the country in terms of animal product (meat and milk) consumption, reflecting the city's above-average per capita income, which is three times the national average (Chavez et al., 2012; GoI, 2022). Delhi is characterized by a network of garbage piles of varying sizes and contents, often associated with impoverished communities residing in slums and engaged in informal waste processing. The presence of refuse-based food subsidies contributes to the distinctive visual and olfactory characteristics of South Asian cities. These subsidies not only attract large congregations of selected species, easily noticeable even to untrained observers, but also support background populations such as rodents that capitalize on these resources (Kumar, 2019c). Together, these communities on detritus, consisting of both human and non-human actors, shape the olfactory landscape and sensory experiences of South Asian urbanization (Doron, 2021).

Second, despite the increasing volumes of unsegregated waste and animal remains, the availability of suitable waste disposal spaces in the twenty-first century is limited. The lack of new locations for large-scale formal garbage disposal has led to management conflicts. Anticipating potential disruptions in livestock rearing to meet the animal protein demands of growing cities, new regulations have been implemented for the disposal of slaughter remains in landfill sites in Delhi. The civic bodies responsible for waste management in the capital face conflicts of interest and jurisdiction: the Delhi Development Authority plans waste spaces every 20 years, municipal corporations handle garbage collection and dumping areas (Talyan et al., 2008), the forest department addresses encroachments by unauthorized dumpers, and animal welfare organizations advocate for animal rights. Despite court orders for decommissioning, all three of Delhi's landfills continue to operate, as finding new sites has proven challenging (Fig. 7.1).

Third, Third, the cultural practice of feeding meat scraps to kites, known as "ritual-feeding," has been passed down for centuries. During this ritual, meat is

Fig. 7.1 Massive flocking of Black Kites is typical for landfills within tropical cities across South Asia. This is a beholding view of >10,000 kites from February 2013 that congregated over and around the *Ghazipur* landfill spread over 70 acres, mostly capitalising over poultry processing waste at and from *Shaheed Ashfaqullah Khan* Chicken, Fish and Egg market (Government of National Capital Territory of Delhi) in the vicinity. Before their local decline, such landfills had massive populations of vulture *spp.* (Photo Credit: Fabrizio Sergio)

thrown into the air for the birds to catch (Fig. 7.2). Ritual-feeding serves various purposes, including seeking blessings, relief from sins, alleviation of worries and philanthropy (Pinault, 2008; Taneja, 2015). In Delhi, this practice is particularly prevalent among members of the Islamic faith, who are concentrated in specific areas known as "Muslim colonies." In these areas, large quantities of meat are tossed to kites at predictable hours each day, leading to the rapid congregation of hundreds of kites within minutes. While Muslims constitute approximately 15% of Delhi's population (Census India, 2011), they are primarily concentrated in less than 10% of the city's total area of 1500 km^2 (Kumar et al., 2018a, 2019b). This concentration creates distinct spatial zones of opportunity for kites, as the meat chunks are precisely cleaved into sizes suitable for aerial consumption (see Fig. 7.2). Consequently, kites become the predominant species benefiting from this resource.

Urban ecosystems are widely recognized for their direct impact on biodiversity through the creation of impervious surfaces during urbanization. However, it is important to acknowledge the powerful influence of cultural factors, such as religion, on the fauna and flora of cities (Aronson et al., 2017). Unfortunately, tropical cities, despite their rich and diverse cultural expressions, have received limited attention in terms of their influence on urban biodiversity. This chapter aims to provide a contextual understanding of these poorly studied aspects by examining the

Fig. 7.2 The Black Kites of Delhi depend heavily on human subsidies offered for religious reasons: Shown here in (**a**) is a man with his sons ritually feeds kites with the typical, compact chunks of red meat (red circle); (**b**) large numbers of kites, sometimes into the hundreds, may congregate at such feeding events; (**c**) the ritual offerings are taken to the nests; (**d**) a parent kite is about to feed its fledgling with a ritual meat chunk. More than 90% of the diet in this population is composed of ritual offerings, which explains the tight link between breeding density and ready access to human cultural subsidies (photo credit for all images: F. Sergio)

human-kite relationships in Delhi. For example, meat tossing is a year-round ritual activity within Delhi, which intensifies after the first morning prayer at mosques. Typically, meat tossing occurs after the first two of the five daily Muslim prayers (Fajr – 05:30 am and Duha – 6:50 am), although individual preferences based on working hours may vary. We obtained currently unpublished data from observations captured by nest cameras (see Kumar et al. 2019a, b). Additionally, ground truthing was conducted based on the recorded ritual-feeding timings and prominent provisioning windows observed in the nest camera footage. Although we do not yet have an exact daily estimate of the total quantity of ritually tossed meat for kites in Delhi, interviews with individual meat sellers indicated that approximately 150–300 kg of meat/seller is sold on Thursday and Jumma, with a daily average of around 50 kg.

Finally, Delhi currently maintains a substantial amount of green cover, which serves as ample nesting habitat for kites. However, there is a pressing concern as the city is experiencing rapid loss of tree cover (Paul & Nagendra, 2015). This necessitates the need to anticipate the potential ecological ramifications of these changes. In the subsequent three sections, we delve into these consequences specifically for kites, thereby expanding the discussion to encompass the broader urban fauna, particularly those species that thrive in urban environments.

3 Urban Wildlife: Habitat Selection and Nesting Patterns

The understanding of habitat selection, particularly in urban environments, remains inadequate in organismal ecology. There is a prevalent misconception that urban changes render landscapes unfit for wildlife (Leong, 2009). While it is true that human food subsidies often attract opportunistic urban exploiters, food availability alone cannot dictate habitat suitability for such species. In our study, we observed significant spatial variation in kite nesting densities across 32 sampling units throughout the city. Although the overall breeding density of kites (~20.1 ± 7.9 nests/km^2) did not show a significant change compared to estimates from the 1970s (Galushin, 1971; Kumar et al., 2014, 2018a, 2019a), the range of nesting densities (0–180 nests/km^2, $n = 32$) indicated distinct spatially explicit patterns in the kites' urban opportunism. To comprehend the factors influencing kite nest substrate selection within the available urban habitat, we conducted a comparative analysis between nesting trees and randomly selected trees near specific point locations generated using the Arc GIS program (Kumar et al., 2018a provides detailed methodology). Anticipating that urban habitat selection is influenced by multiple resources, we examined how nesting trees differed from random trees at the micro (nest-level) and landscape levels, considering various urban and human factors as potential covariates. Furthermore, as part of a parallel study on the movement ecology of kites, we GPS-tagged some individuals from our study population. These tagged kites frequently visited Muslim localities, where we observed large congregations of kites (e.g., >200) during ritual-feeding episodes (Kumar et al., 2018a).

GPS data and on-field observations revealed frequent utilization of specific habitat patches by kites for foraging, even at distant sites. To investigate this phenomenon, we examined the patterns in successive distances from each nest or random location to multiple Muslim colonies. Through exploratory analysis, we compared the proximity of mosques (considered as point locations for meat tossing) to nests and random points. The results indicated that kites selected nesting sites to ensure proximity to multiple meat-tossing locations. Notably, successive distances to point locations stopped showing any significant difference between random points and nest locations after the third closest Muslim colony. Beyond this spatial limit the difference became negligible (mean distance to the third closest colony from 154 nests was <2000 m). To quantify the significance of this foraging resource in the kites' habitat selection, we developed a comprehensive measure by integrating the proximity (distance) values from each sampling location to three successive Muslim colonies. Additionally, we considered the Muslim (human) population density within a 2000 m radial area around the nest/random locations, assuming that denser colonies would have higher rates of refuse and ritualized feeding. Principal Component Analysis (PCA) was employed to analyse the distances to the first, second, and third closest Muslim colonies, as well as their human population densities. The resulting PC1, referred to as "access to Muslim subsidies," explained 65% of the variance and exhibited a strong positive loading on Muslim population density, while displaying high negative loadings on the distances to the three closest Muslim colonies. This composite index of access to abundant "Muslim subsidies" was incorporated into all subsequent models (Kumar et al., 2018a provides detailed methodology).

We successfully differentiated nest locations from random locations using a logistic model based on a training dataset of 100 nests and 100 random locations, employing the maximum log-likelihood approach (Kumar et al., 2018a). The final model, consisting of six variables, performed well when applied to validation datasets, correctly reclassifying 82% of the 108 locations we set-aside for validation (97.4% of 54 nests and 74.1% of 54 random sites). According to the final model, breeding kites showed a preference for parkland or woodland areas near densely populated sections of the city, which provided easy access to ritual food subsidies near mosques. These choices were influenced by cues from human activities and practices related to food provisioning, and kites settled near road configurations that facilitated easy access to ritually tossed meat chunks by Muslims and human refuse. Additionally, we examined whether the logistic model for habitat selection could explain urban adaptability. To investigate this, we explored correlations between the explanatory variables and **(i)** ***occupancy***, representing the repeated use of a nesting territory, and **(ii)** ***overall breeding success***, measured as the average breeding success over the four years of monitoring. This analysis aimed to determine the adaptive nature of the observed habitat choices (Clark & Shutler, 1999; Sergio, 2003). From the six variables in the final model, which explained habitat selection, linear mixed modelling showed that woodland structure and access to Muslim food subsidies (PC1) had significant effects on both the duration of territory occupation and the cumulative number of fledglings produced by each pair between 2013 and 2016. These findings were consistent with studies on other avian species, including other kite populations (Sergio & Newton, 2003), prompting us to further explore how human and urban factors shape behavioural phenotypes. We hypothesized that individuals of higher quality would exhibit a propensity for occupying and exploiting the best habitat patches. However, humans, who serve as both foraging resources and potential threats, create a range of choices in association with infrastructure. Therefore, we delved into these dynamics to gain a comprehensive understanding of behavioural coupling (Kumar et al., 2018a, b).

4 Behavioural Bottleneck for the Non-humans: *To Be or Not to Be Near People*

Organisms exhibit hierarchical behaviours in selecting habitats, involving identification of cues across scales, which are challenged by urban changes (McMahon et al., 2017). Urbanization often disrupts as to how cues from physical habitat features at micro and landscape scales link with foraging opportunities. For urban exploiters, behavioural strategies involve deciding the level of proximity to maximize benefits from humans while minimizing anthropogenic threats. However, mutually exclusive anthropogenic effects, like urban infrastructure growth, food subsidies, and human presence, have been largely confounded in discussions on biotic homogenization in cities. Previous studies have not provided mechanisms for coarse-scale resource identification in urban habitats (crucial for fitness), in combination with intensive resource patch utilization at finer scales (which enhances fitness) (McMahon et al., 2017). Thus, we aimed to investigate whether kites' proximity to humans varies along the urban gradient, considering the changing costs associated with infrastructure and religio-cultural perceptions within heterogeneously developed areas. Assessing tolerance to humans in animals is typically done by studying patterns in Flight Initiation Distance (FID) values. Literature has extensive documentation of variations in FID based on taxa, population, individual life histories, and factors such as sex, age, and behavioural state (Seltmann et al., 2012; Weston et al., 2012). However, such studies have not explored the specific influences of urban gradients and the role of infrastructure and cultural factors in shaping kites' proximity choices to humans. Understanding these dynamics will provide insights into the behavioural bottleneck experienced by non-human species in urban environments and shed light on the interactions between wildlife and human-dominated landscapes.

While FID serves as a useful measure for assessing how individuals, populations, or species cope with anthropogenic environments based on personality traits, it does not account for how animals defend their young in fixed nests or dens. In urban environments, where offspring are immobile (e.g., altricial chicks or canine pups), breeders face a distinct selection pressure that differs from animals relying on flight to evade humans. Understanding the potential threat posed by humans and infrastructure, which varies across space (nesting territory vs. other areas) and time (breeding vs. non-breeding stages), is crucial for the organismic ecology of urban-dwelling species (Sol et al., 2013; Wong & Candolin, 2015). Breeding decisions in urban areas involve a combination of inherited traits and cultural learning, making risk assessment in heterogeneous urban environments a critical limiting factor. Furthermore, as previously described, the suitability of urban spaces is further influenced by human attitudes toward wildlife, resulting in the co-production of a geographic pattern of coexistence between select fauna and humans. Consequently, understanding how animals modulate their behaviours and parental strategies in response to human presence and urban features is essential for comprehending the dynamics of urban wildlife and the challenges they face. By investigating these aspects, we can gain insights into the complex interactions between wildlife and human society within urban settings and develop strategies for promoting harmonious cohabitation.

During our initial monitoring of kite breeding ecology in Delhi (Kumar et al., 2014), we observed a fascinating behaviour: kites attacked people's heads with their open hallux, dive-bombing from above to protect their offspring. This nest defense behaviour within their territory contrasts sharply with their European counterparts, which typically flee to safe locations upon the approach by humans (F. Sergio pers. comm.). However, even within Delhi, this personality trait exhibited high variability both across and within sampling units. Approximately one-third of the monitored nests, where we documented aggressive nest defense, resulted in head injuries and bleeding. Notably, this behaviour was predominantly displayed when females are nest-bound, which accounted for about 60–80 days of the overall breeding effort. The consistent presence of our research team members during nest monitoring, along with the random presence of citizens around the nesting territories, provided a unique opportunity for a natural experiment. We aimed to investigate how kite defense behaviour varied across Delhi's mosaic of urban structures, human densities, and practices, using the same set of environmental, urban, and human variables previously identified as important components of habitat quality and food availability in this population (Kumar et al., 2018a). Additionally, we examined the anthropogenic characteristics that influenced the likelihood of aggression by comparing attacking and non-attacking pairs while controlling for factors such as year, date, breeding stage, previous nest visits, and local habitat structure (see Kumar et al., 2019a for detailed information). By conducting this analysis, we aimed to gain insights into the interplay between kite behaviour and the urban landscape, shedding light on the factors that trigger nest defense and aggression in response to human presence.

In line with our expectations, the likelihood of a breeding pair making investment choices, i.e. ordinal levels of aggression in nest defense was explained by the same variables that distinguished nesting locations from random ones (see Fig. 7.3 for details) (Kumar et al., 2018b). Surprisingly, we found that offspring defense was not influenced by inter-nest spacing, the timing of defense trials, or the characteristics of the visiting team. These observations suggested the possibility of prior assessment of parental investment based on territory quality and the value of the brood. High-ranking and experienced parents seemed capable of identifying the potential benefits of investing in offspring defense early on. To test this hypothesis, we examined whether the intensity of offspring defense during the early incubation stage was linked to subsequent breeding success. Our findings revealed that pairs (likely the highest quality breeders) that chose to invest in aggressive nest defense during incubation (approximately two months before fledging) had significantly higher fledging success during the monitoring period from 2013 to 2018. These results highlight the co-adaptations of urban animals and their ability to perceive and respond to human perceptions, attitudes, and practices. Behavioural studies focusing on hierarchical habitat selection should prioritize understanding the fitness outcomes associated with patch selection. This knowledge will be crucial for

Fig. 7.3 The breeding kite pairs exhibited nest defence, for which we developed an ordinal score of 1, 2, and 3, which signify the increasing cost of alarming and making targeted attacks on potential nest intruders. Our research hypotheses tested the additive as well as interactive impacts of nest content (brood side), nesting stage and urban habitation, which we expected to be positively correlated with higher ordinal scores (for details, see Kumar et al., 2018b, 2019a)

wildlife management strategies aimed at addressing increasing agonistic interactions in multiple cities. By unravelling the intricacies of nest defense behaviours and their impact on breeding success, we can gain valuable insights into the adaptive responses of urban wildlife. This understanding will contribute to effective wildlife management approaches that promote coexistence and mitigate conflicts in urban environments.

One aspect that still required explanation was the variation we observed within a single sampling unit in the propensity of breeding pairs that attacked human intruders. In urban environments that are utilized by humans and animals in multiple dimensions, patterns of interactions select species, defining their urban niche (Lowry et al., 2013). To investigate the factors influencing such variations in behavioral traits among pairs within the same urban setting, we implemented a logistic differentiation of select kite pairs based on whether they had physically attacked humans in the past or not (n = 36 cases each) (Kumar et al., 2019a). This paired comparison design controlled for the effects of year, date, breeding stage, previous nest visits, and local habitat structure, which were already accounted for in previous models.

Our final model revealed that the likelihood of a breeding pair being aggressive attackers increased with: (i) breeding success, (ii) the presence of waste-rich areas near the nest, (iii) local architecture that brought them in proximity with humans (e.g., a balcony adjacent to the nest), (iv) greater access to ritual subsidies from Muslim colonies, and (v) higher human density in the streets within a 200-meter radial area from the nest. The finding that severe aggression in kites was associated with regional architecture, socio-religious, and cultural features, rather than landscape composition or classical ecological variables, introduces new perspectives for urban wildlife management. Studying such dynamics in developing cities has implications for managing species that have reached an advanced stage of synurbization (Francis & Chadwick, 2012). These results also suggest that urban systems can support a gradient of behavioural phenotypes, challenging previous notions that only bolder and more aggressive individuals are successful urban colonizers. Additionally, semi-structured interviews ($n = 279$) provided further insights into aggressive displays exhibited by synurbic species like kites. Interviewees reported that apart from religious and cultural beliefs they tolerated kites due to their aggressiveness being a time-specific behavioral decision during the breeding season (60–90 days), outside of which these birds did not defend their territories. The responses of Delhi's diverse populace demonstrated widespread acknowledgement of the scavenging ecosystem service provided by kites and similar organisms. Such acceptance of urban commensals in a twenty-first-century megacity is unimaginable in Western cities, where raptors and other animals have historically been persecuted by people for creating nuisance, a trend that continues in various parts of the world (Torres et al., 2018).

These studies that examined human-kite interactions have initiated an important discussion that highlights the need to incorporate ethnozoological aspects in understanding the ecology of populations responding to anthropogenic selection pressures. Traditionally, human-wildlife conflict has been approached by monitoring spatial and temporal patterns in animal responses to human-occupied landscapes. However, our findings reveal that human socio-cultural factors play a significant role in shaping the decisions of both humans and animals, highlighting their participation in a "coupled system." In tropical landscapes, people and animals co-shape each other's socio-ecological space through repeated interactions. Specifically, we provide evidence that kite attacks on humans are influenced by the geography of human religion, hygiene, and poverty, demonstrating that coexistence is not a uniform concept. The convergence of human expressions towards animals in shared spaces, particularly as religious expressions among followers of different religions in tropical megacities, opens up new avenues for research on the congruence of human responses to animals. Furthermore, we found that kite behaviour was fine-tuned to humans, tolerating them at close range during feeding but attacking them when provoked. Our evidence suggests that the aesthetic, cultural, spiritual, and ecosystem service benefits offered by kites currently outweigh any local discomfort caused by aggressive individuals. It is crucial to emphasize the intangible benefits of wildlife to humans and incorporate human socio-cultural aspects into disciplines such as behavioural ecology, wildlife management, and urban ecology. In the face of dynamic changes in ethnozoological values within urban social conglomerates like Delhi, research and management activities should consistently evaluate and integrate socio-cultural aspects as essential tools.

In summary, as the human-animal interface undergoes changes due to increasing attraction towards human subsidies and habitat fragmentation, conflicts are likely to escalate globally. Therefore, a comprehensive examination of animal ecology and behaviour using ethnozoological approaches provides an opportunity to develop transferable wildlife management skills, recognizing that biodiversity transcends political boundaries (see Kumar et al., 2019a, b, c for more details).

5 Life History Strategies in an Urbanising World: *Migratory vs. Resident Populations*

Recent advancements in telemetry technology have provided researchers with unprecedented opportunities to uncover migration routes and understand the connectivity between breeding and non-breeding areas for thousands of species in tropical countries (Robinson et al., 2010). However, despite the profound urbanization-related changes taking place in the Global South, we lack baseline estimates from most wintering regions of migratory populations, limiting our ability to assess the full extent of these changes (Newton, 2010). In addition to direct anthropogenic impacts, the proliferation of opportunistic species poses a significant threat to native biodiversity. These indirect effects often manifest as interference competition, but interactions with new or introduced species can also introduce diseases, and contribute to biotic homogenization. Recent studies have shed light on the influence of anthropogenic factors, such as infrastructure development, predictable food subsidies, and artificial light at night (ALAN), on migratory paths, stopover locations, wintering home ranges, and fidelity to these locations in both space and time (Horton et al., 2019; Robinson et al., 2010). Conversely, we have limited information about how resident breeding populations differ from wintering bird populations in terms of trait plasticity and the co-creation of new niches by opportunistic urban fauna that capitalize on predictable anthropogenic resources. Understanding the trait differences between resident and wintering bird populations is crucial, particularly when wintering ecology of migrants is directly linked to the benefits from urban resources, such as predictable food subsidies. This phenomenon represents an eco-evolutionary process that involves rapid changes in the relative importance of wintering strategies. Moreover, studying the human-bird interface becomes vital in such scenarios. Urban food subsidies for wintering migrants offer significant opportunities, but the quality of refuse may be low and contaminated with toxicants and pathogens that ultimately compromise survival and overall fitness. Finally, considering that these urban niches are influenced by human practices and beliefs, which are associated with socio-economic and cultural values, it is essential to understand how food subsidies drive inter- and intra-population dietary differences (Inger et al., 2016; Oro et al., 2013; West & Jones, 2022).

Regarding the points mentioned above, numerous studies have documented the significant congregation of kites in Delhi, particularly around landfill sites, garbage dumps of various sizes, and near slaughterhouses (up to 10,000 kites counted at a regularly monitored single landfill) (e.g., Galushin, 1971) (Fig. 7.1). However, the coexistence of two kite subspecies in urban settings in South Asia, particularly the migratory and wintering patterns of Black-eared Kites (*Milvus migrans lineatus*), lacked comprehensive information. Similar to historical records, our research group initially struggled with the frequent misidentification of migratory kites, which was further complicated by the difficulty of distinguishing Black-eared Kites from Small Indian Kites (*Milvus migrans govinda*). This confusion surrounding subspecies status, especially when observing hundreds or even thousands of individuals simultaneously, had implications for the hypotheses that landfills played a crucial role in the habitat selection of resident kite breeders (Kumar et al., 2018a). However, through GPS tagging of kites from Delhi's largest landfill at Ghazipur during the period of 2014–2018, we discovered that kite congregations over garbage were actually migratory in nature (see Kumar et al., 2020).

The literature on the responses of non-human animals to human presence often focuses on population-level differences between urban and rural environments and overlooks the individual-level versatility exhibited by animals throughout their lifetimes (see Ríos-Chelén et al., 2013). There is substantial evidence challenging the common assumption that variations in animal responses to anthropogenic impacts are discrete and uniform (e.g., Bermúdez-Cuamatzin et al., 2009). In line with the findings of Sergio et al. (2014) on individual improvements in migration, our observations on tagged kites demonstrated how animals likely adapt to the complex and heterogeneous environments of cities through nuanced behavioural adjustments (Kumar et al., 2018b). For example, in the case of resident birds, we discovered that the breeding territories of females were approximately 150 times smaller than their territory size during May to November ($n = 6$). On the other hand, for male birds ($n = 2$) primarily involved in provisioning duties, the variation in their use of urban space was around 5 times. During the breeding season, female activity was largely confined within a 500 m radius, whereas males often ventured up to 2 km from the nest. This pattern can be attributed to the division of labour, with male birds taking responsibility for provisioning while females focus on nest construction and incubation of eggs and chicks until they reach approximately 35 days of age (Malhotra, 2007).

Through regular assessment of how urban management decisions affect the distribution of resources and threats over space and time (see Kumar, 2019c, for the impacts of changing waste management at Ghazipur), we discovered that kites modify their decision parameters for daily movements based on their behavioral states. The significant reduction in breeding home ranges and the seasonality of urban space usage by kites in Delhi suggested that foraging decisions within urban areas involve complex and multi-modal optimization processes (e.g., Shochat et al., 2004). Given the observed differences in urban space utilization between experienced and inexperienced birds of both kite subspecies (Votier et al., 2017; authors' unpublished data), it is crucial to investigate how the spatiotemporal risk gradient in urban areas interacts with individual-level risk-sensitivity, including personality traits. In the Indian context, all tagged kites exhibited specialization in anthropogenic resources, relying almost exclusively on landfills and abattoirs while frequenting urban or suburban sites in Delhi and surrounding cities/towns. Moreover, the breeding ranges of Black-eared Kites were much larger than their non-breeding ranges in India (Fig. 7.4). These findings shed light on the substantial impact of

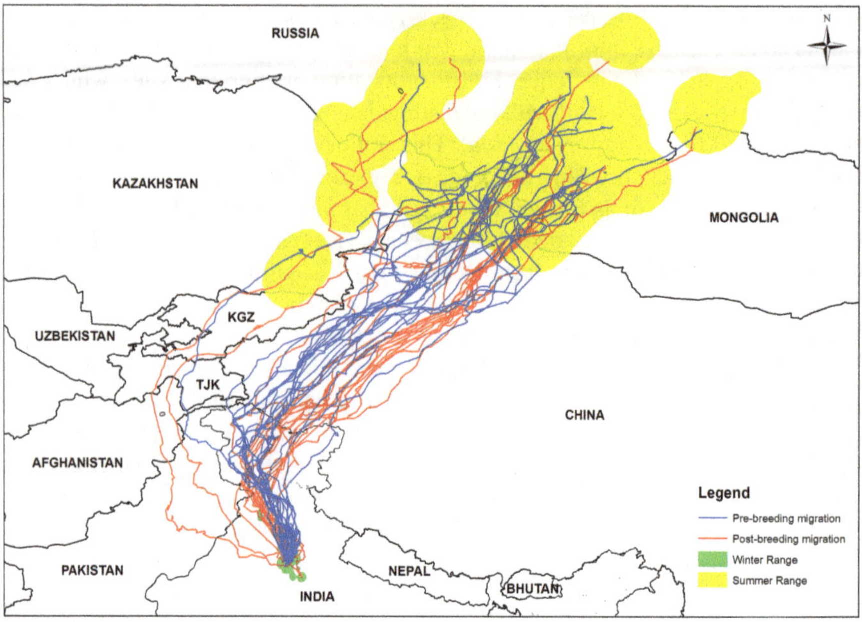

Fig. 7.4 Pre-breeding (blue tracks) and post-breeding (red tracks) migration routes, and breeding and nonbreeding ranges (yellow and green polygons respectively) of Black-eared Kites *Milvus migrans lineatus* GPS-tagged in Delhi (India). The polygons of the ranges represent 99% Kernels built on the cumulative sum of the GPS-locations of all tagged individuals. This map was generated with ArcGIS 10.5

abundant and predictable human food subsidies on migratory avian scavengers during their wintering period. Although our study on kite movement ecology is still in its preliminary stage, it inadvertently provides an opportunity to explore the interactions between life history strategies (resident and migratory) and individual life histories (from juvenile to adult stages, with respect to breeding investment).

6 Conclusion: *The Future of Systemic Links Amongst People, Animals and Waste in Cities*

Our studies provide evidence as to how human practices and perceptions can be incorporated as ecological covariates to understand the human-animal interface. Approaches that consider links between opportunistic availability of anthropogenic resources and organisms' life-history strategies (resident vs. migratory), breeding status, individual experience, and personality traits (e.g., neophobia) will be key to effective management of ecosystem health.

The perspectives and approaches needed to characterize urban ecologies that encompass non-western realities are still underdeveloped, primarily due to a limited understanding of how animal and human ecologies have been integrated over centuries of coexistence. While there has been a notable increase in research on the correlations between the environment, biodiversity, and human-induced effects, we still lack knowledge about the mechanisms that facilitate the coexistence of hundreds of species living in close contact with humans in thousands of non-Western cities. However, even in the absence of a scientific outlook, folk perspectives have played a significant role in governing human-animal interactions within vast and diverse human-dominated urban ecosystems in South Asia.

In South Asian cities, folk perspectives often result in people showing tolerance towards commensals and facultative scavengers. This coexistence is intertwined with rituals supported by mythological beliefs, which explain the offerings of food subsidies to several animals, including cows, dogs, monkeys, crows, kites, and small passerines. The accessibility of food subsidies is mediated by the spatial configuration of built spaces and roads, while its availability is influenced by a complex geography of human density and urban communities practicing their own animal feeding rituals. Therefore, spatial responses not only underlie the increasing global human-animal conflicts but also shed light on why certain individuals are attracted to substandard habitats (ecological traps), necessitating further investigation into the conundrums of habitat choices by taxa affected by urbanizing ecosystems.

Addressing these gaps in understanding will enable effective management of anthropogenic impacts for biodiversity conservation and the mitigation of conflicts and diseases. By incorporating holistic perspectives that consider the intertwined relationships between people, animals, and waste in cities, we can foster sustainable coexistence and ensure the well-being of both human and non-human urban inhabitants.

Acknowledgements We are grateful for the support and contributions from Shri Virendra R Tiwari (Director), Dean and Research Coordinator of Wildlife Institute of India, and various individuals and organizations throughout this study. The discussions during the THINKPAWS (www.thinkPAWS.org) Network's workshops (India-Oxford Initiative's GCRF fund AT13.01) played a crucial role in shaping this chapter. Financial support from the DBT/Wellcome Trust UK India Alliance (Grant number: IA/CPHE/21/1/505985), University of Oxford, including the Travel and Discretionary Grants from Somerville College/Mansfield College and the Felix Scholarship, as well as funding from the Raptor Research and Conservation Foundation, Mumbai, and the Government of India's Ministry of Environment, Forest and Climate Change (MoEF&CC), was instrumental in carrying out the research. We would like to express our appreciation to the Indian Air Force, Delhi Police, Forest Departments (Government of NCT of Delhi and Uttar Pradesh), Municipal corporations in Delhi, civic bodies (PWD, Delhi Transport Corporation and MTNL), and institutions (Delhi University & constituent Colleges) for granting the necessary permits for conducting our study. Special thanks go to Yadvendradev V Jhala, Qamar Qureshi, Fabrizio Sergio, and Andrew Gosler for their supervision of the "Black Kite Project" during our M.Sc./D.Phil. programs at the Wildlife Institute of India and the University of Oxford. We are grateful to our fieldwork team members, Laxmi Narayan, Prince, Zehen, Aparajita, Poonam, Navneet, Abhinandan, Prashant, Ranjan, Hisham, Nikita, and Gunjesh, for their valuable contributions. We would also like to thank colleagues from Dr B R Ambedkar University Delhi, University of Georgia, the citizens of Delhi, Ujjwal Kumar, and the staff at the Wildlife Institute of India and the University of Oxford's Department of Biology for their support.

References

Angeoletto, F., Fellowes, M. D., Essi, L., Santos, J. W., Johann, J. M., da Silva Leandro, D., & Mendonca, N. M. (2019). Ecología urbana y planificación: una convergéncia ineludible. *Revista Electrônica em Gestão, Educação e Tecnologia Ambiental, 23*(1), 1–8.

Angeoletto, F., Tryjanowski, P., Santos, J., Martinez-Miranzo, B., Leandro, D., Bohrer, J., et al. (2022). Will Brazilian City dwellers actively engage in urban conservation? A case study with the charismatic Neotropical blue-and-yellow macaw (Ara ararauna). *Birds, 3*(2), 234–244. https://doi.org/10.3390/birds3020015

Anonymous. (2016). *Use our infographics to explore the rise of the urban planet.* https://www.science.org/content/article/use-our-infographics-explore-rise-urban-planet

Aronson, M. F. J., Lepczyk, C. A., Evans, K. L., Goddard, M. A., Lerman, S. B., MacIvor, J. S., Nilon, C. H., & Vargo, T. (2017). Biodiversity in the city: Key challenges for urban green space management. *Frontiers in Ecology and the Environment., 15*(4), 189–196. https://doi.org/10.1002/fee.1480

Bermúdez-Cuamatzin, E., Garcia, C. M., Ríos-Chelén, A. A., & Gil, D. (2009). Strategies of song adaptation to urban noise in the house finch: Syllable pitch plasticity or differential syllable use? *Behaviour, 146*(9), 1269–1286. https://doi.org/10.1163/156853909X423104

Carey, M. P., Sanderson, B. L., Barnas, K. A., & Olden, J. D. (2012). Native invaders—Challenges for science, management, policy, and society. *Frontiers in Ecology and the Environment, 10*(7), 373–381. https://doi.org/10.1890/110060

Census India. (2011). https://www.census2011.co.in

Champion, H. G., & Seth, S. K. (1968). A revised survey of the forest types of India. *Manager of Publications*. https://dds.crl.edu/crldelivery/23005

Chavez, A., Ramaswami, A., Nath, D., Guru, R., & Kumar, E. (2012). Implementing trans-boundary infrastructure-based greenhouse gas accounting for Delhi, India. *Journal of Industrial Ecology, 16*(6), 814–828. https://doi.org/10.1111/j.1530-9290.2012.00546.x

Clark, R. G., & Shutler, D. (1999). Avian habitat selection: Pattern from process in Nest-site use by ducks? *Ecology, 80*(1), 272–287. https://doi.org/10.1890/0012-9658(1999)080[0272:AHSPFP]2.0.CO;2

Cox, D. T., & Gaston, K. J. (2016). Urban bird feeding: Connecting people with nature. *PLoS One, 11*(7), e0158717. https://doi.org/10.1371/journal.pone.0158717

Doron, A. (2021). Stench and sensibilities: On living with waste, animals and microbes in India. *The Australian Journal of Anthropology, 32*(S1), 23–41. https://doi.org/10.1111/taja.12380

Drewitt, A. L., & Langston, R. H. W. (2008). Collision effects of wind-power generators and other obstacles on birds. *Annals of the New York Academy of Sciences, 1134*(1), 233–266. https://doi.org/10.1196/annals.1439.015

Ferguson-Lees, J., & Christie, D. A. (2001). *Raptors of the world.* Houghton Mifflin Harcourt.

Francis, R. A., & Chadwick, M. A. (2012). What makes a species synurbic? *Applied Geography, 32*(2), 514–521. https://doi.org/10.1016/j.apgeog.2011.06.013

Fuller, R. A., Irvine, K. N., Davies, Z. G., Armsworth, P. R., & Gaston, K. J. (2013). *Interactions between people and birds in urban landscapes.* https://doi.org/10.1525/california%2F9780520273092.001.0001

Galushin, V. M. (1971). A huge urban population of birds of prey in Delhi India. *Ibis, 113*(4), 522–522. https://doi.org/10.1111/j.1474-919X.1971.tb05189.x

Gangoso, L., Agudo, R., Anadón, J. D., de la Riva, M., Suleyman, A. S., Porter, R., & Donázar, J. A. (2013). Reinventing mutualism between humans and wild fauna: Insights from vultures as ecosystem services providers. *Conservation Letters, 6*(3), 172–179. https://doi.org/10.1111/j.1755-263X.2012.00289.x/

GoI. (2022). *Animal Husbandry Statistics (AHS)|department of animal husbandry & dairying.* https://dahd.nic.in/animal-husbandry-statistics

Grimm, N. B., Faeth, S. H., Golubiewski, N. E., Redman, C. L., Wu, J., Bai, X., & Briggs, J. M. (2008). Global change and the ecology of cities. *Science, 319*(5864), 756–760. https://doi.org/10.1126/science.1150195

Hagemeijer, W. J., & Blair, M. J. (1997). *The EBCC atlas of European breeding birds* (p. 479). Poyser.

Horton, K. G., Nilsson, C., Van Doren, B. M., La Sorte, F. A., Dokter, A. M., & Farnsworth, A. (2019). Bright lights in the big cities: Migratory birds' exposure to artificial light. *Frontiers in Ecology and the Environment, 17*(4), 209–214. https://doi.org/10.1002/fee.2029/

Hunold, C., & Mazuchowski, M. (2020). Human–wildlife coexistence in urban wildlife management: Insights from nonlethal predator management and rodenticide bans. *Animals, 10*(11), 11. https://doi.org/10.3390/ani10111983/

Inger, R., Cox, D. T. C., Per, E., Norton, B. A., & Gaston, K. J. (2016). Ecological role of vertebrate scavengers in urban ecosystems in the UK. *Ecology and Evolution, 6*(19), 7015–7023. https://doi.org/10.1002/ece3.2414

Kumar, N., Singh, A., & Harriss-White, B. (2019c). Urban waste and the human–animal interface in Delhi. *Economic and Political Weekly. 54*(47), 42–47.

Kumar, N., Mohan, D., Jhala, Y. V., Qureshi, Q., & Sergio, F. (2014). Density, laying date, breeding success and diet of Black Kites Milvus migrans govinda in the city of Delhi (India). *Bird Study, 61*(1), 1–8. https://doi.org/10.1080/00063657.2013.876972

Kumar, N., Gupta, U., Jhala, Y. V., Qureshi, Q., Gosler, A. G., & Sergio, F. (2018a). Habitat selection by an avian top predator in the tropical megacity of Delhi: Human activities and socio-religious practices as prey-facilitating tools. *Urban Ecosystems, 21*(2), 339–349. https://doi.org/10.1007/s11252-017-0716-8

Kumar, N., Qureshi, Q., Jhala, Y. V., Gosler, A. G., & Sergio, F. (2018b). Offspring defense by an urban raptor responds to human subsidies and ritual animal-feeding practices. *PLoS One, 13*(10), e0204549. https://doi.org/10.1371/journal.pone.0204549

Kumar, N., Jhala, Y. V., Qureshi, Q., Gosler, A. G., & Sergio, F. (2019a). Human-attacks by an urban raptor are tied to human subsidies and religious practices. *Scientific Reports, 9*(1), 1–10. https://doi.org/10.1038/s41598-019-38662-z

Kumar, N., Gupta, U., Malhotra, H., Jhala, Y. V., Qureshi, Q., Gosler, A. G., & Sergio, F. (2019b). The population density of an urban raptor is inextricably tied to human cultural practices. *Proceedings of the Royal Society B, 286*(1900), 20182932. https://doi.org/10.1098/rspb.2018.2932

Kumar, N., Gupta, U., Jhala, Y. V., Qureshi, Q., Gosler, A. G., & Sergio, F. (2020). GPS-telemetry unveils the regular high-elevation crossing of the Himalayas by a migratory raptor: Implications for definition of a "Central Asian Flyway". *Scientific Reports, 10*(1), 1–9. https://doi.org/10.1038/s41598-020-72970-z/

Leong, K. M. (2009). The tragedy of becoming common: Landscape change and perceptions of wildlife. *Society and Natural Resources, 23*(2), 111–127. https://doi.org/10.1080/08941920802438642

Lowry, H., Lill, A., & Wong, B. B. M. (2013). Behavioural responses of wildlife to urban environments. *Biological Reviews, 88*(3), 537–549. https://doi.org/10.1111/brv.12012

Malhotra, A. K. (2007). *Tiger of sky-pariah kite*. PhD Thesis, Shilalekh Publishers.

Margalida, A., & Colomer, M. À. (2012). Modelling the effects of sanitary policies on European vulture conservation. *Scientific Reports, 2*(1), 1–7. https://doi.org/10.1038/srep00753

Marzluff, J. M. (2017). A decadal review of urban ornithology and a prospectus for the future. *Ibis, 159*(1), 1–13. https://doi.org/10.1111/ibi.12430

Marzluff, J. M., McGowan, K. J., Donnelly, R., & Knight, R. L. (2001). Causes and consequences of expanding American crow populations. In J. M. Marzluff, R. Bowman, & R. Donnelly (Eds.), *Avian ecology and conservation in an urbanizing world* (pp. 331–363). Springer. https://doi.org/10.1007/978-1-4615-1531-9_16

McKinney, M. L. (2002). Urbanisation, biodiversity, and conservation: The impacts of urbanisation on native species are poorly studied, but educating a highly urbanised human population about these impacts can greatly improve species conservation in all ecosystems. *Bioscience, 52*(10), 883–890. https://doi.org/10.1641/0006-3568(2002)052[0883:UBAC]2.0.CO;2

McMahon, L. A., Rachlow, J. L., Shipley, L. A., Forbey, J. S., & Johnson, T. R. (2017). Habitat selection differs across hierarchical behaviours: Selection of patches and intensity of patch use. *Ecosphere, 8*(11), e01993. https://doi.org/10.1002/ecs2.1993

Moleón, M., Sánchez-Zapata, J. A., Margalida, A., Carrete, M., Owen-Smith, N., & Donázar, J. A. (2014). Humans and scavengers: The evolution of interactions and ecosystem services. *Bioscience, 64*(5), 394–403. https://doi.org/10.1093/biosci/biu034/

Newton, I. (2010). *The migration ecology of birds*. Elsevier.

Nilon, C. H. (2011). Urban biodiversity and the importance of management and conservation. *Landscape and Ecological Engineering, 7*(1), 45–52. https://doi.org/10.1007/s11355-010-0146-8

Oro, D., Genovart, M., Tavecchia, G., Fowler, M. S., & Martínez-Abraín, A. (2013). Ecological and evolutionary implications of food subsidies from humans. *Ecology Letters, 16*(12), 1501–1514. https://doi.org/10.1111/ele.12187/

Ortlieb, R. (1998). Schwarzmilan. *Westarp*. Wissenschaften.

Parker, T. S., & Nilon, C. H. (2012). Urban landscape characteristics correlated with the synurbization of wildlife. *Landscape and Urban Planning, 106*(4), 316–325. https://doi.org/10.1016/j.landurbplan.2012.04.003

Paul, S., & Nagendra, H. (2015). Vegetation change and fragmentation in the mega city of Delhi: Mapping 25 years of change. *Applied Geography, 58*, 153–166. https://doi.org/10.1016/j.apgeog.2015.02.001

Pinault, D. (2008). *Notes from the fortune-telling parrot: Islam and the struggle for religious pluralism in Pakistan*. Equinox Publ.

Rademacher, A., & Sivaramakrishnan, K. (Eds.). (2013). *Ecologies of urbanism in India: Metropolitan civility and sustainability* (1st ed.). Hong Kong University Press. https://www.jstor.org/stable/j.ctt1ffjpbk/

Richard, E., Zapata, D. I. C., & Angeoletto, F. (2023). Geofagia y plasticofagia en Coragyps atratus. *Ecosistemas: Revista Cietifica y Tecnica de Ecologia y Medio Ambiente, 32*(1), 2482–2482.

Ríos-Chelén, A. A., Quirós-Guerrero, E., Gil, D., & Macías Garcia, C. (2013). Dealing with urban noise: Vermilion flycatchers sing longer songs in noisier territories. *Behavioral Ecology and Sociobiology, 67*(1), 145–152. https://doi.org/10.1007/s00265-012-1434-0/

Robinson, W. D., Bowlin, M. S., Bisson, I., Shamoun-Baranes, J., Thorup, K., Diehl, R. H., Kunz, T. H., Mabey, S., & Winkler, D. W. (2010). Integrating concepts and technologies to advance the study of bird migration. *Frontiers in Ecology and the Environment, 8*(7), 354–361. https://doi.org/10.1890/080179

Santangelo, J. S., Ness, R. W., Cohan, B., Fitzpatrick, C. R., Innes, S. G., Koch, S., et al. (2022). Global urban environmental change drives adaptation in white clover. *Science, 375*(6586), 1275–1281. https://doi.org/10.1126/science.abk0989/

Seltmann, M. W., Öst, M., Jaatinen, K., Atkinson, S., Mashburn, K., & Hollmén, T. (2012). Stress responsiveness, age and body condition interactively affect flight initiation distance in breeding female eiders. *Animal Behaviour, 84*(4), 889–896. https://doi.org/10.1016/j.anbehav.2012.07.012

Sergio, F. (2003). From individual behaviour to population pattern: Weather-dependent foraging and breeding performance in black kites. *Animal Behaviour, 66*(6), 1109–1117. https://doi.org/10.1006/anbe.2003.2303/

Sergio, F., & Newton, I. (2003). Occupancy as a measure of territory quality. *Journal of Animal Ecology, 72*(5), 857–865. https://doi.org/10.1046/j.1365-2656.2003.00758.x

Sergio, F., Blas, J., Forero, M., Fernández, N., Donázar, J. A., & Hiraldo, F. (2005). Preservation of wide-ranging top predators by site-protection: Black and red kites in Doñana National Park. *Biological Conservation, 125*(1), 11–21. https://doi.org/10.1016/j.biocon.2005.03.002

Sergio, F., Tanferna, A., De Stephanis, R., Jiménez, L. L., Blas, J., Tavecchia, G., Preatoni, D., & Hiraldo, F. (2014). Individual improvements and selective mortality shape lifelong migratory performance. *Nature, 515*(7527), 410–413. https://doi.org/10.1038/nature13696

Shochat, E., Lerman, S. B., Katti, M., & Lewis, D. B. (2004). Linking optimal foraging behavior to bird community structure in an Urban-Desert landscape: Field experiments with artificial food patches. *The American Naturalist, 164*(2), 232–243. https://doi.org/10.1086/422222

Shochat, E., Lerman, S. B., Anderies, J. M., Warren, P. S., Faeth, S. H., & Nilon, C. H. (2010). Invasion, competition, and biodiversity loss in urban ecosystems. *Bioscience, 60*(3), 199–208. https://doi.org/10.1525/bio.2010.60.3.6

Soanes, K., Sievers, M., Chee, Y. E., Williams, N. S. G., Bhardwaj, M., Marshall, A. J., & Parris, K. M. (2019). Correcting common misconceptions to inspire conservation action in urban environments. *Conservation Biology, 33*(2), 300–306. https://doi.org/10.1111/cobi.13193

Sol, D., Lapiedra, O., & González-Lagos, C. (2013). Behavioural adjustments for a life in the city. *Animal Behaviour, 85*(5), 1101–1112. https://doi.org/10.1016/j.anbehav.2013.01.023

Strohbach, M. W., Warren, P. S., & Peterson, M. N. (2014). Urban Wildlife Science in Coupled Human–Natural Systems. In R. A. McCleery, C. E. Moorman, & M. N. Peterson (Eds.), Urban wildlife conservation: Theory and practice (pp. 33–53). Springer. https://doi.org/10.1007/978-1-4899-7500-3_3.

Talyan, V., Dahiya, R., & Sreekrishnan, T. (2008). State of municipal solid waste management in Delhi, the capital of India. *Waste Management, 28*(7), 1276–1287. https://doi.org/10.1016/j.wasman.2007.05.017

Taneja, A. V. (2015). Saintly animals: The shifting moral and ecological landscapes of North India. *Comparative Studies of South Asia, Africa and the Middle East, 35*(2), 204–221. https://doi.org/10.1215/1089201x-3138988

Thiollay, J.-M., Newton, I., & Bretagnolle, V. (2004). *Rapaces nicheurs de France: Distribution, effectifs et conservation*. Delachaux et Niestlé.

Torres, D. F., Oliveira, E. S., & Alves, R. R. N. (2018). Conflicts between humans and terrestrial vertebrates: A global review. *Tropical Conservation Science, 11*, 1940082918794084. https://doi.org/10.1177/1940082918794084/

United Nations, Department of Economic and Social Affairs, Population Division. (2019). *World urbanization prospects: The 2018 revision (ST/ESA/SER.A/420)*. United Nations.

Votier, S. C., Fayet, A. L., Bearhop, S., Bodey, T. W., Clark, B. L., Grecian, J., Guilford, T., Hamer, K. C., Jeglinski, J. W. E., Morgan, G., Wakefield, E., & Patrick, S. C. (2017). Effects of age and reproductive status on individual foraging site fidelity in a long-lived marine predator. *Proceedings of the Royal Society B: Biological Sciences, 284*(1859), 20171068. https://doi.org/10.1098/rspb.2017.1068

Watson, V. (2013). Planning and the 'stubborn realities' of global south-east cities: Some emerging ideas. *Planning Theory, 12*(1), 81–100. https://doi.org/10.1177/1473095212446301

West, E. H., & Jones, H. H. (2022). Human food subsidies drive individual specialisation and intrapopulation dietary differences in a generalist predator. *Ecosphere, 13*(9), e4246. https://doi.org/10.1002/ecs2.4246

Weston, M. A., McLeod, E. M., Blumstein, D. T., & Guay, P.-J. (2012). A review of flight-initiation distances and their application to managing disturbance to Australian birds. *Emu-Austral Ornithology, 112*(4), 269–286. https://doi.org/10.1071/MU12026

Wong, B. B. M., & Candolin, U. (2015). Behavioural responses to changing environments. *Behavioral Ecology, 26*(3), 665–673. https://doi.org/10.1093/beheco/aru183

Chapter 8
Urban Green Spaces in Neotropical Cities: Biodiversity Conservation and Menaces

Rosemeri Segecin Moro, Silvia Méri Carvalho, and Taise Ernestina Prestes Nogueira Duarte

Abstract Latin American cities, with few exceptions, have been mostly established within biodiversity hotspots. However, these urban environments and their relationships with high biodiversity boundaries are yet few studied in conservation biology. The potential use of cities by different species requires a detailed understanding at multiple scales of the processes involved. This chapter focuses on the weaknesses and strengths related to biodiversity conservation detected in urban green network such as parks, street afforestation, home gardens, and backyards of Latin American cities. We address factors that influence the urban diversity of tropical and subtropical climate areas, and how Latin American urban public policies are in general inefficient in terms of maintaining local biodiversity, and green infrastructure planning tends to ignore basic concepts of urban ecology. In general, the larger green areas consist of forest remnants spared from urbanization due to land ownership or affective issues, together with restored areas. The rapid urban densification in Neotropical cities grants a low possibility of expanding green spaces, and this constrains the colonization and maintenance of more taxa beyond peri anthropic and synurbic species. Moreover, socioeconomic inequalities also difficult urban environmental preservation since vulnerable populations are bound to occupy risky areas for building. Also, removing the existing vegetation and exploiting natural resources, hence making difficult the compatibility of the urban development with species resilience. Moreover, social housing policies sometimes provide people with such small lots that the destination of space for green private areas is unfeasible. Finally, the attempts to create and optimize functional green corridors does not always follow technical and ecological criteria enough to ensure an effective connection among surroundings and the green areas within the city. Moreover, not all conservation

R. S. Moro (✉) · S. M. Carvalho
Universidade Estadual de Ponta Grossa, Ponta Grossa, Paraná, Brazil
e-mail: rsmoro@uepg.br

Taise Ernestina Prestes Nogueira Duarte
SEDUC MT Secretaria de Estado de Educação de Mato Grosso, Rondonópolis, Mato Grosso, Brazil
e-mail: taise.duarte@edu.mt.gov.br

F. Angeoletto et al. (eds.), *Ecology of Tropical Cities*,
https://doi.org/10.1007/978-3-031-48012-6_8

biology assumptions made for natural environments have been validated for Neotropical urban environments.

Keywords Urban ecology · Backyards · Urban parks · Street afforestation · Urban biodiversity

1 What This Chapter Is About

There was a great conceptual complexity and multidisciplinary involvement in discussing urban green spaces, especially in the Neotropical biogeographical region, which stretches from South America to southern Mexico and the Caribbean islands (Feeley & Stroud, 2018), which comprises Latin America. In this text, we focused our interest on the extra-Amazonian region, since the Amazon and the Caribbean insular environments deserve a very particular analysis due to their specificities.

The global urban population rate is about 67%. In Latin America, this rate increased to 84% (Williams, 2013). Moreover, 90% of the Latin American population is expected to live in cities by 2050, making the Neotropical region one of the most urbanized in the world (Secretariat of the Convention on Biological Diversity, 2012). Latin American cities have been established, with few exceptions, within biodiversity hotspots (Williams, 2013; Pauchard & Barbosa, 2013; MacGregor-Fors & Ortega-Álvarez, 2013; Angeoletto et al., 2023). These environments support an expressive number of rare and endemic species that have been greatly affected by the suppression of habitats due to the urbanization process (Mittermeier et al., 2004). This includes areas as diverse as Atlantic Rainforest and Cerrado in Brazil, Madrean Pine Oak Woodlands in Mexico, Tropical Andes in Colombia, and the Chilean Winter Rainfall-Valdivian Forests in Chile. For example, the Brazilian Cerrado presents around 1500 cities among areas inhabited by more than 4800 endemic plant and vertebrate species (Strassburg et al., 2017; Angeoletto et al., 2022) in which a population density of 6.5 Km^{-2} has grown at 1.5% a year in five years only between 2005–2010 (Williams, 2013). Other Neotropical hotspots except Amazonian and Caribbean showed similar rates—Atlantic Forest in Brazil (115.2 Km^{-2} dens. and 1.1% growing annual rate), Mesoamerica (72.6 Km^{-2} dens. and 1.6%), Tropical Andes (57.9 Km^{-2} dens. and 1.4%), Choco-Darien-Western Ecuador (56.2 Km^{-2} dens. and 1.6%), and Chilean forests (45.8 Km^{-2} dens. and 1.1%).

As remember Hall (2019), urban ecosystems, even those under high human impact, are natural systems equally subject to ecological processes and cycles as, for instance, thermal control by insolation and air currents, material recycling by microorganisms, nutrients competition and so one. However, as pointed by Perrotti and Iuorio (2019), large and medium Neotropical cities exhibit a low degree of homeostasis. That is, they present a low capability to reduce the influence of negative externalities on the urban environment since the productivity-centered logic of consumers demands a lot of natural resources. This can be attributed to the choice of a heterotrophic linear urban metabolism model instead of a circular urban metabolism model with sustainable energy flows and less external dependence (Girardet, 1989; Tainter, 2019).

Specific abiotic factors thought to be novel to cities also have analogs in wildland environments. However, the intensity, duration, periodicity, and combination of these factors in cities present potent selective forces on urban organisms (Faeth et al., 2011). These are the few authors who listed factors that make patterns of urban diversity in tropical and subtropical climate regions different from those established for temperate climate regions, as the major biodiversity, the general lack of severe winter and minor annual variation in the length of days.

The results of no understanding and including natural ecosystem dynamics in urban planning are well-known and studied (Lepczyk et al., 2017, Angeoletto et al., 2019; Ordóñez Barona et al., 2020), together with the associated effects in a variety of ecological, economical and human wellbeing aspects: changes in biodiversity and landscape aesthetics (Bensusan, 2002), interference with groundwater recharge (Costanza et al., 2014; Sun et al., 2019), heat island formation (Heisler & Brazel, 2019; Nowak, 2019), different sources of pollution (Escobedo et al., 2011; Nowak, 2019), emergence of risky areas to urbanization (Smardon, 2019), lack of natural spaces for outdoor activities and the desirable coexistence between people and urban nature (Hartig et al., 2014; Smardon, 2019), and the loss of memory and identity linked to natural spaces (Bensusan, 2002; Smardon, 2019).

Although there are various definitions and typologies related to the terms 'urban green areas', 'urban green spaces', and 'urban green infrastructure', we have chosen to employ those directly focused on the plant and its subordinate animal communities associated with the concept of urban forest. In this sense, according to FAO (2017), urban forests could be defined as networks or systems comprising all woodlands, groups of trees, and individual trees located in urban and peri-urban areas. To the U.S. Department of Agriculture, they include forests remnants, road tree lines, and vegetation in parks and home gardens, backyards, vacant lots, and brownfields regardless of whether trees or not (https://www.fs.usda.gov/managing-land/urban-forests). Urban forests are the backbone of green infrastructure, bridging the gap between rural and urban areas and improving a city's environmental footprint.

According to Perrotti and Iuorio (2019), and Perini and Sabbion (2017), green infrastructure creates a natural or semi-natural spatial/functional network of green and blue areas that aim at increasing urban resilience, adaptative capacity, and mitigation of city anomalies, and that include basic habitat networks for maintaining biodiversity (Tzoulas et al., 2007). Bechara et al. (2016) also emphasized the cultural value of urban biodiversity and how the provision to the population with material and intangible assets increases the need for natural elements in the city together with the urban sprawl.

However, in Latin America such 'void' spaces are under socio-economic pressures for housing and industrial uses, what lead Kattwinkel et al. (2011) to name a 'temporary biodiversity' including private spaces such as backyards, brownfields, and vacant lots, as the vegetation of sidewalks. Another concerning is the progressive reduction of urban lots as the main impediment to the presence of an expressive biodiversity in backyards and the road remodeling with tree suppressions.

In this chapter, we intend to review points that are sometimes not considered in the planning and management of urban green spaces in the Neotropical region. The resilience and sustainability of the biodiversity of urban green spaces are the focus of this discussion although we cannot ignore that socioeconomic inequalities in Latin America are a main factor of most dysfunctional green spaces. In major Neotropical cities, attempts are made to develop programs to improve the quality of life of the population, which, when successfully implemented, lead to gentrification of these areas and socio-spatial exclusion of the initially targeted beneficiaries. Even in projects developed for the upper classes, the lack of knowledge of conservation biology by planners has led to great mistakes, both to people and native species.

2 Spatial Relation Between the Social Context and the Neotropical Urban Green Structure

The spatial configuration of Neotropical cities, from European colonization, follows, with few exceptions, linear models of settlement along a defining economic axis, such as coastlines and riverbanks, railways, and highways (Sedrez, 2013). Alternatively, they may follow a radial model, designed or not, starting from a central district that concentrates commercial, liberal, and residential activities, displacing to the periphery the lower income housing and industries. Finally, cities development may even follow a hybrid model, merging features of both urban designs, with multiple subcenters subordinate to the downtown area (Torres et al., 2003).

Planned green spaces in Latin American cities followed, in the late 19th and early twentieth century, European cultural and aesthetic standards, with the establishment of forested avenues and squares, zoological and botanical gardens, with distribution of the vegetation structurally controlled (Sedrez, 2013). They reflected the desires of the bourgeois elite, with a positivist and hygienist vision, often excluding the population that did not conform to this model (Rolnik, 1997).

From the mid-twentieth century, the rapid intensification of the urbanization process by the migration of rural populations to the most Latin American cities resulted in the creation of new cities and the development of intra-urban growth provoked phenomena that altered and even suppressed the major existing green areas (Torres, 1978, 1993; Delgado, 1991). In addition to this, other urban design ideas have become patent in the last decades: residential and commercial verticalization in the core area of cities, the replacement of existing wide courtyards with built-up spaces, frequent irregular and disorganized occupation of public and/or at-risk areas (e.g. slums), the implementation of huge clusters of low-cost housing projects in the periphery, an increasingly distant of suburbs respect to the city center, and the gentrification of the sectors that maintained larger private areas and a certain proportion of the green infrastructure (Garza Villareal, 2000; Torres et al., 2003; Sedrez, 2013; Duarte et al., 2017; Angeoletto et al., 2017). Municipal urban plans attempt to organize urban land use by promoting spatial segregation of residential, commercial, and industrial areas, for instance, as green spaces as well. However, intentional urban voids are often created primary for real estate speculation instead of vegetation preservation, so posterior conversion to gray infrastructure does not usually face greater difficulties.

Finally, high demographic density and social inequality have created a very particular phenomena in the periphery of cities, the "suburbanisation of the elites" as gated communities, with private neighbourhoods and security systems, where communal areas do not always feature green spaces (Torres, 1997; Torres et al., 2003). At the same time, low-income neighbourhoods are typically interspersed with local ecosystems in peri-urban areas or completely devoid of green spaces (Unikel, 1976; Gorelik & Silvestri, 2000; Moreira, 2002; Hope et al., 2003; Pauchard & Barbosa, 2013).

While increasing population density and traffic has required the removal of trees from streets and channelization of rivers to make way for major avenues, it has also resulted in efforts to create 'environmental' parks, linear parks, 'city' parks, and other designations used for large multifunctional community green areas (Bechara et al., 2016). Many of these were created through the revitalization of stream banks, abandoned mining or inactive industrial areas (as the architect Jaime Lerner established while mayor of Curitiba at the end of the twentieth century). In addition, in recent decades, green spaces disconnected from the local culture, such as university campuses, water parks, and cemetery parks have been introduced in the peripheral area of large or medium-sized Latin American cities (Pauchard & Barbosa, 2013).

The criteria used in the implementation of green spaces integrate social needs in general, but not always environmental values, or when the latter are present, they are very simplified and sometimes misunderstood. Similarly, there are no reliable and systematized inventories of urban green spaces, valuation of their environmental quality, and supported ecosystem services. Government and community initiatives devoted to their creation are partial and isolated, not deriving from an overall plan (Peña Salmón, 2011; MMA, 2021). According to Rastandeh et al. (2017), when analyzing six sites in South America, they observed a general lack of knowledge of the spatial distribution of native species in urban spaces for the development of urban architecture projects and the creation of public policies that favor the urban green system.

When employing the Green View Index (GVI) using Google Street View (GSV) panoramas to four completely different cities in Ecuador, Brazil, and Argentina (http://senseable.mit.edu) one can see almost the same proportion, ranging from only 10 to 15% of greenery (Quito – 10.8%, São Paulo—11.7%, Franca—13.7%, and Buenos Aires—14.5%) and there is a high probability that almost 70% of this vegetation is mainly road afforestation (Gilbert, 1991).

3 What Remains and What Should Return: Resilience and Difficulties for Renaturing Processes

The biggest challenges faced in the conservation of urban green spaces is the suppression of habitats and the barriers created for species movements, so the access to resources could be very limited (Bechara et al., 2016; Leveau et al., 2017), leading to alteration of their structure and quality, as the isolation of remaining populations could cause changes in gene flows, intensified competition, and even the extinction of certain species (Primack, 2001). The reduced connectivity makes it difficult the formation of ecological corridors between the fragmented spaces available for feeding and shelter for the native species (Lopez et al., 2018). In addition to the spatial configuration of the green zones, the distance from preserved rural areas to the city determines the potential colonization of peri anthropic species, whose diversity is notably smaller in central areas of large cities than in their surroundings (Carbó-Ramírez & Zuria, 2011; McKinney, 2008; Faeth et al., 2011).

Urban ecosystems are intentionally maintained at mid-successional stages, in the absence of top-level predators and in the dominance of generalist and omnivorous species (Chollet et al., 2018; Callaghan et al., 2019; Amaya-Espinel et al., 2019). To Lopez et al. (2018), several landscape variables as housing density and large-scale impervious surface cover were each significant predictors of species composition for some exotic and native species groups. As pointed out by Faeth et al. (2011) and McKinney (2008), among others, on a large scale, plant species richness usually increases in cities due to the introduction of exotic species, while animal species in general decrease, even if the abundance of some groups increases, mainly birds and arthropods (Pacheco & Vasconcelos, 2007).

This pattern of diversity increasing, however, may not remain at small local spatial scales and often does not signify an increase in trophic chain diversity once Pautasso (2007) pointed that at small spatial scales increasing human population size tends to eliminate species via extreme human disturbances in the local vicinity. However, at coarser spatial scales, people have preferentially settled in areas of high biodiversity and have enhanced beta diversity even further with many species introductions and increasing habitat heterogeneity via complex land use patterns. However, to Hooper and Vitousek (1997) and Díaz and Cabido (2001), species diversity is often an inadequate surrogate, as there is a growing consensus that functional diversity, or the value and range of species traits, rather than species numbers per se, strongly determines ecosystem functioning.

In renaturing projects, the goal of preserving and increasing existing populations as an incentive for the return of native species is often stated. But, to what extent would this be desirable? Which species should be considered? In what situations and on what scale?

There are many proposals for renaturing processes facilitating the reintroduction of natural elements into cities to reduce the impact of urbanization in the natural communities (Muller & Werner, 2010; Herzog, 2016; Beatley, 2017; United Nations, 2014; Rumble et al., 2019). Renaturing projects in green areas appear at varied scales, in a mosaic of habitats imposed by human occupation in relatively small areas, which does not occur in the natural environments (Tzoulas et al., 2007; Hostetler et al., 2011; Rastandeh et al., 2017). Rehabilitation and revitalization projects aimed to resignify the public use of green spaces and do not satisfy the need for ecological restoration, which presupposes lower levels of anthropization.

The areas available for renaturation projects in general are small and disconnected, imposed by human occupation and sometimes by real estate speculation, in varying conditions but almost always unfavorable to the main assumptions of conservation biology – minimum area, connectivity, biodiversity, and functional redundancy (Tzoulas et al., 2007; Hostetler et al., 2011; Rastandeh et al., 2017). However, Faeth et al. (2011) raised many questions regarding the inadequacy of applying the standards established for natural areas to urban ecosystems. There is a growing understanding that green urban spaces would be new habitats, radically different from more natural ones, with new characteristics to be further investigated, even if the typical environmental factors of cities such as heat islands and pollution, also affect the diversity of natural environments.

Difficulties in this process include high spatial heterogeneity and the rapid colonization rates and growth of exotic species (cultivated, ruderal and/or invasive), and increasing synurbic and peri anthropic populations due, for instance, to warmer temperatures caused by the urban heat island (Velez-Ruiz et al., 2013; Faeth et al., 2011; Lopez et al., 2018; Rumble et al., 2019), by increased stress in hot summers from heat, water loss and/or increased insect herbivory, as warmer winter temperatures potentially promote the expansion of introduced species from warmer climates (Lopez et al., 2018).

Urbanization also increases the number of invasive alien species by the replacement of native species by cosmopolitan ones (Filloy et al., 2014; MacGregor-Fors et al., 2017). To Lopez et al. (2018), urbanization appears to impede dispersal for some native species, especially those with short dispersal distances, and that these effects are likely to increase over time. In contrast, exotic species show spatial patterns consistent with spread along stream corridors from multiple urban source locations. This irruption of exotic species with the potential to spread in non-native areas can impact ecosystems by causing biotic homogenization (Secretariat of the Convention on Biological Diversity, 2012; Santos et al., 2019; Martínez & de la Barrera, 2020). However, the homogenization caused by planting few native species in afforestation projects, for instance, has been questioned. Møller et al. (2012) claim that behavioral plasticity and timing can better explain differences in bird species richness and invasive success in urban environments. To Gong et al. (2013), the establishment of alien plant species instigated by anthropogenic disturbances do cause biotic homogenization in the artificial habitats, but the near-natural habitats show some resistance against the establishment of alien species. Latitude could be a factor for the loss of bird richness in the most subtropical city (Leveau et al., 2017), as well the biome in which the city is located, whether more forested or in grassland biomes (Leveau & Leveau, 2005). Therefore, the settlement of artificial habitats in the cities was proven distinct effects on fauna and flora. Nascimento et al.

(2020) drew attention to the role that herbaceous and shrub species play in providing resources to urban pollinators.

Recent studies have demonstrated that invasive species, no matter if intentionally cultivated or ruderal, exotic or native, can play also a positive role favoring urban and peri-urban birds, bats, and other small urban mammals in the Neotropical climate region (Reis et al., 1993; Galetti & Morellato, 1994; Schulze et al., 2000; Estrada & Coates-Estrada, 2001; Almeida et al., 2005; Leal et al., 2019; Bonthoux et al., 2019). Since many exotic plant species provide the native peri-urban fauna with relevant resources (Zortéa & Chiarello, 1994; Cunha et al., 2006), the need for their eradication has raised intense debates (Schlaepfer et al., 2011; Angeoletto et al., 2020).

Callaghan et al. (2019) stated that the heterogeneity of urban environments results in higher bird diversity in densely populated Neotropical cities and suggested that in addition to maintaining large green areas, small green spaces such as backyards and home gardens could play an equally important role in their conservation. Private gardens provide significant habitats and increase the overall functional dimension of the green space network, acting as corridors and even creating new habitat patches (Rudd et al., 2002; Carbó-Ramírez & Zuria, 2011; Faeth et al., 2011; Ikin et al., 2013; Gonçalves & Toledo, 2016; Amaya-Espinel et al., 2019), Unfortunately, these are spaces susceptible to disturbance, due to their small size and exposure to conditions imposed by the surrounding urban matrix and by the private management of the owners (Fernández et al., 2018).

The replacement of native vegetation could preclude many species from invading cities and even migrate to the outskirts of urban areas (Sol et al., 2013; Brown, 2020). These peri anthropic species, which live in urban forests, vary depending on the latitude and quality of the surrounding rural areas (Filloy et al., 2014; Leveau et al., 2017; Vergara & Ibarra, 2019). Positive and negative consequences can arise once rural species start to exploit cities. They can cause disturbances if they start foraging inside houses, and they are also at risk of being roadkilled or attacked by domestic animals or humans, Synurbic species, on the other hand, are adapted to and dependent on the urban environment (Rumble et al., 2019). To Francis and Chadwick (2012), the term 'synurbic', is sometimes used within the more recent urban ecology literature to refer to a species that colonizes or is found within urban ecosystems. But they argue that this is too simplistic an interpretation and consider that the term should be reserved for species populations that have higher densities in urban compared to rural areas. However, the scarcity of populational studies of the Neotropical mammals and insects, both urban and rural made it unpractical. According to Hostetler et al. (2011), long-term studies based on monitoring are also needed to understand the variety of elements, the demography of species, and the interactions that support green infrastructure and maintain its biological integrity.

Synurbic species include a range of insects, arachnids, mammals, and birds of different sizes, which are important pollinators and disperser of species in the urban forest (Silva & Ferreira, 2011; Lutinski et al., 2013; Estrada et al., 2014; Ribeiro et al., 2018). These groups exhibit distinct behaviors and ecology when compared to non-urban populations, within the plasticity that species exhibits such as higher populational density associated with reduced individual territories (mate, family), reduced migratory behavior, extended reproductive period (with offspring in winter), and other changes in the reproductive ecology (Yeh, 2004; Ibáñez-Alamo & Soler, 2010; Partecke & Gwinner, 2007; Bozek et al., 2007; Davison et al., 2009). However, these competitive advantages can be affected by altered and prolonged circadian activities and changes in diet and foraging behavior (Slabbekoorn & Ripmeester, 2008). The observed increase in life expectancy is generally accompanied by poorer health parameters, an increase in intraspecific aggression, domestication, and various adaptations to human behavior.

In general planners don't forecast the volume of resources offered to and requested by the species that could be attracted by the new landscape configuration. So, it is irresponsible to induce non-urban populations to turn into peri anthropic or synurbic communities (Fig. 8.1). This would increase conflicts with dwellers, as monkey invasions for foraging, birds attacks near nests, or even the occurrence of zoonosis (Rosatte et al., 2010; Arduíno et al., 2020). In this sense, Lepczyk et al. (2017), Lopez et al. (2018) and López-Flores et al. (2009) discuss when green spaces are more likely to act as ecological traps or population sinks, by increasing nest predation rates for instance, or when reproduction and survival rates are unable to sustain a population. Nevertheless, these situations are rarely foreseen and evaluated in projects of renaturing or to improve green areas connection.

In the following sections, we present topics in the two most addressed scales of urban green space analysis currently used in Latin American city planning: (i) the challenges in implementing and maintaining street trees and urban parks, (ii) the most recent discussion on the importance of small private green spaces, mainly home gardens and backyards.

Fig. 8.1 Diversity of peri-anthropic and synurbic species in the urban area of Rondonópolis, Brazil (Cerrado biome): (**a**) Rhea (*Rhea americana*); (**b**) Blue and Yellow Macaw (*Ara ararauna*); (**c**) Howler Monkey (*Alouatta fusca*). (Photos: (**a**) T.E.P.N. Duarte; (**b**, **c**) C. Silva Jr)

4 Urban Forestry: Looking for the Lost Biodiversity

The vegetation cover, formed by different typologies such as forest fragments, public and private green areas, permanent preservation areas, and street trees, constitutes the urban forest (Fig. 8.2). However, especially in Brazil, the term urban afforestation has become more popular (Biondi, 2015).

According to Rossetti et al. (2010), urban afforestation refers to a pattern distribution of trees along with high anthropized areas, such as sidewalks, central beds, crossroads, and roundabouts. Biondi (2015) also includes 'street trees' as a component of the urban forest, along with woodlands, squares, and home gardens.

Trees in the urban environment, whether incorporated into forest fragments or individually placed in squares, parks, and streets, play an important role in the composition and maintenance of the urban ecosystem, representing the habitat of fauna species, corridor, or even steppingstones for animal populations through a hostile urban matrix (Beninde et al., 2015; Fernández et al., 2018). In Latin America and the Caribbean, studies on urban forestry have expanded remarkably in the last

Fig. 8.2 Aspects of the Latin-American urban forest. (**a**) Santiago do Chile; (**b**) Mexicali (Mexico); (**c**) Buenos Aires (Argentina); (**d**) Belo Horizonte (Brazil). (Photos: (**a**) R.S.Moro; (**b**) R.I. Rojas C.; (**c, d**) S.M. Carvalho)

decade, mainly in Brazil, which concentrates about half of these studies, followed by Mexico and Chile (Barona et al., 2020).

An extensive study carried out by Barona et al. (2020) shows that most research developed in Latin America focuses on field surveys that report on the diversity of street trees and isolated trees in public areas, without addressing more deeply the social or management aspects. According to Rossetti et al. (2010), the most technically challenging and quite expressive is the planting of street trees in the design of existing cities. The major focus of street tree implementation could be linked to its role in connectivity between urban green spaces in a network that should be not only structural but also functional (Kirchner et al., 1990).

The main function of tree insertion in cities, besides aesthetics, is to provide a partial balance of urban space (Rossetti et al., 2010), however, the ignorance of urban afforestation, often leads to huge mistakes and endless conflicts, such as the coexistence of trees with overhead and underground cables, excessively paved environments, poor sizing of the canopy projection, among others. As pointed out by Furlan (2004), some factors hinder the presence of street trees, which ends up at disadvantage in relation to other uses of such spaces, mainly sidewalks and flower beds.

Another relevant issue to be taken into consideration when implementing or maintaining trees in the urban environment is the existing hostile environmental conditions. The trees are subjected to air pollutants, contaminated, and poorly developed soils, insufficient surrounding free area, and mechanical injury (Rossetti et al., 2010; Lawrence et al., 2012).

More recently, some debate has been raised about the fact that climate changes may lead to even greater biological limitations than are normally found in the urban environment (Ferrini et al., 2014). Pretzsch et al. (2017) developed a study on ten major cities in the world, including Santiago, Chile, and reported that, in general, urban trees have shown accelerated growth since the 1960s onwards. Even if the benefits of increased surface and underground biomass development, with the resulting carbon sequestration, outweigh the negative effects, accelerated growth may result in faster aging and a shorter lifespan, which would possibly require faster replacement and replanting (Vrecenak et al., 1989).

Another factor to be considered is the value of the spontaneous herbaceous component that grows among the gray system structures present on sidewalks, flower beds, roadsides, and vacant lots (Bonthoux et al., 2019) once these categories are just different scales in the analysis of urban forests. The expansion of afforestation biodiversity suitable for the urban environment is necessary and recognized by the scientific community as well as by technicians involved in public tree management. However, more specific indications are still scarce such as the optimal proportion and distribution of species within a given territory. The species suitable for sidewalks have not yet been fully studied yet, and the choices are sometimes subject to the temporary popularity of certain species. In Mexico City, 72% of the tree population is represented by only nine species (Chacalo et al., 1994). In Brazil, the predominance of three to five species is mainly noted in urban afforestation, with a high incidence of exotic plants (Rossetti et al., 2010; Esteves & Corrêa, 2018).

Esteves and Corrêa (2018), in a study including 55 Brazilian cities, showed that in 78% of them, the number of exotic species on streets exceeded that of the native species. The Amazon biome was the best represented, while native tree species outside the Amazon were the least represented. The authors stressed that the valorization of local flora should be promoted in urban afforestation as a strategy to reduce costs in the maintenance of green infrastructure, and to make Brazilian cities more environmentally balanced. Environments with trees support significant functional bird biodiversity (Beninde et al., 2015) and the use of exotic species in the urban afforestation alters local ecological processes in bird trophic chains. Bechara et al. (2016), and Nascimento et al. (2020) stressed the importance of urban afforestation in a more ecological way, i.e., ecosystem-based management, supported by the floral biology, phenology, and permanent food supply for pollinators and dispersers.

There is rarely an official program to encourage research, selection, genetic improvement, or commercial propagation of native species with ornamental potential for gardening, urban afforestation, or restoration projects for degraded urban areas. Although neotropical urban green spaces should favor the implementation and maintenance of native species in their urban green systems, the maintenance of exotic trees could be admitted since they do not present an invading or toxic behavior to native fauna and offer niche and trophic chain conditions to native species (Angeoletto et al., 2020). No consensus has been reached regarding this issue. A study by Bello et al. (2020) in large Atlantic Rainforest fragments concluded that exotic palms with more than 80% niche overlapping the native palm *Euterpe edulis* could lead to the disruption of its seed dispersion. Campagnoli and Antunes (2017) drew attention to the decrease in bird biodiversity in areas of bio-invasion by exotic palms. Due to the lack of data on this effect in urban fragments, Faeth et al. (2011) pointed out that preserving or rebuilding natural habitats by introducing native and non-native plant species does not guarantee the appearance of other components of the biological community. Finally, they also state that understanding the processes that lead to diversity is essential for effective decision making to preserve urban biodiversity.

5 Urban Parks: Challenges to Maintain and Expand

The permanence of vegetation cover in Neotropical cities is highly dependent on the dynamics of urban populations (Secretariat of the Convention on Biological Diversity, 2012; Williams, 2013; Angeoletto et al., 2020). Urban parks in Latin America, both public and private, are composted mostly by forest remnants embedded in the urban or peri-urban zone, generally in secondary-succession stages (Chollet et al., 2018; Amaya-Espinel et al., 2019) or vegetation that has been replaced in areas spared from urbanization, either due to their unhealthiness (mountains or valley, decommissioned quarries, sand extraction sites, etc.) or left behind for land or economic reasons (around the headquarters of former farms, bankrupt enterprises, unclaimed lands, etc.). But active selective planting can occur, which may be well-intentioned but is often technically flawed, mixing climax native and exotic ornamental species, and often making it difficult to characterize the successional stage of this new habitat (Isernhagen et al., 2009).

So, parks can be strongly structured, designed, and managed (Fig. 8.3a, b, d), or offer a more 'natural' appearance (Fig. 8.3c, e, f). Sometimes, even though parks consist of original forest remnants are currently managed.

As examples of how different origins and purposes Latin American parks and gardens could be, the Botanical Garden of the Universidad Austral de Chile (Valdivia, Chile) and Parque Ducke Botanical Garden (Manaus, Brazil) were managed to document the Valdivian and Amazonian jungle respectively (Figs. 8.3a, e). On the other way, the Tijuca National Park (Rio de Janeiro, Brazil), which was completely deforested to coffee plantations during the eighteenth century, had its native vegetation restored by active planting during the nineteenth century to improve the city water supply. So, currently one hardly could perceive it was totally planted (Fig. 8.3f). Cerro Santa Lucía (Santiago, Chile) in turn, is a human-constructed garden with intentional landscaping and plant arrangements to embellish the wealthy surroundings (Fig. 8.3b), while Las Bellotas (Cuauhtémoc, Mexico) and Cerro Nielol (Temuco, Chile) take advantage of the spontaneous distribution of its vegetation cover for leisure and contemplation purposes (Figs. 8.3c, d).

Though Fernández et al. (2018) did not identify a significant relationship between income and biodiversity in Chile's urban parks, extreme poverty is incompatible with urban environmental preservation, as the homeless are forced to occupy risky areas, clearing of existing vegetation and creating precarious housing clusters (Williams, 2013; Caputo et al., 2019). Those who manage to obtain a regular piece

Fig. 8.3 Urban parks in Latin-American hotspots: (**a**) UACH Botanical Garden, Valdivia (Chile); (**b**) Cerro Santa Lucía, Santiago (Chile); (**c**) Cerro Nielol, Temuco (Chile); (**d**) Las Bellotas, Cuauhtémoc (Mexico); (**e**) Parque Ducke Botanical Garden, Manaus (Brazil); (**f**) Tijuca National Park, Rio de Janeiro (Brazil). (Photos: R.S. Moro)

of land usually end up occupying a lot so small that no green space is feasible (Smardon, 2019). In a scenario of social inequality and latent violence in large Latin American cities the use of these spaces for social or cultural functions, considering their suitability, can be difficult and even dangerous (Donoso, 2017). In the absence of the public power, and lack of a culture of communitarian enjoyment of public green spaces, no rare they are quickly occupied for other interests, sometimes illegal. Conversely, within the current context of a society surrendered to a pervasive neoliberal doctrine, green areas are used as an added financial value for real estate, thus increasing social-spatial segregation (Crompton, 2001; Bensusan, 2002; Caputo et al., 2019).

Depending on their size, urban parks can harbor resident animal populations, or become the foraging, roosting, or nesting area for peri-urban species (Scherer et al., 2012; Barros et al., 2006). They are also important ecological steppingstones for species moving between green patches. In this sense, Fernández et al. (2018) and Hostetler et al. (2011) have pointed out the influence of the environmental quality of the urban matrix in the surroundings of woods and parks on the ecological processes and the ecosystem services they provided.

Beninde et al. (2015), about the Brazilian Cerrado, reported that to maintain a higher level of urban diversity it would be necessary to increase the potential habitats in urban parks to at least one large block of 50 hectare, in addition to the existence of a network of functional corridors. Threlfall et al. (2016) also pointed out the need for large areas to ensure the integrity of bird and bat communities.

To evaluate the difference in species richness between urban parks and rural areas, Estevo et al. (2017) compared the distribution of functional trophic guilds of birds in a 54-hectare urban park and a 38,000-hectare rural area in the Atlantic Forest. They found no significant differences between them and concluded that urban bird communities may have sufficient functional redundancy to maintain the biological functions in less preserved areas, which somewhat disagrees with the conclusions of Faeth et al. (2011) about species richness decreasing in urban areas. Plant communities however may have difficulties in reaching more mature successional stages in small areas, as observed by Santana et al. (2019) when analyzing a 150-year-old 8-hectare urban park, and Fonseca et al. (2013) in a 40- years-old-1.2-hectare urban green space.

Nonetheless, fragmentation alone does not result in the loss of trophic or demographic mechanisms that justify a decrease in biodiversity. Many abiotic factors and interactions between competing urban species also occur in the natural environment, but not with the same intensity, duration, and frequency seen in cities, which act as selective forces in the urban environment.

Rebuilding natural or semi-natural environments also requires large inputs of energy, time, and resources, often dictated by law, and public opinion and perception. Failures in the supply of energy and materials, usually result in rapid successional changes, immigration, and dominance of invasive ruderal species, to the point where the desired community disappears. Natural processes such as immigration and emigration, predation and herbivory, disease, parasitism, competition, succession, local extinction, and natural selection are profoundly altered by human decision-making (Faeth et al., 2011; Lopez et al., 2018). Thus, not all urban parks are functional and there should be a cost-benefit assessment of the implementation of new areas capable of increasing the resilience of natural communities based in field research evidence.

The lack of knowledge of the ecological processes that occur in subtropical urban forests isolated in the city matrix is one of the challenges of applying ecological principles to the issue of urban planning (Dislich & Pivello, 2002; Roy et al., 2012; Fernández et al., 2018). Along with the effects of pollution, disease, mechanical injury, and microclimate conditions, there is a lack of landscape-scale information on minimum functional area, shape, and distance from the nearest native patches (Faeth et al., 2011; Lawrence et al., 2012; Matthies et al., 2015; Lopez et al., 2018), edges (Hardt et al., 2013), and others (see Chap. 17). In addition, for instance, little is known about the inextricable mycorrhizal association between fungi and roots that allows most Neotropical plants to survive. There is a lack of funding for the study of urban biological structures, their functions, and problem-solving based on animal, plant, and local ecosystem characteristics (Pauchard & Barbosa, 2013). In their review, these authors show that despite the increase of papers related to the urban ecosystem in the last 40 years, less than 2% address urban wildlife, and only 3.7% address Latin America (most of them refer to megacities in Brazil, Chile, Argentina, Colombia, and Mexico).

6 Importance of Home Gardens and Backyards in the Urban Forest

Defined as private unbuilt spaces adjacent to dwellings, home gardens and backyards can present varied forms and uses depending on geographical, climate, social, cultural, economic, and even gender factors (Eyzaguirre & Linares, 2004). Galluzzi et al. (2010) defined urban backyards as micro-environments or small-scale socio-ecological systems, aiming at socio-economic compatibility with the potential to support biodiversity in urban environments.

Backyards and home gardens proportion varies greatly depending on the urban development context and Rudd et al. (2002) consider that even though backyards are apparently too small to be biologically significant, when they are added together, sometimes they reach areas of considerable size. As example, in León (Nicaragua), the area covered by backyards represents 86% of the urban green space surface (González-García & Gómez Sal, 2008). Meanwhile in the densely populated North Zone of Rio de Janeiro (Brazil), backyards represent only 40% of the urban green area (Ferreira, 2016).

FAO estimative for 2020 indicates that 85% of low-income people in Latin America live in urban areas and this lower classes generally use their backyards to produce food and generate income (Rayol & Miranda, 2017). Thus, these spaces are used to produce goods for sale, store materials either for construction or disposal, or even to build rooms for rent or to add family members. But in the medium and big cities downtown, when backyards exist, they often consist of small patches of grass, rarely trees (Bensusan, 2002; Angeoletto et al., 2017). Angeoletto et al. (2017) have found that in the low-income neighborhoods prevailed utilitarian species (i.e., medicinal and food species) whereas in the backyards of the upper-middle class, ornamental species prevailed. But Pauchard and Barbosa (2013), based on evidence from Rio de Janeiro (Brazil) and Valdivia (Chile), and by Fernández et al. (2018) in Santiago (Chile), found no significant relationship between income and biodiversity. To Williams (2013), studies link habitat loss to urban and agricultural expansion and the effects of globalization more than to population growth. In some cases, local low-income communities manage to reintroduce urban green spaces in their neighborhood; in others, maintaining vacant lot vegetation results in increased species number, most generalist native species. Finally, in some affluent neighborhoods, the quality of the habitat has been improved by replacing non-native vegetation by native ornamental species.

Regarding the plants in backyards, the European colonization introduced several species from a matrix of African, Asian, and Mediterranean plant species (Eyzaguirre & Linares, 2004). Beyond that, the Iberian heritage is observed in the distribution of green areas in homes, so in Spanish-speaking countries backyards follow the ancestral model of a Roman village which denotes a 'closed' house, with the rooms facing a small inner courtyard, sometimes just ornamental. Conversely, in the Portuguese culture, which predominates in Brazil, houses tend to be 'open' to the outside through several windows, with a functional backyard with fruit trees, spices, and medicinal plants. Both traditions however do not favor ornamental gardens in front of the dwellings, as they face directly onto the street. On the contrary, immigrants from other parts of Europe (mainly established in South Brazil, and Chile) brought homesteads with a clear functional division between ornamental front home gardens, and multifunctional backyards. Finally, Afro-Latin American descendants tended to reproduce the ancestral African pattern of small houses, clustered around multifunctional communal courtyards, valuing outdoor work segregated from the surrounding green areas (Pereira, 2011; Weimer, 2020). These arrangements can be seen where urban planning does not exist, such as in the favelas in the hills of Rio de Janeiro and in the traditional rural settlements of Afro-descendants throughout Brazil, the quilombos.

According to Lope-Alzina and Howard (2012) in the last years home gardens have been the object of numerous studies, mainly in the field of agroecology, agroforestry, and ethnobotany; lesser studied is the backyards contribution to urban biodiversity. Almeida et al. (2005) investigate the dwellers knowledge about its backyard's birds and Banaszak-Cibicka et al. (2016) has pointed how urban green space favorable for large and diverse bee populations. Nascimento et al. (2020) pointed out that research of urban pollinators on green infrastructure privileges areas such as parks and university campuses (69%), followed by squares and streets (11%), as vacant lots, gardens, and roadside areas did not deserve more than 5% of the studies analyzed.

Being tropical and subtropical environments in Latin America locations of high biodiversity, it is natural that in urban areas the diversity of species of fauna and flora is higher when compared to other regions, and this includes the potential plant species observed in backyards. But MacGregor-Fors and Ortega-Álvarez (2013) cautioned that in some cases, backyards are points of introduction of exotic species that become invasive in pristine environments. In São Luís (Maranhão, northeastern Brazil) 57% of the backyards have more than 20 plant species. Similar diversity was observed in Boa Vista (Roraima, northern Brazil) and Santarém (Pará, northern Brazil), where 43 and 98 species of plants were identified, respectively (Akinnifesi et al., 2010). Meanwhile Angeoletto (2012) found 900 plant species grown in around 300 urban backyards in Maringá (Paraná, southern Brazil), including endangered species. To Lope-Alzina and Howard (2012) home gardens in Yucatán (Mexico) presented from 5 to 387 plants. Regarding the number of trees in urban backyards, Duarte et al. (2017) found marked regional differences in Brazil ranging from 37% in the North and 72 to 74% in the South and Southeast regions, respectively. Meanwhile Lope-Alzina and Howard (2012) have found 31 to 58% of trees in the Yucatán home gardens. The large qualitative and quantitative differences found are due to the continental proportions of the region Neotropical, where climatic, economic, and cultural conditions are considerably varied.

It is very common that areas near forest parks, squares, and well-wooded streets would be more valorized places, occupied by higher-income families, while for the poorest population backyards may be the only possibility of access to green spaces (Crompton, 2001; Duarte et al., 2017). Even though backyards are fundamental for the quality of life of this population, in terms of their income supplementation, food production and thermal comfort, increasingly dense settlements have progressively reduced backyards throughout Latin America (Fig. 8.4) and the paved backyards predominate (Angeoletto et al., 2017).

Fig. 8.4 Condominium Melchiades Figueiredo in Rondonópolis, Brazil. (Photos: T.E.P.N. Duarte, and FV Group, respectively)

On the other hand, some public policies have been successful in seeking to increase biodiversity in backyards, such as adopting tax incentives for maintaining green areas (Carvalho & Amaral, 2020).

Social housing projects are usually located in low value-added areas in the suburbs. These areas become a transition zone between the countryside and the city and should be better planned to preserve local diversity, with a greater variety of plant species attractive to native bird species, for instance. So, it is necessary to plan these spaces achiving maximum performance of their ecological function as well considering their multiple uses within each social reality.

Considering urban backyards for the preservation of biodiversity is an emerging need, however, allying it with urban development is an even greater challenge. Elaborating viable alternatives to increase urban biodiversity in tropical cities is as difficult a challenge as reducing deforestation in the countryside. However, the dissemination of knowledge about the importance of urban backyards for the preservation of biodiversity in urban areas, as well as the dissemination of affirmative experiences, are already the first step.

7 Final Remarks

Finally, what are the factors of biodiversity conservation and menaces to urban green spaces in Neotropical cities? First, the conservation of Neotropical urban green spaces is closely linked to the socioeconomic aspects of each city but subject to a frequent omission of the state, followed by the lack of skilled human resources for urban planning, as well as lack of mobilization of the stakeholders.

The sprawl dynamic in most Latin American cities did not deserve larger areas to biodiversity, and in most cases, large green spaces were only possible due to the presence of unhealthy areas for building, limiting the possibility of urban development. Latin American public urban policies have generally proven be inefficient in terms of maintaining regional biodiversity, and green infrastructure planning seldom apply conservation biology concepts. What is generally observed is little likelihood of expanding green spaces into continuous blocks, which would ensure biodiversity beyond generalist, peri anthropic, or synurbic species. In addition, street tree planting does not always follow technical and ecological criteria that allow functional connectivity between green spaces.

Cities should be designed with a minimum urban forest area capable to host the widest possible variety of taxa with viable populations, together with the functions they play and the ecosystems services they offer. However, very few green space projects in Latin America are comprehensive, to the point of aiming not only networks of ecological corridors and the connectivity between public spaces but also the quality of the green matrix at multiple scales. To achieve this, it would include yet the most numerous small private spaces as possible.

Some key concepts of conservation biology have not yet been tested in Neotropical urban green spaces, such as the role of exotic species in the ecosystem functioning, as well as measuring functional groups, and detecting its key species. In most studies, species richness in urban areas remains the focus in biodiversity assessment—a species-based approach that focuses on the unique role of each species in the community. More studies with a functional-type approach, based on common solutions presented by various species to the same environmental challenge are still needed.

References

Akinnifesi, F. K., Sileshi, G. W., Ajayi, O. C., Akinnifesi, A. I., Moura, E. G., et al. (2010). Biodiversity of the urban home gardens of São Luís city, North-Eastern Brazil. *Urban Ecosystem, 13*, 129–146. https://doi.org/10.1007/s11252-009-0108-9

Almeida, C. G., Moro, R. S., & Zanon, C. M. V. (2005). Dieta de duas espécies de morcegos frugívoros (Chiroptera, Phyllostomidae) em remanescentes florestais alterados na área urbana de Ponta Grossa, PR. *Publicatio UEPG, 11*(3/4), 15–21.

Amaya-Espinel, J. D., Hostetlewr, M., Henríquez, C., & Bonacic, C. (2019). The influence of building density on Neotropical bird communities found in small urban parks. *Landscape and Urban Planning, 190*, 103578. https://doi.org/10.1016/j.landurbplan.2019.05.009

Angeoletto, F. H. S. (2012). *Planeta ciudad: ecología urbana y planificación de ciudads medias de Brasil*. Thesis, Universidad Autónoma de Madrid.

Angeoletto, F. H. S., Ruiz Sanz, J. P., Albertin, R. M., & Da Silva, F. F. (2017). The grass is always greener on the other side of the fence: The flora in urban backyards of different social classes. *Ambiente e Sociedade, 20*(1), 1–20. https://doi.org/10.1590/1809-4422ASOC141293V2012017

Angeoletto, F., Fellowes, M. D., Essi, L., Santos, J. W., Johann, J. M., da Silva Leandro, D., & Mendonca, N. M. (2019). Ecología urbana y planificación: una convergéncia ineludible. *Revista Electrônica em Gestão, Educação e Tecnologia Ambiental, 23*(1), 1–8.

Angeoletto, F. H. S., Leite, L. B., Duarte, T. E. P. N., Johann, J. M., & Negri, I. (2020). The role played by exotic plants in urban ecosystems: Comments on Spennemann. *European Journal of Ecology, 6*(1), 13–15. https://doi.org/10.17161/eurojecol.v6i2.13432

Angeoletto, F., Tryjanowski, P., Santos, J., Martinez-Miranzo, B., Leandro, D., Bohrer, J., et al. (2022). Will Brazilian city dwellers actively engage in urban conservation? A case study with the charismatic neotropical Blue-and-yellow macaw (*Ara ararauna*). *Birds, 3*(2), 234–244.

Angeoletto, F. H. S., Richard, E., Santos, J., Guedes, N., Oliveira, S. L., Albertín, R. M., et al. (2023). When we think of urban afforestation in Brazil, what comes to mind is a mouth that has lost its teeth. *Terr@ Plural, 17*, e2321654.

Arduíno, M. B., Mucci, L. F., Santos, L. M., & Soares, M. F. S. (2020). Importance of micro-environment to arbovirus vector distribution in an urban area, São Paulo, Brazil. *Revista da Sociedade Brasileira de Medicina Tropical, 53*, e20190504. https://doi.org/10.1590/0037-8682-0504-2019

Banaszak-Cibicka, W., Raty'nska, H., & Dylewski, L. (2016). Features of urban green space favourable for large and diverse bee populations (Hymenoptera: Apoidea: Apiformes). *Urban Forestry & Urban Greening, 20*(1), 448–452. https://doi.org/10.1016/j.ufug.2016.10.015

Barona, C. O., Devisscher, T., Dobbs, C., Aguilar, L. O., Baptista, M. D., et al. (2020). Trends in urban forestry research in Latin America and the Caribbean: A systematic literature review and synthesis. *Urban Forestry & Urban Greening, 47*, 126544. https://doi.org/10.1016/j.ufug.2019.126544

Barros, R. S. M., Bisaggio, E. L., & Borges, R. C. (2006). Morcegos (Mammalia, Chiroptera) em fragmentos florestais urbanos no município de Juiz de Fora, Minas Gerais, Sudeste do Brasil. *Biota Neotropica, 6*(1), BN02206012006.

Beatley, T. (2017). Biophilic cities and healthy societies. *Urban Planning, 2*(4), 1–4. https://doi.org/10.17645/up.v2i4.1054

Bechara, F. C., Topanotti, L. R., & Silva, L. M. (2016). Aspectos da arborização urbana ecológica. *Revista Ibero-Americana de Ciências Ambientais, 7*(1), 49–55. https://doi.org/10.6008/SPC2179-6858.2016.001.0004

Bello, C., Cintra, A. L. P., Barreto, E., Vancina, M. H., Sobral-Souza, T., et al. (2020). Environmental niche and functional role similarity between invasive and native palms in the Atlantic Forest. *Biological Invasions, 13*, 741–754. https://doi.org/10.1007/s10530-020-02400-8

Beninde, J., Veith, M., & Hochkirch, A. (2015). Biodiversity in cities needs space: A meta-analysis of factors determining intra-urban biodiversity variation. *Ecology Letters, 18*(6), 581–592. https://doi.org/10.1111/ele.12427

Bensusan, N. (2002). *Seria melhor mandar ladrilhar? Biodiversidade: como, para que, por quê*. ISA/UNB Press.

Biondi, D. 2015. Floresta urbanas: conceitos e terminologias.

Bonthoux, S., Voisin, L., Vouché-Pillon, S., & Chollet, S. (2019). More than weeds: Spontaneous vegetation in streets as a neglected element of urban biodiversity. *Landscape and Urban Planning, 185*, 163–172. https://doi.org/10.1016/j.landurbplan.2019.02.009

Bozek, C. K., Prange, S., & Gehrt, S. D. (2007). The influence of anthropogenic resources on multi-scale habitat selection by raccoons. *Urban Ecosystem, 10*, 413–425. https://doi.org/10.1007/s11252-007-0033-8

Brown, A. (2020). Deer, Bear and everywhere: animals move into the city. *Stateline*. https://www.pewtrusts.org/en/research-and-analysis/blogs/stateline/2020/03/09/deer-bear-and-everywhere-animals-move-into-the-city

Callaghan, C. T., Bino, G., Major, R. E., Martin, J. M., Lyons, M. B., & Kingsford, R. T. (2019). Heterogeneous urban green areas are bird diversity hotspots: Insights using continental-scale citizen science data. *Landscape Ecology, 34*, 1231–1246. https://doi.org/10.1007/s10980-019-00851-6

Campagnoli, M. L., & Antunes, A. Z. (2017). Density of invasive exotic palms affecting the understory avifauna in the Atlantic Forest, Southeastern Brazil. *Neotropical Biology & Conservation, 12*(1), 37–47. https://doi.org/10.4013/nbc.2017.121.05

Caputo, S., Donoso, V. G., Izaga, F., & Britto, P. (2019). The democracy of green infrastructure: Some examples from Brazil and Europe. In F. Lemes de Oliveira & I. Mell (Eds.), *Planning cities with nature: Theories, strategies and methods* (pp. 137–152). Springer. https://doi.org/10.1007/978-3-030-01866-5_10

Carbó-Ramírez, P., & Zuria, I. (2011). The value of small urban greenspaces for birds in a Mexican city. *Landscape and Urban Planning, 100*(3), 213–222. https://doi.org/10.1016/j.landurbplan.2010.12.008

Carvalho, F. T., & Amaral, C. T. G. (2020). A extrafiscalidade tributária como instrumento para a concretização de políticas públicas: A construção de cidades sustentáveis e o estudo de caso IPTU Verde. *Revista de Direito à Cidade, 12*, 214–255. https://doi.org/10.12957/rdc.2020.46124

Chacalo, A., Aldama, A., & Grabinsky, J. (1994). Street tree inventory in Mexico City. *Journal of Arboriculture, 20*(4), 222–226.

Chollet, S., Brabant, C., Tessier, S., & Jung, V. (2018). From urban lawns to urban meadows: Reduction of mowing frequency increases plant taxonomic, functional, and phylogenetic diversity. *Landscape and Urban Planning, 180*, 121–124. https://doi.org/10.1016/j.landurbplan.2018.08.009

Costanza, R., de Groot, R., Sutton, P., der Ploeg, S., Anderson, S. J., et al. (2014). Changes in the global value of ecosystem services. *Global Environmental Change, 26*, 152–158. https://doi.org/10.1016/J.GLOENVCHA.2014.04.002

Crompton, J. L. (2001). The impact of parks on property values: A review of the empirical evidence. *Journal of Leisure Research, 33*(1), 1–31. https://doi.org/10.1080/00222216.2001.11949928

Cunha, A. A., Vieira, M. V., & Grelle, C. E. V. (2006). Preliminary observations on habitat, support use and diet in two nonnative primates in an urban Atlantic Forest fragment: The capuchin monkey (*Cebus* sp.) and the common marmoset (*Callithrix jacchus*) in the Tijuca Forest, Rio de Janeiro. *Urban Ecosystem, 9*(4), 351–359. https://doi.org/10.1007/s11252-006-0005-4

Davison, J., Huck, M., Delahay, R. J., & Roper, T. J. (2009). Restricted ranging behavior in a high-density population of urban badgers. *Journal of Zoology, 277*, 45–53.

Delgado, J. (1991). *Las nuevas periferias de la Ciudad de México*. Cocodes.

Díaz, S., & Cabido, M. (2001). Vive la différence: plant functional diversity matters to ecosystem processes. *Trends in Ecology & Evolution, 16*(11), 646–655. https://doi.org/10.1016/S0169-5347(01)02283-2

Dislich, R., & Pivello, V. R. (2002). Tree structure and species composition changes in an urban tropical forest fragment (São Paulo, Brazil) during a five-year interval. *Boletim de Botânica da Universidade de São Paulo, 20*, 1–11. https://doi.org/10.11606/issn.2316-9052.v20i0p1-11

Donoso, V. G. (2017). *Paisagem e cotidiano em habitação social nas Regiões Metropolitanas de São Paulo e Santiago do Chile*. PhD Thesis. University of São Paulo, São Paulo. https://www.teses.usp.br/teses/disponiveis/16/16135/tde-09062017-110211/en.php. Accessed 7 Feb 2022.

Duarte, T. E. P. N., Angeoletto, F. H. S., Richard, E., Vacchiano, M. C., Leando, D. S., et al. (2017). Urban afforestation in Brazil: A reflection of environmental injustice. *Terr@ Plural, 11*(2), 291–303. https://doi.org/10.5212/TerraPlural.v.11i2.0008

Escobedo, F. J., Kroeger, T., & Wagner, J. E. (2011). Urban forests and pollution mitigation: Analyzing ecosystem services and disservices. *Environmental Pollution, 159*(8–9), 2078–2087. https://doi.org/10.1016/j.envpol.2011.01.010

Esteves, M. C., & Corrêa, R. S. (2018). Natividade da flora usada na arborização de cidades brasileiras. *Cadernos de Arquitetura e Urbanismo, Paranoá, 22*, 159–171. https://doi.org/10.18830/issn.1679-0944.n22.2018.11

Estevo, C. A., Nagy-Reis, M. B., & Silva, W. R. (2017). Urban parks can maintain minimal resilience for Neotropical bird communities. *Urban Forestry & Urban Greening, 27*, 84–89. https://doi.org/10.1016/j.ufug.2017.06.013

Estrada, A., & Coates-Estrada, R. (2001). Species composition and reproductive phenology of bats in a tropical landscape at Los Tuxtlas, México. *Journal of Tropical Ecology, 17*, 672–646. https://doi.org/10.1017/S026646740100147X

Estrada, M. A., Coriolano, R. E., Santos, N. T., Caixeiro, L. R., Barbosa, A., et al. (2014). Influence of Urban green areas on the Ant Fauna. *Floresta e Ambiente, 21*(2), 162–169. https://doi.org/10.4322/floram.2014.035

Eyzaguirre, P. B., & Linares, O. F. (2004). *Home gardens and agrobiodiversity*. Smithsonian Press.

Faeth, S. H., Bang, C., & Saari, S. (2011). Urban biodiversity: Patterns and mechanisms. *Annals of the New York Academy of Sciences, 1223*, 69–81. https://doi.org/10.1111/j.1749-6632.2010.05925.x

Feeley, K. J., & Stroud, J. T. (2018). Where on Earth are the "tropics"? *Frontiers of Biogeography, 10*(1–2), 1–7. https://doi.org/10.21425/F5101-238649

Fernández, I. C., Wu, J., & Simonetti, J. A. (2018). The urban matrix matters: Quantifying the effects of surrounding urban vegetation on natural habitat remnants in Santiago de Chile. *Landscape and Urban Planning, 187*, 181–190. https://doi.org/10.1016/j.landurbplan.2018.08.027

Ferreira, J. R. P. (2016). *Uma nova variável no planejamento para o desenvolvimento urbano sustentável: Áreas verdes em quintais no subúrbio da cidade do Rio de Janeiro*. Thesis, Universidade Federal do Rio de Janeiro. http://www.ppe.ufrj.br/index.php/en/publicacoes/teses-e-dissertacoes/2016/206-uma-nova-variavel-no-planejamento-para-o-desenvolvimento-urbano-sustentavel-areas-verdes-em-quintais-no-suburbio-da-cidade-do-rio-de-janeiro. Accessed 7 Feb 2022.

Ferrini, F., Bussotti, F., Tattini, M., & Fini, A. (2014). Trees in the urban environment: response mechanisms and benefits for the ecosystem should guide plant selection for future plantings. *Agrochimica, 58*(3), 234–246. https://doi.org/10.12871/0021857201432

Filloy, J., Grosso, S., & Bellocq, M. (2014). Urbanization altered latitudinal patterns of bird diversity-environment relationships in the southern Neotropics. *Urban Ecosystem, 18*(3), 777–791. https://doi.org/10.1007/s11252-014-0429-1

Fonseca, S. N., Ribeiro, J. H., & Carvalho, F. A. (2013). Estrutura e Diversidade da Regeneração Arbórea em uma Floresta Secundária Urbana (Juiz de Fora, MG, Brasil*). Floresta e Ambiente, 20*(3), 307–315. https://doi.org/10.4322/floram.2013.031

Food and Agriculture Organization of United Nations (FAO). (2017). Urban and Peri-urban Forestry. http://www.fao.org/forestry/urbanforestry/87025/en/#:~:text=Urban%20forests%20can%20be%20defined,and%20trees%20in%20derelict%20corners. Accessed 12 Apr 2021.

Francis, R. A., & Chadwick, M. A. (2012). What makes a species synurbic? *Applied Geography, 32*(2), 514–521. https://doi.org/10.1016/j.apgeog.2011.06.013

Furlan, S. A. (2004). Paisagens sustentáveis: São Paulo e sua cobertura vegetal. In A. P. A. Carlos & A. U. Oliveira (Eds.), *Geografias de São Paulo* (pp. 255–283). Contexto.

Galetti, M., & Morellato, L. P. C. (1994). Diet of the large fruit-eating bat *Artibeus lituratus* in a forest fragment in Brasil. *Mammalia, 58*(4), 661–665. https://doi.org/10.1515/mamm.1994.58.4.657

Galluzzi, G., Eyzaguirre, P., & Negri, V. (2010). Home gardens: neglected hotspots of agrobiodiversity and cultural diversity. *Biodiversity and Conservation, 19*, 3635–3654. https://doi.org/10.1007/s10531-010-9919-5

Garza Villareal, G. (org.). (2000). *La Ciudad de México en el fin del segundo milenio*. Colegio de México/Gobierno del Distrito Federal.

Gilbert, O. L. (1991). *The ecology of urban habitats*. Chapman & Hall.

Girardet, H. (1989). *Ciudades*. Gayam.

Gonçalves, S. F., & Toledo, M. A. B. (2016). Aves em quintais e a relação com os proprietários de residências em área urbana da cidade de Jacareí, São Paulo, Brasil. *Revista Ambiente e Água, 11*(5), 1149–1162. https://doi.org/10.4136/ambi-agua

Gong, C., Chen, J., & Yu, S. (2013). Biotic homogenization and differentiation of the flora in artificial and near-natural habitats across urban green spaces. *Landscape and Urban Planning, 120*, 158–169. https://doi.org/10.1016/j.landurbplan.2013.08.006

González-García, A., & Gómez Sal, A. (2008). Private urban greenspaces or "patios" as a key element in the urban ecology of tropical Central America. *Human Ecology, 36*, 291–300. https://doi.org/10.1007/s10745-007-9155-0

Gorelik, A., & Silvestri, G. (2000). Ciudad y cultura urbana, 1976-99: el fin de la expansión. In J. L. Romero & L. A. Romero (Eds.), *Buenos Aires: historia de cuatro siglos*. Editorial Abril. v.2.

Green View Index (GVI). Disponible at http://senseable.mit.edu. Accessed 20 Apr 2021.

Hall, C. A. S. (2019). Urban ecology from a biophysical and systems perspective. In M. H. P. Hall & S. B. Balogh (Eds.), *Understanding urban ecology: An interdisciplinary systems approach* (pp. 27–57). Springer Nature. https://doi.org/10.1007/978-3-030-11259-2

Hardt, E., Pereira-Silva, E. F. L., Santos, R. F., Tamashiro, J. Y., Ragazzi, S., & Lins, D. B. S. (2013). The influence of natural and anthropogenic landscapes on edge effects. *Landscape and Urban Planning, 120*, 59–69. https://doi.org/10.1016/j.landurbplan.2013.08.014

Hartig, T., Mitchell, R., de Vries, S., & Frumkin, H. (2014). Nature and health. *Annual Review of Public Health, 35*, 207–228. https://doi.org/10.1146/annurev-publhealth-032013-182443

Heisler, G. M., & Brazel, A. J. (2019). The climate system. In M. H. P. Hall & S. B. Balogh (Eds.), *Understanding urban ecology: An interdisciplinary systems approach* (pp. 137–173). Springer Nature. https://doi.org/10.1007/978-3-030-11259-2_7

Herzog, C. P. (2016). A multifunctional green infrastructure design to protect and improve native biodiversity in Rio de Janeiro. *Landscape and Ecological Engineering, 12*(1), 141–150. https://doi.org/10.1007/s11355-013-0233-8

Hooper, D. U., & Vitousek, P. M. (1997). The effects of plant composition and diversity on ecosystems processes. *Science, 277*, 1302–1305. https://doi.org/10.1126/science.277.5330.1302

Hope, D., Gries, C., Zhu, W., Fagan, W. F., Redman, C. L., et al. (2003). Socioeconomics drive urban plant diversity. *Proceedings of the National Academy of Sciences, 100*, 8788–8792. https://doi.org/10.1073/pnas.153755710

Hostetler, M., Allen, W., & Meurk, C. (2011). Conserving urban biodiversity? Creating green infrastructure is only the first step. *Landscape and Urban Planning, 100*, 369–371. https://doi.org/10.1016/j.landurbplan.2011.01.011

Ibáñez-Alamo, J. D., & Soler, M. (2010). Does urbanization affect selective pressures and life-history strategies in the common blackbird (*Turdus merula* L.)? *Biological Journal of the Linnean Society, 101*, 759–766. https://doi.org/10.1111/j.1095-8312.2010.01543.x

Ikin, K., Beaty, R. M., Lindenmayer, D. B., Knight, E., Fischer, J., et al. (2013). Pocket parks in a compact city: How do birds respond to increasing residential density? *Landscape Ecology, 28*(1), 45–56. https://doi.org/10.1007/s10980-012-9811-7

Isernhagen, I., Le Bourlegat, J. M. G., & Carboni, M. (2009). Trazendo a riqueza arbórea regional para dentro das cidades: possibilidades, limitações e benefícios. *Revista da SBAU, 4*(2), 117–138. https://doi.org/10.5380/revsbau.v4i2.66277

Kattwinkel, M., Biedermann, R., & Kleyer, M. (2011). Temporary conservation for urban biodiversity. *Biological Conservation, 144*, 2335–2343. https://doi.org/10.1016/j.biocon.2011.06.012

Kirchner, F. F., Detzel, V. A., & Mitishita, E. A. (1990). *Mapeamento da vegetação urbana*. Anais do III Encontro Nacional Sobre Arborização Urbana.

Lawrence, A. B., Escobedo, F. J., Staudhammer, C., & Zipperer, W. (2012). Analyzing growth and mortality in a subtropical urban forest ecosystem. *Landscape and Urban Planning, 104*, 85–94. https://doi.org/10.1016/j.landurbplan.2011.10.004

Leal, E. S. B., Guerra Filho, D. Q., Ramalho, D. F., Silva, J. M., Bandeira, R. S., et al. (2019). Bat Fauna (Chiroptera) in an urban environment in the Atlantic Forest, northeastern Brazil. *Neotropical Biology & Conservation, 14*(1), 55–82. https://doi.org/10.3897/neotropical.14.e34837

Lepczyk, C. A., Aronson, M. F. J., Evans, K. L., Goddard, M. A., Lerman, S. B., & MacIvor, J. S. (2017). Biodiversity in the city: Fundamental questions for understanding the ecology of urban green spaces for biodiversity conservation. *Bioscience, 67*, 799–807. https://doi.org/10.1093/biosci/bix079

Leveau, C. M., & Leveau, L. M. (2005). Avian community response to urbanization in the Pampean Region, Argentina. *Ornitología Neotropical, 16*, 503–510.

Leveau, L. M., Leveau, C. M., Villegas, M., & Cursach, J. A. (2017). Bird communities along urbanization gradients: A comparative analysis among three Neotropical cities. *Ornitología Neotropical, 28*, 77–87.

Lope-Alzina, D. G., & Howard, P. L. (2012). The structure, composition, and functions of homegardens: Focus on the Yucatán Peninsula. *Etnoecológica, 9*(1), 17–41.

Lopez, B. E., Urban, D., & White, P. S. (2018). *Global Ecology and Biogeography*, 1–14. https://doi.org/10.1111/geb.12760

López-Flores, V., MacGregor-Fors, I., & Schondube, J. E. (2009). Artificial nest predation along a Neotropical urban gradient. *Landscape and Urban Planning, 92*(2), 90–95. https://doi.org/10.1016/j.landurbplan.2009.03.001

Lutinski, J. Á., Lopes, B. C., & Morais, A. B. B. (2013). Urban ant diversity (Hymenoptera: Formicidae) in ten cities of southern Brazil. *Biota Neotropica, 13*(3), 332–342. https://doi.org/10.1590/S1676-06032013000300033

MacGregor-Fors, I. J., & Ortega-Álvarez, R. (Eds.). (2013). *Ecología urbana: experiencias en América Latina*. http://www1.inecol.edu.mx/libro_ecologia_urbana/

MacGregor-Fors, I. J., Quesada, J., Lee, J. G.-H., & Yeh, P. J. (2017). Space invaders: House Sparrow densities along three urban-agricultural landscapes. *Avian Conservation and Ecology, 12*(2), 11. https://doi.org/10.5751/ACE-01082-120211

Martínez, D. N., & de la Barrera, E. (2020). Ecofisiología de la germinación de tres malezas efímeras periurbanas en Morelia, Michoacán, México. *La Granja, 31*(1), 47–55. https://doi.org/10.17163/lgr.n31.2020.03

Matthies, S. A., Rüter, S., Prasse, R., & Schaarschmidt, F. (2015). Factors driving the vascular plant species richness in urban green spaces: Using a multivariable approach. *Landscape and Urban Planning, 134*, 177–187. https://doi.org/10.1016/j.landurbplan.2014.10.014

McKinney, M. (2008). Effects of urbanization on species richness: A review of plants and animals. *Urban Ecosystem, 11*, 161–176. https://doi.org/10.1007/s11252-007-0045-4

Ministério de Meio Ambiente (MMA). (2021). Programa Cidades + Verdes. https://antigommagovbr/images/agenda_ambiental/areas/ProgramaCidadesVerdespdf. Accessed 20 Apr 2021.

Mittermeier, R. A., Gill, P. R., Hoffmann, M., & Pilgrim, J. (2004). *Hotspots revisited: Earth's biologically richest and most endangered terrestrial ecoregions*. CEMEX.

Møller, A. P., Diaz, M., Flensted-Jensen, E., Grim, T., Ibáñez-Álamo, J. D., et al. (2012). High urban population density of birds reflects their timing of urbanization. *Oecologia, 170*, 867–875. https://doi.org/10.1007/s00442-012-2355-3

Moreira, T. (2002). *A política habitacional e fundiária no Brasil. Seminário "Terra urbana para políticas sociais"*. LabHab e Lincoln Institute of Land Policy.

Muller, N., & Werner, P. (2010). Urban biodiversity and the case for implementing the convention on biological diversity in towns and cities. *Urban Biodiversity and Design*, 1–33. https://doi.org/10.1002/9781444318654.ch1

Nascimento, V. T., Agostini, K., Souza, C. S., & Maruyama, P. K. (2020). Tropical urban areas support highly diverse plant-pollinator interactions: An assessment from Brazil. *Landscape and Urban Planning, 198*, 103801. https://doi.org/10.1016/j.landurbplan.2020.103801

Nowak, D. J. (2019). The atmospheric system: Air quality and greenhouse gases. In M. H. P. Hall & S. B. Balogh (Eds.), *Understanding urban ecology: An interdisciplinary systems approach* (pp. 175–199). Springer Nature. https://doi.org/10.1007/978-3-030-11259-2_8

Pacheco, R., & Vasconcelos, H. L. (2007). Invertebrate conservation in urban areas: Ants in the Brazilian Cerrado. *Landscape and Urban Planning, 81*, 193–199. https://doi.org/10.1016/j.landurbplan.2006.11.004

Partecke, J., & Gwinner, E. (2007). Increased sedentariness in European Blackbirds following urbanization: A consequence of local adaptation? *Ecology, 88*(4), 882–890. https://www.jstor.org/stable/27651179

Pauchard, A., & Barbosa, O. (2013). Regional assessment of Latin America: Rapid urban development and social economic inequity threaten biodiversity hotspots. In T. Elmqvist et al. (Eds.), *Urbanization, biodiversity and ecosystem services: Challenges and opportunities. A global assessment* (pp. 589–608). Springer.

Pautasso, M. (2007). Scale-dependence of the correlation between human population presence and vertebrate and plant species richness. *Ecology Letters, 10*(1), 16–24. https://doi.org/10.1111/j.1461-0248.2006.00993.x

Peña Salmón, C. A. (2011). *Medotología para la planificación de áreas verdes urbanas: El caso de Mexicali, Baja California*. UABC Press.

Pereira, V. M. T. B. (2011). A herança da arquitetura africana nas comunidades quilombolas. Anais do XXVI Simpósio Nacional de História, 1–15 July, São Paulo, Brasil. http://encontro2014rjanpuhorg/resources/anais/14/1308185752_ARQUIVO_herancadaarquiteturaafricanapdf. Accessed 18 Apr 2021.

Perini, K., & Sabbion, P. (2017). *Urban sustainability and river restoration: Green and blue infrastructure*. John Wiley & Sons. https://doi.org/10.1002/9781119245025

Perrotti, D., & Iuorio, O. (2019). Green infrastructure in the space of flows: An urban metabolism approach to bridge environmental performance and user's wellbeing. In F. Lemes de Oliveira & I. Mell (Eds.), *Planning cities with nature: Theories, strategies and methods* (pp. 265–277). Springer. https://doi.org/10.1007/978-3-030-01866-5_18

Pretzsch, H., Bider, P., Uhl, E., Schütze, G., Perkins, D., et al. (2017). Climate change accelerates growth of urban trees in metropolises worldwide. *Science Reporter, 7*, 15403. https://doi.org/10.1038/s41598-017-14831-w

Primack, R. B. (2001). *Essentials of conservation biology*. Sinauer Associates.

Rastandeh, A., Brown, D. K., & Zari, M. P. (2017). *Biodiversity conservation in urban environments: A review on the importance of spatial patterning of landscapes*. Ecocity World Summit, 12–14 July. Melbourne.

Rayol, B. P., & Miranda, I. S. (2017). Influence of socioeconomic factors on plant species richness of home gardens of the Central Amazon. *Revista Espacios, 38*(46), 28.

Reis, N. R., Peracchi, A. L., & Onuki, M. K. (1993). Quirópteros de Londrina, Paraná, Brasil (Mammalia, Chiroptera). *Revista Brasileira de Zoologia, 10*(3), 371–381. https://doi.org/10.1590/S0101-81751993000300001

Ribeiro, C. L., Melo Jr, J. C. F., Mouga, D. M. D. S., & Carneiro, E. (2018). Phenology and floral visitors of *Psychotria nuda* (Rubiaceae) in an urban fragment of the Atlantic Forest in South Brazil. *Neotropical Biology & Conservation, 13*(3), 192–201. https://doi.org/10.4013/nbc.2018.133.02

Rolnik, R. (1997). *A cidade e a lei: legislação política urbana e territórios na cidade de São Paulo*. Studio Nobel.

Rosatte, R., Ryckman, M., Ing, K., Proceviat, S., Allan, M., et al. (2010). Density, movements, and survival of raccoons in Ontario, Canada: Implications for disease spread and management. *Journal of Mammalogy, 91*, 122–135. https://doi.org/10.1644/08-MAMM-A-201R2.1

Rossetti, A. I. N., Pellegrino, P. R. M., & Tavares, A. R. (2010). As árvores e suas interfaces no ambiente urbano. *Revista da SBAU, 5*(1), 1–24. https://doi.org/10.5380/revsbau.v5i1.66231

Roy, S., Byrne, J., & Pickering, C. (2012). A systematic quantitative review of urban tree benefits, costs, and assessment methods across cities in different climatic zones. *Urban Forestry & Urban Greening, 11*, 351–363. https://doi.org/10.1016/j.ufug.2012.06.006

Rudd, H., Vala, J., & Schaefer, V. (2002). Importance of backyard habitat in a comprehensive biodiversity conservation strategy: A connectivity analysis of urban green spaces. *Restoration Ecology, 10*(2), 368–375. https://doi.org/10.1046/j.1526-100X.2002.02041.x

Rumble, H., Angeoletto, F. H. S., Connop, S., Goddard, M. A., & Nash, C. (2019). Understanding and applying ecological principles in cities. In F. Lemes de Oliveira & I. Mell (Eds.), *Planning cities with nature: Theories, strategies and methods* (pp. 217–234). Springer. https://doi.org/10.1007/978-3-030-01866-5_15

Santana, L. D., Fonseca, C. R., & Carvalho, F. A. (2019). Ecological aspects of regenerating species of an urban forest with 150 years old of forest succession: The risk of exotic species. *Ciencia Florestal, 29*(1), 1–13. https://doi.org/10.5902/1980509830870

Santos, G. S., Ribeiro, I. C., Centoducatte, L. D., & Mendes, S. L. (2019). Bird breeding biology and homogenization process in an urban green area at Atlantic rainforest of southeastern Brazil. *Neotropical Biology & Conservation, 14*(1), 83–98. https://doi.org/10.3897/neotropical.14.e34838

Scherer, A. L., Scherer, J. F. M., & Petry, M. V. (2012). Distribution and trophic structure of birds in a petrochemical industrial gradient in southern Brazil. *Biological Conservation, 7*(3), 162–170. https://doi.org/10.4013/nbc.2012.73.03

Schlaepfer, M. A., Sax, D. F., & Olden, J. D. (2011). The potential conservation value of non-native species. *Conservation Biology, 25*(3), 428–437. https://doi.org/10.1111/j.1523-1739.2010.01646.x. PMid:21342267.

Schulze, M. D., Seavy, N. E., & Whitacre, D. F. (2000). A comparison of the phyllostomid Bat assemblages in undisturbed neotropical forest and in forest fragments of a slash-and-burn farming mosaic in Petén, Guatemala. *Biotropica, 32*(1), 174–184.

Secretariat of the Convention on Biological Diversity. (2012). Cities and biodiversity outlook: Actions and policy. Montreal. https://www.cbd.int/doc/health/cbo-action-policy-en.pdf. Accessed 22 Apr 2021.

Sedrez, L. (2013). Natureza urbana na América Latina: cidades diversas e narrativas comuns. In C. Leal, J. A. Pádua, & J. Soluri (Eds.), *Novas Histórias Ambientais da América Latina e do Caribe* (pp. 59–66). Rachel Carson Center.

Silva, M. S., & Ferreira, R. L. (2011). Succession of the richness, abundance and species composition of frugivorous insects in *Psidium guajava* (L.). *Neotropical Biology & Conservation, 6*(2), 103–111. https://doi.org/10.4013/nbc.2011.62.05

Slabbekoorn, H., & Ripmeester, E. A. P. (2008). Birdsong and anthropogenic noise: Implications and applications for conservation. *Molecular Ecology, 17*(1), 72–83. https://doi.org/10.1111/j.1365-294X.2007.03487.x

Smardon, R. C. (2019). Social processes, Urban ecosystems, and sustainability. In M. H. P. Hall & S. B. Balogh (Eds.), *Understanding urban ecology: An interdisciplinary systems approach* (pp. 59–82). Springer Nature. https://doi.org/10.1007/978-3-030-11259-2_3

Sol, D., Lapiedra, O., & Lagos, C. G. (2013). Behavioural adjustments for a life in the city. *Animal Behaviour, 85*(5), 1101–1112. https://doi.org/10.1016/j.anbehav.2013.01.023

Strassburg, B. B. N., Brooks, T., Fektran-Barbieri, R., Iribarrem, A., Crouzeilles, R., et al. (2017). Moment of truth for the Cerrado hotspot. *Nature Ecology & Evolution, 1*, 0099. https://doi.org/10.1038/s41559-017-0099

Sun, N., Limburg, K. E., & Hong, B. (2019). The urban hydrological system. In M. H. P. Hall & S. B. Balogh (Eds.), *Understanding urban ecology: An interdisciplinary systems approach* (pp. 119–136). Springer Nature. https://doi.org/10.1007/978-3-030-11259-2_6

Tainter, J. (2019). Scale and metabolism in ancient cities. In M. H. P. Hall & S. B. Balogh (Eds.), *Understanding urban ecology: An interdisciplinary systems approach* (pp. 85–99). Springer Nature. https://doi.org/10.1007/978-3-030-11259-2_4

Threlfall, C. G., Williams, N. S. G., Hahs, A. K., & Livesley, S. J. (2016). Approaches to urban vegetation management and the impacts on urban bird and bat assemblages. *Landscape and Urban Planning, 153*, 28–39. https://doi.org/10.1016/j.landurbplan.2016.04.011

Torres, H. (1978). El mapa social de Buenos Aires en 1943, 1947 y 1960. *Desarollo Económico, 18*(70), 163.

Torres, H. (1993). *El mapa social de Buenos Aires (1940–1990)*. FADUUBA.

Torres, H. (1997). *Procesos recientes de fragmentación socioespacial en Buenos Aires: la suburbanización de las elites. Seminário "El nuevo milenio y lo urbano"*. Universidad de Buenos Aires.

Torres, H. G., Marques, E., Ferreira, M. P., & Bitar, S. (2003). Pobreza e espaço: padrões de segregação em São Paulo. *Estudos Avançados, 17*(47), 97–128. https://doi.org/10.1590/S0103-40142003000100006

Tzoulas, K., Korpela, K., Venn, S., Pelkonnen, V. Y., Kazmierczak, A., et al. (2007). Promoting ecosystem and human health in urban areas using green infrastructure: A literature review. *Landscape and Urban Planning, 81*(3), 167–178. https://doi.org/10.1016/j.landurbplan.2007.02.001

Unikel, L. (1976). *El desarollo urbano de México*. El Colegio de México.

United Nations. (2014). *World urbanization prospects: The 2014 revision, Highlights* (ST/ESA/SER.A/352).

Velez-Ruiz, R. I., Gonzalez, V. G., & Engel, M. S. (2013). Observations on the urban ecology of the Neotropical stingless bee *Tetragonisca angustula* (Hymenoptera: Apidae: Meliponini). *Journal of Melittology, 15*, 41647. http://orcid.org/0000-0003-3067-077X

Vergara, G., & Ibarra, J. T. (2019). Paisajes en transición: Gradientes urbano-rurales y antropización del bosque templado andino del sur de Chile. *Revista de Geografía Norte Grande, 74*, 139–158.

Vrecenak, A. J., Vodak, M. C., & Fleming, L. E. (1989). The influence of site factors on the growth of urban trees. *Journal of Arboriculture, 15*, 206–209.

Weimer, G. (2020). Arquitetura popular afro-brasileira. *Em Questão, 26*, 291–316. https://doi.org/10.19132/1808-5245260.291-316

Williams, J. N. (2013). Humans and biodiversity: Population and demographic trends in the hotspots. *Population and Environment, 34*(4), 510–523. https://doi.org/10.1007/s11111-012-0175-3

Yeh, P. J. (2004). Rapid evolution of a sexually selected trait following population establishment in a novel habitat. *Evolution, 58*(1), 166–174. https://doi.org/10.1111/j.0014-3820.2004.tb01583.x

Zortéa, M., & Chiarello, A. G. (1994). Observations on the big fruit-eating bat, *Artibeus lituratus*, in an urban reserve of Southeast Brazil. *Mammalia, 58*(4), 665–670.

Chapter 9
Green and Gray Grids: Urban Network, Natural Resources and the Construction of a Sustainable Landscape in Northern Paraná State, Brazil

Karin Schwabe Meneguetti, Renato Leão Rego, Gislaine Elizete Beloto, and Mayara Henriques Coimbra

Abstract The network of planned new towns in the tropical northern Paraná state, Brazil, shows a rather surprising functional coexistence between human occupation and environmental flows, a current pressing concern in urban planning and design. Stemming from an overseas speculative enterprise conducted on privately owned land and oriented by machine-age reasoning, the urban network originally depicted organic urban grids positioned on the main ridges, as well as the regular parceling of rural areas along natural lines. Despite the massive deforestation that took place for agricultural purposes, this venture created a regional landscape based on the garden city model, which has led to an amicable relationship between built-up areas and the natural environment. This chapter explores the historical events behind the establishment of that urban network and its relationship with natural resources, so as to trace the formation and evolution of green patches and green corridors, not only in the countryside but also within cities. Thus, the potentialities of connected green spaces in these urban forms will be addressed, since they are known to affect the quantity and quality of animal species that inhabit urban spaces.

Keywords New towns · Urban network · Regional landscape · Garden-city · Green corridors · Sustainability, Northern Paraná state

This chapter is based upon a research financed in part by the Coordenação de Aperfeiçoamento de Pessoal de Nível Superior—Brasil (CAPES)

K. S. Meneguetti (✉) · R. L. Rego · G. E. Beloto · M. H. Coimbra
Departamento de Arquitetura e Urbanismo, Universidade Estadual de Maringá, Maringá, PR, Brazil
e-mail: ksmeneguetti@uem.br

F. Angeoletto et al. (eds.), *Ecology of Tropical Cities*,
https://doi.org/10.1007/978-3-031-48012-6_9

1 Introduction

The survival of wildlife species depends on the quality of their habitat, the supply of food and water, and the primordial ability to move around the landscape (Angeoletto et al., 2022; Moroń et al., 2024). Movement is required to provide favorable environments and access to food, shelter, and adequate reproduction areas. Urbanization can negatively interfere with this dynamic balance, as it fragments landscapes and changes natural processes. Cities conform a favorable habitat for human development that has stimulated the progressive urbanization of populations. Nonetheless, while cities have become the problem, it is within them that solutions for natural imbalances should be sought.

Environmentally-based urban planning focused on sustainability can make use of historical references and success stories in the attempt to find the balance between urbanization and environmental protection and to predict better scenarios. There is a consensual perception that landscape connectivity is a determining factor in achieving such balance. While gray connections allow man to appropriate the territory to develop their activities, green connections provide for natural processes across the landscape.

Odum (2007) and Lyle (1994) state that, by establishing an interconnected network of green spaces along the landscape, a system capable of dealing with important environmental and ecosystem issues, such as environmental conservation, sustainable development, water and air quality, as well as the flow of species is intuitively formed (Ribeira, 2019). Since the nineteenth century, landscape architects, such as Frederick Law Olmsted, Calvert Vaux and Horace Cleveland, have taken advantage of the concept of connectivity to create urban leisure systems through the integration of open spaces (Ribeira, 2019). The connectivity created by these systems has proven to be effective in maintaining ecological movement and flows at various levels, with minimal or no disruption caused by the road system. Connectivity allows not only the link between natural portions of the territory, but also between city and nature (Magalhães, 2001).

In this sense, the urbanization of northern Paraná State, Brazil, is presented in this chapter as a noteworthy instance. From a commercial enterprise aiming at profit, an example of urban and rural development was born with an environmental quality that outshines other similar activities in Brazil. This chapter presents the foundations on which the enterprise was formed and the premises that have enabled the construction of an exemplary landscape. At both regional and urban scales, decisions based on environmental issues, or just arbitrarily proposed, were responsible for the configuration of the landscape. Thus, it would be important to understand those decisions that have resulted in the outcomes that are more relevant to current issues, and how they can support a more intelligent future scenario that takes into consideration all animal species and their permanence in the territory.

2 Regional Planning in Northern Paraná—The Design of Grey and Green Grids

The design of gray and green grids in northern Paraná state can be traced back to the British garden-city movement. The private development and colonization of this region was initiated by a British company in the 1920s, when the idea of urban and regional planning, then intensely debated in England, was translated into the garden-city framework of social cities.

An experienced and visionary entrepreneur commanded the initiative in northern Paraná after visiting the region in 1924. Lord Lovat (Simon Fraser, 1871–1933) was a Scottish aristocrat, member of the British parliament and enthusiastic imperialist. He had already pioneered in colonization and settlement in Swaziland, and was one of the investors in cotton farming in the Ghezira scheme, in Sudan. His visit to Brazil was the result of a mission by British bankers to investigate the country's economic situation in response to a request for a large loan by the Brazilian government. As an explorer, Lord Lovat soon realized the potential of the red lands in northern Paraná, which were still covered by native forest (Rego, 2009).

The initial idea of growing cotton in the region was abandoned due to the low competitiveness of the Sudanese produce output. But coffee eventually became the solution to the problem. Coffee growing was expanding into the hinterlands of the state of São Paulo, reaching the banks of the Paranapanema River on the border with Paraná. Notably, the history of northern Paraná begins as "a chapter of São Paulo's life in Paraná territory" (Müller, 1956: 55). In fact, although administratively belonging to the state of Paraná, this region was considered under the economic influence of the state of São Paulo (Monbeig, 1984: 221). Coffee plantation in the neighboring state triggered the colonization of the left bank of the Paranapanema River (Fresca, 2007: 212; Luz, 1997: 13; Katzman, 1977: 69), and all coffee production, whether from northern Paraná or the west of São Paulo, converged to a technical, financial and mercantile hub—the city of São Paulo—and then to the port of Santos. Lord Lovat knew how to make use of the pioneering development of private colonization schemes in association with the British imperialist experience, and systematically planned the occupation of northern Paraná.

Lord Lovat acquired from the impoverished Paraná government an area of approximately 372 million m^2, equivalent to Kent, Sussex and Hampshire together. The development required the parceling of the territory into rural lots, the provision of infrastructure, opening roads, the occupation by settlers, and obviously the creation of towns. A railroad was also essential to ensure that the agricultural production reached the consuming markets. Due to the lack of adequate physical-territorial planning, Lord Lovat had faced several problems in his previous colonization venture in southern Africa in early twentieth century. That experience motivated him to listen to the recently emerging science of town planning, especially to its main movement in England. It is not unreasonable to imagine that the ideas proposed by Ebenezer Howard, a Parliament official, contained in his book "Garden Cities of Tomorrow", published in 1902, served as a reference for the colonization plan of northern Paraná. The idea was to order the territory and plan new and better towns to attract people back to the countryside.

The cities idealized by Howard would add the advantages of the urban life within the rural area by connecting one to the other. Low population density and abundant green areas should provide the environmental quality capable of solving many of the problems that affected the post-industrial revolution British cities and the late Victorian society. Called garden cities, together they would function as social cities, sharing opportunities and services. According to Howard, social cities would be

> a carefully-planned cluster of towns, so designed that each dweller in a town of comparatively small population is afforded, by a well-devised system of railways, waterways, and roads, the enjoyment of easy, rapid, and cheap communication with a large aggregate of the population, so that the advantages which a large city presents in the higher forms of corporate life may be within the reach of all, and yet each citizen of what is destined to be the most beautiful city in the world may dwell in a region of pure air and be within a very few minutes' walk of the country (Hall and Ward, 1998: 23).

A leading British planner recognized in 1925 that the two cities created by Howard—Letchworth Garden City and Welwyn Garden City—

> were undertaken as home colonization schemes. They ought to be considered as providing models for the organization of the planned transfer of communities to Australia, New Zealand, Canada, and the African colonies (…). If emigration were to take the form of the establishment of new industrial towns, not villages but real towns, each with its agricultural belt, we should have the basis of a new population policy for the British Commonwealth (…) The two garden cities show how it could be done (Purdom, 1949: 483).

Lord Lovat's company developed a regional planning of unprecedented scale and nature in Brazil. The deliberate process of systematic colonization involved the parceling of the soil, the establishment of new towns, the construction of a railway, and the provision of infrastructure. Although Lovat never came to see the final result of his initiative, when the British company was liquidated in 1944, nine cities had been founded along approximately 80 km of railway. They were cities for approximately 30,000 inhabitants, equivalent in form and infrastructure, independent and equidistant, separated by a green belt and connected by the railway, which thus resembled the Howardian scheme of social cities (Rego, 2011).

The threat of nationalization of the railway—which eventually happened—was responsible for the liquidation of the British company. The company's new Brazilian directors, including two polytechnic Engineers from São Paulo, adapted the colonization scheme, which then no longer had control over the railway. Intending at a quicker and less costly regional occupation, the company's management adopted the British satellite city model, widely discussed in São Paulo, especially in the Polytechnic School of Engineering and Architecture.

Thus, from mid-1940s, the British scheme for population decentralization and occupation of the territory through the creation of satellite cities was applied to further the colonization of northern Paraná. Numerous small, interconnected and dependent cities were then founded establishing a regional urban hierarchy (Beloto, 2015; Macedo, 2011; Rego & Meneguetti, 2010).

A report from the colonizing company states that four main cities with intermediate urban centers were planned. The main cities, arranged at every 100 km along the railway line, would function as regional centers. In addition to the nine existing cities, others were created, each with approximately 10,000 inhabitants. Small urban centers were implanted between them at a distance of 10 to 15 km. These urban centers were designed for 1,000 or 5,000 people and, like the larger cities, also had a green belt. Together, major cities, towns, and villages created an interdependent and service sharing urban region (Companhia, 1975: 77–79; Macedo, 2011). Such structure ensured the continuity of the close city-countryside relationship that had marked the new towns since their origin.

The cities of northern Paraná were implanted along roads on the main ridges. They were located on prominent points of the terrain and their original green belt—consolidated by small farms circumscribing the urban form—coincided with the nearby springs. They were, therefore, cities on the high ground, never cities in the valleys, to avoid problems resulting from the proximity to watercourses, such as seasonal floods, insects, and diseases. The railway station, almost always positioned on a straight stretch, indicative of flatter terrain, was the starting point for the layout of the future city. Thus accommodated, the cities assumed irregular shapes, according to the sites they occupied, limited by steeper slopes and springs, which would be protected by the green belt to be created around the urban form. The green belts, made up of farms that would supply each city with fresh produce, not only limited the size of each urban area, but also protected the surrounding springs (Rego, 2019: 96: 115).

Although the grid was the recurring pattern in the design of these cities, it was always adapted to the relief of the site and customized with characteristic formal motifs in order to generate unique urban forms. The orthogonality that marks the layout of so many towns is not only a simple formula for the urban arrangement, it is also a way of creating order. Since the Renaissance, it has been understood that cities with regularly arranged streets, blocks and buildings would have order. But this principle of order, regularity and simplicity provided by the rectilinear street layout at right angles can also become rigid and dull if applied mechanically. Because the grid is neutral and non-hierarchical, it seems monotonous until it is imbued with specific contents in order to become unique, and have a spatially attractive hierarchy. The orthogonal layout of cities in northern Paraná acquired this specific content and, therefore, abandoned neutrality by being diligently adapted to the site, having a peculiar center, and sensibly incorporating artistic arrangements of elementary morphological elements, such as blocks, tree-lined streets and open spaces, as well as the implantation of buildings formatted and grouped according to classic precepts of urban composition (Rego, 2019: 113–114). Furthermore, when adapting the layout to the terrain, urbanization slightly gave way to natural elements, reducing its interference on watercourses.

The practicality of the grid also served for the parceling of rural areas, following a 'long plot' approach, supposedly of German origin, introduced in southern Brazil. This system has the advantage of using the natural lines of the terrain for land division into plots: long and, in theory, rectangular, delimiting strips of land that go from the watercourses in the valleys to the ridges. The roads were designed to run between the watercourses similarly to the railway, which ran along the main ridges. Along the watercourses, riparian forest would be maintained and subsistence orchards planted. As a result, an organic system generated from orthogonal lines adapted to topographic circumstances can be observed.

Small lots favored the occupation by family farmers, targeted by the colonizing company, to result in a larger rural population that generated demand for urban services. Lord Lovat calculated that, by allowing eight families per square kilometer of agricultural land, with at least half of this population installed in cities and towns tending to agriculture auxiliary services, this rural parceling system would provide the region the capacity to support a population between one and two million inhabitants when fully developed (Rego, 2019: 96).

The connection between cities and countryside was evident in the regular proximity between urban centers. These nuclei were, on average, fifteen kilometers apart, so that, at midpoint, an individual could easily reach a neighboring city on horseback or on foot. Although the regularity of short intervals between railway stations facilitated refueling, the briefly spaced stops were not directly related to the railway's technical demands. Actually, they responded to territorial planning. This is evidenced by the fact that in railway stretches outside the colonizing company lands, the train did not stop at the same regular short intervals. The proximity between cities constrained the possibility of large rural properties being formed, and their connections facilitated rural life, the use of urban services, and the flow of rural production (Rego, 2019: 97).

The individualized urban forms, separated by green belts, close to each other and regularly spaced, connected to the countryside and among them, made the northern Paraná colonization scheme a successful case of urban and regional planning. The colonization numbers in northern Paraná demonstrates the success of the enterprise. From a technical point of view, this positive evaluation stems from the successful colonization experience and coordinated regional planning (Dozier, 1956; Kaztman, 1977, 1978). Almost a century after the British enterprise began, northern Paraná is still seen as an exemplary case of planned urbanization and systematic colonization.

With the expansion of capitalism in Brazil in the first half of the twentieth century, new cities were created to promote national development, advance the occupation of the national territory, and to support the emerging rural economy. In the transformation from a natural into an urban landscape, the city-nature binomial relationship played a role that was different from that seen in northern Paraná. The preservation of green areas as well as the idea of sustainability was secondary to development. On the new agricultural frontiers, colonization efforts were often directed at the unconditional transformation of natural landscapes. Progress, always viewed positively, manifested itself as the natural environment was transformed into cultivable land, and even more so into urbanized areas (Rego, 2019: 76–77). In contrast, what was seen in northern Paraná, was an amicable relationship between built-up areas and nature, despite the massive deforestation that had initially taken place for agricultural purpose.

3 Sustainable Landscape, Not So Sustainable Development

With an urbanization rate averaging 97%, the axis connecting the two largest cities in northern Paraná—Londrina and Maringá—housed an urban population of 1,688,636 inhabitants in the year 2020 (IBGE, 2020). The number of people living in the cities that make up the backbone of this urban network is 15 times larger than that expected in the 1950s, when the regional conformation was planned. Certainly, this population growth has led to significant changes in the shape of the cities as a whole. Until the 1960s, the polynucleated urban structure with defined boundaries, together with the railroad and the highways, created a connected regional infrastructure, with intense economic and social relationships among the cities. The clusters of cities benefited from the regional structure and expanded along it (Ribeiro & Beloto, 2020). Changes in agricultural practices, however, had a direct impact on this balance.

The rural matrix, in principle intended to support coffee production by family members, consisted of a mosaic of rural plots of regular shape and size. Mainly from the 1970s onwards, that scenario was significantly altered by mechanization and the monoculture of soybeans and wheat, which resulted in the enlargement of rural properties. The rural population started to migrate from smaller urban settlements to larger cities, significantly altering relations of interdependency and the balance within the network.

Between 1960 and 1990, both Londrina and Maringá expanded in a north-south direction, transversally to the axis that connects them, through the establishment of large neighborhoods. However, encouraged by the concept of development corridors by Charles Frederick John Whebell, from the late 1960s, the plans elaborated by the State government during the next decade focused on industrialization as a solution for urban development, as well as an economic response to the import model (Beloto, 2015). As a result, the urban form also began to spread along the logistical axis, where areas for industrial use and storage of grains were concentrated.

Metronor—Northern Paraná Linear Metropolis (*Metrópole Linear do Norte do Paraná, in Portuguese*)—was the name given to the successive plans developed by the State government during the 1970s for the Londrina-Maringá axis. They envisaged the complete conurbation of all the entities involved in the urban and regional dynamics along the railway structure at a time when most prominent urban forms on the axis were still concise and fragmented only at the edges (Coimbra & Beloto, 2020; Ribeiro & Beloto, 2020). It was believed at the time, that the force emanating from the transport structure responsible for the main connection between the cities, would be sufficient to form an urban continuum from Londrina to Maringá. Thus, from the 1990s onwards, another territorial arrangement has emerged, different from the polynuclear structure initially planned by the colonizing company. Nevertheless, despite the strength of the road structure envisaged by Metronor, the topographical constraints and the shape of the lots that made up the rural matrix remained as morphological determinants of expansion.

Suburbanization has become evident from the year 2000. Urban fragments not only spread along the main transport routes, but also scattered over increasingly larger areas away from the main cities. The effect of cultural movements such as the 'gated condominiums' and 'neo-ruralism' can be clearly observed in these changes. The expansion, eminently directed by the road structure between Londrina and Maringá, also corresponds to the expansion of the other cities on the axis. Albeit to a lesser extent, the constitution of linear urban continuum can also be seen throughout the region. The expansion of cities to the north of the traditional axis and, above all, the suburbanization resulting from the cities on the axis, should consolidate a new regional arrangement in northern Paraná in the coming years, transforming the Londrina-Maringá axis into an urban ring (Fig. 9.1).

Despite the transformations observed in the regional organization, the network of green corridors is the ecological fabric that still environmentally sustains the region. The State of Paraná was extensively deforested for agricultural exploitation, and the vegetation that covered northern Paraná was decimated in the first half of the twentieth century, with the introduction and expansion of agriculture. The 20% of the area of the rural plots that should be reserved for the maintenance of the native forest, initially established by the colonizing company, was progressively extinguished in the region over the last decades. The replacement of the forest by monocultures "created the conditions for Antarctic polar air fronts that wrecked the coffee plantations, causing a deep economic crisis" (Meneguetti, 2009: 74). After that, the extensive and mechanized agricultural model adopted from the 1970s in replacement to coffee cultivation, exposed the soil to erosive processes. These events forced new soil conservation policies, among which the protection of river basins and the replanting of the riparian forest (Fleischfresser, 1999). Basically, what stands out in the natural landscape nowadays is the riparian vegetation along the hydrographic network, showing the dendritic pattern of natural drainage (Figs. 9.2 and 9.3).

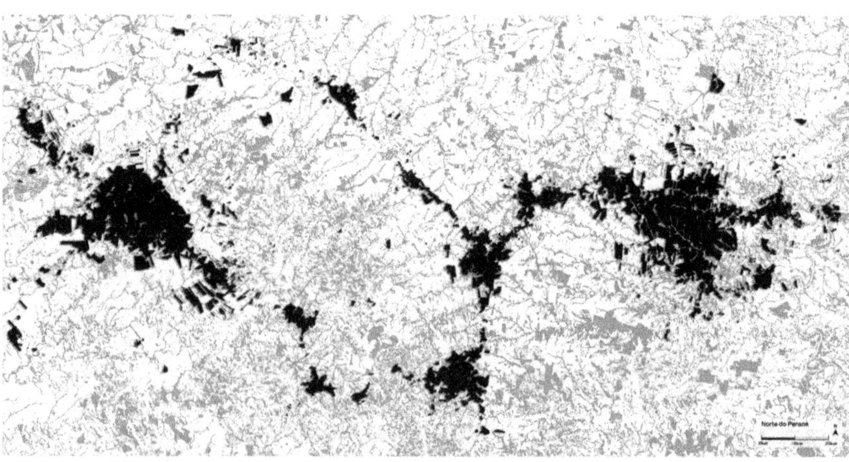

Fig. 9.1 Londrina-Maringá axis and urban ring. (Organised by the authors. Base: Googlemaps)

Fig. 9.2 Green corridor in the rural area between Londrina and Maringá. (Source: Grupo de Pesquisa Cidade e Paisagem [City and Landscape Research Group])

Fig. 9.3 Green corridor within the city of Maringá. (Source: Grupo de Pesquisa Cidade e Paisagem [City and Landscape Research Group])

In addition to the reforestation along watercourses, northern Paraná cityscapes also present small urban parks, significant vegetation along the streets shaping the urban landscape, and several remnant areas covered with the original vegetation scattered in the countryside. Thus, the landscape resulting from the combined forces at play reveals a mosaic composed of cultivated, urban and forested areas. While the urban ring occupies the ridges, the valleys shelter the forested reserves, in a condition similar to the model Forman (1995: 309) calls 'interdigitated'.

In general, the rural matrix as it stands—geometrically defined rural plots, direct access to the central areas of all cities in the Londrina-Maringá axis, and minimal natural obstacles—facilitates the expansion of urban forms. Similarly, small forested areas in the rural environment can also attract urban development, as natural features enhance the landscape and so, increase real estate value. These natural environments might thus be maintained over time, which does not necessarily mean that urbanization will not eventually affect them.

4 Green and Gray Urban Grids—The City of Maringá

The city of Maringá stands out among the other cities in the northern Paraná network for the environmental quality of its urban layout. Maringá was designed when the colonization enterprise began to demonstrate its economic viability, and marketing actions catapulted the success of sales of rural and urban plots. The colonization company hired an engineer and town planner already known for urbanization projects in the State of São Paulo, Jorge de Macedo Vieira, to design the new city. In Maringá, Vieira could practice the same 'hybrid' urban planning he had employed to a lesser extent in previous neighborhood and city layouts, adapting English garden city formal principles to local conditions.

The colonizing company realized the relevance of an urban form that considered natural features and the presence of urban green to sell the dream of quality of life in the new agricultural frontier. Three fundamental points were taken as reference in the choice of the site and later in the layout of the city: the future railway line crossing in the east-west direction, and two small valleys to the south. The natural conditions determined the urban form: parks were created around the two springs in the valleys to the south, and the residential streets were defined by the topographic characteristics of the terrain (Meneguetti, 2007, 2009, 2017; Meneguetti & Rego, 2008; Angeoletto, 2012).

Surrounding the urban layout, the rural area was divided into farms of approximately 24,200 m^2, with an orthogonal configuration reminiscent of the green belts advocated by Ebenezer Howard in the garden city movement (Howard, 1965). Although the green belt was also a feature around the other cities of the Company, in Maringá a more evident formal intention can be observed, with the size and format of plots compatible with the expected crop production.

As one of the two cities that concentrated the rural exodus and large real estate developments, Maringá has experienced moments of great urban expansion, both horizontal and vertical. Vertical buildings of up to forty floors were first built in the city center, and later in adjacent neighborhoods. Yet, town planning in Maringá has been able to maintain the city's environmental quality and guarantee minimum urbanization standards in expansion areas (Fig. 9.4).

The city's Master Plan drawn up in 1967 determined areas of environmental protection along urban streams, many of which responsible for connecting the parks in the city. Then, the 1979 Street Guidelines Plan established strips of vegetation 120 meters wide along all watercourses, corresponding to green corridors. These corridors ensured that green areas previously separated from each other in the original project were connected, as well as the conservation and protection of environmentally fragile areas, and above all, survival prospects of protected zones of the initially designed area (Meneguetti, 2007, 2009, 2017; Meneguetti & Rego, 2013).

Along the lines of urban green infrastructure theories, the system of open spaces created by the initial design and later augmented by these two city plans has resulted in an efficient ecological structure (Meneguetti, 2007). The establishment of an interconnected green infrastructure within the city of Maringá has been extremely important since the fragmentation of the landscape is recognized as one of the most

Fig. 9.4 Green and Gray urban grids of Maringá. (Source: LURe - Laboratório de Urbanismo Regional [Regional Urbanism Lab])

incisive factors that incur in the loss of the quality of natural habitats and, consequently, the diversity of urban environment. Additionally, the fragmentation caused by urbanization can also prevent or hinder natural flows and movements, shrinking habitats and species that are unable to live isolated in patches (Jongman & Pungetti, 2004; Odum, 2007).

The impact of landscape fragmentation has been studied in a significant way in environmental and ecosystem issues linked to the studies of the cities. Studies presented by Forman and Godron (1986), Lyle (1994), Ahern (1995), and Odum (2007) suggest that isolation and fragmentation must give way to connectivity, and that this connectivity should progress from a local to a global scale. Urban green corridors can be effective in this mission: "These corridors have greatly influenced the migration and perpetuation of wildlife in cities" (Hough, 1995: 172) (Fig. 9.5).

In the case of the city of Maringá, a green infrastructure is well established, with public open spaces (mainly riparian corridors and parks) protected by urban legislation in the form of "protection zones", and tree-lined streets and avenues with central islands forming a continuous canopy. This system of open spaces, in its current status, can be enhanced and complemented by the activity of ecological planning, in order to face the challenges, not always friendly, imposed by the relationship between human occupation and natural processes. The idea of a network connecting green corridors, forested areas, and other punctual interventions, dealing with water

Fig. 9.5 Green corridors in northern Paraná cities. (Source: LURe - Laboratório de Urbanismo Regional [Regional Urbanism Lab])

management and enhancing open spaces system to provide environmental services, is a way to face this delicate relationship (Meneguetti, 2017).

Although the riparian corridors have evolved significantly with increasing vegetation cover over the last decades, the green corridors formed by the street canopy have lost quality with decreased tree coverage. Alongside with the progressive loss of permeability in squares, open-space subsystems suffer from changes in the flow of matter and energy (Ribeira, 2019).

5 Lessons Learned

In a changing world, where the possibility of life is continuously and progressively being challenged, landscape planners play a unique role. It is paramount that knowledge gathered from sciences such as History, Geography, Biology, Ecology, Sociology and others are combined together in the definition of human activities that are more compatible with the maintenance of the balance of natural processes.

Despite the inevitability of urbanization for the accommodation of human beings, cities do not always need to be seen as an antagonist to the environment. Processes, changes and interventions need to be clearly understood in order that smarter and more sustainable cities with more diversity of life can be planned.

The examples presented in this text are the result of experiences initiated in the last century based on foreign ideas and concepts adapted to local conditions. Although considered a success in many fronts, results have not always been so positive. By looking at the examples presented, important lessons can be drawn when planning the cities of the future.

One of the aspects that stands out is adaptation. The import of ideas is only valid when selectively adapted to local conditions. Regional planning needs to be adjusted to the morphology of the territory. City layouts must naturally emerge from the location, and enhance the qualities of the landscape. Planning should provide for amendments over time due to economic and social changes without compromising its initial qualities. Because of the dynamic and evolutionary characteristics of the landscape, flexibility and adaptability to changes are essential to sustainability-based planning (Ahern, 2013).

However, the pragmatism of establishing a cluster of towns in northern Paraná has brought some negative consequences as well. These happened over the last decades based on an urban plan designed to occupy the land and establish an uninterrupted network of urban roads. In an attempt to ensure urban mobility, road planning neglected natural structures, fragmenting some of the green corridors. The cities of the cluster were distributed on ridges to provide the best terrain and conditions for the implantation of the railway line, and avoid problems with floods. Although areas where the land is flat or present low declivity favor the reticulated urban layout with less environmental impact, paving the soil in these areas can prevent stormwater from returning to the subsoil.

Another relevant aspect is connectivity. Clear understanding of natural processes is crucial. Planning should consider movements and flows (of water, wildlife, people) to avoid the fragmentation of the landscape. In this sense, the concept of green grid stands out, because narrow green corridors can be economically more viable to be created and maintained and yet act properly if extensively connected (Jongman & Pungetti, 2004; Lemes de Oliveira, 2017). In the cases presented, the green and gray grids have played a crucial role in the sustainability of the initial plan and can lead to adequate expansions provided they are understood and reinforced in subsequent plans.

Nonetheless, connectivity is also about to face a new challenge: the advance of the rapidly expanding urban tissue over the remaining green areas in the rural territory. Unforeseen by the colonizing company that operated in the region, this is happening in the light of real estate interests, rather than based on socio-environmental plans. This problem may be dealt with the application of the 'green wedge' concept (Lemes de Oliveira, 2017), by linking the green grids to a more robust structure in rural areas, which should constrain and drive urban expansion, shaping it to the morphology of the terrain (Fig. 9.6).

Fig. 9.6 Urban grid meeting the green grid. (Source: LURe - Laboratório de Urbanismo Regional [Regional Urbanism Lab])

6 Conclusion

It is essential that urban and regional planning is governed by the objectives of sustainable development. The sustainability paradigm has gradually provided planning proposals with a systemic and more integrated view, rather than the univocal view of modernist planning. It is about incorporating relational factors and no longer dealing with pre-established standards. From sustainability, socioenvironmental issues emerge, which are much broader than merely urban or rural issues.

The planning of an urban territory under the sustainability paradigm faces multiple challenges at different time and space levels. The key to responding to these challenges lies on the legacy of previous plans and proposals implemented in different cities or regions. The landscape in northern Paraná, for example, is the result of the strength planning has had since the initial and systematic occupation of the territory. What we have as a result are urban forms that are still relatively compact, the prevalence of road and street structures in the formation and expansion of the urban territory in association with a discreet historical environmental relationship that culminated in the constitution of green corridors and the permanence of potential forested patches in the rural area.

Based upon the history of the occupation of northern Paraná, it is possible to list both positive and negative aspects arising from the implementation of different urban planning practices. Considering the singularity of the subdivision and occupation processes in this region, some of the points raised here may become design references for urban territories in expansion, or even re-naturalization processes of areas that are environmentally fragile or necessary for a balanced environment. The interface of green and gray grids has occurred both at the regional scale—the rural matrix and the network of cities—as well as the urban scale—urban green patches and corridors.

The rural matrix, as implemented in northern Paraná, supports the socioeconomic requirements for sustainable regional development based on small and medium-sized properties, access to distribution logistics, and the maintenance of part the rural plots as vegetated areas. This matrix could be used as the basis for shaping cities in the future, taking into consideration the quality of the expanding urban area *per se,* as well as in relation to the remaining rural vegetation.

The green patches and corridors inside urban tissues have been formed over the decades with the help of urban elements and instruments. The permanence of these green structures within the urban fabric is the main point in ensuring environmental sustainability based on a green infrastructure. The potential for the integration of green corridors and patches with the urban fabric is robust. Currently, they interpenetrate the urban area with an ecological function, with great potential for leisure and recreation to be further explored. Only with the complete social participation within these structures—the rapprochement between man and nature—can the maintenance of physical structures, and the improvement of habitats and connectivity be guaranteed to maintain fundamental flows and movements for a richer biodiversity.

The challenges of planning cities and regions in developing countries are enormous, mainly due to the speed urban growth takes place and the financial constraints to sustain an effective technical staff in the management of the territory. Sustainability-based planning actions in the tropics are not widely spread out but good examples as the colonization experience and urbanization presented here can be spotted.

References

Ahern, J. (1995). Greenways as a planning strategy. *Landscape and Urban Planning, 33*, 131–155.

Ahern, J. (2013). Urban landscape sustainability and resilience: The promise and challenges of integrating ecology with urban planning and design. *Landscape Ecology, 28*, 1203–1212. https://doi.org/10.1007/s10980-012-9799-z

Angeoletto, F. H. S. (2012). *Planeta Ciudad: ecología urbana y planificación de ciudades medias de Brasil*. Universidad Autónoma de Madrid.

Angeoletto, F., Tryjanowski, P., Santos, J., Martinez-Miranzo, B., Leandro, D., Bohrer, J., ... & Fellowes, M. D. (2022). Will Brazilian city dwellers actively engage in urban conservation? A case study with the charismatic neotropical Blue-and-yellow macaw (Ara ararauna). *Birds, 3*(2), 234–244.

Beloto, G. E. (2015). *From region to Metropolis. The territory designed by conceptual models*. (Unpublished PhD Thesis) FAUUSP. [in Portuguese].

Coimbra, M. H., & Beloto, G. E. (2020). Urban expansion: From compact to fragmented form. *Acta Scientiarum Technology, 42*, 1–8. https://doi.org/10.4025/actascitechnol.v42i1.42407

Companhia Melhoramentos Norte do Paraná. (1975). *Colonization and development of northern Paraná*. S.l.: CMNP. [in Portuguese].

Lemes de Oliveira, F. (2017). *Green wedge urbanism*. Bloomsbury Academic.

Dozier, C. L. (1956). Northern Paraná, Brazil: An example of organized regional development. *Geographical Review, 46*(3), 318–333.

Fleischfresser, V. (1999). Public policies and the formation of conservationist networks in hydrographic basins—The example of rural Paraná. *Revista Paranaense de Desenvolvimento, 95*, 61–77. [in Portuguese].

Forman, R. T. T. (1995). *Land mosaics: The ecology of landscapes and regions*. Cambridge University Press.

Forman, R. T. T., & Godron, M. (1986). *Landscape ecology*. IE-Wiley.

Fresca, T. M., & Carvalho, M. S. (2007). *Geography and the north of Paraná: A historical retrieve*. Humanidades. [in Portuguese].

Hall, P., & Ward, C. (1998). *Sociable cities. The legacy of Ebenezer Howard*. John Wiley & Sons.

Hough, M. (1995). *Cities and natural process*. Routledge.

Howard, E. (1965). *Garden cities of to-morrow*. The MIT Press.

IBGE. (2020). *Population estimation*. [in Portuguese] https://www.ibge.gov.br/estatisticas/sociais/população/9103-estimativas-de-populacao.html?=&t=downloads. Accessed 21 Jan 2021.

Jongman, R. H. G., & Pungetti, G. (2004). *Ecological networks and greenways: Concept, design, implementation*. Cambridge University Press.

Katzman, M. T. (1977). *Cities and frontiers in Brazil: Regional dimensions of economic development*. Harvard University Press.

Katzman, M. T. (1978). Colonization as an approach to regional development: Northern Paraná, Brazil. *Economic Development and Cultural Change, 26*(4), 709–724.

Luz, F. (1997). *The urban phenomenon in a pioneering zone: Maringá*. A Prefeitura. [in Portuguese].

Lyle, J. T. (1994). *Regenerative design for sustainable development*. Wiley.
Macedo, J. (2011). Maringá: A British garden city in the tropics. *Cities, 28*(4), 347–359.
Magalhães, M. R. (2001). *Landscape architecture, morphology and complexity*. Editorial Estampa. [in Portuguese].
Meneguetti, K. S. (2007). *From garden city to sustainable city: Potentialities for an ecological urban structure and the city of Maringá—PR*. (Unpublished PhD Thesis), FAUUSP. [in Portuguese].
Meneguetti, K. S. (2009). *Garden city, sustainable city: The ecological urban structure and the city of Maringá*. EDUEM. [in Portuguese].
Meneguetti, K. S. (2017). Before the green infrastructure: The layout of Maringá. In P. Pellegrino (Ed.), *Strategies for a green infrastructure* (pp. 101–118). Manole. [in Portuguese].
Meneguetti, K. S. & Rego, R. L. (2008). British urban form in twentieth-century Brazil. *Urban Morphology, 12*, 25–34.
Meneguetti, K. S., & Rego, R. L. (2013). Improving open spaces for a sustainable city. *Journal of Civil Engineering and Architecture, 7*, 1150–1156.
Monbeig, P. (1984). *Pioneers and farmers of São Paulo*. Hucitec: Polis. [in Portuguese] [1st edition 1935].
Moroń, D., Beim, M., Gudowska, A., Angeoletto, F., Celary, W., Cwajna, A., ... & Tryjanowski, P. (2024). Evaluating tramway infrastructure on biodiversity and ecosystem services. *Scientific Reports, 14*(1), 9394.
Müller, N. L. (1956). Contribution to the study of northern Paraná. *Boletim Paulista de Geografia, 22*, 55–97. [in Portuguese].
Odum, E. P., & Barret, G. W. (2007). *Fundamentals of ecology*. Thomson Learning. [in Portuguese].
Purdom, C. B. (1949). *The building of satellite towns. A contribution to the study of town development and regional planning*. J. M. Dent & Son Ltd. [1st edition 1925].
Rego, R. L. (2009). *Planted towns. Britons and the construction of Northern Paraná landscape*. Humanidades. [in Portuguese].
Rego, R. L. (2011). A tropical enterprise: British planning ideas in a private settlement in Brazil. *Planning Perspectives, 26*(2), 261–282. https://doi.org/10.1080/02665433.2011.550448
Rego, R. L. (2019). *Ideas for new towns. Architecture and Urbanism in the interior of Brazil in the 20th century*. Kan. [in Portuguese].
Rego, R. L., & Meneguetti, K. S. (2010). Planted towns and territorial organization: The morphology of a settlement process in Brazil. *Urban Morphology, 14*(2), 101–109.
Ribeira, A. B. (2019). *Abordagem sistêmica na Arquitetura da Paisagem*. (Unpublished Master's dissertation) UEM. [in Portuguese].
Ribeiro, R. R., & Beloto, G. E. (2020). Compacity and urban dispersion from the regional structures. *Arquitetura Revista, 16*(1), 63–77. [in Portuguese]. https://doi.org/10.4013/arq.2020.161.04

Chapter 10
Loss of Seaweed Biodiversity in Marine Tropical Regions Caused by Unplanned Coastal Urbanization

Fernando Scherner

Abstract Marine biodiversity is facing increasing threats from unplanned coastal urbanization worldwide. Poor coastal planning and pollution are typically among the main factors contributing to these impacts, which affect the physiology of benthic community structuring groups, such as seaweeds. This leads to changes at both population and community levels, resulting in substantial alterations to coastal biological communities in urban areas. While some groups are severely affected, others tend to benefit from this scenario. Calcareous algae, an important foundation group for benthic communities, are among those most impacted. Conversely, ephemeral species thrive in such environments, as they are physiologically adapted to these conditions, including several green seaweed species, among other groups. These differences in performance lead to significant shifts in marine benthic community structures, typically resulting in major biodiversity losses in the affected areas. Adequate coastal planning, enhancement, management, and restoration techniques can help mitigate the loss of biodiversity in coastal urban areas. Here, we present the main causes of the problem and a summary of possible mitigation and restoration techniques.

Keywords Marine biodiversity · Coastal urbanization · Pollution · Seaweed · Biodiversity loss

1 Introduction

Urban centers are disproportionately located along rivers and coastlines, with 40% of the world's population occupying 7% of the Earth's land area located on narrow coastal zones (Sterzel et al., 2020). Among these, over 1.3 billion people live on tropical coasts, primarily in developing countries (Sale et al., 2014). Population

F. Scherner (✉)
Pell Frischmann, London, UK
e-mail: Fscherner@pellfrischmann.com

F. Angeoletto et al. (eds.), *Ecology of Tropical Cities*,
https://doi.org/10.1007/978-3-031-48012-6_10

density at the coast (≤100 km from the coast and ≤100 m above sea level) is approximately three times higher than the global average and is increasing (Small & Nicholls, 2003). The pace at which the worlds coastal areas are being populated is very intense. The world's population is urbanizing quickly (Angeoletto et al., 2022; Bloom, 2011; Goddard et al., 2021; Santangelo et al., 2022; Seto et al., 2011; Beridze et al., 2023; Moroń et al. 2024) while mass migration is observed towards coastlines (Creel, 2003; McGranahan et al., 2007).

Urban development on the coasts is changing coastal environments worldwide through unmanaged human population growth, decreases in vegetation cover and replacement with impervious surfaces (Yuan & Bauer, 2007). This is increasing human pressure and the run-off of polluted waters into coastal oceans (Bay et al., 2003; Marsalek et al., 1999) and impacting the biodiversity of marine coastal areas, leading to biodiversity declines and community shifts (Scherner et al., 2013). Furthermore, recent studies suggest that coastal urbanization may have an evolutionary effect on marine organisms by creating new selective regimes or influencing evolution by impacting gene flows or partitioning of genetic diversity across seascapes (Alter et al., 2020).

Halpern et al. (2008) developed an ecosystem-specific, multiscale spatial model to synthesize 17 global data sets of anthropogenic drivers of ecological change for 20 marine ecosystems. The results indicate that no area is unaffected by human influence and that a large fraction (41%) is strongly affected by multiple drivers. The study predicted that human impact on the oceans shows strong spatial heterogeneity but with every square kilometer affected by some anthropogenic driver of ecological change. Over a third (41%) of the world's oceans have medium high to very high impact scores with a small fraction (0.5%) but relatively large area (~2.2 million km^2) experiencing very high impact. Most of the highest predicted cumulative impact is in areas of continental shelf and slope, which are subject to both land- and ocean-based anthropogenic drivers. Large areas of high predicted impact occur in all the continents and include tropical areas where systems of high marine biodiversity are present.

As described by Todd et al. (2019), coastal urbanization can lead to consequences which can be broken down to three main groups: 1. Exploitation of living and non-living marine resources, including recreational, subsistence and commercial fishing, and mineral extraction; 2. Pollution, including urban runoff, sediments, industrial and municipal waste, untreated or partially treated domestic waste, animal/slaughterhouse waste, and other contaminant sources, and well as light and noise pollution; 3. The alteration of natural habitats into a different state, such as beach replenishment, seawalls, jetties, piers, marinas, breakwaters, bridges, port and harbor infrastructure. Obviously, the three groups can interact, resulting in potential synergistic effects for marine ecosystems. Other factors which have a global nature such as climate alterations may also combine with the effects above adding additional pressure to these systems.

Coastal tropical regions comprise highly diverse and productive ecosystems such as coral reefs, mangroves and seagrass meadows. Tropical coastal seas are predominantly situated within developing countries, where more than one fifth of the global population inhabits and where anthropogenic pressure is expected to increase in the coming decades (Sale et al., 2014). Thus, it is important to identify how coastal urbanization affects biodiversity in these areas so that adequate coastal management with a focus on marine biodiversity mitigation can be implemented in these regions. However, although urban ecology has advanced rapidly in recent decades, advances have primarily focused on terrestrial and freshwater systems. Urban marine ecology, in comparison to the terrestrial and freshwater systems, can be considered a field in its early developments and still requires the theoretical and empirical foundations underpinning urban ecosystem science on land (Todd et al., 2019).

Here, the main impacts of coastal urbanization on seaweed communities in tropical regions were reviewed and recommendations for the management of these areas are provided. The implementation of appropriate management practices in coastal urban areas could help to mitigate the loss of biodiversity in these areas and potentially reinstate to a certain degree the baseline conditions.

2 Coastal Urbanization Pollution and Disturbances

2.1 Urban Runoff

Urban runoff is among the most deleterious consequences of costal urbanization and probably the most important factor to be considered concerning the impacts of coastal urbanization on marine coastal ecosystems. Contaminants derived from urban runoff have been shown to accumulate in marine sediments, reaching concentrations capable of causing biological effects. With respect to toxicity, intermittent pulses of pollutants characterize urban runoff discharges (Marsalek et al., 1999).

According to Marsalek et al., 1999, urban traffic is an important source of runoff pollutants. It has been reported that urban highways runoff pollution is of concern in highways with more than 30,000 vehicles a day (Ellis et al., 1997) and impervious drainage elements (Barrett et al., 1998), although attenuation of runoff toxicity during transport is be expected to occur in areas such as grassed swales (Marsalek et al., 1999).

Göbel et al. (2007) points out that the characteristics of the urban surface is determined by the composition of the surface, type and degree of utilization, weathering processes, development age, surface slope, exposure, and spatial location. The authors divide urban surface storm water runoff into three main types, as follows: 1. partly sealed surfaces (e.g. overgrown soil in backyards, urban green spaces and porous paving); 2. impermeable roof surfaces; and 3. impermeable road surfaces. The authors treat other spaces such as storage and trade centers in commercial and industrial areas, agricultural property and airports as a special category, which may contribute additional pollutant loading and remark these should be studied individually to define specific pollutant loading.

In coastal urban environments, numerous pollutants are normally present in combination when discharged from sewages and pluvial channels. Inputs of several toxic compounds have been reported for urban environments (Ellis et al., 1997; Haiyan & Stuanes, 2003; Marsalek et al., 1997; Sansalone & Buchberger, 1997) in intensities that vary along time and according to weather conditions (Göbel et al., 2007), which makes urban coastal environments subjected to extensive variations in seawater quality. The pollutants found in coastal urban environments include nutrients, heavy metals, polycyclic aromatic hydrocarbons (PAHs), herbicides and sediments (Göbel et al., 2007). Seawater salinity is also affected by coastal urbanization, being significantly more variable in urban environments in comparison to non-urban or less urban environments (Scherner et al., 2012a, b).

2.2 Other Sources of Pollution

Although less important than urban, runoff, other sources of pollution may also be present in coastal urban areas. For instance, areas where there is substantial traffic of boats and ferries are subjected to oil leaks, as well as gray and black water dumped into the sea, among others.

Solid waste pollution is also conspicuous is most coastal urban environments. Organic and inorganic solid wastes can reach marine habitats thought littering and dumping but can also be carried by the urban runoff via drainage systems. Plastic is probably one of the most important solid pollutant in marine environments these days. According to Gallo et al., 2018, persistent plastics, with estimated lifetime for degradation of hundreds of years, can break up into micro and nano elements over shorter timescales, which facilitates their uptake by marine biota throughout the food chain. These polymers may contain chemical additives and contaminants that may be harmful for marine biota, presenting potential risks to marine ecosystems, biodiversity and food availability.

2.3 Physical Disturbance

The intertidal areas in coastal urban zones are subject to trampling, especially if located in touristic regions. Furthermore, development and infrastructure can also directly affect coastal biological communities through the construction of urban ports and bays and sea defense and enhancement structures such as groynes, seawalls, gabions, breakwater, and beach replenishments (Fig. 10.1).

Fig. 10.1 Two intertidal areas in Pernambuco, Northeastern Brazil. (**a**) a non-urban area dominated by perennial brown seaweed; (**b**) a heavily urbanized area in Recife, dominated by ephemeral green seaweed

3 Seaweed Diversity Declines in Coastal Urban Areas

Several studies have shown that coastal urbanization leads to changes in biological communities, typically reducing species richness by the removal of the most sensitive species and changing the structure of the communities.

3.1 Seaweed Community Changes

Seaweeds are key primary producers in coastal regions, and in combination with seagrasses they support a range of goods and services provided by costal environments (Beaumont et al., 2007; Harley et al., 2012). For instance, these primary producers provide habitat, nursery and food for marine fauna (Grall et al., 2006), removal of pollutants from seawater (Wang & Zhao, 2007), food for humans as well as having a diversity of biotechnological applications in agriculture, medicine, cosmetics and the food industry (Zemke-White & Ohno, 1999). For instance, brown seaweed of the genera *Sargassum* typically form seaweed beds in tropical and temperate regions, being amongst the most important constituents of the phytobenthos worldwide (Mattio et al., 2008). *Sargassum* species form canopy-dominated systems, which are productive and structurally complex habitats that act as foundations to entire systems (Russell et al., 2009). Established canopies preserve the recruitment of associated fauna, providing habitat and energy for several fish and invertebrate species, supporting ecosystem services (Rönnbäck et al., 2007; Vitousek et al., 1997) and fisheries (Russell et al., 2009). Furthermore, calcareous algae play a key role in the carbon and carbonate budgets of coral reef ecosystems (Littler & Littler, 1988; Sinutok et al., 2012). Ultimately, these primary producers play a fundamental role in the maintenance of coastal marine life. However, this group as well as several phaeophyceae species are sensitive to pollutants. Calcareous algae are known to be negatively affected by excess nutrients (Bjork et al., 1995), while phaeophyceae such as *Sargassum* species have their reproduction and physiology impacted by urban pollution (Kevekordes, 2001; Scherner et al., 2012a), which ultimately lead to their removal from polluted areas. Such process occurs while other opportunistic seaweed groups, such as certain species of chlorophyta benefit from this scenario (Bjork et al., 1995; Liu et al., 2009; Scherner et al., 2013, 2012a; Teichberg et al., 2010). Several studies have reported that human induced stress has been causing declines of brown seaweeds (Airoldi et al., 2008; Connell et al., 2008; Russell et al., 2009) while opportunistic seaweed such as those of the genera *Ulva* thrive (Teichberg et al., 2010).

The different responses of different seaweed groups and species to the effects of coastal urbanization ultimately leads to major seaweed communities shifts. Such shifts normally swings from systems dominated by perennial species to systems dominated by ephemeral ones, and this process normally brings with it significant declines in biodiversity. Significant loss of primary producers is greatly disadvantageous to the marine environment since it cascades through the food chain, causing further loss of fauna at different levels and altering the trophic structure of marine communities (Airoldi et al., 2008; Haddad et al., 2009).

3.2 Ecophysiology

The underlying mechanisms by which environmental pollutants ultimately impact seaweeds at the community and population levels are normally a result of the alterations that occur at the physiological level. These changes impact on the individuals affected and cascade through populations and communities, modifying ecological interactions and indirectly affecting entire marine food webs. Such physiological alterations comprise changes to the photosynthetic performance and enzymatic alterations.

It has been shown that species normally absent from impacted coastal urban areas such as *Sargassum* species have their photosynthetic performance significantly reduced in such conditions, which demonstrates its vulnerability to urban stressors (Scherner et al., 2012a). Not surprisingly, opportunistic seaweeds such as *Ulva* species normally have their physiological fitness improved towards urban areas. Thus, different physiological strategies are used by these groups. In general, opportunistic species lose less energy via non-regulated mechanisms than perennial species, and thus they are able to convert more energy into effective quantum yield. Therefore, it is assumed that the presence of pollutants accentuates the loss of energy by perennial seaweeds such as *Sargassum* species, while ephemeral seaweeds such as *Ulva* species increase the conversion of energy into photochemistry in such situations. This eventually leads to a loss of biomass of perennial species and expansion of the ephemeral ones. The negative effects of urban sewage to the biomass and species composition of seaweed communities have been reported by different studies (Liu et al., 2007; Scherner et al., 2013; Terlizzi et al., 2002). For instance, Tewari and Joshi (1988) demonstrated the tolerance of *U. lactuca* and the sensitiveness of *Sargassum* species to waters contaminated by industry effluents and domestic sewage using biomass as a descriptor.

Heavy metal pollution is also commonly present in coastal urban areas. Heavy metals are known to affect the physiology and cellular structure of seaweeds. Changes to metabolism and cellular organization have been observed in different seaweed species, promoting alterations in seaweed metabolism and stimulating protective mechanisms against oxidative stress (e.g. Costa et al., 2016).

The effects of heavy metals pollution can be particularly damaging to seaweed species when it occurs on their distribution limits. Tropical seaweeds living at their temperature limits of distribution are naturally exposed to physiological stressors, and additional stress is added when they are exposed to coastal pollution. For instance, *Halimeda jolyana*, an important calcareous component of tropical southwestern Atlantic reefs presented negative physiological effects and ultrastructural alterations when exposed to trace metals at concentrations observed in polluted regions at their lowest temperature limits of distribution (Scherner et al., 2018). Additionally, biomass yield and photosynthetic performance declined substantially in samples exposed to metal, although partial photosynthetic recovery was observed at the highest temperature when metal enrichment was ceased, and exposure to higher metal concentrations caused substantial ultrastructural alterations to chloroplasts regardless of temperatures. It was suggested that the lack of photosynthetic recovery at the lower temperatures indicates a higher vulnerability of the species at its temperature limits of distribution in the tropical southwestern Atlantic.

Variations on seawater salinity is also among the effects of costal urbanization. Due to the large areas covered by impervious surfaces in urban areas, the urban runoff at high intensities can cause significant variations of salinity on these areas. Such variations may be attenuated in non-urban areas as a good amount of the rainwater will be penetrate the bare ground. Different levels of salinity are known to determine species occurrence and distribution (Martins et al., 1999; Wilkinson, 1981; Wilkinson et al., 2007). Ephemeral seaweed normally have higher tolerance to the exposure to low salinities, which may contribute to the shift of perennial seaweed dominated into ephemeral seaweed dominated ecosystems (Scherner et al., 2012b) in coastal urban areas. The above is supported by physiological aspects of brown seaweeds exposed to salinity variations, such as photosynthetic performance (Scherner et al., 2012b), the reproduction patterns of *Sargassum muticum* (Hales & Fletcher, 1990; Norton, 1977; Steen, 2004), and growth rates of *Laminaria digitata, Fucus vesiculosus, F. serratus* (Gordillo et al., 2002) and *S. muticum* (Steen, 2004).

4 How Can Urban Coastal Zones Be Managed to Minimize Impacts on Coastal Biodiversity?

In this context, more efficient land use in urban areas is important to mitigate the impacts of coastal urbanization and human population growth on marine biodiversity (Sushinsky et al., 2013). Considering the results presented here and the current worldwide fast growing coastal urbanization process, public policies should consider a combination of expanded conservation efforts of coastal land-sea systems, and techniques to mitigate the contamination of the marine environment by anthropogenic pollutants in urban areas, as well as restoration possibilities.

4.1 Green Spaces and Biodiversity Enhancement

In general, it is recognized that urbanization affects local climate, hydrology, air quality and biodiversity (Angeoletto et al., 2018, 2019; Fellowes et al., 2020; Rumble et al., 2019; Santangelo et al., 2022; Souza et al., 2023; Whitford et al., 2001). The proportion of green space and tree cover have a fundamental role in improving all of the above. It is also important to improve quality of life and leisure opportunities for local communities. Regarding the impacts of runoff to marine coastal biodiversity, it is substantially lower in areas with higher green coverage (Scherner et al., 2013; Whitford et al., 2001). Green areas have the capacity to attenuate the effects of urban runoff by retaining pollutants and slowing down the runoff flow into the sea. Thus, local planning policies for the development of new urban areas should encourage the presence of significant areas of green spaces and biodiversity enhancement. The installation of green roofs on buildings should also be encouraged whenever possible, as this would also improve the total green cover in newly developed or existing urban areas.

4.2 Sewage Treatment

The lack of appropriate sewage treatment may be one of the most important factors affecting coastal biodiversity. Most of the tropical coastal areas are situated in developing or deprived countries, where sewage treatment is not universal. For instance, in Brazil, public policies have failed to address this issue. Only 46% of the sewage is treated in the country (SNIS, 2018), and much of this is on coastal cities. For decades states has been responsible to provide sewage treatment to the population through public companies. However, the laws that regulate the sector have recently changed, allowing the private sector to invest and carry out sewage treatment projects. This is an important progress, which may accelerate the sewage treatment in the country, with optimistic predictions for the following years.

Excessive nutrients, particularly nitrogen (N), is responsible for a large fraction of marine biodiversity degradation (Howarth, 2000). As marine waters are normally N limited, and human derived activities result in rich N outputs. Thus, removal of N during sewage treatment is an important measure. However, for many coastal regions, human sewage is not the major source of nutrients. Animal wastes from livestock feedlot operations are one of the biggest sources of nutrients to coastal waters (Howarth, 2000). Thus, such operations should have N removed to acceptable levels before emission and if possible should be avoided in coastal urban areas and in coastal areas in general.

4.3 Urban Design

It has been suggested that cities with lower per capita ecological impact are characterized by high residential density, with large interstitial green spaces and small backyards. This compact urban development model minimizes ecological disruption while maintaining human access to public green spaces (Sushinsky et al., 2013). This could also minimize the toxicity of urban runoffs as runoffs from mixed urban land use has been shown to be substantially less toxic and rarely severely toxic (Marsalek et al., 1999).

Besides increasing green space areas and enhancing biodiversity in urban areas as discussed above, local plans for coastal urban areas should avoid the presence of busy roads along the coast whenever possible. As the busier the road the worse the runoff pollution (Ellis et al., 1997).

Local plans should consider activities that can potentially increase the runoff pollution near the coast and such activities should preferably be located away from the coasts. Furthermore, tertiary treatment of activities that generate effluent pollution should be implemented effectively in order to reduce the input of nutrients into the sea, particularly of N.

Input and accumulation of contaminants from runoff in receiving environments can be reduced by treatment technology, such as detention ponds (Morrisey et al., 2003). These ponds could be implemented in conjunction with other technologies to slow down the speed in which runoff enters the marine systems, giving the ecosystem time to process the load of nutrients and fresh water, and helping to reduce the variations of such pulses. This is also an effective measure against flooding, as surface water attenuation systems store and slowly release surface water run-off as the surface water enters sewers or water courses, reducing the risk of downstream flooding.

4.4 The Biodiversity Net Gain Approach

Biodiversity Net Gain is an approach that encourages developers to leave biodiversity in a better condition than previously. Development proposals that may have an impact on biodiversity are encouraged to provide an increase in suitable habitat and ecological features in a way that it not only replaces what has been lost in terms of biodiversity but also actually achieves a biodiversity gain. This way, it is expected that the loss of biodiversity through developments would be fully offset and ecological networks could then be restored (Baker et al., 2019). This approach is currently being implemented in the English planning system, where the national planning policy frameworks already encourage the biodiversity net gain approach. Similar approaches could be implemented on tropical coastal areas and could include marine systems as well as the adjacent terrestrial area.

The Biodiversity Net Gain approach comprises the use of a metric for calculating the negative impacts on habitats caused by developments and calculating how much new or restored habitat, and of what types is required to deliver satisfactory biodiversity net gain. However, it still relies on the application of the mitigation hierarchy to avoid, mitigate or compensate for biodiversity losses. Similarly, the Environmental Net Gain (ENG) is based on the same rationale although it comprises a wider range of environmental aspects such as air quality and flood risk management (CIEEM, CIRIA, IEMA, 2016).

4.5 Solid Wastes

The implementation or improvement of environmentally efficient waste collection and management of urban waste is an essential step to reducing solid wastes inputs into marine systems, particularly in developing countries. Recycling systems should be encouraged locally and regionally, which could generate jobs and benefit deprived communities in coastal areas of developing countries and least developed countries. This should be supported by technical assistance and capacity-building activities to stakeholders in implementing efficient waste collection and management measures.

Environmental awareness programs is an important tool to help reduce the amount of littering on public areas, which would eventually end up in the sea. This can be implemented with the dissemination, information and training activities to local communities and tourists to improve awareness on the causes and consequences of littering.

The above measures could be complemented by the development and installation of innovative systems to retain solid waste on the end of drainage systems. Again, such a system would require adequate management with constant removal of wastes to avoid the saturation and damage of the structures involved.

Further actions should be implemented regarding plastic pollution. Innovative alternatives have been developed; however, some are not degradable in the marine environment. Marine biodegradable polymers could have an important role to play in reducing the environmental damage of plastics in marine systems (e.g. polyhydroxyalkanoates, PHAs). However, their biodegradability would require further study and validation under a range of conditions in seawater, and internationally accepted certification (Gallo et al., 2018). The current issues caused by plastic in the oceans should encourage the continuous development of alternative and economically viable biodegradable polymers.

The following measures have been suggested to reduce marine plastic litter in National Implementation Plans for the Stockholm Convention on Persistent Organic Pollutants, as summarized by Gallo et al. (2018): a. Promoting substitution and green chemistry to avoid POPs and other harmful chemicals in plastics, especially EDCs; b. Encouraging plastic waste prevention and supporting development and implementation of safer or more benign alternatives to persistent plastics in the marine environment; c. Supporting research on environmental and health impacts of marine plastics, microplastics and nanoplastics and related fate of EDCs and POPs; d. Encouraging ecodesign for better packaging recyclability; e. Encouraging plastic waste recycling when feasible; f. Promoting BATs to reduce plastic leakage to oceans and improving information on input loads, sources and originating sectors; g. Encouraging the improvement and efficiency of collection and sound environmental management of waste; h. Encouraging changes in consumption and littering behavior.

5 Restoring Impacted Marine Ecosystems

The term restoration describes the intervention in the recovery process of a marine ecosystem, with the purpose to enhance the process, but not necessarily leading the system to recover to its original condition (Elliott et al., 2007; Hawkins et al., 1999). In most cases, impacted ecosystems should be able to recover naturally when the impacting pressure (e.g. pollution) has been removed, especially if no physical damage has occurred. In fact, the most cost effective approach to restore an open marine ecosystems is to stop the damaging activity and let recolonization follow naturally (Hawkins et al., 1999). However, some areas may require the use of techniques to accelerate this recovery. In any case, the appropriate detection of impacts precedes the recovery or restoration of the ecosystem, which should be followed by actions to stop or minimize the risks of occurrence of further impacts. Intervention may be needed to remove undesired materials (i.e. mud, litter) from the area to allow the natural recolonization by propagules and larvae.

Once impacts have been stopped or at least reduced then natural recovery can occur. This should be monitored to record the progress of recovery. However, active restoration may be required to speed up the process. Transplantation or enhancement of ecosystem engineers (Lawton, 1994) can be employed. Hard substrate communities can be created to provide substrate for the development of fouling communities, creating artificial rocky shores and artificial reefs.

Several methods have been used to restore seaweed beds, including employing artificial reefs, seaweed ropes, spore bags, and transplanted cultures. Regarding brown seaweed beds restoration, a number of techniques have been proposed. For instance, Terawaki et al. (2003) have proposed the following techniques: (1) Creation of shallow and moderate sloping bottom substrata, which have been shown to be effective for the reestablishment of management-free seagrass and Sargassum beds on urban coasts; (2) Seeding or transplanting using artificial substratum for extension of nursery and fishing grounds around natural Sargassum beds; and (3) Periodic transplantation of Sargassum individuals using artificially produced seedlings, which is an effective method to produce niche, allowing faunal re-colonization in polluted areas with sparse vegetation cover. Another recent study examined the use of a new technique using encapsulated seaweed zygotes with polysaccharide-like alginates to improve their attachment using the brown alga *Sargassum fulvellum*, which plays an important role in seaweed forests. The technique increased the germination density and growth rates of encapsulated zygotes up to four and seven times, respectively, when compared to un-encapsulated zygotes, as well as providing better germination density and growth rates (Jung et al., 2020).

There are indications on the literature that habitat rehabilitation costs can vary substantially between and within ecosystems. A review of the economics based on various coastal habitats considering the Total Economic Value approach, which accounts for direct and indirect uses and non-uses, indicated that such projects could have significantly high costs (Spurgeon, 1998), which may be an obstacle to implementing such plans.

References

Airoldi, L., Balata, D., & Beck, M. W. (2008). The gray zone: Relationships between habitat loss and marine diversity and their applications in conservation. *Journal of Experimental Marine Biology and Ecology, 366*, 8–15.

Alter, S. E., Tariq, L., Creed, J. K., & Megafu, E. (2020). Evolutionary responses of marine organisms to urbanized seascapes. *Evolutionary Applications, 14*, 210–232.

Angeoletto, F., Fellowes, M. D., & Santos, J. W. (2018). Counting Brazil's urban trees will help make Brazil's urban trees count. *Journal of Forestry, 116*(5), 489–490.

Angeoletto, F., Fellowes, M. D., Essi, L., Santos, J. W., Johann, J. M., da Silva Leandro, D., & Mendonca, N. M. (2019). Ecología urbana y planificación: una convergéncia ineludible. *Revista Electrônica em Gestão, Educação e Tecnologia Ambiental, 23*(1), 1–8.

Angeoletto, F., Tryjanowski, P., Santos, J., Martinez-Miranzo, B., Leandro, D., Bohrer, J., et al. (2022). Will Brazilian city dwellers actively engage in urban conservation? A case study with the charismatic neotropical Blue-and-yellow macaw (Ara ararauna). *Birds, 3*(2), 234–244.

Baker, J., Hoskin, R., & Butterworth, T. (2019). *Biodiversity net gain: Good practice principles for development. A practical guide*. CIRIA. https://cieem.net/wp-content/uploads/2019/02/C776a-Biodiversity-net-gain.-Good-practice-principles-for-development.-A-practical-guide-web.pdf

Barrett, M. E., Irish, L. B., Malina, J. F., & Charbeneau, R. J. (1998). Characterization of highway runoff in/Austin, Texas, area. *Journal of the Environmental Engineering Division ASCE, 124*, 131–137.

Bay, S., Jones, B. H., & Schiff, K. (2003). Water quality impacts of stormwater discharges to Santa Monica Bay. *Marine Environmental Research, 56*, 205–223.

Beaumont, N. J., Austen, M. C., Atkins, J. P., Burdon, D., Degraer, S., Dentinho, T. P., Derous, S., Holm, P., Horton, T., van Ierland, E., Marboe, A. H., Starkey, D. J., Townsend, M., & Zarzycki, T. (2007). Identification, definition and quantification of goods and services provided by marine biodiversity: Implications for the ecosystem approach. *Marine Pollution Bulletin, 54*, 253–265.

Beridze, B., Mahamoud-Issa, M., Zmudzinski, M., Bednarz, P. A., Filutowska, Z. K., Sikora, B., ... & Angeoletto, F. (2023). Welcome to the (urban) jungle: notes on new science, biological evolution in cities. *European Journal of Ecology, 9*(2).

Bjork, M., Mohammed, S. M., Bjorklund, M., & Semesi, A. (1995). Coralline algae, important coral-reef builders threatened by pollution. *Ambio, 24*, 502–505.

Bloom, D. E. (2011). 7 billion and counting. *Science, 333*, 562–569.

CIEEM, CIRIA, IEMA. (2016). *Biodiversity Net Gain: Good practice principles for development. Chartered Institute of Ecology and Environmental Management*. Institute of Environmental Management and Assessment, Construction Industry Research and Information Association, UK. https://cieem.net/wp-content/uploads/2019/02/Biodiversity-Net-Gain-Principles.pdf

Connell, S. D., Russell, B. D., Turner, D. J., Shepherd, S. A., Kildea, T., Miller, D., Airoldi, L., & Cheshire, A. (2008). Recovering a lost baseline: Missing kelp forests from a metropolitan coast. *Marine Ecology Progress Series, 360*, 63–72.

Costa, G. B., de Felix, M. R. L., Simioni, C., et al. (2016). Effects of copper and lead exposure on the ecophysiology of the brown seaweed *Sargassum cymosum*. *Protoplasma, 253*, 111–125. https://doi.org/10.1007/s00709-015-0795-4

Creel, L. (2003). *Ripple effects: Population and coastal regions* (pp. 1–7). Population Reference Bureau.

Elliott, M., Burdon, D., Hemingway, K. L., & Apitz, S. E. (2007). Estuarine, coastal and marine ecosystem restoration: Confusing management and science – A revision of concepts. *Estuarine, Coastal and Shelf Science, 74*, 349e366.

Ellis, J. B., Revitt, D. M., & Llewellyn, N. (1997). Transport and the environment: Effects of organic pollutants on water quality. *Water and Environmental Management, 11*, 170–177.

Fellowes, M. D., Acquaah-Harrison, K., Angeoletto, F., Santos, J. W., da Silva Leandro, D., Rocha, E. A., et al. (2020). Map-a-mole: Greenspace area influences the presence and abundance of the European mole Talpa europaea in urban habitats. *Animals, 10*(6), 1097.

Gallo, F., Fossi, C., Weber, R., et al. (2018). Marine litter plastics and microplastics and their toxic chemicals components: The need for urgent preventive measures. *Environmental Sciences Europe, 30*, 13. https://doi.org/10.1186/s12302-018-0139-z

Göbel, P., Dierkes, C., & Coldewey, W. G. (2007). Storm water runoff concentration matrix for urban areas. *Journal of Contaminant Hydrology, 91*, 26–42.

Goddard, M. A., Davies, Z. G., Guenat, S., Ferguson, M. J., Fisher, J. C., Akanni, A., et al. (2021). A global horizon scan of the future impacts of robotics and autonomous systems on urban ecosystems. *Nature Ecology & Evolution, 5*(2), 219–230.

Gordillo, F. J. L., Dring, M. J., & Savidge, G. (2002). Nitrate and phosphate uptake characteristics of three species of brown algae cultured at low salinity. *Marine Ecology Progress Series, 234*, 111–118.

Grall, J., Le, F., Guyonnet, B., & Riera, P. (2006). Community structure and food web based on stable isotopes (δ 15 N and δ 13 C) analysis of a North Eastern Atlantic maerl bed. *Journal of Experimental Marine Biology and Ecology, 338*, 1–15.

Haddad, N. M., Crutsinger, G. M., Gross, K., Haarstad, J., Knops, J. M. H., & Tilman, D. (2009). Plant species loss decreases arthropod diversity and shifts trophic structure. *Ecology Letters, 12*, 1029–1039.

Haiyan, W., & Stuanes, A. O. (2003). Heavy metal pollution in air-water-soil-plant system of Zhuzhou City, Hunan Province, China. *Water, Air, and Soil Pollution, 147*, 79–107.

Hales, J. M., & Fletcher, R. L. (1990). Studies on the recently introduced brown alga *Sargassum muticum* (Yendo) Fensholt. V. Receptacle initiation and growth, and gamete release in laboratory culture. *Botanica Marina, 33*, 241–249.

Halpern, B. S., Walbridge, S., Selkoe, K. A., Kappel, C. V., Micheli, F., D'Agrosa, C., Bruno, J. F., Casey, K. S., Ebert, C., Fox, H. E., Fujita, R., Heinemann, D., Lenihan, H. S., Madin, E. M. P., Perry, M. T., Selig, E. R., Spalding, M., Steneck, R., & Watson, R. (2008). A global map of human impact on marine ecosystems. *Science, 319*, 948–952.

Harley, C. D. G., Anderson, K. M., Demes, K. W., Jorve, J. P., Kordas, R. L., Coyle, T. A., & Graham, M. H. (2012). Effects of climate change on global seaweed communities. *Journal of Phycology, 48*, 1064–1078.

Hawkins, S. J., Allen, J. R., & Bray, S. (1999). Restoration of temperate marine and coastal ecosystems: Nudging nature. *Aquatic Conservation: Marine and Freshwater Ecosystems, 9*, 23–46.

Howarth, R. W. (2000). Nutrient pollution of coastal rivers, bays, and seas. *Issues in Ecology, 7*, 1–15.

Jung, S. M., Lee, J. H., Han, S. H., et al. (2020). A new approach to the restoration of seaweed beds using Sargassum fulvellum. *Journal of Applied Phycology, 32*, 2575–2581. https://doi.org/10.1007/s10811-020-02054-y

Kevekordes, K. (2001). Toxicity tests using developmental stages of *Hormosira banksii* (Phaeophyta) identify ammonium as a damaging component of secondary treated sewage effluent discharged into Bass Strait. *Marine Ecology Progress Series, 219*, 139–148.
Lawton, J. H. (1994). What do species do in ecosystems. *Oikos, 71*, 367–374.
Littler, M. M., & Littler, D. S. (1988). Structure and role of algae in tropical reef communities. In C. A. Lembi & J. R. Waaland (Eds.), *Algae and human affairs* (Vol. 28, pp. 29–56). Cambridge University Press.
Liu, D., Bai, J., Song, S., Zhang, J., Sun, P., Li, Y., & Han, G. (2007). The impact of sewage discharge on the macroalgae community in the Yellow Sea coastal area around Qingdao, China. *Water, Air, and Soil Pollution, 7*, 683–692.
Liu, D., Keesing, J. K., Xing, Q., & Shi, P. (2009). World's largest macroalgal bloom caused by expansion of seaweed aquaculture in China. *Marine Pollution Bulletin, 58*, 888–895.
Marsalek, J., Brownlee, B., Mayer, T., Lawal, S., & Larkin, G. A. (1997). Heavy metals and PAHs in stormwater runoff from the Skyway Bridge in Burlington, Ontario. *Water Quality Research Journal of Canada, 32*, 815–827.
Marsalek, J., Rochfort, Q., Brownlee, B., Mayer, T., & Servos, M. (1999). An exploratory study of urban runoff toxicity. *Water Science and Technology, 39*, 33–39.
Martins, I., Oliveira, J. M., Flindt, M. R., & Marques, J. C. (1999). The effect of salinity on the growth rate of the macroalgae *Enteromorpha intestinalis* (Chlorophyta) in the Mondego estuary (West Portugal). *Acta Oecologica, 20*, 259–265.
Mattio, L., Dirberg, G., Payri, C. E., & Andréfouët, S. (2008). Diversity, biomass and distribution pattern of Sargassum beds in the South West lagoon of New Caledonia (South Pacific). *Journal of Applied Phycology, 20*, 811–823.
McGranahan, G., et al. (2007). The rising tide: Assessing the risks of climate change and human settlements in low elevation coastal zones. *Environment and Urbanization, 19*, 17–37.
Moroń, D., Beim, M., Gudowska, A., Angeoletto, F., Celary, W., Cwajna, A., ... & Tryjanowski, P. (2024). Evaluating tramway infrastructure on biodiversity and ecosystem services. *Scientific Reports, 14*(1), 9394.
Morrisey, D. J., Turner, S. J., Mills, G. N., Williamson, R. B., & Wise, B. E. (2003). Factors affecting the distribution of benthic macrofauna in estuaries contaminated by urban runoff. *Marine Environmental Research, 55*, 113–136.
Norton, T. A. (1977). Ecological experiments with *Sargassum muticum. Journal of Marine Biological Association of the United Kingdom, 57*, 33–43.
Rönnbäck, P., Kautsky, N., Pihl, L., Troell, M., Söderqvist, T., & Wennhage, H. (2007). Ecosystem goods and services from Swedish coastal habitats: Identification, valuation, and implications of ecosystem shifts. *Ambio, 36*, 534–544.
Rumble, H., Angeoletto, F., Connop, S., Goddard, M. A., & Nash, C. (2019). Understanding and applying ecological principles in cities. In *Planning cities with nature: Theories, strategies and methods* (pp. 217–234).
Russell, B., Thompson, J. A. I., Falkenberg, L. J., & Connell, S. D. (2009). Synergistic effects of climate change and local stressors: CO2 and nutrient-driven change in subtidal rocky habitats. *Global Change Biology, 15*, 2153–2162.
Sale, P. F., Agardy, T., Ainsworth, C. H., et al. (2014). Transforming management of tropical coastal seas to cope with challenges of the 21st century. *Marine Pollution Bulletin, 85*, 8–23.
Sansalone, J. J., & Buchberger, S. G. (1997). Partitioning and first flush of metals in urban roadway storm water. *Journal of Environmental Engineering, 123*, 134–143.
Santangelo, J. S., Ness, R. W., Cohan, B., Fitzpatrick, C. R., Innes, S. G., Koch, S., et al. (2022). Global urban environmental change drives adaptation in white clover. *Science, 375*(6586), 1275–1281.
Scherner, F., Barufi, J. B., & Horta, P. A. (2012a). Photosynthetic response of two seaweed species along an urban pollution gradient: Evidence of selection of pollution-tolerant species. *Marine Pollution Bulletin, 64*, 2380–2390.

Scherner, F., Ventura, R., Barufi, J. B., & Horta, P. A. (2012b). Salinity critical threshold values for photosynthesis of two cosmopolitan seaweed species: Providing baselines for potential shifts on seaweed assemblages. *Marine Environmental Research*, 14–25. https://doi.org/10.1016/j.marenvres.2012.05.007

Scherner, F., Horta, P. A., Oliveira, E. C., Simonassi, J. C., Hall-Spencer, J., Chow, F., Nunes, J. M. C., & Pereira, S. M. B. (2013). Coastal urbanization leads to remarkable seaweed species loss and community shifts in the SW Atlantic. *Marine Pollution Bulletin, 76*, 106–115.

Scherner, F., Bastos, E., Rover, T., Oliveira, E. M., Almeida, R., Itokazu, A. G., Bouzon, Z. L., Rörig, L. R., Pereira, S. M. B., & Horta, P. A. (2018). *Halimeda jolyana* (Bryopsidales, Chlorophyta) presents higher vulnerability to metal pollution at its lower temperature limits of distribution. *Environmental Science and Pollution Research, 25*, 11775–11786. https://doi.org/10.1007/s11356-018-1395-6

Seto, K. C., et al. (2011). A meta-analysis of global urban land expansion. *PLoS One, 6*, e23777.

Sinutok, S., Hill, R., Doblin, M. A., Kühl, M., & Ralph, P. J. (2012). Microenvironmental changes support evidence of photosynthesis and calcification inhibition in *Halimeda* under ocean acidification and warming. *Coral Reefs, 31*, 1201–1213.

SNIS. (2018). *Ranking Do Saneamento Instituto Trata Brasil 2020 (SNIS 2018)* (p. 2020). GO Associados.

Souza, F. L., Fenoglio, M. S., & Angeoletto, F. (2023). To be a Brazilian City dweller, sometimes we must learn to say enough! *Sustainability, 15*(4), 3699.

Spurgeon, J. (1998). The socio-economic costs and benefits of coastal habitat rehabilitation and creation. *Marine Pollution Bulletin, 37*, 373–382.

Steen, H. (2004). Effects of reduced salinity on reproduction and germling development in *Sargassum muticum* (Phaeophyceae, Fucales). *European Journal of Phycolohy, 39*, 293–299.

Sterzel, T., Lüdeke, M. K. B., Walther, C., Kok, M. T., Sietz, D., et al. (2020). Typology of coastal urban vulnerability under rapid urbanization. *PLoS One, 15*(1), e0220936. https://doi.org/10.1371/journal.pone.0220936

Sushinsky, J. R., Rhodes, J. R., Possingham, H. P., Gill, T. K., & Fuller, R. A. (2013). How should we grow cities to minimize their biodiversity impacts? *Global Change Biology, 19*, 401–410. https://doi.org/10.1111/gcb.12055

Teichberg, M., Fox, S. E., Olsen, Y. S., Valiela, I., Martinetto, P., Iribarne, O., Muto, E. Y., Petti, M. A. V., Corbisier, T. N., Soto-Jiménez, M., Pães-Osuna, F., Castro, P., Freitas, H., Zitelli, A., Cardinaletti, M., & Tagliapietra, D. (2010). Eutrophication and macroalgal blooms in temperate and tropical coastal waters: Nutrient enrichment experiments with *Ulva* spp. *Global Change Biology, 16*, 2624–2637.

Terawaki, T., Yoshikawa, K., Yoshida, G., Uchimura, M., & Iseki, K. (2003). Ecology and restoration techniques for Sargassum beds in the Seto Inland Sea, Japan. *Marine Pollution Bulletin, 47*, 198–201.

Terlizzi, A., Fraschetti, S., Guidetti, P., & Boero, F. (2002). The effects of sewage discharge on shallow hard substrate sessile assemblages. *Marine Pollution Bulletin, 44*, 544–550.

Tewari, A., & Joshi, H. V. (1988). Effect of domestic sewage and industrial effluents on biomass and species diversity of seaweeds. *Botanica Marina, 31*, 389–397.

Todd, P. A., Heery, E. C., Loke, L. H. L., Thurstan, R. H., Kotze, D. J., & Swan, C. (2019). Towards an urban marine ecology: Characterizing the drivers, patterns and processes of marine ecosystems in coastal cities. *Oikos, 128*, 1215–1242.

Vitousek, P. M., Mooney, H. A., Lubchenco, J., & Melillo, J. M. (1997). Human domination of Earth's ecosystems. *Science, 277*, 494–499.

Wang, X., & Zhao, H. (2007). Uptake and biodegradation of polycyclic aromatic hydrocarbons by marine seaweed. *Journal of Coastal Research, 50*, 1056–1061.

Whitford, V., Ennos, A. R., & Handley, J. F. (2001). "City form and natural process" – Indicators for the ecological performance of urban areas and their application to Merseyside, UK. *Landscape and Urban Planning, 57*, 91–103.

Wilkinson, M. (1981). Survival strategies of attached algae in estuaries. In N. V. Jones & W. J. Wolff (Eds.), *Feeding and survival strategies of estuarine organisms* (pp. 29–38). Plenum Publishing Company.

Wilkinson, M., Wood, P., Wells, E., & Scanlan, C. (2007). Using attached macroalgae to assess ecological status of British estuaries for the European Water Framework Directive. *Marine Pollution Bulletin, 55*, 136e150.

Yuan, F., & Bauer, M. E. (2007). Comparison of impervious surface area and normalized difference vegetation index as indicators of surface urban heat Island effects in Landsat imagery. *Remote Sensing of Environment, 106*, 375–386.

Zemke-White, W. L., & Ohno, M. (1999). World seaweed utilisation: An end-of-century summary. *Journal of Applied Phycology, 11*, 369–376.

Chapter 11
Urban Biological Evolution in Tropical Cities

Diego Carmona, Mariana Chávez-Pesqueira, Diego F. Angulo, Fabio Angeoletto, Luis Abdala-Roberts, Xoaquín Moreira, Paula Sosenski, and Victor Parra-Tabla

Abstract In the last decade, urban evolution has gained relevance by showing that in few generations evolutionary changes can occur in organisms living in cities. Despite the ample evidence showing that ecology and evolution operate in different ways in temperate and tropical latitudes, most of the urban evolution studies have been conducted in temperate cities. Here, we review arguments that support that such strong biogeographic bias is constraining our understanding and capacity to achieve broader inferences across latitudes on the effects of the worldwide urban ecosystem expansion on the evolutionary process. We also highlight that despite plants, and its ecological interactions, have played a key role in developing ecological and (co)evolutionary theory, there are relative few urban evolution studies on plants in contrast with those focusing on animals. Based on this bias and the lack of studies in the tropics, we review plant-mutualistic and antagonistic interactions and

D. Carmona (✉) · L. Abdala-Roberts · P. Sosenski · V. Parra-Tabla
Departamento de Ecología Tropical, Campus de Ciencias Biológicas y Agropecuarias, 17 Universidad Autónoma de Yucatán, Mérida, Yucatán, Mexico
e-mail: diego.carmona@correo.uady.mx

M. Chávez-Pesqueira
Centro de Investigación Científica de Yucatán, Unidad de Recursos Naturales, Mérida, Yucatán, Mexico

D. F. Angulo
Interdisciplinary Center for Biological and Human Research, Faculty of Chemical Sciences, Benito Juárez Autonomous University of Oaxaca, Oaxaca, Mexico

F. Angeoletto
Programa de Pós-Graduação em Geografia da Universidade Federal de Rondonópolis, Mato Grosso, Brasil

Programa de Pós-Graduação em Gestão e Tecnologia Ambiental, Universidade Federal de Rondonópolis, Rondonópolis, Brasil

PUMAS, Pessoas Unidas pelo Meio Ambiente, Agricultura & Sociedade, Chapada, Brasil

X. Moreira
Misión Biológica de Galicia (MBG), Consejo Superior de Investigaciones Científicas, Pontevedra, Spain

F. Angeoletto et al. (eds.), *Ecology of Tropical Cities*,
https://doi.org/10.1007/978-3-031-48012-6_11

latitudinal patterns to suggest how plant-ecological interactions can evolve in tropical cities. While it has been suggested that urban biological evolution can play a key role to design sustainable and resilient cities, it can be challenging or unfruitful to apply standardized strategies in developed countries to a socio-eco-evolutionary context not yet studied. We need to put tropical urban evolution on the map.

Keywords Gene flow · Natural selection · Mutation rates · Habitat fragmentation · Genetic drift · Biological homogenization · Urban expansion · Plant ecological interaction · Plant reproductive systems · Latitudinal gradient

1 Urban Evolution in the Tropics

The urban ecosystems are rapidly growing (Seto et al., 2010). Currently, ca. 0.5% of Earth's land is covered by urban areas (Liu et al., 2020), which is equivalent to approximately 10% of the intact tropical forest coverage on Earth (~5–7%; Wilson, 1988; Watson et al., 2018). In 2050 the urban ecosystems are predicted to triple its current area (Seto et al., 2012), supporting 70% of the human population (United Nations, 2018). Considering jointly tropical deforestation rates (1.4% between 2010–2017; State of the Tropics, 2020) and urban expansion, the urban ecosystem coverage could equal or exceed the coverage of tropical forests in years to come.

Close to 43% of the world's population currently lives in the tropics (United Nations 2019), and it is expected that in 2050 tropical regions will contribute with 60% of the total urban land coverage around the world, in which more than half of the world's population will reside (State of the Tropics, 2020; United Nations, 2018). In 2018, the United Nations (World Urbanization Prospect database) identified the existence of 1248 temperate and 612 tropical cities with more than 300,000 inhabitants (Fig. 11.1). Interestingly, the vast majority of tropical cities occur in developing countries (97.5%; only 15 tropical cities are associated to high income countries; United Nations, 2018), and 14 of the 30 mega cities in the world occur in the tropics as well. In regard to mega cities, it is predicted that in the year 2100 71% of them (51 of 70; population $\geq$ 10 million) will occur in the tropics reaching top populations of close to 90 million people in some cities (Hoornweg & Pope, 2017).

In the last decade, the study of biological evolution in urban environments has become a rapidly growing field (Fig. 11.2a). The Fig. 11.2a was created based on a literature survey including studies on urban evolution reported in Johnson and Munshi-South (2017), in Miles et al. (2019a), as well as studies that hit the following terms in the Web of Science: urban*ecol*genet, "urban*ecol*genom, urb*ecol*evol, urb*ecol*pheno, and urban evolution. Studies were included in the database if selection, mutation, genetic drift and/or gene flow were considered (our review included articles published until mid 2021). From this literature review, we estimated a mean publication of 41.6 ± s.e. 3.44 (between 2016–2020), evaluating whether evolution can be altered by urbanization (e.g., mutation, selection, gene flow, and genetic drift, Fig. 11.2b) (Donihue & Lambert, 2015; Johnson & Munshi-South, 2017; Diamond & Martin, 2021). In general, a great amount of the reviewed

Latitude	Cities (> 300,000 inhabitants)	Cities where urban evolution has been studied	% of studied cities
Tempered	1,248	177	14%
Tropical	612	38	6%

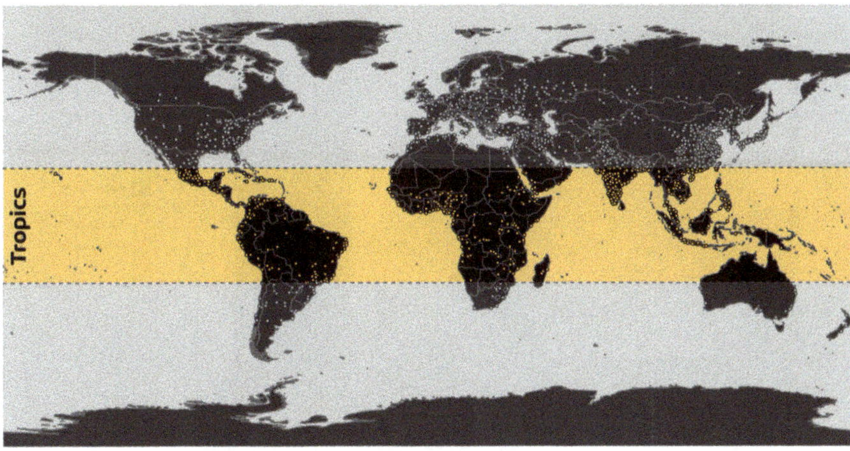

Fig. 11.1 Global distribution of cities (> 300,000 inhabitants). Data obtained from World Urbanization Prospect database; United Nations, 2018). Data for numeralia about cities where urban evolution have been conducted were obtained from the urban evolution literature survey

literature provide evidence that rapid evolution can be shaped by urban environments (Hendry, 2016), suggesting that evolution should be considered in the design of sustainable and resilient urban ecosystems (Donihue & Lambert, 2015). In this way, a general goal of urban evolutionary biology is to understand how the interplay between evolutionary forces facilitates or constrains the adaptive response of organisms to urban environments (Szulkin et al., 2020b).

Interestingly, within this growing body of research, there is a strong bias towards temperate latitudes (74.2%, 297 out of 400 studies of urban evolution; see Supplementary material S1) (Fig. 11.2a). Such bias leaves almost unexplored the evolutionary response of species to urbanization in the tropics (11.2%; 45 out of 400 studies), a region that paradoxically harbors more than 50% of the described and undescribed species (Wilson, 1988; Dirzo & Raven, 2003). Currently, studies of urban evolution have been performed in 14% of the worldwide temperate cities, but only in 6% of tropical ones (cities >300,000 inhabitants sensu United Nations, 2019) (Fig. 11.1). This bias constrains our capacity to achieve broader inferences across ecological communities, taxonomic groups, cities, cultures, and environmental conditions to better understand how urbanization drives biological evolution as well as the viability to apply socio-eco-evolutionary strategies to bring sustaintability to tropical cities (Alberti et al., 2020; Des Roches et al., 2021).

It is widely recognized that biodiversity increases towards tropical latitudes (i.e., the latitudinal biodiversity gradient) and several theories explain this pattern (reviewed by Willig et al., 2003, Mittelbach et al., 2007, Brown, 2014; Fine, 2015; Lawrence & Fraser, 2020). Recurrently, these theories invoke that evolutionary

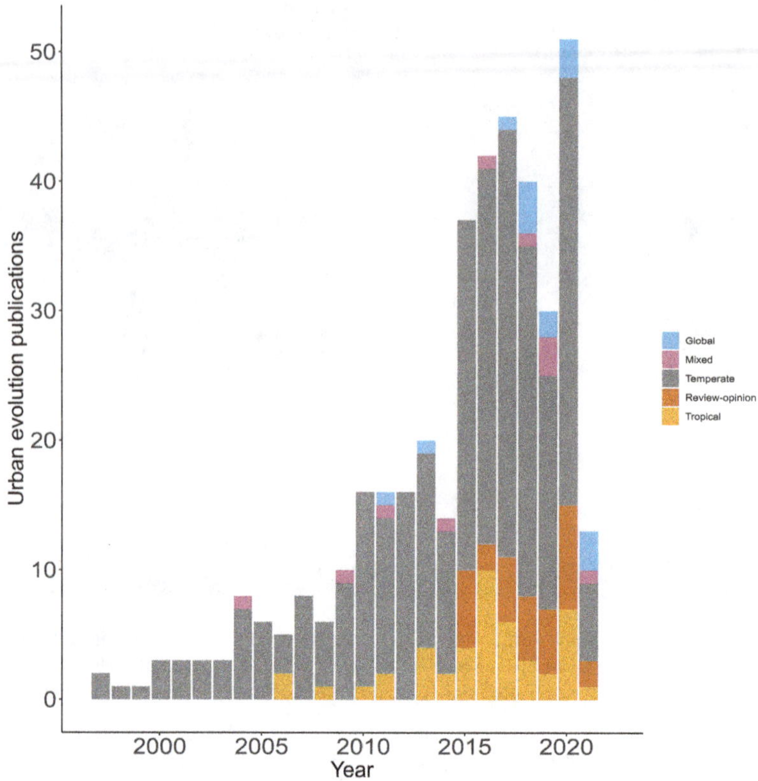

Fig. 11.2 Iincrease in publication rate of urban evolution studies from 1996 to 2021. Earlier sporadically publications were removed for the sake of visualization; the earlier work was published in 1955 (Kettlewell, 1955). Data are provided in the supplementary material S1

forces in combination with ecological processes should operate differently and are expected to cause contrasting speciation, extinction, and migration along the latitudinal gradient (Dowle et al., 2013). For instance, higher mutation rates are expected to occur in the tropics due to the kinetic effect of temperature on biological processes (Rohde, 1992; Wright et al., 2003, 2006; Davies et al., 2004; Orton et al., 2019), higher genetic drift in the tropics is also expected due to smaller populations (Fedorov, 1966; Stevens, 1989; Currie et al., 2004), and stronger selective pressures promoted by more intense biotic interactions are also predicted (Dobzhansky, 1950; Schemske et al., 2009; Brown, 2014). Moreover, higher recombination rates are expected in the tropics, which in turn can lead to faster adaptive changes (Levin, 1975) and favour greater selective divergence in the face of gene flow (Moritz et al., 2000). Together, all these differences in evolutionary processes between tropical and temperate latitudes would be expected to promote contrasting diversification rates (species origin-extinction) and contribute to the high biodiversity observed in the tropics (Dowle et al., 2013; Fine, 2015). While there is evidence supporting higher speciation rates in the tropics (Rolland et al., 2014; Pyron & Wiens, 2013; Condamine et al., 2012) there are other examples showing equal or higher speciation rates in temperate regions (see Mittelbach et al., 2007). For instance, a recent study based on 60,000 angiosperm species found that speciation in this group cools down in the tropics (Igea & Tanentzap, 2020) supporting the alternative hypothesis of "empty niches" (Schluter, 2016; Schluter & Pennell, 2017; Lawrence & Fraser, 2020). Despite that the debate about the causes of latitudinal biodiversity is still open, the evidence that evolutionary processes and patterns are different between temperate and tropical latitudes suggest that the evolutionary implications of urbanization in the tropics might not be directly predicted from results reported in temperate regions, given intrinsic differences in the pace, mode, and underlying factors driving evolution across latitudes (see above). Moreover, this loss of inference can also be increased if patterns of urban expansion (e.g. growth rate and form) are also contrasting between temperate and tropical regions.

Urbanization generates a mosaic of heterogeneous habitats from which complex environmental gradients emerge (McDonnell & Hahs, 2013; Szulkin et al., 2020a), defining the urban ecological scenario where evolutionary process will occur (Hutchinson, 1959). In this context, it might be relevant to ask not only if evolutionary forces act differently between tropical and temperate latitudes but if patterns of urban expansion are also contrasting between these two regions. Different patterns of urban expansion such as growth rate and urban landscape complexity (e.g. complexity, centrality, porosity, and density; Huang et al., 2007; Fig. 11.3) between tropical and temperate cities can affect the inner connectivity between patches within cities and non-urban environments, conditioning gene flow, genetic drift, as well as the adaptive process and eco-evolutionary dynamics to urban conditions (Alberti et al., 2020). Unfortunately, studies focused on urban expansion do not test for the tropical/temperate contrast (Seto et al., 2010; Liu et al., 2014; Xie et al., 2020), making it difficult to evaluate urban expansion across the latitudinal gradient (but see the project State of the Tropics, 2020).

Despite the reduced amount of systematic information about urban expansion in the tropics (see table 6 in Xie et al., 2020) there are some key studies that give some insight about urbanization patterns between temperate and tropical cities (reviewed in Xie et al., 2020); however, none of them explicitly tested for contrast between temperate and tropical latitudes because they are focused on economical regions. Using the supplementary data shared by Huang et al. (2007), which included 77 cities around the world, we found that tropical cities are less complex, less spread, less porous and denser than cities from temperate latitudes, attributes that can importantly affect urban evolutionary processes ($n_{\text{tropical}} = 18$; $n_{\text{temperate}} = 59$; Fig. 11.3a–f). In the same way, the rate of urban expansion is different between tropical and temperate cities. Using data from the World Urbanization Prospect (United Nations, 2018), we estimated that tropical cities grow 1.24 faster than temperate cities (2.79 and 2.25 average annual rate of change in tropical and temperate cities considering the period between 1950 and 2020, respectively) (see Fig. 11.3g and figure legend for the statistics). Beyond answering what are the driving factors of such contrasts, the fact is that such differences in urban dynamic expansion and urban shape could importantly alter the evolutionary dynamics occurring in temperate and tropical cities.

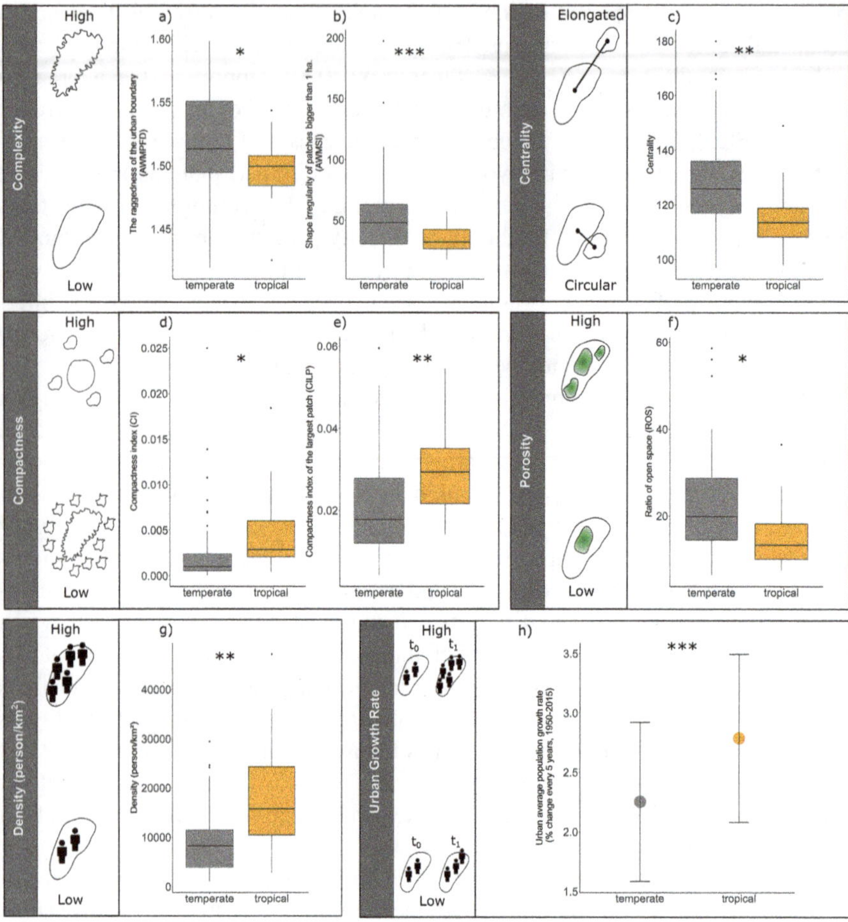

Fig. 11.3 Reanalysed data from Huang et al. (2007) to explore contrast between temperate and tropical cities (59 and 18 cities, respectively) in four main indices describing pattern of urban form. Conclusions about urban complexity incorporate two indices (**a**) Area Weighted Mean Patch Fractal Dimension (AWMPFD) which describes the raggedness for the urban boundary, and (**b**) Area Weighted Mean Shape Index (AWMSI) represents the shape irregularity of the urban patches (> 1 ha.). These two indices show that tropical cities are less complex than tempered cities since have more regular patches with less ragged boundaries ($T_{\text{AWMPFD}} = 2.590$, $df = 37.781$, $P < 0.05$; $T_{\text{AWMSI, unequal variances}} = 3.903$, $df = 72.981$, $P < 0.001$). Tropical cities also are less elongated than temperate cities as indicates the centrality index (**c**), which measures the average distance of the dispersed parts to the centroid of the largest urban patch, considered the city center. ($T_{\text{Centrality}} = 2.84$, $df = 75$, $P < 0.001$). Based on the Compactness Index (CI; C) and Compactness Index of the Largest Patch (CLIP; (**d**) tropical cities resulted more compact than temperate cities because in the tropics there are more regular patches and less number of patches ($T_{\text{CI}} = -2.245$, $df = 75$, $P < 0.05$); and the largest patches in tropical cities are also more compact ($T_{\text{CILP}} = -2.904$, $df = 75$, $P < 0.001$). Porosity analysis indicate that temperate cities have 32.4% higher ratio of open space (ROS), including vegetation and water bodies, than tropical cities. This open spaces are pores or "holes" in relation to the calculates entire urbanized area ($T_{\text{ROS}} = 2.509$, $df = 75$, $P < 0.014$). Also, population density in tropical cities is 2.07 higher than in temperate cities (**h**) ($T_{\text{unequal variances}} = -3.275$,

The idea of using cities as experimental replicates is grounded on the expectation that urban homogenization could favor that even distant cities would share more common features than with nearby undisturbed areas (Donihue & Lambert, 2015). Based on this logic it is possible to test for evolutionary predictions and to explore how replicable the outcomes are (see Santangelo et al., 2020a, b, c, 2022). One way to assess repeatability on evolutionary patterns is to test how frequent the same evolutionary pattern occurs across multiple urban populations of the same specie (one-species-many-cities) (parallel evolution *sensu* Santangelo et al., 2020a), or how frequent such outcome occurs across different species facing urban environments (see Borowy & Swan, 2020; several-species-at-least-one-city) while controlling, ideally, for phylogenetic relationships (convergent evolution at species level *sensu* Bolnick et al., 2018; see Snell-Rood & Wick, 2013; Winchell et al., 2020; Martin et al., 2021; Kunc & Schmidt, 2021). The latter case, in our view, would correspond to the study of urban evolution at the macro evolutionary scale, which has been out of the radar in multiple opinion and review articles (Donihue & Lambert, 2015; Johnson et al., 2015; Szulkin et al., 2020b). This approach could be used to test if preadaptations or exaptations to urban environments occur at the same rate in tropical and temperate latitudes. In either of both approaches, the bias toward temperate cities remain. Only 12 one-species-many-cities studies were performed in tropical latitudes (i.e. studies including more than one city) while in temperate cities we detected 80 studies. In the other hand, only 2 studies including several-species-at-least-one-city were conducted in tropical cities, 11 in temperate latitudes, and two studies included tropical and temperate cities simultaneously. Finally, when we conducted the literature survey we found 18 studies exploring macro evolutionary patterns in the cities, 11 of which included data on functional traits.

2 Urban Evolution of Plant-Associated Species Interactions

Besides the latitudinal gradient bias in the study of urban evolution, we detected a couple more of important biases. First, the number of urban evolution studies focusing on animals was 8.6 times higher than on plants (from the 435 studies of cases only 45 reports were performed in plants; Fig. 11.4), and secondly only 22% of those studies focusing on plants were framed within the context of the evolution of biotic interactions (only one study was performed in tropical cities and nine in temperate cities). Observing that research on animals either in temperate and tropical

Fig.11.3 (continued) *df* = 20.078, $P < 0.01$). **Finally, using data from World Urbanization Prospect (2018) and a linear-mixed model, we found that** tropical cities showed a 1.24 times higher urban average population growth rate (% change every 5 years) than temperate cities (**g**; marginal means and 95% CI) (tropical = 2.79, temperate = 2.25; $F_{1,\ 1808.31} = 16.352$, $P < 0.001$) [2]. The model included *Latitude* (tropical/temperate) and *Country Developed Level* as fixed factors, and *City*, *City nested in Country*, and *Year* as random factors. The model was run with the lme4 package in R (Bates et al., 2015)

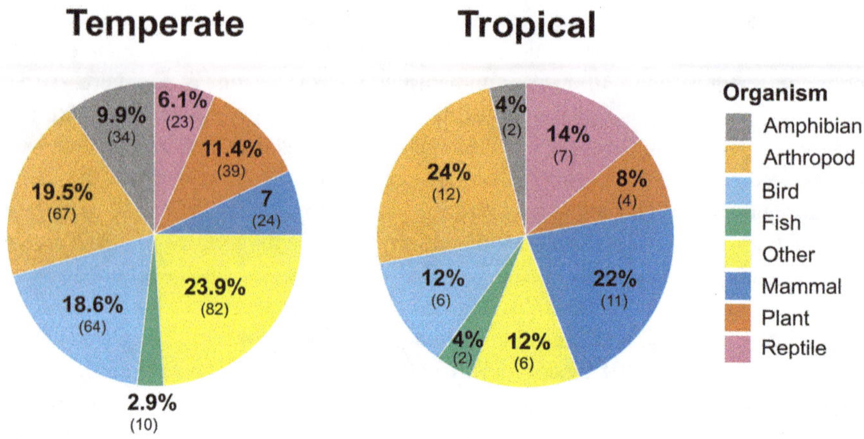

Fig. 11.4 Organisms used in urban evolution studied in temperate and tropical cities (386 studies). Data are provided in the supplementary material S1

cities (78%; Fig. 11.4) is higher than in plants is at odds since plants have played a key role to developing and testing evolutionary theory (Briggs & Walters, 2016) as well as to building the coevolutionary theory in which ecological interactions are a major driver for the evolution of biodiversity (Ehrlich & Raven, 1964; Carmona et al., 2015). There is thus an open field to develop plant-centered research in urban evolution and to use cities as an experimental opportunity to confirm evolutionary predictions using field sampling, combined with common garden and reciprocal transplants experiments making use of genomic tools. In the next section we will focus on plant adapting to urban environments and the importance of using them to study the role of biotic interactions focusing in the available data from temperate cities and discussing what could be expected in tropical latitudes.

3 Plant-Pollinator Interactions and the Evolution of Plant Reproductive Systems in Urban Environments

3.1 Urban Evolution of Plant-Pollinator Interactions

Urbanization leads to the emergence of fragmented and disturbed landscapes dominated by spatial discontinuity, vegetation loss, as well as stressful environmental conditions such as increases in temperature and higher concentrations of greenhouse gases (Johnson & Munshi-South, 2017; see above). These conditions disrupt mutualistic associations such as those between plants and their pollinators (Kiers et al., 2010; Irwin et al., 2020). Like other biotic interactions (*see section: Urbanization effects on plant-herbivore interactions*), there is more information on the effects of urbanization on the disruption of plant-pollinator interactions in temperate than in tropical regions (Silva et al., 2020). Even though it is possible to suggest some general predictions regarding adaptive outcomes due to selective pressures from urbanization (Willig et al., 2003; Chartier et al., 2021), it is also possible to suggest some differences in the strength of the effects of urbanization between temperate and tropical regions through knowledge of patterns of plant and pollinator diversity distribution and plant reproductive systems.

The most drastic effect that urbanization may have in disrupting plant-pollinator interactions, is the loss of mutualistic partners (Kiers et al., 2010). Thus, if the level of urbanization is a *proxy* for the decline in pollinator abundance, one would expect to find a gradient of reduction in pollinator richness and floral visits from wild to urban areas. This decline would translate into pollen limitation that can affect adaptive and non-adaptive evolutionary processes (e.g. genetic drift). For instance, decreasing gene flow in plant species with cross-pollination or mixed reproductive systems (Johnson & Munshi-South, 2017; Miles et al., 2019a; Bennett et al., 2020), accordingly, such pollen limitation can also increase selective pressures on traits that attract new pollinator partners or favor the evolution of autogamy or autonomous selfing. The evolution of these traits, aimed at achieving reproductive assurance in plants (Irwin et al., 2020), would depend on the cost imposed by inbreeding depression (see section on reproductive systems). In this way, the reduction of pollination services in extreme urban fragmented landscapes are central to mediate the dynamics of (co)evolutionary process occurring on the cities.

Although a large body of evidence has shown that fragmentation significantly reduces pollinator richness and pollination services (Aguilar et al., 2006; Eckert et al., 2010), studies of the effect of urbanization on those aspects are still scarce (Silva et al., 2020; Wenzel et al., 2020). Nonetheless, the evidence so far allow us to visualize some general patterns. For example, Wenzel et al. (2020) found in a review that a high level (i.e., percentage of impervious surface) of urbanization is associated with an overall decrease in pollinating insects (importantly, very few of the studies were performed in tropical regions). Another study of Deguines et al. (2016) observed that urbanization results in an over-representation of generalist insects. Furthermore, other authors have found that cities can be refuges for a substantial diversity of insect species, particularly bees (Baldock et al., 2015; Hall et al., 2017). However, one important bias in most of the studies on this topic is that they have focused on insects, primarily on bees without considering other pollinator groups (but see Dzul-Cauich & Munguía-Rosas, 2022 for an study with bats), or without assessing differences between native and introduced pollinators in tropical and temperate cities (Wenzel et al., 2020; Orr et al., 2021; Silva et al., 2020). A more realistic interpretation of the overall effect of urbanization on pollination services needs to consider a broad analysis that encompasses the greatest number of pollinator taxa and their functional diversity, since the effects of urbanization seem not to be uniform (Cane, 2001; Deguines et al., 2016). For example, results of a recent meta-analysis about the effects of urbanization on terrestrial arthropod communities, showed that diversity and abundance of Coleoptera and Lepidoptera were more affected by urbanization than any other arthropods (Fenoglio et al., 2020). Interestingly, analysis of arthropods as functional groups detected that urbanization negatively affected the diversity and abundance of pollinators and natural enemies among other functional groups (Fenoglio et al., 2020), thus suggesting that species with particular traits and their ecological role are key in their capacity to respond to novel urban environments.

Temperate regions are generally characterized by lower pollinator species richness than tropical ones (Ollerton, 2017), and the latter also have higher proportion of plant species specialized in insect pollinators (i.e., bees, butterflies, beetles, wasps and thrips) and a lower proportion of species specialized in fly and bird pollinators (Ollerton et al., 2006). Considering this, it could be expected that urbanization would affect certain pollinator taxa or functional groups and their associated plant partners differentially, depending on their geographical distribution (but see Ollerton, 2017 and Vizentin-Bugoni et al., 2018) and their potential to adapt to particular urban conditions. For example, Ericales species are characterized as being pollinated by members of several taxa (i.e., bees, flies, moths, hummingbirds, and flying and non-flying mammals) that are not evenly distributed (or even absent) between tropical and temperate regions (Chartier et al., 2021), suggesting that some of them would be more affected by urbanization depending on the degree of disturbance associated with the tropical and temperate cities they inhabit.

One of the main biological features of the difference between temperate and tropical cities are the quantitative (species diversity) and qualitative (species identity) aspects of biological communities, which determine the interaction network structure and dynamics. These contrasting patterns may impose different conditions for the complexity of selective responses, driving the evolution of phenotypic traits involved in pollination (Strauss & Irwin, 2004). The higher species richness, and diversity of taxonomic and functional pollination vectors expected in tropical cities will be affected by the higher complexity, lower spread, lower porosity and density than temperate cities (Fig. 11.3).

3.2 Urban Evolution of Plant Reproductive Systems

While urbanization can importantly affect the evolution of the interaction between plants and pollinators acting on traits involved in the interaction, special attention must be given to how this can affect the evolution of mating systems and reproductive attributes of plants under a scenario of pollinator limitation (Irwin et al., 2018, 2020; Cheptou, 2019). As in the previous section was stated, there is global evidence that urbanization increases pollen limitation which can influence the evolution of mating system strategies. For example, using a phylogenetically controlled meta-analysis that included more than 1200 plants species, Bennett et al. (2020) found that plants relying on pollinators in urban settings were more limited by pollen than similar pollinator-dependent plants in natural or managed landscapes. In this study, Bennett et al. (2020) concluded that urbanization-related factors such as fragmentation, pollution and traffic are highly disruptive of plant–pollinator interactions imposing strong selective pressures on the sexual reproductive assurance of the plants. In this context, autogamy has been interpreted as a strategy that would provide self-compatible plants with reproductive assurance in the absence or limitation of pollinators and/or low availability of mates, conditions that would increase pollen limitation (Thomann et al., 2013; Bennett et al., 2020). Several studies have tested this hypothesis by trying to explain the evolution of autogamy in self-compatible species inhabiting alpine communities (Zhang & Li, 2008), species colonizing islands (Barrett, 1996), and species found in fragmented or disturbed habitats (Cheptou & Avendaño, 2006). For example, Eckert et al. (2010) analyzed the effect of anthropogenic habitat disturbance on the mating system of 27 species and found a lower rate of outcrossing in disturbed than on undisturbed sites. Taken together, these studies suggest that urbanization can lead to changes in plant mating systems driven by pollinator limitation, promoting the evolution of autogamy or in the case of mixed mating systems favoring higher selfing rates. However, to our knowledge, few experimental studies (e.g., reciprocal transplant between plants from urban and natural environments, common garden settings) have tested whether urban environments promote the evolution of autogamy. For instance, using the self-compatible species *Crepis sancta*, Cheptou and Avendaño (2006) found that in urban environments that this species had a marginally higher rate of autogamy than in rural environments (Cheptou & Avendaño, 2006). Interestingly, this result was not due to an increase in autonomous selfing, but to a reduction in outcrossed seeds because of reduced pollinator availability (Cheptou & Avendaño, 2006). In addition, the divergent evolution of mating system strategies between non-urban and urban environments can importantly affect the evolutionary dynamics of populations leading to speciation. For instance, evolution toward autogamy or increasing selfing rates could promote the evolution of reproductive isolation by increasing genetic drift fixing incompatible alleles in hybrids (e.g. drift base speciation; Thompson et al., 2018).

In agreement with the evidence that ecological and evolutionary process are different between temperate and tropical regions, it is important to highlight that also plant mating systems respond to a latitudinal gradient. Outcrossing systems are more frequent at lower latitudes while autogamy are more common at higher latitudes (but Moeller et al., 2017). Therefore, it is expected more outcrossing species should be affected by urbanization in the tropics. However, this pattern seems to be related not only to differences in plant species and pollinator richness, but also to a greater diversity of pollination systems in the tropics (Ollerton et al., 2006). Therefore, it could be expected that selection on the plant mating system, favoring autogamy, would be stronger in tropical than temperate urban environments (Desaegher et al., 2019), or that the local or global extinction rates of outcrossing plants facing urbanization in the tropics will be higher. The occurrence of both possible outcomes would depend on the evolutionary ability (evolvability) of mating systems of each taxa.

One central mechanism that will favor the evolution of autogamy to evolve is the loss of self-incompatible alleles (SI). In particular, the incompatibility system of plants inhabiting cities represents a good example of a reproductive trait that could be used to test the prediction that urbanization will reduce allele richness (Miles et al., 2019a), because the optimal functioning of the incompatibility system depends on the population's richness of incompatibility alleles (S-locus). In fragmented urban landscapes, both the size and fragmentation of plant populations and the activity of pollinators could affect the availability of mates, limiting the possibilities for mating. Thus, if plants in self-incompatible small populations (bearer of the Self-Incompatible alleles; SI) receive few floral visits, pollinators will transport pollen from a reduced number of donors, which in turn, could result in a reduction in gene flow and a high genetic structure (Cheptou & Avendaño, 2006). Consequently, a low representation of the pool of S alleles could be expected to be found in populations, since only some of these alleles would remain after an expected bottleneck effect in urban environments. The loss of a significant number of S alleles would favor the appearance and spread of new self-compatibility (SC) mutations that would break or modify incompatibility responses (Busch & Schoen, 2008). These mutations would promote the occurrence of autogamy or a mixed mating system when outcrossing becomes a disadvantageous mating strategy if pollinators are scarce. To test the hypothesis that SC mutations should spread more rapidly in populations with low visitation rates, Voillemot et al. (2019) introduced individuals from a self-compatible population of the species *Linaria cavanillesii* to experimental self-incompatible arrays with different levels of pollinator availability. The authors found that self-compatibility fixed within a few generations, particularly in those populations in which pollinators were partially or totally excluded (but see Layman et al., 2017). Although this case helps explaining the evolution of partial self-incompatibility or self-compatibility in self-incompatible species under scenarios of pollinator limitation, as those expected for urban environments, experimental studies evaluating the effect of urbanization on the level of compatibility within and between urban and natural populations are lacking. The latter is also true for studies comparing the evolutionary shift from self-incompatibility to self-compatibility systems in species distributed in tropical and temperate cities. Interestingly, self-incompatible, including dioecious species, are more frequent in tropical than temperate regions (Igic et al., 2008), suggesting that the breakdown of self-incompatibility in response to urbanization could be also more frequent in species of tropical than temperate cities.

The adaptive responses of populations to changes in selective pressures imposed by a new pollination environment will depend on the evolutionary ability of floral attributes related to mating systems (i.e. reduction in herkogamy and/or flower size when autogamy is favored, Opedal, 2019). It has been shown, for instance, that adaptive evolution of certain floral traits in response to pollinator limitation can occur in few generations (Brys & Jacquemyn, 2012; Gervasi & Schiestl, 2017). For example, in just five generations in the absence of pollinators, Roels and Kelly (2011) detected a significant reduction in anther-length distance (herkogamy) associated to an increase in seed production from autonomous selfing in experimental populations of *Mimulus gutattus*. Although the evolvability of the attributes is determined by their genetic variance and heritability (Thomann et al., 2013; Cheptou, 2021), it must be considered that certain groups of attributes will have greater constraints (genetical or developmental) on adaptive evolution than others. For example, Opedal (2019) analyzed the evolvability of animal-pollinated flowers in novel pollinator communities and found that floral attributes associated with mating system (e.g. herkogamy) and attraction (e.g. floral display and volatile compound concentrations) have higher evolvability than those related to the mechanical fit between flowers and pollinators (e.g. shape), which had a lower-than-average evolvability value.

So far, there is insufficient evidence to discuss general patterns of the effect of urbanization on the evolution of floral attributes mediating pollination and influencing the evolution of mating system. Therefore, it is not yet known how rapidly these attributes evolve in urban environments. Considering this, predicting the evolutionary trajectory of floral attributes in the cities seems not feasible yet. For example, contrary to expectations of selection for smaller floral displays, and shorter and wider flowers in plants of the species *Gelsemium sempervirens* in urban compared to non-urban sites, Irwin et al. (2018) detected that selection was 28% stronger in urban sites on floral display size, with no differences in selection for flower size or flower shape (Irwin et al., 2018). In contrast, a recent analysis of phenotypic variation in reproductive attributes in 366 plant species distributed along an urbanization gradient, suggested that reproductive strategies such as self-pollinated, self-compatible and non-entomophilous are favored in urban environments (Desaegher et al., 2019). Also, the same study detected that tubular and narrow flowers were less frequent on those scarce entomophilous species founded in urban environments (Desaegher et al., 2019). These results highlight other important aspect associated with urban environments, besides the apparent advantage of autogamy, the disadvantage of specialization due to constraints of specialized pollination when pollinators are scarce.

If plant species are pollinator-limited in urban environments, it would be expected that species with specialized floral designs would receive even fewer visits than in wild populations, or than generalist species in urban communities. In other words, urbanization would be expected to favor not only autogamy, but also generalization in plant-pollinator interactions. Interestingly, Doré et al. (2021) analyzed 295 pollination networks in 123 sites around the world and detected a positive association between anthropogenic pressures (assessed through different variables, including the level of urbanization) and the increase in the level of generalization in plant-pollinator networks. These results are consistent with the higher proportion of generalist insects found in urban landscapes (e.g., Deguines et al., 2016), suggesting then that one of the results of urbanization is the loss of specialized interactions in pollination interactions (see Bennett et al., 2020). Furthermore, urban environments are home of a large number of alien plant species that have been introduced for ornamental purposes, and invasive species that have migrated to cities by various means (Alberti et al., 2017). These species have been shown to be highly generalist in their use of the resident pollinators of the communities they invade, which allows them to easily integrate into plant-pollinator networks (Parra-Tabla & Arceo-Gómez, 2021). For example, many invasive species, and particularly ornamental species, produce flowers that are highly attractive to floral visitors, so their presence could rapidly change the dynamics of pollination networks where they are introduced (Johnson et al., 2017). Other aspects of urbanization that we have not addressed here, but which deserve attention, is to try to understand how rising temperatures (e.g. heat islands) and greenhouse gases affect floral traits that mediate plant-pollinator interactions, such as nectar and floral scents (see Irwin et al., 2020). For instance, it has been found that plant phenology is been altered in urban cities, likely due to the appearance of heat islands (Neil & Wu, 2006). This phenological asynchrony between urban and non-urban habitats could cause impede outcrossing, contributing to the aforementioned process of divergent evolution.

There is growing interest in the evolutionary consequences of urbanization, but it is not yet known how often the changes that species undergo are adaptive and whether these changes differ between tropical and temperate regions. So far, in mutualist plant-pollinator interactions, there are plausible predictions suggesting that urbanization results in pollinator limitation, more generalized interactions, and a shift towards an autogamy mating system regardless of geographic region. However, differences in the richness of interactions between plants and their pollinators and in the diversity of plant mating systems, suggests that the changes resulting from urbanization would be more dramatic in the tropics than in temperate regions. Yet, it is important to underline that there is still a lack of experimental evidence to shed light on specific environmental effects associated with urbanization that may influence the plant-pollinator interactions on the evolution of the plant's mating systems, especially in the tropics.

4 Urbanization Effects on Plant-Herbivore Interactions

4.1 Ecological Factors Driving Urbanization Effects on Plant-Herbivore Interactions

A growing number of studies have reported on changes in trophic species interactions between natural or rural environments and urban settings (Turrini et al., 2016; Kozlov et al., 2017; El-Sabaawi, 2018; Moreira et al., 2019; Fenoglio et al., 2020; Theodorou, 2022). One such interactions of widespread importance which are affected by urbanization are those involving plants and herbivores, mainly insects (Raupp et al., 2010; Miles et al., 2019b). Altered patterns of herbivory in urban areas have been attributed to changes in plant and herbivore species composition (via environmental filters) determining which interactions take place and how they play out in urban settings (Faeth et al., 2005; Diamond et al., 2015). Likewise, a number of abiotic factors (e.g. temperature, atmospheric CO_2, pollutants), physical conditions (e.g. spatial connectivity), and biotic factors (e.g. competition, predation) affect the ways in which planted/rewilded or persisting plant and herbivore species interact in urban sites (Kozlov & Zvereva, 2016; reviewed by Miles et al., 2019b). Accordingly, recent studies have found differences in the amount and type of herbivore damage on plants in natural vs. urban areas or along rural-urban gradients (Kozlov et al., 2017; Moreira et al., 2019). In turn, these changes in herbivory (together with other factors, e.g. abiotic conditions) are expected to lead to changes in herbivore selection on plants, ultimately determining the evolution of plant defences and plant-herbivore co-evolution (Johnson & Munshi-South, 2017; Santangelo et al., 2022).

Studies reporting on changes in herbivory, particularly by insects, between urban and natural settings or along transitions between urban and rural habitats have received increasing attention in last decades, mainly in temperate regions (Dreistadt et al., 1990; Raupp et al., 2010; Turrini et al., 2016). Evidence for directionality of urbanization effects on herbivory is mixed: some have reported increased insect herbivory in urban areas relative to rural habitats (e.g. Cuevas-Reyes et al., 2013, Turrini et al., 2016), whereas others have found the opposite pattern (e.g. Bode & Gilbert, 2016; Kozlov et al., 2017; Moreira et al., 2019). While outcomes appear somewhat idiosyncratic and species-specific (Diamond et al., 2015; Miles et al., 2019b), some consistency can be observed when grouping by herbivore guild with patterns generally countering the prediction of increased herbivory in urban areas. For example, leaf herbivory by chewing insects (e.g. caterpillars, beetles) on temperate trees (e.g. Quercus, Populus, Betula) is generally lower in urban than in rural sites (Bode & Gilbert, 2016; Kozlov et al., 2017; Moreira et al., 2019). In addition, leaf-miner damage has been shown to also decrease in urban settings for several temperate tree species (Kozlov et al., 2017) though in one other case no difference was found (Moreira et al., 2019). Although these results suggest reduced herbivory in urban areas is more common, a greater number of studies reporting on herbivore guild-specific differences and their underlying causes are needed to uncover mechanisms.

Changes in herbivory patterns in urban vs. natural setting may take place via different mechanisms, including direct abiotic or physical effects on herbivores and also indirect effects via changes in plant traits such as chemical or physical defences and/or nutritional quality (i.e. bottom-up effects; Thompson et al., 2016) and/or in the diversity and abundance of herbivore enemies (i.e., top-down effects; Turrini et al., 2016; Kozlov et al., 2017). These drivers of urbanization effects on herbivory have unfortunately been seldom tested. Nonetheless, the few studies that have done so provide a first glimpse at the relative importance of biotic and abiotic drivers and hint at underlying mechanisms. For example, it has been proposed that increased resource availability due to nutrient deposition (e.g. nitrogen) or increased CO_2 availability may result in higher growth and lowered plant defences in urban environments, and, in turn, expectedly higher herbivory (note also that in some respects urban environment may be more stressful for plants; Johnson et al., 2015; Miles et al., 2019b). However, in a recent study (Moreira et al., 2019) found that increased CO2 emissions were correlated with lowered levels of leaf defences as expected (and higher leaf nutrients as well) for oak trees in urban sites relative to those in rural sites. However, contrary to expectations, leaf chewer damage was lower in urban sites and did not relate to the measured plant defensive traits (Moreira et al., 2019). Likewise, despite some evidence of differences in leaf physical traits between urban and rural sites, (Kozlov et al., 2017) found lower leaf miner and chewer damage in urban sites for several tree species and leaf traits apparently did not explain these differences in herbivory. Combined, these studies suggest that other unmeasured plant traits could have driven these herbivory patterns, or simply that bottom-up effects of urbanization on herbivory via plant traits are often weak.

Studies addressing altered top-down effects have found some evidence that changes in predation in urban environments could underlie herbivory differences between natural and urban settings. While predator and parasitoid diversity and abundance are usually lower in urban areas (Burkman & Gardiner, 2014), specific guilds or species of natural enemies can in some cases exert strong control on phytophagous insects in urban areas (Faeth et al., 2005; Philpott et al., 2020). However, as for herbivores, natural enemy responses also are highly variable as well as guild-specific. Levels of parasitism on insect herbivores are variable, including higher (Sumoski et al., 2009) or lower (Nelson & Forbes, 2014; Kozlov et al., 2017) parasitism in urban areas, as well no response to urbanization (Fenoglio et al., 2009). For insectivorous birds, patterns are also mixed with evidence for both increased (Kozlov et al., 2017) and decreased predation in urban areas, whereas for ants at least one study has found increased predation in urban sites (Kozlov et al., 2017). More studies addressing simultaneously changes in bottom-up and top-down effects are clearly needed to increase inference and better understand the mechanisms underlying changes in plant-herbivore interactions in urban settings.

4.2 Urban Evolution on Plant-Herbivore Interactions

While the effects of urbanization on plant evolution have been addressed by a relatively small number of studies (Fig. 11.2), few of these have explicitly considered the importance ecological interactions (Johnson & Munshi-South, 2017; 22% 10 out of 45 studies in plants; see previous section on plant-pollinator interactions). Indeed, the few studies that have tested for changes in plant evolution due to urbanization have focused on phenological, physiological and other life-history traits in relation to selection by abiotic factors, whereas work on plant defense evolution is scarce (e.g. Johnson et al., 2015; Pisman et al., 2020; Qu et al., 2022). Similarly, evolutionary effects of urbanization on herbivore traits (e.g. feeding behavior, host location, feeding physiology and performance) associated with herbivory have not been commonly studied (Miles et al., 2019b; Theodorou, 2022).

As described in the previous section, urbanization alter the ecological context (drivers) in which plant-herbivores interact, this can potentially affect the adaptive evolution of plant defensive traits as has been previously shown (Thompson et al., 2016; Santangelo et al., 2020c). Basically, it is expected that natural selection will optimize (cost/benefit relationship) the expression of defense to attain maximal fitness (Simms, 1992). While the benefit to allocate resources to defense depends on the amount and probability of damage, the cost of defenses are expected to depend on the availability of limiting resources (Rhoades, 1979; Maschinski & Whitham, 1989; Miles et al., 2019b), the direct cost of the defensive trait, and the indirect costs emerged due to pleiotropic effects (e.g. ecological costs; Strauss et al., 2002; Koricheva, 2002). Such optimal approximation (Rhoades, 1979; Stamp, 2003) set the basis to predict that maximal fitness in cities, in which herbivory could be expected to be low (Raupp et al., 2010; Miles et al., 2019b), can be attained by reducing resources allocation to defenses, and potentially explaining the higher levels of defense in rural areas than in cities, as it has been detected for *Trifolium repens* across the globe (Santangelo et al., 2022). Nevertheless, such adaptive response can be constrained by other evolutionary forces such as gene flow and genetic drift (Johnson & Munshi-South, 2017; Miles et al., 2019a; Alberti et al., 2020), which can play a very relevant role in the highly fragmented urban (Santangelo et al., 2018; Johnson et al., 2015).

Even though urbanization occur globally, and it can be considered the best and larger experiment in the history to study adaptation to new environments (see Santangelo et al., 2022), the study of how urbanization affect plant defense evolution is taking time to flourish. Particularly, the use of the rural-urban experimental contrast or gradient (see previous section) can be used to explore changes in the adaptive value of plant traits. Also, these experiments can help to assess how the costs and benefits of defensive strategies (e.g. tolerance and resistance, induced and constitutive) and their adaptive value, change along resource availability (Simms & Triplett, 1994; Maschinski & Whitham, 1989; Miles et al., 2019b), and whether changes in the complexity of herbivore community favor the evolution of alternative or mixed defensive strategies (Carmona & Fornoni, 2013). From a multivariate perspective (Strauss et al., 2004; Wise & Rausher, 2013; Santangelo et al., 2020b) changes in the defensive phenotype between urban and rural environments can be used to explore the relevance of genetic and ecological constraints for the maintenance of phenotypic variance in the expression of defense (e.g. the role of plant reproductive systems on the evolution of plant defense; Bello-Bedoy & Núñez-Farfán, 2010; Núñez-Farfán et al., 2007; Johnson et al., 2015) as well as to search for recognizable defensive syndromes evolving in urban environments (Agrawal & Fishbein, 2006; Santangelo et al., 2020b).

Currently, only a handful of studies have examined the urban evolution of plant defense. One of such is work on white clover, *Trifolium repens,* which has been shown to have evolved decreased levels of hydrogen cyanides in response to urbanization (Thompson et al., 2016; Johnson et al., 2018; Santangelo et al., 2020c, 2022). In particular, such pattern might be due to the combined selection of drought stress and herbivory (see Santangelo et al., 2022). Recently Santangelo et al. (2022) searched for patterns of parallel evolution on adaptive clines in the production of hydrogen cyanides at worldwide scales. This research was launched as part of a global project called Global Urban Evolution (GLUE) that managed to integrated 287 scientists from 26 countries and 160 cities around the globe. Interestingly very few researchers from tropical latitudes got involved, which favored the strong temperate bias discussed above. In this study only 10 out of 160 cities were sampled in tropical cities (6.2%). Using Santangelo et al. (2022) supplementary data we found that the probability of detecting adaptive clines (positive or negative) was similar between temperate and tropical cities ($\chi^2_1 = 1.41$, $P = 0.235$). Some other studies have folded in tri-trophic interactions, starting by classical work on *Biston betularia* showing that the evolution of this moths' coloration is driven by plant-mediated crypsis to avian predators (Kettlewell, 1955; Cook & Saccheri, 2013).

4.3 Urbanization Effects on Plant-Herbivore Evolution in the Tropics

While most of the above studies looking at urbanization-driven ecological and evolutionary patterns in plant-herbivore interactions come from temperate zones (Johnson & Munshi-South, 2017; Miles et al., 2019b). A good example in the tropics was done in an insular system found that urbanization altered selection by Darwin's finches on fruit traits of *Tribulus cistoides* in the Galapagos Islands (Rivkin et al., 2020). Seed and fruit removal rates were higher in urban compared to natural populations and urbanization modified selection on fruit traits that confer defence against avian predation. In particular, they found that seed or fruit removal imposed stronger selection for large and poorly defended mericarps in urban populations than in non-urban populations (Rivkin et al., 2020). In addition, lower seed and fruit removal rates in urban populations appear to be at least partly mediated by reduced finch diversity and alternative food sources in urban areas which mediates reduced attack rates on fruits (Rivkin et al., 2020). To our knowledge, there are no studies comparing evolutionary effects of urbanization on plant-herbivore interactions in temperate vs. tropical regions.

Although there is a large amount of heterogeneity in city landscapes which results in concomitant variation in abiotic, biotic and physical conditions, several factors vary relatively consistently along urban to rural/natural gradients in tropical and temperate regions. For instance, cities generally exhibit higher mean daily surface and air temperature compared with nearby rural areas (Diamond et al., 2015), as well as higher levels of some nutrients (e.g. nitrogen deposition) and pollutants or gases (e.g., CO_2) (Moreira et al., 2019), which directly or indirectly influence plants and herbivores. These factors may differ in the degree to which they vary along urban-rural gradients across temperate and tropical regions. For example, effects of urbanization on herbivory due to warming might vary across latitudes (Kozlov et al., 2017), as differences in temperature between urban and nearby rural sites are expected to be smaller in tropical compared to temperate sites and because thermal tolerances are narrower for insect species at low relative to high latitudes (Diamond et al., 2015). Similarly, effects of water availability may vary less markedly between urban and natural sites in tropical areas due to increased and less variable rainfall (Miles et al., 2019b). These examples point at plausible differences in how abiotically determined effects of urbanization on plant-herbivore interactions playout across latitudes. Presumably, stronger urban to rural/natural habitat differences suggests these climate-related urbanization effects will be stronger in temperate regions. However, reductions in diversity between background natural communities relative to urban settings will be greater in tropical regions, potentially leading to stronger effects of reduced diversity on herbivory rates (Diamond et al., 2015) at low relative to high latitudes. In addition, outcomes will also depend on species plasticity. While theory predicts that plasticity will be greater in temperate than tropical taxa and therefore buffer temperate taxa against abiotically-driven urbanization effects (Diamond et al., 2015), recent evidence indicates that plasticity in some traits (e.g. associated with thermal acclimation) increases at lower latitudes (Seebacher et al., 2015). Therefore, outcomes are likely to depend on the individual or combined effects of multiple biotic or abiotic drivers and plasticity responses of specific taxa or guilds of plants and herbivores.

Due to the paucity of tropical studies, there is expectedly no basis for a robust comparison of temperate vs. tropical differences in urbanization effects on plant-herbivore interactions and their evolutionary implications. However, two recent studies tested for latitudinal variation in urbanization effects on plant-herbivore interactions within temperate regions of Europe (Kozlov et al., 2017; Moreira et al., 2019). Both studies found no evidence that urbanization effects on herbivory or plant defenses on a several tree taxa were contingent on latitude (Kozlov et al., 2017, Moreira et al., 2019). Yet, being restricted to a temperate region these findings might not be representative of broader latitudinal gradients that encompass tropical regions (Carmona et al., 2020), thus calling for studies testing for latitudinal variation in urbanization effects that span both temperate and tropical sites.

5 Urban Evolution on Belowground-Plant Interactions

Soils are importantly affected by urbanization (Grimm et al., 2008), and such alterations can affect the evolution of below and aboveground plant traits (Geem et al., 2013; Mejia-Alva et al., 2018; Colom & Baucom, 2020). However, to our knowledge, there are no studies yet on belowground urban evolution. For instance, in the case of plant-mycorrhizal fungi (MF) interactions there are few studies evaluating the effect of urbanization on MF communities and root infection and if such alterations could affect urban tree health or ecosystem services (Baxter et al., 1999; Ochimaru & Fukuda, 2007; Egerton-Warburton & Allen, 2000; Stabler et al., 2001; Cousins et al., 2003; Lin et al., 2021; Buil et al., 2021); however, the evolutionary perspective is lacking (Newbound et al., 2010; Irwin et al., 2020). Interestingly, in contrast to plant-ectomycorrhizal fungi associations the interaction between plant and arbuscular mycorrhizal fungi is widely distributed in the tropics (Allen et al., 1995). Such biogeographic patterns can favor different ecological and evolutionary dynamics between temperate and tropical environments (Averill et al., 2019) including cities as well as the process of invasion of the urban ecosystem and worldwide spread of tropical plants in a warming world (Pringle et al., 2009; Borden & Flory, 2021). In general, there is a great opportunity to assess the impact of urbanization affecting soil (biotic and abiotic component) and its (co)evolutionary implications occurring belowground across temperate as tropical cities.

6 Conclusions and Perspectives

Urbanization effects are rapidly expanding across the globe and we are only starting to explore how organisms adapt to these new environments. Even though the study of plants has played a key role to understand ecosystem functioning and services (Hooper et al., 2005), as well as to develop the (co)evolutionary theory (Ehrlich & Raven, 1964; Briggs & Walters, 2016), there is still few research with a phytocentric point of view in the study of urban evolution, as well as a very limited amount of work assessing the impact of urbanization on plant-associated ecological interactions, particularly in tropical regions. Ecological interactions are fundamental to understand the process that generate biodiversity (Ehrlich & Raven, 1964) and they need to be studied in the context of urbanization and in tropical regions.

Even though cities occur at the planetary scale, currently our understanding about urban evolution is based on evidence generated mainly in temperate cities. Given that evolutionary processes unfold differently in tropical regions compared to temperate latitudes, and that urban expansion (along with socio-economic and cultural drivers) follows distinct patterns, there is a pressing need to study these effects across latitudes in order to develop a more nuanced, mechanistic understanding of regional differences. Recently, it has been shown that systemic racism and socio-economical segregation within cities can drive contrasting ecological and evolutionary responses in different organisms (Schell et al., 2020; Schmidt & Garroway, 2022; Lima et al., 2023). A global view of such inequality suggest divergent evolutionary processes also between temperate and tropical cities since they are commonly associated with developed and developing countries, respectively. In the tropics, due to weak governance, limited resources and ineffective policy implementations create significant challenges in studying urban tropical systems. As a result, the development and implementation of appropriate policies often lags behind, contributing to the emergence of complex and challenging urban environments in tropical cities (Reiner Jr et al., 2015; Fang et al., 2020; Leandro et al., 2025; Lima et al., 2023; Souza et al., 2023). While it is suggested that urban evolution can be play a key role to design sustainable and resilient cities (Donihue & Lambert, 2015; Beridze et al. 2023; Leandro et al., 2025), it can be challenging or unfruitful to apply standardized strategies in developed countries (Dobbs et al., 2019) to socio-eco-evolutionary context not yet studied. We call to study urban evolution in the tropics and create and being part of global projects (e.g. GLUE project; Santangelo et al., 2022), to put tropical urban evolution on the map.

Supplementary Data
S1: Literature survey database

Acknowledgements This work was supported by the CONAHCyT grant 'Ciencia de Frontera 2019' (Grant No. 568482)

References

Agrawal, A. A., & Fishbein, M. (2006). Plant defense syndromes. *Ecology, 87*, S132–S149.

Aguilar, R., Ashworth, L., Galetto, L., & Aizen, M. A. (2006). Plant reproductive susceptibility to habitat fragmentation: Review and synthesis through a meta-analysis. *Ecology Letters, 9*, 968–980.

Alberti, M., Correa, C., Marzluff, J. M., Hendry, A. P., Palkovacs, E. P., Gotanda, K. M., et al. (2017). Global urban signatures of phenotypic change in animal and plant populations. *Proceedings of the National Academy of Sciences, 114*, 8951–8956.

Alberti, M., Palkovacs, E. P., Roches, S. D., Meester, L. D., Brans, K. I., Govaert, L., et al. (2020). The complexity of urban eco-evolutionary dynamics. *Bioscience, 70*, 772–793.

Allen, E. B., Allen, M. F., Helm, D. J., Trappe, J. M., Molina, R., & Rincon, E. (1995). Patterns and regulation of mycorrhizal plant and fungal diversity. *Plant and Soil, 170*, 47–62.

Averill, C., Bhatnagar, J. M., Dietze, M. C., Pearse, W. D., & Kivlin, S. N. (2019). Global imprint of mycorrhizal fungi on whole-plant nutrient economics. *Proceedings of the National Academy of Sciences, 116*, 23163–23168.

Baldock, K. C., Goddard, M. A., Hicks, D. M., Kunin, W. E., Mitschunas, N., Osgathorpe, L. M., et al. (2015). Where is the UK's pollinator biodiversity? The importance of urban areas for flower-visiting insects. *Proceedings of the Royal Society B: Biological Sciences, 282*(1803), 20142849.

Barber, N. A. (2018). Cross-compartment herbivory effects on antagonists and mutualists and their consequences for plant fitness. In T. Ohgushi, S. Wurst, & S. N. Johnson (Eds.), *Aboveground–Belowground Community Ecology* (pp. 247–269). Springer.

Barrett, S. C. H. (1996). The reproductive biology and genetics of Island plants. *Philosophical Transactions of the Royal Society of London. Series B: Biological Sciences, 351*, 725–733.

Bates, D., Mächler, M., Bolker, B., & Walker, S. (2015). Fitting linear mixed-effects models using lme4. ArXiv preprint arXiv 1406.5823.

Baxter, J. W., Pickett, S. T., Carreiro, M. M., & Dighton, J. (1999). Ectomycorrhizal diversity and community structure in oak forest stands exposed to contrasting anthropogenic impacts. *Canadian Journal of Botany, 77*, 771–782.

Bello-Bedoy, R., & Núñez-Farfán, J. (2010). Cost of inbreeding in resistance to herbivores in Datura stramonium. *Annals of Botany, 105*(5), 747–753.

Bennett, J. M., Steets, J. A., Burns, J. H., Burkle, L. A., Vamosi, J. C., Wolowski, M., et al. (2020). Land use and pollinator dependency drives global patterns of pollen limitation in the Anthropocene. *Nature Communications, 11*, 1–6.

Beridze, B., Mahamoud-Issa, M., Zmudzinski, M., Bednarz, P. A., Filutowska, Z. K., Sikora, B., Winiarska, D., Carmona, D., & Angeoletto, F. (2023). Welcome to the (urban) jungle: notes on new science, biological evolution in cities. *European Journal of Ecology, 9*(2).

Bode, R. F., & Gilbert, A. B. (2016). Seed predators, not herbivores, exert natural selection on Solidago spp. in an urban archipelago. *Environmental Entomology, 45*, 150–154.

Bolnick, D. I., Barrett, R. D., Oke, K. B., Rennison, D. J., & Stuart, Y. E. (2018). (Non) parallel evolution. *Annual Review of Ecology, Evolution, and Systematics, 49*, 303–330.

Borden, J. B., & Flory, S. L. (2021). Urban evolution of invasive species. *Frontiers in Ecology and the Environment, 19*, 184–191.

Borowy, D., & Swan, C. M. (2020). A multi-trait comparison of an urban plant species pool reveals the importance of intraspecific trait variation and its influence on distinct functional responses to soil quality. *Frontiers in Ecology and Evolution, 8*, 68.

Briggs, D., & Walters, S. M. (2016). *Plant variation and evolution*. Cambridge University Press.

Brown, J. H. (2014). Why are there so many species in the tropics? *Journal of Biogeography, 41*, 8–22.

Brys, R., & Jacquemyn, H. (2012). Effects of human-mediated pollinator impoverishment on floral traits and mating patterns in a short-lived herb: An experimental approach. *Functional Ecology, 26*, 189–197.

Buil, P. A., Renison, D., & Becerra, A. G. (2021). Soil infectivity and arbuscular mycorrhizal fungi communities in four urban green sites in Central Argentina. *Urban Forestry & Urban Greening, 64*, 127285.

Burkman, C. E., & Gardiner, M. M. (2014). Urban greenspace composition and landscape context influence natural enemy community composition and function. *Biological Control, 75*, 58–67.

Busch, J. W., & Schoen, D. J. (2008). The evolution of self-incompatibility when mates are limiting. *Trends in Plant Science, 13*, 128–136.

Cane, J. H. (2001). Habitat fragmentation and native bees: A premature verdict? *Conservation Ecology, 5*, 3.

Carmona, D., & Fornoni, J. (2013). Herbivores can select for mixed defensive strategies in plants. *New Phytologist, 197*, 576–585.

Carmona, D., Fitzpatrick, C. R., & Johnson, M. T. J. (2015). Fifty years of co-evolution and beyond: Integrating co-evolution from molecules to species. *Molecular Ecology, 24*, 5315–5329.

Carmona, D., Moreira, X., & Abdala-Roberts, L. (2020). Latitudinal and elevational gradients in plant defences and herbivory in temperate trees: Recent findings, underlying factors, and an evolutionary perspective using genomic tools. In J. Núñez-Farfán & P. L. Valverde (Eds.), *Evolutionary ecology of plant-herbivore interaction* (pp. 343–368). Springer.

Chartier, M., von Balthazar, M., Sontag, S., Löfstrand, S., Palme, T., Jabbour, F., et al. (2021). Global patterns and a latitudinal gradient of flower disparity: Perspectives from the angiosperm order Ericales. *New Phytologist, 230*, 821–831.

Cheptou, P. O. (2019). Does the evolution of self-fertilization rescue populations or increase the risk of extinction? *Annals of Botany, 123*, 337–345.

Cheptou, P. O. (2021). Pollination strategies in the face of pollinator decline. *Botany Letters, 168*, 316–323.

Cheptou, P. O., & Avendaño, V. L. G. (2006). Pollination processes and the Allee effect in highly fragmented populations: Consequences for the mating system in urban environments. *New Phytologist, 172*, 774–783.

Cheptou, P. O., Hargreaves, A. L., Bonte, D., & Jacquemyn, H. (2017). Adaptation to fragmentation: Evolutionary dynamics driven by human influences. *Philosophical Transactions of the Royal Society B: Biological Sciences, 372*(1712), 20160037.

Colom, S. M., & Baucom, R. S. (2020). Belowground competition can influence the evolution of root traits. *The American Naturalist, 195*, 577–590.

Condamine, F. L., Sperling, F. A., Wahlberg, N., Rasplus, J. Y., & Kergoat, G. J. (2012). What causes latitudinal gradients in species diversity? Evolutionary processes and ecological constraints on swallowtail biodiversity. *Ecology Letters, 15*, 267–277.

Cook, L. M., & Saccheri, I. J. (2013). The peppered moth and industrial melanism: Evolution of a natural selection case study. *Heredity, 110*, 207–212.

Cousins, J. R., Hope, D., Gries, C., & Stutz, J. C. (2003). Preliminary assessment of arbuscular mycorrhizal fungal diversity and community structure in an urban ecosystem. *Mycorrhiza, 13*, 319–326.

Cuevas-Reyes, P., Gilberti, L., González-Rodríguez, A., & Fernandes, G. W. (2013). Patterns of herbivory and fluctuating asymmetry in Solanum lycocarpum St. Hill (Solanaceae) along an urban gradient in Brazil. *Ecological Indicators, 24*, 557–561.

Currie, D. J., Mittelbach, G. G., Cornell, H. V., Field, R., Guégan, J. F., Hawkins, B. A., et al. (2004). Predictions and tests of climate-based hypotheses of broad-scale variation in taxonomic richness. *Ecology Letters, 7*, 1121–1134.

Davies, T. J., Savolainen, V., Chase, M. W., Moat, J., & Barraclough, T. G. (2004). Environmental energy and evolutionary rates in flowering plants. Proceedings of the Royal Society of London. *Series B: Biological Sciences, 271*, 2195–2200.

Deguines, N., Julliard, R., De Flores, M., & Fontaine, C. (2016). Functional homogenization of flower visitor communities with urbanization. *Ecology and Evolution, 6*, 1967–1976.

Des Roches, S., Brans, K. I., Lambert, M. R., Rivkin, L. R., Savage, A. M., Schell, C. J., et al. (2021). Socio-eco-evolutionary dynamics in cities. *Evolutionary Applications, 14*, 248–267.

Desaegher, J., Nadot, S., Machon, N., & Colas, B. (2019). How does urbanization affect the reproductive characteristics and ecological affinities of street plant communities? *Ecology and Evolution, 9*, 9977–9989.

Diamond, S. E., & Martin, R. A. (2021). Evolution in cities. *Annual Review of Ecology, Evolution, and Systematics, 52*, 519–540.

Diamond, S. E., Dunn, R. R., Frank, S. D., Haddad, N. M., & Martin, R. A. (2015). Shared and unique responses of insects to the interaction of urbanization and background climate. *Current Opinion in Insect Science, 11*, 71–77.

Dirzo, R., & Raven, P. H. (2003). Global state of biodiversity and loss. *Annual Review of Environment and Resources, 28*, 137–167.

Dobbs, C., Escobedo, F. J., Clerici, N., de la Barrera, F., Eleuterio, A. A., MacGregor-Fors, I., et al. (2019). Urban ecosystem services in Latin America: Mismatch between global concepts and regional realities? *Urban Ecosystems, 22*, 173–187.

Dobzhansky, T. (1950). Evolution in the tropics. *American Scientist, 38*, 209–221.

Donihue, C. M., & Lambert, M. R. (2015). Adaptive evolution in urban ecosystems. *Ambio, 44*, 194–203. https://doi.org/10.1007/s13280-014-0547-2

Doré, M., Fontaine, C., & Thébault, E. (2021). Relative effects of anthropogenic pressures, climate, and sampling design on the structure of pollination networks at the global scale. *Global Change Biology, 27*, 1266–1280.

Dowle, E. J., Morgan-Richards, M., & Trewick, S. A. (2013). Molecular evolution and the latitudinal biodiversity gradient. *Heredity, 110*, 501–510.

Dreistadt, S. H., Dahlsten, D. L., & Frankie, G. W. (1990). Urban forests and insect ecology. *Bioscience, 40*, 192–198.

Dzul-Cauich, H. F., & Munguía-Rosas, M. A. (2022). Negative effects of light pollution on pollinator visits are outweighed by positive effects on the reproductive success of a bat-pollinated tree. *The Science of Nature, 109*, 1–11.

Eckert, C. G., Kalisz, S., Geber, M. A., Sargent, R., Elle, E., Cheptou, P. O., et al. (2010). Plant mating systems in a changing world. *Trends in Ecology & Evolution, 25*, 35–43.

Egerton-Warburton, L. M., & Allen, E. B. (2000). Shifts in arbuscular mycorrhizal communities along an anthropogenic nitrogen deposition gradient. *Ecological Applications, 10*, 484–496.

Ehrlich, P. R., & Raven, P. H. (1964). Butterflies and plants: A study in coevolution. *Evolution, 18*, 586–608.

El-Sabaawi, R. (2018). Trophic structure in a rapidly urbanizing planet. *Functional Ecology, 32*, 1718–1728.

Faeth, S. H., Warren, P. S., Shochat, E., & Marussich, W. A. (2005). Trophic dynamics in urban communities. *Bioscience, 55*, 399–407.

Fang, S. C., Rodrigues, E. G., & Christiani, D. C. (2020). Environmental health hazards in the tropics. In E. T. Ryan, D. R. Hill, T. Solomon, N. E. Aronson, & T. P. Endy (Eds.), *Hunter's tropical medicine and emerging infectious diseases* (pp. 200–208). Elsevier.

Fedorov, A. A. (1966). The structure of the tropical rain forest and speciation in the humid tropics. *The Journal of Ecology, 54*, 1–11.

Fenoglio, M. S., Salvo, A., & Estallo, E. L. (2009). Effects of urbanisation on the parasitoid community of a leafminer. *Acta Oecologica, 35*, 318–326.

Fenoglio, M. S., Rossetti, M. R., & Videla, M. (2020). Negative effects of urbanization on terrestrial arthropod communities: A meta-analysis. *Global Ecology and Biogeography, 29*, 1412–1429.

Fine, P. V. (2015). Ecological and evolutionary drivers of geographic variation in species diversity. *Annual Review of Ecology, Evolution, and Systematics, 46*, 369–392.

Geem, M. V., Gols, R., Dam, N. M. V., der Putten, W. H. V., Fortuna, T., & Harvey, J. A. (2013). The importance of aboveground–belowground interactions on the evolution and maintenance of variation in plant defense traits. *Frontiers in Plant Science, 4*, 431.

Gervasi, D. D., & Schiestl, F. P. (2017). Real-time divergent evolution in plants driven by pollinators. *Nature Communications, 8*, 1–8.

Grimm, N. B., Faeth, S. H., Golubiewski, N. E., Redman, C. L., Wu, J., Bai, X., & Briggs, J. M. (2008). Global change and the ecology of cities. *Science, 319*, 756–760.

Hall, D. M., Camilo, G. R., Tonietto, R. K., Ollerton, J., Ahrné, K., Arduser, M., et al. (2017). The city as a refuge for insect pollinators. *Conservation Biology, 31*, 24–29.

Hendry, A. P. (2016). *Eco-evolutionary dynamics*. Princeton University Press.

Hooper, D. U., Chapin III, F. S., Ewel, J. J., Hector, A., Inchausti, P., Lavorel, S., Lawton, J. H., Lodge, D. M., Loreau, M., Naeem, D., Schmid, B., Setälä, H., Symstad, A. J., Vandermeer, J., & Wardle, D. A. (2005). Effects of biodiversity on ecosystem functioning: a consensus of current knowledge. *Ecological Monographs, 75*(1), 3–35.

Hoornweg, D., & Pope, K. (2017). Population predictions for the world's largest cities in the 21st century. *Environment and Urbanization, 29*, 195–216.

Huang, J., Lu, X. X., & Sellers, J. M. (2007). A global comparative analysis of urban form: Applying spatial metrics and remote sensing. *Landscape and Urban Planning, 82*, 184–197.

Hutchinson, G. E. (1959). Homage to Santa Rosalia or why are there so many kinds of animals? *The American Naturalist, 93*, 145–159.

Igea, J., & Tanentzap, A. J. (2020). Angiosperm speciation cools down in the tropics. *Ecology Letters, 23*, 692–700.

Igic, B., Lande, R., & Kohn, J. R. (2008). Loss of self-incompatibility and its evolutionary consequences. *International Journal of Plant Sciences, 169*, 93–104.

Irwin, R. E., Warren, P. S., & Adler, L. S. (2018). Phenotypic selection on floral traits in an urban landscape. *Proceedings of the Royal Society B, 285*(1884), 20181239.

Irwin, R. E., Youngsteadt, E., Warren, P. S., & Bronstein, J. L. (2020). The evolutionary ecology of mutualisms in urban landscapes. In M. Szulkin, J. Munshi-South, & A. Charmantier (Eds.), *Urban evolutionary biology* (pp. 111–129). Oxford University Press.

Johnson, M. T. J., & Munshi-South, J. (2017). Evolution of life in urban environments. *Science, 358*(6363), eaam8327.

Johnson, M. T. J., Thompson, K. A., & Saini, H. S. (2015). Plant evolution in the urban jungle. *American Journal of Botany, 102*, 1951–1953.

Johnson, A. L., Fettersand, A. M., & Ashman, T.-L. (2017). Considering the unintentional consequences of pollinator gardens for urban native plants: Is the road to extinction paved with good intentions? *New Phytologist, 215*, 1298–1305.

Johnson, M. T. J., Prashad, C. M., Lavoignat, M., & Siani, H. S. (2018). Contrasting the effects of natural selection, genetic drift and gene flow on urban evolution in white clover (*Trifolium repens*). *Proceedings of the Royal Society B: Biological Sciences, 285*(1883), 20181019.

Kettlewell, H. B. D. (1955). Selection experiments on industrial melanism in the Lepidoptera. *Heredity, 9*, 323–342.

Kiers, T. E., Palmer, T. M., Ives, A. R., Bruno, J. F., & Bronstein, J. L. (2010). Mutualisms in a changing world: An evolutionary perspective. *Ecology Letters, 13*, 1459–1474.

Koricheva, J. (2002). Meta-analysis of sources of variation in fitness costs of plant antiherbivore defenses. *Ecology, 83*, 176–190.

Kozlov, M. V., & Zvereva, E. L. (2016). Industrial pollution affects behaviour of the leafmining moth Stigmella lapponica (Lepidoptera: Nepticulidae). *Entomologia Experimentalis et Applicata, 158*, 69–77.

Kozlov, M. V., Lanta, V., Zverev, V., Rainio, K., Kunavin, M. A., & Zvereva, E. L. (2017). Decreased losses of woody plant foliage to insects in large urban areas are explained by bird predation. *Global Change Biology, 23*, 4354–4364.

Kunc, H. P., & Schmidt, R. (2021). Species sensitivities to a global pollutant: A meta-analysis on acoustic signals in response to anthropogenic noise. *Global Change Biology, 27*, 675–688.

Lawrence, E. R., & Fraser, D. J. (2020). Latitudinal biodiversity gradients at three levels: Linking species richness, population richness and genetic diversity. *Global Ecology and Biogeography, 29*, 770–788.

Layman, N. C., Fernando, M. T. R., Herlihy, C. R., & Busch, J. W. (2017). Costs of selfing prevent the spread of a self-compatibility mutation that causes reproductive assurance. *Evolution, 71*, 884–897.

Leandro, D., Johnson, M., et al. (2025). GLUEing the world: The GLobal urban evolution project and its importance to urban biodiversity conservation. In F. Angeoletto, P. Tryjanowski, & M. Fellowes (Eds.), *Ecology of tropical cities: Natural and social sciences applied to the conservation of urban biodiversity*. Springer Nature.

Levin, D. A. (1975). Pest pressure and recombination systems in plants. *The American Naturalist, 109*, 437–451.

Lima, G. F., Correa Santos, J. W. M., Albertin, R. M., Martínez-Miranzo, B., Souza, F. L., & Angeoletto, F. (2023). Backyards are a way to promote environmental justice and biodiversity conservation in Brazilian cities. *Diversity, 15*(7), 815.

Lin, L., Chen, Y., Xu, G., Zhang, Y., Zhang, S., & Ma, K. (2021). Impacts of urbanization undermine nestedness of the plant–arbuscular mycorrhizal fungal network. *Frontiers in Microbiology, 12*, 250.

Liu, Z., He, C., Zhou, Y., & Wu, J. (2014). How much of the world's land has been urbanized, really? A hierarchical framework for avoiding confusion. *Landscape Ecology, 29*, 763–771.

Liu, X., Huang, Y., Xu, X., Li, X., Li, X., Ciais, P., et al. (2020). High-spatiotemporal-resolution mapping of global urban change from 1985 to 2015. *Nature Sustainability, 3*, 564–570.

Martin, P. R., Burke, K. W., & Bonier, F. (2021). Plasticity versus evolutionary divergence: What causes habitat partitioning in urban-adapted birds? *The American Naturalist, 197*, 60–74.

Maschinski, J., & Whitham, T. G. (1989). The continuum of plant responses to herbivory: The influence of plant association, nutrient availability, and timing. The American Naturalist 134:1–19.

McDonnell, M. J., & Hahs, A. K. (2013). The future of urban biodiversity research: Moving beyond the 'low-hanging fruit'. *Urban Ecosystems, 16*, 397–409.

Mejia-Alva, B., Ramos-Zapata, J., Abdala-Roberts, L., & Parra-Tabla, V. (2018). Effects of arbuscular mycorrhizal fungi on above-ground tri-trophic interactions are contingent upon plant genetic effects of cross type in the perennial herb *Ruellia nudiflora* (B Silliman, Ed.). *Journal of Ecology, 106*, 1133–1141.

Miles, L. S., Rivkin, L. R., Johnson, M. T. J., Munshi-South, J., & Verrelli, B. C. (2019a). Gene flow and genetic drift in urban environments. *Molecular Ecology, 28*, 4138–4151.

Miles, L. S., Breitbart, S. T., Wagner, H. H., & Johnson, M. T. J. (2019b). Urbanization shapes the ecology and evolution of plant-arthropod herbivore interactions. *Frontiers in Ecology and Evolution, 7*, 310.

Mittelbach, G. G., Schemske, D. W., Cornell, H. V., Allen, A. P., Brown, J. M., Bush, M. B., et al. (2007). Evolution and the latitudinal diversity gradient: Speciation, extinction and biogeography. *Ecology Letters, 10*, 315–331.

Moeller, D. A., Briscoe Runquist, R. D., Moe, A. M., Geber, M. A., Goodwillie, C., Cheptou, P. O., et al. (2017). Global biogeography of mating system variation in seed plants. *Ecology Letters, 20*, 375–384.

Moreira, X., Abdala-Roberts, L., Berny Mier y Teran, J. C., Covelo, F., de La Mata, R., Francisco, M., et al. (2019). Impacts of urbanization on insect herbivory and plant defences in oak trees. *Oikos, 128*, 113–123.

Moritz, C., Patton, J. L., Schneider, C. J., & Smith, T. B. (2000). Diversification of rainforest faunas: An integrated molecular approach. *Annual Review of Ecology and Systematics, 31*, 533–563.

Neil, K., & Wu, J. (2006). Effects of urbanization on plant flowering phenology: A review. *Urban Ecosystems, 9*, 243–257.

Nelson, A. E., & Forbes, A. A. (2014). Urban land use decouples plant-herbivore-parasitoid interactions at multiple spatial scales. *PLoS One, 97*, e102127.

Newbound, M., Mccarthy, M. A., & Lebel, T. (2010). Fungi and the urban environment: A review. *Landscape and Urban Planning., 96*, 138–145.

Núñez-Farfán, J., Fornoni, J., & Valverde, P. L. (2007). The evolution of resistance and tolerance to herbivores. *Annual Review of Ecology, Evolution and Systematics*, 541–566.

Ochimaru, T., & Fukuda, K. (2007). Changes in fungal communities in evergreen broad-leaved forests across a gradient of urban to rural areas in Japan. *Canadian Journal of Forest Research, 37*, 247–258.

Ollerton, J. (2017). Pollinator diversity: Distribution, ecological function, and conservation. *Annual Review of Ecology, Evolution, and Systematics, 48*, 353–376.
Ollerton, J., Johnson, S. D., & Hingston, A. B. (2006). Geographical variation in diversity and specificity of pollination systems. In N. M. Waser & J. Ollerton (Eds.), *Plant-pollinator interactions from specialization to generalization* (pp. 283–308). University of Chicago Press.
Opedal, Ø. H. (2019). The evolvability of animal-pollinated flowers: Towards predicting adaptation to novel pollinator communities. *New Phytologist, 221*, 1128–1135.
Orr, M. C., Hughes, A. C., Chesters, D., Pickering, J., Zhu, C. D., & Ascher, J. S. (2021). Global patterns and drivers of bee distribution. *Current Biology, 31*, 451–458.
Orton, M. G., May, J. A., Ly, W., Lee, D. J., & Adamowicz, S. J. (2019). Is molecular evolution faster in the tropics? *Heredity, 122*, 513–524.
Parra-Tabla, V., & Arceo-Gómez, G. (2021). Impacts of plant invasions in native plant–pollinator networks. *New Phytologist, 230*, 2117–2128.
Philpott, S. M., Lucatero, A., Bichier, P., Egerer, M. H., Jha, S., Lin, B., & Liere, H. (2020). Natural enemy-herbivore networks along local management and landscape gradients in urban agroecosystems. *Ecological Applications, 30*, e02201.
Pisman, M., Bonte, D., & de la Peña, E. (2020). Urbanization alters plastic responses in the common dandelion *Taraxacum officinale*. *Ecology and Evolution, 10*, 4082–4090.
Pringle, A., Bever, J. D., Gardes, M., Parrent, J. L., Rillig, M. C., & Klironomos, J. N. (2009). Mycorrhizal symbioses and plant invasions. *Annual Review of Ecology, Evolution, and Systematics, 40*(1), 699–715.
Pyron, R. A., & Wiens, J. J. (2013). Large-scale phylogenetic analyses reveal the causes of high tropical amphibian diversity. *Proceedings of the Royal Society B: Biological Sciences, 280*(1770), 20131622.
Qu, J., Bonte, D., & Vandegehuchte, M. L. (2022). Phenotypic and genotypic divergence of plant–herbivore interactions along an urbanization gradient. *Evolutionary Applications, 15*, 865–877.
Raupp, M. J., Shrewsbury, P. M., & Herms, D. A. (2010). Ecology of herbivorous arthropods in urban landscapes. *Annual Review of Entomology, 55*, 19–38.
Reiner, R. C., Jr., Smith, D. L., & Gething, P. W. (2015). Climate change, urbanization and disease: Summer in the city…. *Transactions of the Royal Society of Tropical Medicine and Hygiene, 109*, 171–172.
Rhoades, D. F. (1979). Evolution of plant chemical defense against herbivores. In G. A. Rosenthal & D. H. Janzen (Eds.), *Herbivores: Their interaction with secondary plant metabolites* (pp. 3–54). Academic Press.
Rivkin, L. R., Johnson, R. A. Chaves, J. A., & Johnson, M. T. J. (2020). Urbanization alters ecological and evolutionary interactions between Darwin's finches and Tribulus cistoides on the Galápagos Islands. bioRxiv.
Roels, B. S. A., & Kelly, J. K. (2011). Rapid evolution caused by pollinator loss in *Mimulus guttatus*. *Evolution, 65*, 2541–2552.
Rohde, K. (1992). Latitudinal gradients in species diversity: The search for the primary cause. *Oikos, 65*, 514–527.
Rolland, J., Condamine, F. L., Jiguet, F., & Morlon, H. (2014). Faster speciation and reduced extinction in the tropics contribute to the mammalian latitudinal diversity gradient. *PLoS Biology, 12*, e1001775.
Santangelo, J. S., Johnson, M. T. J., & Ness, R. W. (2018). Modern spandrels: The roles of genetic drift, gene flow and natural selection in the evolution of parallel clines. *Proceedings of the Royal Society B: Biological Sciences, 285*(1878), 20180230.
Santangelo, J. S., Miles, L. S., Breitbart, S. T., Murray-Stoker, D., Rivkin, L. R., & Johnson, M. T. J. (2020a). Urban environments as a framework to study parallel evolution. In M. Szulkin, J. Munshi-South, & A. Charmantier (Eds.), *Urban evolutionary biology* (pp. 36–53). Oxford University Press.
Santangelo, J. S., Rivkin, L. R., Advenard, C., & Thompson, K. A. (2020b). Multivariate phenotypic divergence along an urbanization gradient. *Biology Letters, 16*, 20200511.

Santangelo, J. S., Thompson, K. A., Cohan, B., Syed, J., Ness, R. W., & Johnson, M. T. J. (2020c). Predicting the strength of urban-rural clines in a Mendelian polymorphism along a latitudinal gradient. *Evolution Letters, 4*, 212–225.

Santangelo, J. S., Ness, R. W., Cohan, B., Fitzpatrick, C. R., Innes, S. G., Koch, S., et al. (2022). Global urban environmental change drives adaptation in white clover. *Science, 375*, 1275–1281.

Schell, C. J., Dyson, K., Fuentes, T. L., Des Roches, S., Harris, N. C., Miller, D. S., et al. (2020). The ecological and evolutionary consequences of systemic racism in urban environments. *Science, 369*, eaay4497.

Schemske, D. W., Mittelbach, G. G., Cornell, H. V., Sobel, J. M., & Roy, K. (2009). Is there a latitudinal gradient in the importance of biotic interactions? *Annual Review of Ecology, Evolution and Systematics, 40*, 245.

Schluter, D. (2016). Speciation, ecological opportunity, and latitude: (American Society of Naturalists Address). *The American Naturalist, 187*, 1–18.

Schluter, D., & Pennell, M. W. (2017). Speciation gradients and the distribution of biodiversity. *Nature, 546*, 48–55.

Schmidt, C., & Garroway, C. J. (2022). Systemic racism alters wildlife genetic diversity. *Proceedings of the National Academy of Sciences, 119*(43), e2102860119.

Seebacher, F., White, C. R., & Franklin, C. E. (2015). Physiological plasticity increases resilience of ectothermic animals to climate change. *Nature Climate Change, 5*, 61–66.

Seto, K. C., Sánchez-Rodríguez, R., & Fragkias, M. (2010). The new geography of contemporary urbanization and the environment. *Annual Review of Environment and Resources, 35*, 167–194.

Seto, K. C., Güneralp, B., & Hutyra, L. R. (2012). Global forecasts of urban expansion to 2030 and direct impacts on biodiversity and carbon pools. *Proceedings of the National Academy of Sciences, 109*, 16083–16088.

Silva, J. L. S., de Oliveira, M. T. P., Cruz-Neto, O., Tabarelli, M., & Lopes, A. V. (2020). Plant–pollinator interactions in urban ecosystems worldwide: A comprehensive review including research funding and policy actions. *Ambio, 50*, 1–17.

Simms, E. L. (1992). Costs of plant resistance to herbivory. In R. S. Fritz & E. L. Simms (Eds.), *Plant resistance to herbivores and pathogens: Ecology, evolution and genetics* (pp. 392–425). University of Chicago Press.

Simms, E. L., & Triplett, J. (1994). Costs and benefits of plant responses to disease: Resistance and tolerance. *Evolution, 48*, 1973–1985.

Snell-Rood, E. C., & Wick, N. (2013). Anthropogenic environments exert variable selection on cranial capacity in mammals. *Proceedings of the Royal Society B: Biological Sciences, 280*(1769), 20131384.

Somers, C. M., McCarry, B. E., Malek, F., & Quinn, J. S. (2004). Reduction of particulate air pollution lowers the risk of heritable mutations in mice. *Science, 304*, 1008–1010.

Souza, F. L., Fenoglio, M. S., & Angeoletto, F. (2023). To be a Brazilian City dweller, sometimes we must learn to say enough! *Sustainability, 15*(4), 3699.

Stabler, L. B., Martin, C. A., & Stutz, J. C. (2001). Effect of urban expansion on arbuscular mycorrhizal fungal mediation of landscape tree growth. *Journal of Arboriculture, 27*, 193–202.

Stamp, N. (2003). Out of the quagmire of plant defense hypotheses. *The Quarterly Review of Biology, 78*, 23–55. https://doi.org/10.1086/367580

State of the Tropics. (2020). *State of the tropics 2020 report*. James Cook University. https://www.researchgate.net/publication/343441325_State_of_the_Tropics_2020_Report

Stevens, G. C. (1989). The latitudinal gradient in geographical range: How so many species coexist in the tropics. *The American Naturalist, 133*(2), 240–256.

Strauss, S. Y., & Irwin, R. E. (2004). Ecological and evolutionary consequences of multispecies plant-animal interactions. *Annual Review of Ecology, Evolution, and Systematics, 35*, 435–466.

Strauss, S. Y., Rudgers, J. A., Lau, J. A., & Irwin, R. E. (2002). Direct and ecological costs of resistance to herbivory. *Trends in Ecology & Evolution, 17*, 278–285. https://doi.org/10.1016/S0169-5347(02)02483-7

Sumoski, S. E., Johncox, A. J., Althoff, D. M., & Segraves, K. A. (2009). Impact of urbanization on tri-trophic interactions in an endemic scrub community. *Florida Entomologist, 92*, 582–587.

Szulkin, M., Garroway, C. J., Corsini, M., Kotarba, A. Z., Dominoni, D., & Szulkin, M. (2020a). How to quantify urbanization when testing for urban evolution. In M. Szulkin, J. Munshi-South, & A. Charmantier (Eds.), *Urban evolutionary biology* (pp. 13–35). Oxford University Press.
Szulkin, M., Munshi-South, J., & Charmantier, A. (2020b). *Urban evolutionary biology*. Oxford University Press.
Theodorou, P. (2022). The effects of urbanisation on ecological interactions. *Current Opinion in Insect Science, 52*, 100922. https://doi.org/10.1016/j.cois.2022.100922
Thomann, M., Imbert, E., Devaux, C., & Cheptou, P. O. (2013). Flowering plants under global pollinator decline. *Trends in Plant Science, 18*, 353–359.
Thompson, K. A., Renaudin, M., & Johnson, M. T. J. (2016). Urbanization drives the evolution of parallel clines in plant populations. *Proceedings of the Royal Society B: Biological Sciences, 283*(1845), 20162180.
Thompson, K. A., Rieseberg, L. H., & Schluter, D. (2018). Speciation and the city. *Trends in Ecology & Evolution, 33*, 815–826.
Turrini, T., Sanders, D., & Knop, E. (2016). Effects of urbanization on direct and indirect interactions in a tri-trophic system. *Ecological Applications, 26*, 664–675.
United Nations, Department of Economic and Social Affairs, Population Division. (2018). *World urbanization prospects: The 2018 revision*, Online Edition.
Vizentin-Bugoni, J., Maruyama, P. K., Souza, C. S. D., Ollerton, J., Rech, A. R., & Sazima, M. (2018). Plant-pollinator networks in the tropics: A review. *Ecological Networks in the Tropics*, 73–91.
Voillemot, M., Encinas-Viso, F., & Pannell, J. R. (2019). Rapid loss of self-incompatibility in experimental populations of the perennial outcrossing plant *Linaria cavanillesii*. *Evolution, 73*, 913–926.
Watson, J. E., Evans, T., Venter, O., Williams, B., Tulloch, A., Stewart, C., et al. (2018). The exceptional value of intact forest ecosystems. *Nature Ecology & Evolution, 2*, 599–610.
Wenzel, A., Grass, I., Belavadi, V. V., & Tscharntke, T. (2020). How urbanization is driving pollinator diversity and pollination–A systematic review. *Biological Conservation, 241*, 108321.
Willig, M. G., Kaufman, D. M., & Stevens, R. D. (2003). Latitudinal gradients of biodiversity: Pattern, process, scale, and synthesis. *Annual Review of Ecology, Evolution, and Systematics, 34*, 273–309.
Wilson, E. O. (1988). The current state of biological diversity. *Biodiversity, 521*, 3–18.
Winchell, K. M., Schliep, K. P., Mahler, D. L., & Revell, L. J. (2020). Phylogenetic signal and evolutionary correlates of urban tolerance in a widespread neotropical lizard clade. *Evolution, 74*, 1274–1288.
Wise, M. J., & Rausher, M. D. (2013). Evolution of resistance to a multiple-herbivore community: Genetic correlations, diffuse coevolution, and constraints on the plant's response to selection. *Evolution, 67*, 1767–1779. https://doi.org/10.1111/evo.12061
Wright, S. D., Gray, R. D., & Gardner, R. C. (2003). Energy and the rate of evolution: Inferences from plant rDNA substitution rates in the western Pacific. *Evolution, 57*, 2893–2898.
Wright, S., Keeling, J., & Gillman, L. (2006). The road from Santa Rosalia: A faster tempo of evolution in tropical climates. *Proceedings of the National Academy of Sciences, 103*, 7718–7722.
Xie, H., Zhang, Y., & Duan, K. (2020). Evolutionary overview of urban expansion based on bibliometric analysis in Web of Science from 1990 to 2019. *Habitat International, 95*, 102100.
Zhang, Z. Q., & Li, Q. J. (2008). Autonomous selfing provides reproductive assurance in an alpine ginger *Roscoea schneideriana* (Zingiberaceae). *Annals of Botany, 102*, 531–538.

Part II
Natural and Social Sciences Applied to the Conservation of Urban Biodiversity

Chapter 12
GLUEing the World: The GLobal Urban Evolution Project and Its Importance to Urban Biodiversity Conservation

Deleon Leandro, Marc T. J. Johnson, Adam C. Schneider, Adrián Lázaro-Lobo, Aude E. Caizergues, Carlos Lara, Carlos Iñiguez-Armijos, César González-Lagos, Christopher Frost, Diana Rennison, Ek Del-Val, Glen Ray Hood, Ilga Porth, Jalene M. LaMontagne, James Santangelo, Juraj Paule, Kaitlin Stack Whitney, Karen Dyson, Kiyoko M. Gotanda, Mattheau Comerford, Michael Just, Philips Akinwole, Spyros Gkelis, Thomas Merritt, William Godsoe, Vera Pfeiffer, and Fabio Angeoletto

> *Understanding the implications of urban development on evolutionary dynamics will provide new insights for conservation and urban planning and for designing strategies that will help species adapt to urbanization, improve prediction of population declines, and will help to develop appropriate conservation and management plans.*
>
> *Marina Alberti, 2020*

Abstract In recent years, there has been a notable increase in research projects aimed at understanding the ecology of cities. How and to what extent urbanization is responsible for species' evolution is historically a much less understood topic,

---

D. Leandro
Instituto de Biociências (IB), Universidade Federal de Mato Grosso, Cuiabá, Brasil

M. T. J. Johnson
Department of Biology, Centre for Urban Environments University of Toronto Mississauga, Mississauga, ON, Canada

A. C. Schneider
Department of Biology, University of Wisconsin-La Crosse, La Crosse, WI, USA

A. Lázaro-Lobo
Biodiversity Research Institute (IMIB), CSIC-University of Oviedo-Principality of Asturias, Mieres, Spain

F. Angeoletto et al. (eds.), *Ecology of Tropical Cities*,
https://doi.org/10.1007/978-3-031-48012-6_12

because urban evolutionary biology is a nascent science. Most of the empirical and theoretical investigation of this new branch of evolution has been published recently, that is within the last 15 years. Studies on urban evolutionary biology have relevant implications for the environmental design of cities and for conserving the evolutionary potential and ecological success of urban biota. There is a strong bias towards studies conducted in North America and Europe, and the literature lacks empirical

A. E. Caizergues
Department of Biology, University of Toronto Mississauga, Mississauga, ON, Canada

C. Lara
Departamento de Ecología, Facultad de Ciencias, Universidad Católica de la Santísima Concepción, Concepción, Chile

C. Iñiguez-Armijos
Laboratorio de Ecología Tropical y Servicios Ecosistémicos (EcoSs-Lab), Universidad Técnica Particular de Loja, Loja, Ecuador

C. González-Lagos
Departamento de Ciencias, Facultad de Artes Liberales, Universidad Adolfo Ibañéz, Santiago, Chile

Centro de Ecología Aplicada y Sustentabilidad (CAPES), Santiago, Chile

C. Frost
BIO5 Institute, University of Arizona, Tucson, AZ, USA

D. Rennison
School of Biological Sciences, University of California San Diego, La Jolla, CA, USA

E. Del-Val
Instituto de Investigaciones en Ecosistemas y Sustentabilidad, Universidad Nacional Autónoma de México, México City, México

G. R. Hood
Department of Biological Sciences, Wayne State University, Detroit, MI, USA

I. Porth
Centre d'étude de la forêt (CEF), Université Laval, Quebec City, QC, Canada

J. M. LaMontagne
Department of Biology, University of Missouri-St. Louis, St. Louis, USA

Whitney R. Harris World Ecology Center, University of Missouri-St. Louis, St. Louis, USA

Science & Conservation Division, Missouri Botanical Garden, St. Louis, USA

J. Santangelo
Department of Integrative Biology, University of California Berkeley, Berkeley, CA, USA

J. Paule
Botanischer Garten und Botanisches Museum Berlin, Freie Universität Berlin, Berlin, Germany

K. S. Whitney
Science, Technology & Society Department, Rochester Institute of Technology, Rochester, NY, USA

K. Dyson
Dendrolytics, Seattle, WA, USA

studies that investigate how the greater biodiversity observed in the tropics responds evolutionarily to urbanization processes. Expanding urban evolutionary biology in tropical cities is a priority for urban evolutionary biologists. This perspective chapter explores how the Global Urban Evolution Project, the world's largest scale study of urban evolutionary ecology, is contributing to understanding how urbanization is affecting the ecology and evolution of life, and how this knowledge can be applied to the conservation of the biological diversity of tropical cities.

Keywords Urban evolutionary biology · Urban ecology · Urban biodiversity · Urban ecosystems · Trifolium repens · Urban evolutionary biology applied to biodiversity conservation · Evolutionary processes · Urban processes · Urban evolutionary biology applied to environmental management · Tropical cities

Urbanization is a global phenomenon, whereby anthropogenically mediated environmental change creates new artificial ecosystems. Urban areas are unlike any other ecosystem on Earth, involving rapid dramatic environmental change over space and time, including bringing humans and thousands of species together to

K. M. Gotanda
Department of Biological Sciences, Brock University, St. Catharines, ON, Canada

M. Comerford
Department of Biology, University of Massachusetts Boston, Boston, MA, USA

M. Just
Training Lands and Heritage Branch, Engineer Research and Development Center, Vicksburg, United States of America

P. Akinwole
Biology Department, DePauw University, Greencastle, IN, USA

S. Gkelis
Department of Botany, School of Biology, Aristotle University of Thessaloniki, Thessaloniki, Greece

T. Merritt
School of Natural Sciences, Laurentian University, Ontario, Canada

W. Godsoe
Department of Pest Management and Conservation Lincoln University, Lincoln, New Zealand

V. Pfeiffer
College of Veterinary Medicine, Washington State University, Washington, DC, USA

F. Angeoletto (✉)
Programa de Pós-Graduação em Geografia da Universidade Federal de Rondonópolis, Mato Grosso, Brasil

Programa de Pós-Graduação em Gestão e Tecnologia Ambiental, Universidade Federal de Rondonópolis, Rondonópolis, Brasil

PUMAS, Pessoas Unidas pelo Meio Ambiente, Agricultura & Sociedade, Chapada, Brasil
e-mail: angeoletto@ufr.edu.br

potential interact in new ways. Scientists are just beginning to understand the ecological and evolutionary consequences of urban systems (Grimm et al., 2008; Johnson & Munshi-South, 2017; Fellowes, et al. 2020; Angeoletto, et al. 2022). This perspective chapter explores how the *Global Urban Evolution Project*, the world's largest scale study of urban evolutionary ecology, is contributing to understanding how urbanization is affecting the ecology and evolution of life, and how this knowledge can be applied to the conservation of the biological diversity of cities.

Urban ecosystems comprise a relatively small amount of Earth's surface, but their impact can be large in magnitude and scale. Although urban development covers just 0.44% of the Earth's terrestrial land surface (Liu et al., 2020), cities influence the entire biosphere, through their immense inflows and outflows of energy, resources and pollution (i.e. telecoupled effects sensu Alberti, 2015). The percentage of land surfaces that collectively constitute telecoupled environments has been estimated to be between 50% and 90% (Pimentel et al., 1992; Vitousek et al., 1997) Like other ecosystems, urban ecological systems are coupled to input environments (i.e., landscapes that provide inputs of organisms, energy, and materials) and output environments (i.e., areas of the biosphere that receive waste from urban metabolism, Odum, 1988; Rees, 1997). However, there are two important differences. First, cities are net consumers of energy and resources, consuming much more energy and resources per unit area than they produce. For example, urban areas account for ≥70% of all anthropogenic energy consumption and resource extraction (Un-Habitat, 2016). Second, technological and cultural innovations drive urban lifestyles that are dependent on the input of energy and materials obtained in their environments, such as fossil fuels, and whose consumption demands greatly amplify output environments. As one example, cities account for ≥70% of all CO_2 emissions (Un-Habitat, 2016). The global scale of the ecological footprint of cities (i.e., the extent of their input and output environments) has decoupled cities from their ecological processes, in that a city's ecological processes extend beyond its administrative or political boundaries (Wackernagel & Rees, 1998). Likewise, evolutionary processes of urban ecological systems may have a biospheric reach, meaning that the contemporary evolution of a vast number of species is directly or indirectly influenced by cities (Miles et al., 2021; Caizergues et al. 2024). Studies on biological evolution in urban ecosystems therefore have potentially large implications for the conservation of local and global biodiversity.

Ecology was consolidated as a science through a combination of field studies and laboratory experiments to better understand the mechanisms that drive ecological processes in natural environments (Begon & Townsend, 2021; Moron et al. 2024). For most of the twentieth century, ecologists largely avoided urban ecosystems (Marzluff et al., 2001). Yet even a the time there was an understanding that cities were shaping wildlife. For example, eastern gray squirrels (*Sciurus carolinensis*) were a forest and rural dwelling animal until the late nineteenth century. After their introduction to cities, their populations grew and become ubiquitous in urban areas, such that they are now a model of how native species can thrive in the unique conditions of cities (Benson, 2013).

In recent years, there has been a notable increase in research projects aimed at understanding the ecology of cities (Karmanova et al., 2022), although most of these studies have taken place in cities in the Northern Hemisphere (Shackleton et al., 2021). How and to what extent urbanization is responsible for species' evolution is historically a much less understood topic (Donihue & Lambert, 2015; Johnson et al., 2015; Szulkin et al., 2020; Diamond & Martin, 2021; Beridze et al., 2023), because urban evolutionary biology is a nascent science (Szulkin et al., 2020). Most of the empirical and theoretical investigation of this new branch of evolution has been published recently, that is within the last 15 years (Rivkin et al., 2019).

This new subdiscipline of evolutionary ecology has many unanswered questions that have obvious connections with the conservation of biodiversity in cities (Lambert & Donihue, 2020). Several recent perspective papers and reviews (Johnson & Munshi-South, 2017; Rivkin et al., 2019; Szulkin et al., 2020; Verrelli et al., 2022) list many of these questions, and we provide a summary of these questions in Table 12.1.

In the case of tropical cities, additional unique challenges arise that add important aspects for considerations when answering the questions described in Table 12.1 (from Verrelli et al. 2022). Some of these challenges include: (i) accelerated urbanization in cities with little capacity for planning and environmental management of pollution (atmospheric, light, noise, water), resource use, and biodiversity management; (ii) elevated environmental injustices (poor neighborhoods have little tree cover); (iii) few researchers in urban ecology and urban evolutionary ecology in tropical regions; and, (iv) weak municipal governments that are susceptible to pressures from development and unable to provide adequate support for environmental protection and conservation (Verrelli et al. 2022; Angeoletto et al., 2023; Lima et al., 2023; Souza et al., 2023).

Despite being recent, urban evolutionary biology has developed rapidly, in part due to initiatives such as the *Global Urban Evolution* (GLUE) *Project*, an international consortium of 287 scientists from 26 countries in North and South America, Europe, Oceania, Asia and Africa. Since 2018, GLUE scientists have been studying white clover (*Trifolium repens*), a cosmopolitan plant found in most cities around the world. This species is a herbaceous member of the legume family (Fabaceae), outcrossing, present in a wide range of habitats, and introduced worldwide because it fixes nitrogen. Two questions guide research of this project: is urbanization causing parallel evolution globally; and, what are the features of cities and the climatic factors that determine whether or not populations adapt to urban environments?

Scientists working on the GLUE Project have conducted the most comprehensive, global-scale study of biological evolution in cities ever performed. They collected more than 110,000 samples of white clover from more than 6000 populations in 160 cities and their surrounding rural areas on five continents. They assayed from each plant whether or not plants produce the highly toxic chemical hydrogen cyanide (HCN), a defense mechanism against herbivores that also increases a plant's tolerance to water stress. The presence or absence of HCN is controlled by just two Mendelian inherited loci (Corkill, 1942; Olsen et al., 2007; Olsen et al., 2008), such

Table 12.1 Emerging themes and questions in Urban Evolutionary Biology

Themes	Questions
Fundamental processes and mechanisms (Processes)	What are the effects of urbanization on somatic and gametic mutation rates? Under what conditions will urbanization promote or impede speciation, and how does urbanization impact speciation rates? How does urbanization influence the (co)evolution of species interactions, and do species interactions exacerbate or ameliorate selective pressures associated with urbanization? To what extent has urbanization led to the evolution of new commensal species, including those that are self-domesticating? How can cities enhance global eco-evolutionary potential by altering intraspecific and interspecific interactions? How do urban areas act as genetic sources or sinks in preserving biodiversity? How is sexual selection altered by urbanization?
Spatial and temporal scales of drivers and responses (Scales)	What is the relative importance of evolution in cities in affecting the magnitude and direction of ecological dynamics and patterns, especially compared with nonevolutionary drivers in cities? What is the spatial and temporal scale of eco-evolutionary dynamics, including in deep time, and how does it differ between urban habitats compared with other nonurban habitats? How does rapid evolution differ between established and newly developing cities? What are the relative strengths of individual and interactive effects of drivers of evolution in cities, both genetic and cultural, and do they vary among taxa and across cities? How does heterogeneity both within the urban environment and between urban and nonurban environments influence eco-evolutionary dynamics? To what extent can eco-evolutionary signatures that we observe in cities tell us about eco-evolutionary trends at the global scale?
Sustainability, health, and well-being (Sustainability)	How do pathogens that cause human disease adapt to the urban environment and how does rewilding and restoration alter our ability to fight disease outbreaks? How can urban microbiomes be harnessed for human well-being, for improved soil health and productivity, and to create more sustainable approaches to inform eco-evolutionary dynamics in terrestrial and aquatic systems? How can eco-evolutionary understanding contribute to the implementation of the Sustainable Development Goals (SDGs), and how will the SDGs influence eco-evolutionary dynamics in cities in different global contexts?
Impacts and interactions with climate change (Climate change)	How will climate change and urbanization interact to alter eco-evolutionary dynamics in terms of extreme weather effects on biodiversity and societal consequences in the past and future? What is the effect of urbanization on ecology and evolution in aquatic environments, especially in the context of artificial aquatic infrastructure such as harbors and submersion of cities due to rising sea levels?

(continued)

Table 12.1 (continued)

Themes	Questions
Politics, governance, culture, their interactions, and their ethical considerations (Sociopolitical)	What is the magnitude of the effect of social inequality and systemic oppression in driving ecological and evolutionary dynamics in cities, and how do we ethically and empirically quantify and test this? To what extent does variation among political systems and governance predict evolutionary responses to, and biodiversity of, urban environments? How does both biological and cultural evolution in humans differ in the past, present, and future, and how does that human evolution feedback to affect the development of cities and cultures? How does expanding/enhancing empathy for nature and blue-green infrastructure alter eco-evolutionary processes in urban environments? How do human diets and behavior impact food webs in urban environments that alter eco-evolutionary dynamics? How can we more effectively collaborate across disciplines and with communities to incorporate ethics and human cultural sensitivity into the study of urban evolutionary ecology and the implementation and dissemination of research findings? What role or responsibility do we have as humans for seeking to influence urban eco-evolutionary outcomes? What role does human society and culture, underpinning the preference and artificial selection for novel traits in pets and ornamentals, have on evolutionary ecology within and among cities?
Innovation in technology and methodology (Technology)	What role do genetic manipulations, including genetic engineering, gene editing, and artificial selection, have on the evolution of organisms including humans, and their impact on society and the urban environment? Will there be synthetic organisms (e.g., drones, robots) that are capable of adapting to the urban environment independent of humans, and if so, how will they reshape the natural environment? How can automated, high-throughput data collection in cities be used as input to artificial intelligence (AI) to enable discovery and implementation of nature-based solutions? How can mathematical theory and data synthesis be used to predict urban eco-evolutionary feedbacks?

Source: Verrelli et al. (2022)

that changes in HCN within and between populations indicates evolutionary change in a genetically controlled plant defence. GLUE's scientists discovered that on average, white clover populations growing in cities produce less HCN than rural populations, although the strength and direction of this relationship varied among cities. Nearly one-half of cities (47%) showed evidence of a significant cline in HCN with distance from the city center, with 39% of cities showing evolution of decreased HCN in cities, and 8% showing the opposite pattern. In some of the cities, the evolution was consistent with herbivores selecting for more HCN in rural areas, particularly in cities and rural areas with an abundance of vegetation that could support herbivores. In other cities, evolution of HCN was more related to changes in drought stress along urban rural gradients, which selected for higher HCN in urban areas that experienced higher potential for drought in cities. Thus, GLUE has helped us to understand when populations are expected to evolve in response to environmental stressors in cities, and why this evolution may vary between cities (Santangelo et al., 2022).

In addition to examining the evolution of HCN in white clover, GLUE sequenced the genomes of over 2700 of the plants sampled, selected from a globally representative set of 26 cities across 5 continents. With these data they were able to detect signatures of selection acting directly on the genes controlling HCN production (Santangelo et al., 2022), ruling out the possibility that the patterns they observed in HCN evolution along urban-rural gradients were caused by neutral evolution or selection on tightly linked genes (Santangelo et al., 2018). Moreover, they were able to provide genome-wide estimates of genetic diversity within populations and genetic differentiation between populations on a global scale. These results showed that genetic diversity was very high ($\pi \sim 0.019$) and on average equivalent between urban and rural populations. Moreover, there was very weak genetic differentiation (mean $F_{ST} = 0.012$) between urban and rural populations in a given city. These results are consistent with white clover populations maintaining high effective population sizes in rural and urban habitats, and there being ongoing rampant gene flow among populations in these different habitats.

GLUE's whole-genome dataset provides a model for answering questions relevant to the conservation of organisms in cities, and how species succeed during global invasions (see GLUE 2.0 projects, below). The genome-wide dataset allows for a thorough analysis of how urbanization influences changes in population size and the movement of individuals between populations. Does the size of a city influence how and when populations change in size and the movement of individuals between urban and rural habitats? Does the age of a city influence when a population starts to expand or contract in size? Does the movement of individuals within or between cities depend on the amount of greenspace or socioeconomic variation in a city? These are just some of the questions that are be answered in combining GLUE's genomic and environmental data.

Since white clover has invaded the globe from its native habitats of Europe (Burdon, 1983; Kjaergaard, 2003), GLUE's global genomic dataset also makes it possible to understand the evolutionary and ecological processes that promote global biological invasions. Using landscape and population genomic methods, it is possible to understand the timing of invasions on different continents, whether invasions involved one or many independent invasions, and how populations have adapted to their new environments.

These future questions are being tackled in ten new GLUE 2.0 projects. This next generation of GLUE projects leverages the existing environmental, phenotypic and genomic data of the original GLUE project, often in combination with newly collected data. For example, one project uses the global genomic dataset to estimate genome-wide signatures of selection to understand whether urbanization results in similar signatures of selection across the genome in all cities, and not just on HCN. Another project uses the genomic dataset to perform demographic modeling to understand whether there are parallel nonadaptive signatures of evolution and historical demography in response to urbanization. Yet another project involved many of the original participants, as well as newly recruited scientists, revisiting many of the original populations to measure how herbivore communities and damage change along urban-rural gradients to explain evolution of HCN. GLUE 2.0 projects are expanding beyond questions related to cities, to understand how anthropogenic activities more generally influence the ecology and evolution of species. As one final example, scientists are using GLUE's genomic data to understand the ecological and evolutionary processes global invasion of diverse habitats. These examples are just a handful of the current GLUE 2.0 projects underway, yet still more are possible. Our larger hope is that these data and results inspire projects in other organisms, so we can gain a more complete understanding of how urbanization and invasions affect the ecology, evolution and conservation of life in the Anthropocene.

Research in urban evolutionary biology such as that developed by the GLUE Project has a clear potential for application in the planning of urban green areas. In fact, evolutionary biologists have highlighted the need to incorporate their discoveries into urban planning to promote connectivity between landscapes and prevent the loss of genetic diversity in urban populations of plants and animals, including those threatened with extinction (Alberti et al., 2020; Beridze et al., 2023; Rivkin et al., 2019). As Santangelo et al. (2022) emphasized, "*this knowledge could help conserve some of Earth's most vulnerable species, mitigate the impacts of pests, improve human wellbeing, and contribute to understanding fundamental eco-evolutionary processes*". Projects like GLUE help us to understand the features of urban landscapes that have the largest positive and negative influence on genetic diversity, gene flow and adaptive evolution, fundamental evolutionary processes that affect the ecological success of species. Generalizing these results to other species will require other large-scale multi-city studies in a diversity of species to build a more complete understanding of the similarities and differences in the evolutionary responses of species to urbanization.

The ongoing massive extinction of species and the accelerated destruction of their habitats impose the need to incorporate cities into conservation strategies. Urban ecosystems are viable habitats for thousands of species, and many of these species are threatened with extinction (Rumble et al., 2019; Soanes & Lentini, 2019). Given that urban ecosystems are the fastest growing ecosystem on Earth, the evolutionary and ecological future of many species that inhabit cities depends on the decisions we are making today (Dunn et al., 2022).

Urban evolutionary biology is an interdisciplinary science (Beridze et al., 2023; Schell et al., 2020), in that researchers must incorporate social sciences into approaches for investigations on biological evolution in cities to understand the processes and patterns associated with urban evolutionary ecology (Szulkin et al., 2020). Several social factors are expected to have considerable influence on the evolution of species occurring in cities. For example, the density of people, buildings and roads, policies around waste removal and environmental protection, and the distribution and heterogeneity of parks are just some of the features that influence the socio-eco-evolutionary dynamics of urban systems (Des Roches et al., 2021). Addressing these problems demands research that examines how planning practices, environmental management and urban design intersect to influence eco-evolutionary processes.

Political decisions greatly influence the ecology of cities and the evolution of species within them (Schell et al., 2020; Uchida et al., 2021; Verrelli et al., 2022; Beridze et al., 2023; Gonzaga et al. 2024). For example, federal housing policies in the USA implemented beginning in 1933, segregated urban residential neighborhoods mainly by race, intentionally prioritizing housing availability and quality for white families (Rothstein, 2017). This practice known as 'redlining', wherein racial and ethnic minorities were denied or provided substandard services and access to housing and banking, led to raciallly segregated communities that varied substantially in their socio-ecological characteristics. These discriminatory practices have been discontinued, but the implementation of such practices across decades shape many urban communities even today. For example, recent studies of hundreds of US cities showed that previously redlined neighborhoods are more likely to have more impervious surfaces, lower tree canopy cover, and higher land surface temperatures (Hoffman et al., 2020), as well as higher levels of air pollution (Lane et al., 2022). Fewer trees result in less support for biodiversity and lower species richness of birds, mammals, arthropods and other taxa (Schell et al., 2020). Lower tree cover in racially segregated neighborhoods in the United States has resulted in lower gene flow of wild species occurring in those cities (Schell et al., 2020). There is a similar situation of environmental injustice in Latin American and tropical cities, where the household income of its residents is strongly correlated with a greater number of trees and green areas in neighborhoods (MacGregor-Fors & Escobar-Ibáñez, 2017; Rumble et al., 2019). It is likely that the same evolutionary impacts are also underway in tropical cities (Beridze et al., 2023), and studies with interdisciplinary methodologies will have greater potential to produce relevant information for the planning and conservation of the biological diversity of those cities.

Studies on urban evolutionary biology have relevant implications for the environmental design of cities. Sets of private spaces such as gardens and backyards can connect fragments of native forests inserted in an urban matrix, influencing the gene flow between them (Johnson & Munshi-South, 2017). In tropical cities, it is not uncommon to have several forest fragments that are relatively isolated from each other by barriers such as avenues or neighborhoods. For example, in Rondonópolis, Brazil, there are hundreds of fragments of the Cerrado biome (a global biodiversity hotspot), with areas ranging from less than one hectare to hundreds of hectares. The accelerated devastation of that biome makes the conservation of those urban fragments even more relevant (Oliveira, 2020). However, to what extent the vegetation present in backyards and on squares promote gene flow between forest fragments in tropical cities represents an unanswered and important question for future research. Results from such research could help guide strategies for conserving the evolutionary potential and ecological success of urban biota (Alberti et al., 2020).

Unfortunately, there is a strong bias towards studies conducted in North America and Europe (Verrelli et al., 2022). Thus, the literature lacks empirical studies that investigate how the greater biodiversity observed in the tropics responds evolutionarily to urbanization processes (Carmona et al., 2025; Szulkin et al., 2020). Expanding urban evolutionary biology in tropical cities is a priority for urban evolutionary biologists. The GLUE project included sampling many tropical cities, which has helped to mobilize researchers to expand the knowledge on this topic to other biological systems native to those environments. The GLUE project also provides a valuable blueprint on how largescale collaborations can be performed to understand the evolutionary ecology of tropical urban cities.

Conservation and management decisions in urban landscapes must take into account evolutionary processes, otherwise they risk of producing limited conservation strategies that could result in the decline of target species (Lambert & Donihue, 2020). We mentioned corridors formed by backyard vegetation can establish gene flow corridors between urban forest fragments, but in some instances, the opposite is desirable (e.g. disease vectors). In the hypothetical situation discussed by Lambert and Donihue (2020), an urban amphibian population adapted to some level of water pollution is connected to the rural populations that are not adapted to contaminants. The result could be the unintentional introduction of maladaptive genes into the population of pollution-resistant amphibians.

Unintentional consequences of urban management which can be detrimental to urban biodiversity conservation are an eloquent testimony to the complexity of urban ecosystem conservation. Furthermore, tropical cities typically have limited planning and management capacity, meaning they are often unable to develop specific environmental policies to conserve urban biological diversity (Rumble et al., 2019). Tropical cities have not been able to incorporate information from urban ecology into planning (Angeoletto, 2018; Angeoletto et al. 2019; Souza et al., 2023) and considerable effort is needed to expand dialogue between scientists and urban managers, in the cities of megadiverse tropical countries, to use urban evolutionary biology as an effective environmental management tool. It is our hope that the GLUE project can provide a framework to move tropical urban research forward, both in terms of the results in generates, but also the method it employs for mobilizing collaborative largescale studies of urban evolutionary ecology across cities.

References

Alberti, M. (2015). Eco-evolutionary dynamics in an urbanizing planet. *Trends in Ecology & Evolution, 30*(2), 114–126.

Alberti, M., Palkovacs, E. P., Roches, S. D., Meester, L. D., Brans, K. I., Govaert, L., et al. (2020). The complexity of urban eco-evolutionary dynamics. *Bioscience, 70*(9), 772–793.

Angeoletto, F. (2018). A busca por cidades saudáveis. *Estudos Avançados, 32*(93), 255–259.

Angeoletto, F., Leandro, D. D. S., & Fellowes, M. D. (2019). The consequences of Brazil's lack of transport planning is written in the blood of sparrows. *Urban Geography, 40*(8), 1191–1197.

Angeoletto, F., Tryjanowski, P., Santos, J., Martinez-Miranzo, B., Leandro, D., Bohrer, J., ... & Fellowes, M. D. (2022). Will Brazilian city dwellers actively engage in urban conservation? A case study with the charismatic neotropical Blue-and-yellow macaw (Ara ararauna). *Birds, 3*(2), 234–244.

Angeoletto, F., Richard, E., Santos, J., et al. (2023). When we think of urban afforestation in Brazil, what comes to mind is a mouth that has lost its teeth. *Terr Plural, 17*, e2321654–e2321654.

Begon, M., & Townsend, C. R. (2021). *Ecology: From individuals to ecosystems*. John Wiley & Sons.

Benson, E. (2013). The urbanization of the eastern gray squirrel in the United States. *The Journal of American History*, 691–710. https://doi.org/10.1093/jahist/jat353

Beridze, B., Mahamoud-Issa, M., Zmudzinski, M., Bednarz, P. A., Filutowska, Z. K., Sikora, B., Winiarska, D., Carmona, D., & Angeoletto, F. (2023). Welcome to the (urban) jungle: Comments on a new science, biological evolution in cities. *European Journal of Ecology*. https://doi.org/10.17161/eurojecol.v9i2.19518

Caizergues, A. E., Santangelo, J. S., Ness, R. W., Angeoletto, F., Anstett, D. N., Anstett, J., ... & Johnson, M. T. (2024). Does urbanisation lead to parallel demographic shifts across the world in a cosmopolitan plant?. *Molecular Ecology, 33*(7), e17311.

Burdon, J. (1983). Trifolium repens L. *Journal of Ecology, 71*, 307–330.

Carmona, D., Chávez-Pesqueira, M., Ângulo, D. F., Angeoletto, F., Abdala-Roberts, L., Moreira, X., Sosenski, P., & Parra-Tabla, V. (2025). Urban evolution in tropical cities. In F. Angeoletto, P. Tryjanowski, & M. Fellowes (Eds.), *Ecology of tropical cities: Natural and social sciences applied to the urban conservation*. Springer Nature.

Corkill, L. (1942). Cyanogenesis in white clover (Trifolium repens L.) V. The inheritance of cyanogenesis. *New Zealand JOURNAL of science and Technology. Section B, 23*, 178–193.

Des Roches, S., Brans, K. I., Lambert, M. R., Rivkin, L. R., Savage, A. M., Schell, C. J., ... & Alberti, M. (2021). Socio-eco-evolutionary dynamics in cities. *Evolutionary Applications, 14*(1), 248–267.

Diamond, S. E., & Martin, R. A. (2021). Evolution in cities. *Annual Review of Ecology, Evolution, and Systematics, 52*(1), 519–540.

Donihue, C. M., & Lambert, M. R. (2015). Adaptive evolution in urban ecosystems. *Ambio, 44*, 194–203.

Dunn, R. R., Burger, J. R., Carlen, E. J., Koltz, A. M., Light, J. E., Martin, R. A., Munshi-South, J., Nichols, L. M., Vargo, E. L., Yitbarek, S., Zhao, Y., & Cibrián-Jaramillo, A. (2022). A theory of City biogeography and the origin of urban species. *Frontiers in Conservation Science, 3*. https://doi.org/10.3389/fcosc.2022.761449

Fellowes, M. D., Acquaah-Harrison, K., Angeoletto, F., Santos, J. W., da Silva Leandro, D., Rocha, E. A., ... & Thomas, R. L. (2020). Map-a-mole: Greenspace area influences the presence and abundance of the European mole Talpa europaea in urban habitats. *Animals, 10*(6), 1097.

Grimm, N. B., Faeth, S. H., Golubiewski, N. E., Redman, C. L., Wu, J., Bai, X., & Briggs, J. M. (2008). Global change and the ecology of cities. *Science, 319*(5864), 756–760.

Gonzaga, C. A. C., Roquette, J. G., Angeoletto, F., Silgueiro, V. D. F., Macedo, L. O. B., Valdiones, A. P., & da Silva, N. M. (2024). The Public Prosecutor's Office's experience using Global Forest Watch to monitor and deter deforestation in the Cerrado. *Environmental Conservation*, 1–6.

Hoffman, J. S., Shandas, V., & Pendleton, N. (2020). The effects of historical housing policies on resident exposure to intra-urban heat: A study of 108 US urban areas. *Climate, 8*(1), 12. https://doi.org/10.3390/cli8010012

Johnson, M. T., & Munshi-South, J. (2017). Evolution of life in urban environments. *Science, 358*(6363), eaam8327.

Johnson, M. T. J., Thompson, K. A., & Saini, H. S. (2015). Plant evolution in the urban jungle. *American Journal of Botany, 102*(12), 1951–1953.

Karmanova, T. N., Feoktistova, N. Y., Fetisova, E. E., Mosalov, A. A., & Surov, A. V. (2022). Urban ecology: Retrospective and research prospects. *Biology Bulletin Reviews, 12*(1), 94–105.

Kjaergaard, T. (2003). A plant that changed the world: The rise and fall of clover 1000-2000. *Landscape Research, 28*, 41–49.

Lambert, M. R., & Donihue, C. M. (2020). Urban biodiversity management using evolutionary tools. *Nature Ecology & Evolution, 4*(7), 903–910.

Lane, H. M., et al. (2022). Historical redlining is associated with present-day air pollution disparities in U.S. Cities. *Environmental Science & Technology Letters, 9*(4), 345–350. https://doi.org/10.1021/acs.estlett.1c01012

Lima, G. F., Correa Santos, J. W. M., Albertin, R. M., Martínez-Miranzo, B., Souza, F. L., & Angeoletto, F. (2023). Backyards are a way to promote environmental justice and biodiversity conservation in brazilian cities. *Diversity, 15*(7), 815.

Liu, X., Huang, Y., Xu, X., Li, X., Li, X., Ciais, P., Lin, P., Gong, K., Ziegler, A. D., & Chen, A. (2020). High-spatiotemporal-resolution mapping of global urban change from 1985 to 2015. *Nature Sustainability, 3*, 564–570.

MacGregor-Fors, I., & Escobar-Ibáñez, J. F. (Eds.). (2017). *Avian ecology in Latin American cityscapes*. Springer.

Marzluff, J. M., Bowman, R., & Donnelly, R. (2001). A historical perspective on urban bird research: Trends, terms, and approaches. In *Avian ecology and conservation in an urbanizing world* (pp. 1–17). Springer.

Miles, L. S., Carlen, E. J., Winchell, K. M., & Johnson, M. T. (2021). Urban evolution comes into its own: Emerging themes and future directions of a burgeoning field. *Evolutionary Applications, 14*(1), 3–11.

Moroń, D., Beim, M., Gudowska, A., Angeoletto, F., Celary, W., Cwajna, A., ... & Tryjanowski, P. (2024). Evaluating tramway infrastructure on biodiversity and ecosystem services. *Scientific Reports, 14*(1), 9394.

Odum, E. P. (1988). *Ecologia*. Editora Guanabara Kogan.

Oliveira, C. M. (2020). *O Potencial das Manchas de Habitat Preservadas no Espaço Urbano, Peri-urbano e Rural de Rondonópolis–MT para a Conservação da Biodiversidade*. Dissertação de Mestrado. Programa de Pós-Graduação em Geografia da UFR, Brasil.

Olsen, K. M., Sutherland, B. L., & Small, L. L. (2007). Molecular evolution of the Li/li chemical defence polymorphism in white clover (Trifolium repens L.). *Molecular Ecology, 16*, 4180–4193.

Olsen, K. M., Hsu, S., & Small, L. L. (2008). Evidence on the molecular basis of the ac/ac adaptive cyanogenesis polymorphism in white clover (Trifolium repens L.). *Genetics, 179*, 517–526.

Pimentel, D., Acquay, H., Biltonen, M., Rice, P., Silva, M., Nelson, J., et al. (1992). Environmental and economic costs of pesticide use. *Bioscience, 42*(10), 750–760.

Rees, W. E. (1997). Urban ecosystems: The human dimension. *Urban Ecosystems, 1*, 63–75.

Rivkin, L. R., Santangelo, J. S., Alberti, M., Aronson, M. F., de Keyzer, C. W., Diamond, S. E., Fortin, M. J., Frazee, L. J., Gorton, A. J., Hendry, A. P., Liu, Y., Losos, J. B., MacIvor, J. S., Martin, R. A., MsDonnel, M. J., Miles, L. S., Munshi-South, J., Ness, R. A., Newman, A. E. M., et al. (2019). A roadmap for urban evolutionary ecology. *Evolutionary Applications, 12*(3), 384–398.

Rothstein, R. (2017). *A forgotten history of how our government segregated America*. WW Norton. https://wwnorton.com/books/the-color-of-law/

Rumble, H., Angeoletto, F., Connop, S., Goddard, M. A., & Nash, C. (2019). Understanding and applying ecological principles in cities. In *Planning cities with nature: Theories, strategies and methods* (pp. 217–234).

Santangelo, J. S., Johnson, M. T. J., & Ness, R. W. (2018). Modern spandrels: The roles of genetic drift, gene flow and natural selection in the evolution of parallel clines. *Proceedings of the Biological Sciences, 285*, 20180230.

Santangelo, J. S., Ness, R. W., Cohan, B., Fitzpatrick, C. R., Innes, S. G., Koch, S., et al. (2022). Global urban environmental change drives adaptation in white clover. *Science, 375*(6586), 1275–1281.

Schell, C. J., Dyson, K., Fuentes, T. L., Des Roches, S., Harris, N. C., Miller, D. S., et al. (2020). The ecological and evolutionary consequences of systemic racism in urban environments. *Science, 369*(6510), eaay4497.

Shackleton, C. M., Cilliers, S. S., Davoren, E., & du Toit, M. J. (Eds.). (2021). *Urban ecology in the global South*. Springer Nature.

Soanes, K., & Lentini, P. E. (2019). When cities are the last chance for saving species. *Frontiers in Ecology and the Environment, 17*(4), 225–231.

Souza, F. L., Fenoglio, M. S., & Angeoletto, F. (2023). To be a Brazilian City dweller, sometimes we must learn to say enough! *Sustainability, 15*(4), 3699.

Szulkin, M., Munshi-South, J., & Charmantier, A. (2020). *Urban evolutionary biology*. Oxford University Press.

Uchida, K., Blakey, R. V., Burger, J. R., Cooper, D. S., Niesner, C. A., & Blumstein, D. T. (2021). Urban Biodiversity and the Importance of Scale. *Trends in Ecology & Evolution, 36*, 123–131.

Un-Habitat. (2016). *World Cities Report 2016: Urbanization and development-emerging futures*. UN, 2016.

Verrelli, B. C., Alberti, M., Des Roches, S., Harris, N. C., Hendry, A. P., Johnson, M. T., et al. (2022). A global horizon scan for urban evolutionary ecology. *Trends in Ecology & Evolution, 37*, 1006–1019.

Vitousek, P. M. (1994). Beyond global warming: Ecology and global change. *Ecology, 75*(7), 1861–1876.

Vitousek, P. M., Mooney, H. A., Lubchenco, J., & Melillo, J. M. (1997). Human domination of Earth's ecosystems. *Science, 277*(5325), 494–499.

Wackernagel, M., & Rees, W. (1998). *Our ecological footprint: reducing human impact on the earth* (Vol. 9). New society publishers.

Chapter 13
A Low-Cost Method to Detect Invasive Species in the Urban Environment—An Example of a Drone-Based Workflow Application

Normandes Matos da Silva, Domingos Sávio Barbosa, Marcin K. Dyderski, Dhonatan Diego Pessi, Eliane Guaraldo, Antonio Conceição Paranhos Filho, Larissa Dall Agnol, Cláudio Angelo Corrêa Gonzaga, Camila Leonardo Mioto, and Fabio Angeoletto

Abstract Invasive species pose a significant threat to biological diversity. Exotic and invasive grass species negatively affect urban biodiversity in Brazilian cities, since they occupy open spaces, compete with the native plants and eventually displace them. These species also increase the probability of wildfires in urban areas. Resources allocated for biodiversity conservation and invasive plants' management purposes remain scarce in Brazilian cities. Therefore, we have developed a low-cost methodology to support the detection of invasive grass species based on using drones, as well as two frameworks for data processing, model development, and

N. M. da Silva (✉) · D. S. Barbosa · D. D. Pessi · L. D. Agnol · C. L. Mioto
Programa de Pós-Graduação em Gestão e Tecnologia Ambiental, Universidade Federal de Rondonópolis, Rondonópolis, MT, Brazil
e-mail: normandes@ufr.edu.br

M. K. Dyderski
Institute of Dendrology, Polish Academy of Sciences, Kórnik, Poland

E. Guaraldo · A. C. P. Filho
Universidade Federal de Mato Grosso do Sul, Campo Grande, MS, Brazil

C. A. C. Gonzaga
Ministério Público do Estado de Mato Grosso, Itiquira, MT, Brazil

Programa de Pós-Graduação em Ciência Ambiental, Universidade de São Paulo, São Paulo, Brasil

F. Angeoletto
Programa de Pós-Graduação em Geografia da Universidade Federal de Rondonópolis, Mato Grosso, Brasil

Programa de Pós-Graduação em Gestão e Tecnologia Ambiental, Universidade Federal de Rondonópolis, Rondonópolis, Brasil

PUMAS, Pessoas Unidas pelo Meio Ambiente, Agricultura & Sociedade, Chapada, Brasil

F. Angeoletto et al. (eds.), *Ecology of Tropical Cities*,
https://doi.org/10.1007/978-3-031-48012-6_13

validation, that can be used anywhere in the world. Local governments could set up a basic workstation and generate geomatic products to enable the environmental management of invasive species by investing approximately 2000$. Another advantage of the presented monitoring methodology is that the approach can be adapted to demands associated with biodiversity conservation in urban ecosystems, e.g. monitoring endangered species.

Keywords Biological invasions · Exotic species · Remote sensing · Unmanned aerial vehicles · Urban land management · Invasive plants' management · Urban biodiversity conservation · Urban ecosystems · Brazil

1 Introduction

Agile and consistent procedures for analyzing urban landscape are critical in the management and planning of cities that rapidly extend their sizes. In this context, information on the spatial arrangement of towns can be used to aid in the decision-making process in areas such as health, education, and security (Lajo, 2021). Urban areas must be managed from the perspective of an urban ecosystem accounting for multiple elements: soil, climate, water, fauna, flora, built environments such as roads, buildings, and water supply, as well as the system's socioeconomic dimension (UEMRI, 2008). Several studies emphasize the importance of urban environments in biodiversity conservation (e.g. Alvey, 2006; Angeoletto et al., 2017; Clark et al., 1997; Lima et al., 2023; Miller & Hobbs, 2002; Santangelo et al., 2022). These ecosystems serve as refuges for endangered species and habitats, allowing humans to interact with nature (Angeoletto et al., 2022; Lambert & Donihue, 2020; Shaffer, 2018; Soanes et al., 2018; Soanes & Lentini, 2019). However, due to the strong transformation of land cover and severe disturbances, cities are particularly vulnerable to biodiversity loss (Angeoletto et al., 2023; Gilbert, 1989; Mckinney, 2002; Rumble et al., 2019; Souza et al., 2023; Williams et al., 2015; Moroń et al., 2024; Caizergues et al., 2024). This is especially driven by the retreat of species typical to natural ecosystems, habitat fragmentation, degradation of natural ecosystems (wetlands and forests), lowering of the groundwater table level, nitrogen deposition, disturbances, and pollution (Kowarik, 2011; Kowarik et al., 2019; Szlavecz et al., 2011).

Another important threat to biodiversity are invasive species, which comprise a considerable part of urban biotas (Dyderski et al., 2015; Dyderski & Jagodziński, 2019; Kowarik, 1995; Pyšek, 2016). Invasive species are a significant threat to biodiversity and could cause multiple environmental impacts that result in changes in community composition, biotic interactions, and other ecosystem processes (Bartz & Kowarik, 2019). Despite these threats, resources for biodiversity conservation and for the management of invasive plants are usually limited (Mostert et al., 2018) and this scarcity of resources is a reality in cities in Brazil (Angeoletto et al., 2016). Invasive species in cities take advantage of disturbances and high propagule pressure, contributing to biotic homogenization (Knapp et al., 2010; Kühn et al., 2004; Mckinney, 2006).

The incorporation of technological solutions that support urban planning practices has become easier in recent years due to the development of novel tools, usually accessible directly via the Internet and smartphones. Their emergence is usually connected with a personalization of technological solutions for urban planning, drawing inspiration from sectors such as leisure and adapting them for professional use (Bakogiannis et al., 2019; Gallacher, 2016; Goddard et al., 2021; Ramana et al., 2016). The use of drones, technically known as Unmanned Aircraft Vehicles (UAVs), can be a low-cost and efficient work tool alternative for enabling products aimed at the provision of public services and assisting in city management decision-making. Also, using published data on species distributions and land-use classification can help in the prioritization of sites for alien species monitoring (Paź-Dyderska et al., 2020)

The difficult task of monitoring a city, which includes places with challenging land access or that are dangerous due to violence, is often assigned to urban managers. Hence, UAV aerial photographs and videos enable a remote analysis of urban areas, which can result in a favorable benefit-cost ratio of UAV-generated products have been used to inspect buildings, flood-prone areas, and criminal expertise services (Chen et al., 2014), as well as to monitor urban traffic and plan land use and coverage (Alexander & Jenkins, 2015). UAVs have been explored in smart cities to acquire information that supports urban management and decision-making on a variety of issues related to the proper operation of the municipality. Aerial images with high spatial and temporal resolution obtained by such equipment can be used to identify degraded urban sectors with illegal solid waste dumping and the presence of invasive species, as well as areas with irregular constructions (Bakogiannis et al., 2019; Bolognez et al., 2021; da Silva et al., 2021; Pessi et al., 2020).

Low-cost aerial UAV photographs, obtained by multirotor equipped with cameras that capture the RGB color range, would be an inexpensive way of collecting remote sensing data that can be processed in the cloud or through free software, providing mosaics and digital elevation models necessary for the decision-making process in urban management. In this sense, UAV equipped with a high-resolution digital camera reduces the time and cost of acquiring images for urban mapping (Colomina & Molina, 2014; Laliberte et al., 2011; Sandino et al., 2020). Therefore, we aimed to present here two geoprocessing and remote sensing routines that use UAVs to aid in the identification of target invasive grass in public spaces and abandoned land in urban areas. We hypothesized that using UAVs and open-source software will allow to obtain fair quality discrimination between target species and background. This will support the development of similar workflows for further studies in other parts of the world, using low-cost equipment and supporting environmental management. We also hypothesized that machine-learning approaches will provide higher accuracy than methods based only on two thresholds of remotely sensed variables.

2 Methodology

2.1 *Studied Taxon*

We choose invasive perennial grasses *Urochloa* spp. as target taxon, that have been widely introduced in the tropical zone. In Brazil *Urochloa* grasses were introduced as pasture species in the 2nd half of the twentieth century, due to high biomass production in agriculture (Baptistella et al., 2020; Foxcroft et al., 2010; Riet-Correa et al., 2011). Genus *Urochloa* is considered to be at high risk for biological invasions (FATMA, 2016). Their species can outcompete native species in savannas grasslands due to efficient photosynthetic rates, high water and nutrient use efficiency, and relatively higher tolerance to abiotic stresses, as well as a high rate of biomass production (Foxcroft et al., 2010; Pivello et al., 1999). In Brazil, the most frequent are *U. decumbens*, *U. brizantha*, and *U. humidicola* (Riet-Correa et al., 2011). Those species comprise a serious threat to biodiversity, as well as to the economy, especially in soybean and sugarcane crops, reducing the crops of target species and generating control and removal costs (Horus, 2015).

2.2 *Study Area*

The study was conducted in the municipalities of Rondonópolis and Itiquira, both located in the southeast of the Brazilian state of Mato Grosso. We choose two study sites located on the outskirts of the urban areas, covering 19.34 ha (Fig. 13.1).

The study sites are relatively flat, with slopes ranging from 0% to 10%. The climate in Rondonópolis (study area I) is classified as Cwa (humid subtropical climate), with an average annual rainfall of 1500 mm and an average temperature of 25 °C. (Pessi et al., 2020). Itiquira (study area II) is classified as Aw (tropical with a dry season), with an annual average precipitation of 1600–1900 mm and an average annual temperature of 22 °C to 24 °C (Vendruscolo et al., 2015). Both study sites represent a typical landscape of urban spaces in medium-sized Brazilian cities, mainly on their outskirts, with heterogeneous soil use and coverage, primarily composed of buildings, exposed soil, and a few patches of vegetation typical of the Cerrado biome (characterized by trees, shrubs, and grasses), including the study's target taxon, i.e. *Urochloa* spp.. Invasive grasses impede native species regeneration by forming dense patches dominating vegetation, competitively excluding native species (Caramaschi, 2015), as well as increasing the risk of fires during drought periods trough fuel acummulation (Accatino et al., 2010; Beerling & Osborne, 2006; Keeley & Rundel, 2005). Patches of *Urochloa* spp. are also favorable habitats for insect vectors of infectious diseases such as malaria and dengue fever.

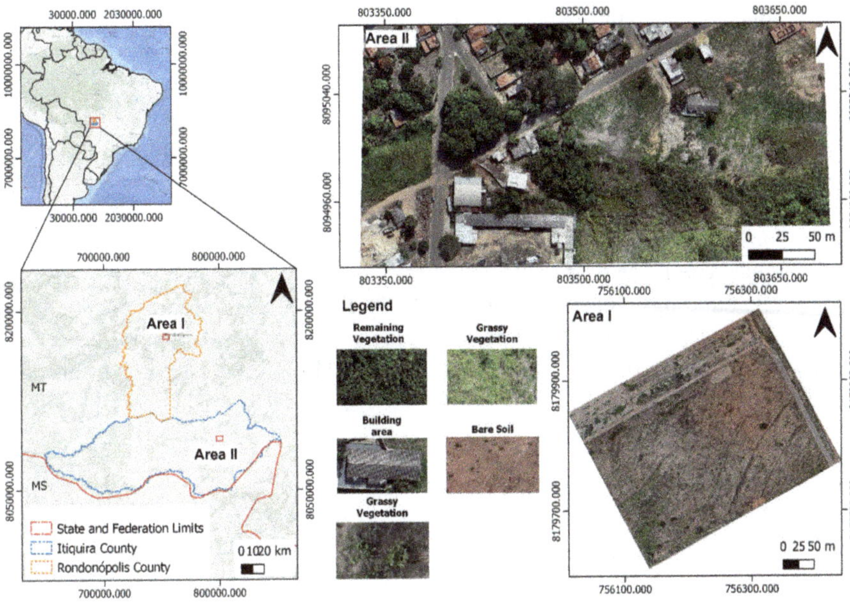

Fig. 13.1 Locality of the study sites. Area I is a vacant lot in the Granville neighborhood of Rondonópolis, MT, Brazil, with invasive grasses and uncovered soil. Area II is located in the urban region of Itiquira, MT, Brazil, and contains invasive grasses, remnant vegetation, uncovered soil, and built areas. Source of vector data for the state and municipal limits—IBGE (2020). Source of images (mosaics) from aerial photographs obtained by UAV

2.3 *Acquisition of Aerial Images*

The suborbital images were captured using a UAV, multirotor model DJI Phantom 4 Pro, equipped with a camera with a CMOS sensor, 20 megapixels, sensitive to the RGB visible spectrum. The automated mission parameters for image collection were planned using the freeware DroneDeploy software, (available at: https://support.dronedeploy.com/lang-pt-BR). The user manual for the UAV platform specifications used in this study is available at: (https://dl.djicdn.com/downloads/phantom_4_pro/20170719/Phantom_4_Pro_Pro_Plus_User_Manual_ES.pdf).

The images were georeferenced using the equipment's GNSS system (GPS and GLONASS), with a positional accuracy error ranging from 5 to 10 m. The images' average frontal and lateral overlap was 75%, and the estimated spatial resolution was 2 cm. The average speed of the UAV along the flight lines varied between 6 and 10 m s^{-1}. Due to the low overlap of images in these sectors of the polygon, the surrounding polygons of the flight lines presented an area larger than the target to be imaged, allowing the edge bands, which have greater distortion, to be discarded.

The images were collected in the field and processed on the OpenDroneMap Web (Webodm) platform, which can be found at: https://www.opendronemap.org/. Webodm (https://webodm.net/) is a cloud processing software that allows the user to configure several parameters for the generation of orthomosaics and digital surface and terrain models. The user obtains processing credits beginning at 5.00$. This investment may render the high costs of purchasing a physical processing station obsolete. Webodm was used to create an orthomosaic of RGB photos, the Digital Surface Model (DSM), and the Digital Elevation Model (DEM), all in geoTIFF format. The DSM represents laser beam reflectance from the top point of the terrain, e.g. buildings, vegetation, or bridges. In contrast, DEM represents the laser beam reflectance from the bottom point, i.e. the ground relief (Paranhos Filho et al., 2021).That way difference between DSM and DEM represents the height of measured object.

2.4 Data Processing

The products generated in Webodm (orthomosaic, DSM, and DEM) were exported to the free and open-source geographic information system (GIS) software QGIS 3.16 (QGIS Development Team, 2021). These files were initially redesigned for the coordinate reference system (CRS) EPSG: 31981, UTM projection (Universal Transverse Mercator), zone 21 S, and Datum SIRGAS 2000. The vegetation index for the RGB orthomosaic layer was calculated using the QGIS raster calculator, to create a new classified image file that highlighted the ground cover. Excess of Green Vegetation Index (ExG) was calculated using the RGB bands:

$$\mathrm{ExG} = 2^{*}\mathrm{G} - \mathrm{R} - \mathrm{B} \tag{13.1}$$

Where: ExG—Excess of Green; G—green band; R—red band; B—blue band (Mao et al., 2003; Meyer & Camargo Neto, 2008).

This vegetation index distinguishes between living vegetation cover, buildings, various residues, and exposed soil. The ExG layer, which is made up of a wide gradient of pixel values, was reclassified by a table in a binary file, with 0 (zero) corresponding to others or no vegetation and 1 (one) corresponding to vegetation cover.

The difference between the DSM and the DEM was calculated using the QGIS raster calculator (Eq. 13.2), providing information about objects (building or vegetation) height.

$$\mathrm{Height} = \mathrm{DSM} - \mathrm{DEM} \tag{13.2}$$

Where: Height—Calculated height [m] of the elements on the ground; DSM—Digital Surface Model; DEM—Digital Elevation Model (Wasser, 2018).

2.5 Data Analysis

We tested two methodological routines: one based on expert knowledge of spectral indices thresholds, and another based on machine learning. In the first, we used two variables to distinguish invasive grasses from background vegetation and we set up their levels manually. In the second, we trained predictive models using the machine learning approach.

2.6 Expert-Based Classification

Based on field observations, we choose a threshold for categorizing the height layer. Small stratum corresponds to elements with a height <2 m (0), including plants that can be represented by studied invasie grasses and shrubs from vacant urban land. Elements with a height ≥2 m were used to represent the layer of high stratum (1). This file was then reclassified in the same manner as the ExG layer, with 0 (zero) indicating a small stratum and 1 (one) indicating a high stratum.

These two generated binary layers (ExG and height) can be merged in the raster calculator to identify tall plants and small plants, which would contain our target invasive grasses. Similar approaches using orthomosaics and DEM/DSM were reviewed by Possoch et al. (2016) and Themistocleous (2019). Then, we used the "reclassify by table" tool in QGIS for reclassification of two binary layers. The values used to set the ExG and Height layers' thresholds were derived from a collection of 25 pixels representing classes 0 and 1. For each layer we calculated mean values. We reclassificated them using a QGIS raster calculator tool, and a binary height map with the following categories: (a) without vegetation; (b) with vegetation; (c) elements of intermediate size, smaller than 2 m; and (d) tall elements, greater than 2 m. (Fig. 13.2).

We determined the overall classification accuracy using the QGis AcATaMa plugin AcaTama (Llano, 2019). We used 100 samples for two classes: non-target (large vegetation or other large structure, greater than 2 m) and target (intermediate or low vegetation, smaller than 2 m), in two study areas (Fig. 13.1).

2.7 Machine Learning Classification

We used R version 4.0.1. (R Core Team, 2024) for data analysis. We exported the RGB mosaic, DEM, and DSM into R, next, we calculated two derivative layers: vegetation height (VH) and excess of green index (ExG). To increase the discrimination accuracy between background vegetation and the invasive species studied we narrowed the area of analysis, calibrating the mask to exclude pixels with VH > 2 m

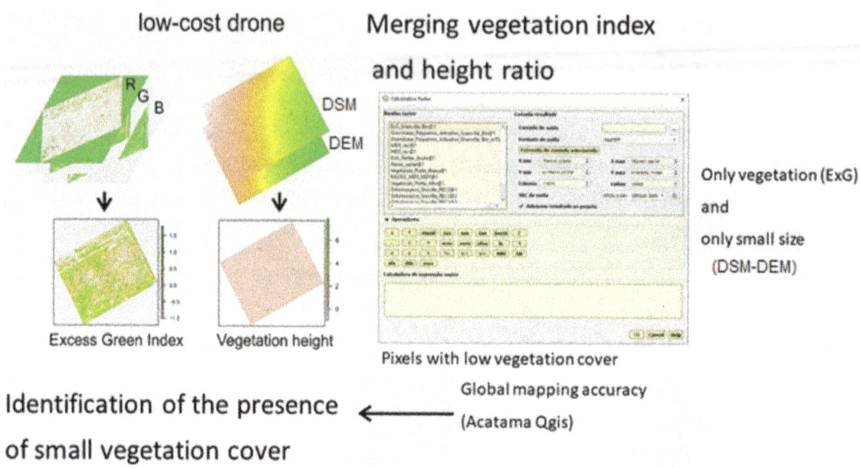

Fig. 13.2 Scheme of the routine used in the QGIS program for aerial image classification and digital height mode. On the left are the DSM and DEM models, as well as the result of the height binary layer and the application of the ExG index based on RGB bands. The QGIS raster calculator is shown on the right, where both procedures were used

height and non-vegetated areas (with ExG < 0.5). Within this mask we selected training polygons for supervised binary classification.

For each study area we randomly selected 10,000 pixels representing invasive grass and 10,000 pixels with background vegetation. Then, we randomly selected 80% of the observations to develop models, and the remaining 20% was used as an independent validation dataset. For each pixel we extracted values of red, green, and blue reflectance, ExG index, and VH. We tested four modeling approaches: logistic regression (GLM; Generalized Linear Model assuming the binomial distribution of the dependent variable), random forest from the randomforest package (Breiman, 2001), support vector machine with linear kernel from the e1037 package (Meyer et al., 2018), and gradient boosting modelling from the GBM package (Ridgeway, 2017). For RF, we tested models with mtry (number of predictors in a single tree) from 1 to 5; for SVM—cost parameters 0.0001, 0.0005, 0.001, 0.005, 0.01, 0.05, and 0.1; and for GBM we tested interaction depths of 1 to 3, number of trees from 50 to 1500 with 50 intervals, shrinkages of 0.01, 0.05, and 0.1, and minimum size of the final node of five.

To standardize development and apply similar internal, repeated cross-validation (ten times by tenfold split), and using centering and scaling for data preprocessing, we developed each machine learning technique using the caret package (Kuhn, 2008). We evaluated model performance using a variety of metrics based on the confusion matrix, i.e., the result of comparing model predictions for independent datasets with real classification. The confusion matrix returns four comparison outcomes: true positive (TP) when both prediction and reference are positive, true negative (TN) when both are negative, false positive (FP) when prediction is positive, but reference is negative, and false negative (FN) when prediction is negative,

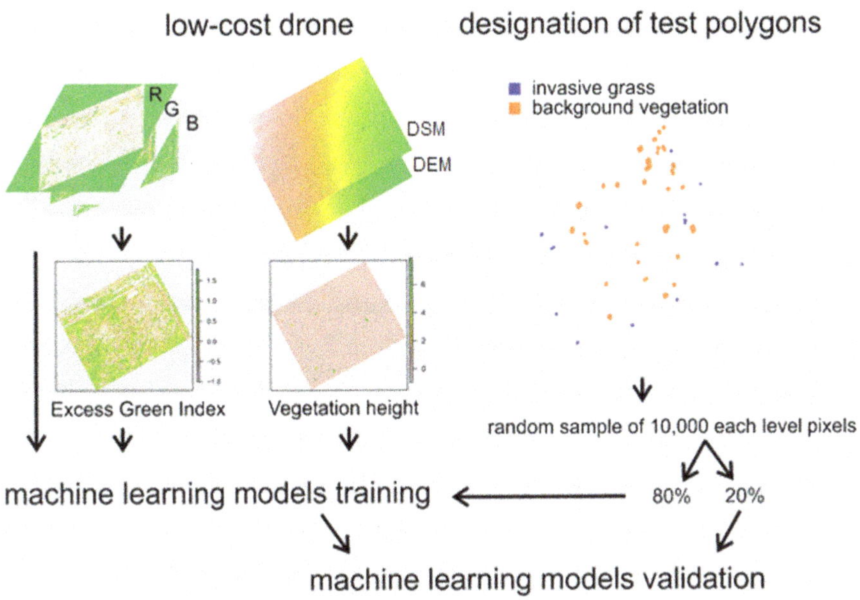

Fig. 13.3 Scheme of the analytical approach used for model development. On the left, are the DSM and DEM models, as well as the result of the DSM-DEM height assessment, and the application of the ExG index based on the RGB bands. On the right, we have polygons that represent the mapped classes, grass samples in blue, and the non-target background in orange

but reference is positive. Comparing the validation dataset with model outcomes, the confusion matrices were obtained and used to calculate the accuracy (proportion of correctly predicted responses), Cohen's kappa (agreement between predictions and reference, referred to expected random guess), specificity (true negative rate, TN/(TN + FP)), sensitivity (true positive rate, TP/(TP + FN)), and true skill statistics (TSS; specificity+sensitivity-1). We also calculated the area under the receiver-operator curve (AUC). These statistics use sensitivity and 1-specificity to demonstrate trade-offs in TP and TN proportions in models. An AUC of 0.5 is comparable to a random guess (it serves as a reference point). Using the same statistics, we tested the transferability of the models developed for one area to another (Fig. 13.3).

3 Results

3.1 Expert-Based Classification

After generating the layer that combined pixels with plants (ExG = 1) and pixels with small elements (height $\leq$ 2 m), it was possible to obtain an accuracy of 0.989 for the mask pixels (ExG = 0 and height >2 m) and of 0.916 for the target pixels

Table 13.1 Granville area: Error matrix (confusion matrix) obtained by the AcaTama plugin—QGIS

		Classified values				
	Study area	1 (Non-target) ExG=0 and height > 2m	2 (Gramineae) ExG=1 and height <= 2m	Total	User accuracy	Total class area (m²)
Thematic raster classes	1	91	9	100	0.91	90028.84037
	2	1	99	100	0.99	2894.60805
	Total	92	108	200		92923.44842
	Producer accuracy	0.98913	0.91667		0.95	
	Overall Accuracy:					

(ExG = 1 and height ≤ 2 m) for the area located in Rondonópolis, Mato Grosso. We obtained an overall accuracy of 91.2%, which demonstrates the routine's robustness. A total of 92.923,44 square meters were mapped (Table 13.1 and Fig. 13.4).

After generating the layer that combined pixels with plants (ExG = 1) and pixels with small elements (Height ≤ 2 m), it was possible to obtain an accuracy of 0.923 for the mask pixels (ExG = 0 and Height > 2 m) and of 0.958 for the target pixels (ExG = 1 and Height ≤ 2 m) for the area located in Itiquira, Mato Grosso. We obtained an overall accuracy of 95.5%, demonstrating the routine's assertiveness also for this area, which is located 150 kilometers from Rondonópolis(Table 13.2 and Fig. 13.5).

3.2 *Machine Learning Classification*

We obtained a comparable performance of GBM and RF for the Granville test site, with the highest specificity and sensitivity, as well as AUC, kappa, and TSS (Table 13.3). Binomial GLM and SVM performed similarly poorly, but still satisfactorily (AUC of 0.80 and TSS of 0.51). All models demonstrated a similar specificity and sensitivity. The predictions showed that the coverage of invasive species studied was more or less similar for all models tested (Fig. 13.6). Binomial GLM, RF, SVM, and GBM correctly predicted the presence of invasive grass in 15.6%, 14.3%, 15.8%, and 14.3% of cases, respectively. In 15.8% of the pixels, at least one model predicted the presence of an invasive species. Only 1.1% of the pixels were chosen by a single model, 8.5% by two models, 0.2% by three models, and 90.2% by four models, indicating a high model convergence (Fig. 13.7).

We found the highest performance of GBM and RF at the Itiquira test site (AUC of 0.881 and 0.755 respectively). Binomial GLM and SVM performed extremely poorly. All models demonstrated a good balance of specificity and sensitivity.

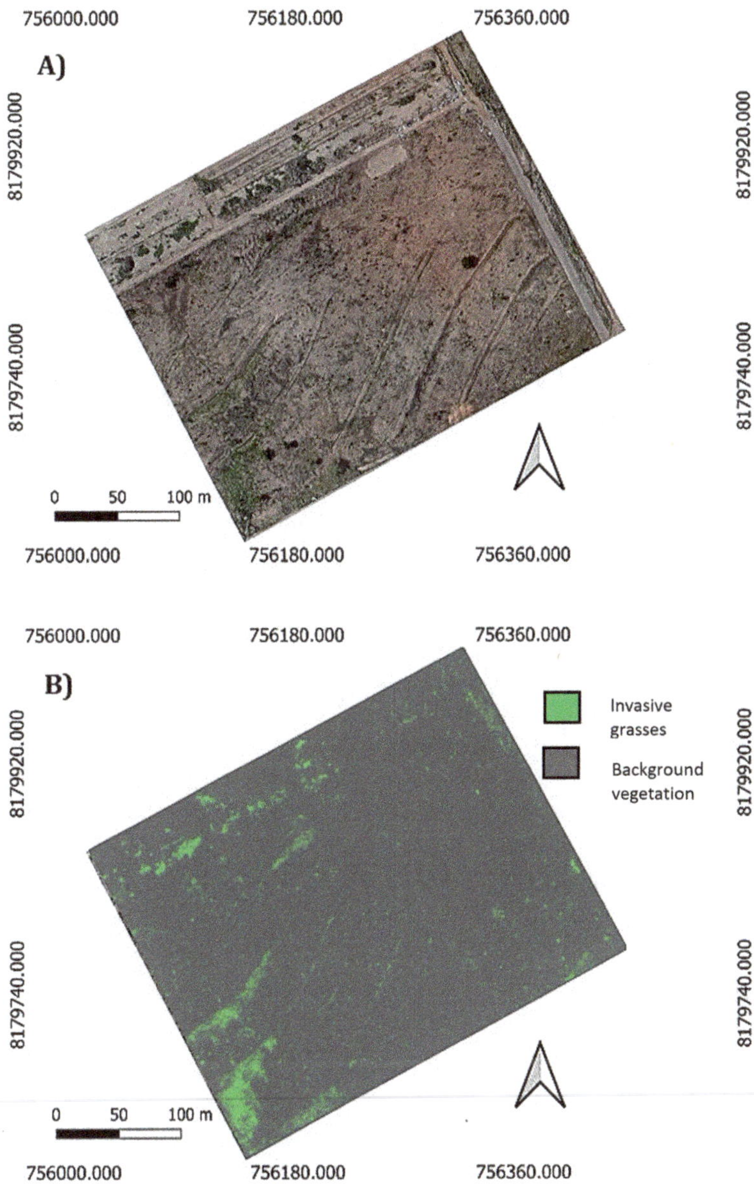

Fig. 13.4 Result of the hybrid classification between the vegetation indexes and the height ratio for the Granville area: a mosaic of images from the Granville area with studied invasive grass *Urochloa* spp. and bare soil (**a**) and the result of classification based on the sum of the two models used to identify grasses (green, **b**)

Table 13.2 Itiquira area: Error matrix (confusion matrix) obtained by the AcaTama plugin—QGIS

		Classified values					
	Study area	1 (Non-target) ExG=0 and height > 2m	2 (Gramineae) ExG=1 and height <= 2m	Total	User accuracy	Total class area (m²)	Wi
Thematic raster classes	1	96	4	100	0.96	55943.53125	0.88204
	2	8	92	100	0.92	7481.795	0.11796
	total	104	96	200		63425.32625	
	Producer accuracy	0.92308	0.95833		0.94		
	Overall Accuracy:			0.95528			

Predictions showed that all models tested had a similar coverage of the invasive species studied (Fig. 13.8). Binomial GLM, RF, SVM, and GBM correctly predicted the presence of invasive grass in 15.8%, 18.5%, 15.8%, and 19.4% of cases, respectively. In 23.3% of pixels, at least one model predicted the presence of an invasive species. Among them, 6.4% of pixels were chosen by only one model, 38.2% by two models, 6.0% by three models, and 49.4% by four models, indicating a small model convergence. (Fig. 13.9).

Applications of the models developed in Granville to Itiquira, and models developed in Itiquira to Granville showed a decrease in the model performance metrics (Table 13.3). Only the weakest models developed for Itiquira (binomial GLM and SVM) performed well in Granville, whereas models developed for Granville performed poorly in Itiquira (with AUC of 0.60).

4 Discussion

4.1 Advantages and Disadvantages of UAVs Application

The aim of our study was to evaluate the effectiveness of using RGB images and digital elevation models obtained from UAVs to distinguish invasive grasses from other objects. We accomplished the goal, showing usefulness of both proposed analytical approaches. In terms of grass identification, the utilization of UAV images proved to be efficient. By employing software and routines capable of analyzing UAV images, we successfully obtained data that revealed areas with the presence of grassy vegetation and other arboreal areas. Our primary objective in this study was to identify invasive species represented by grassy vegetation in the selected areas. As a result, our system demonstrated the ability to differentiate between various types of plant species and showed potential in discriminating between invasive and

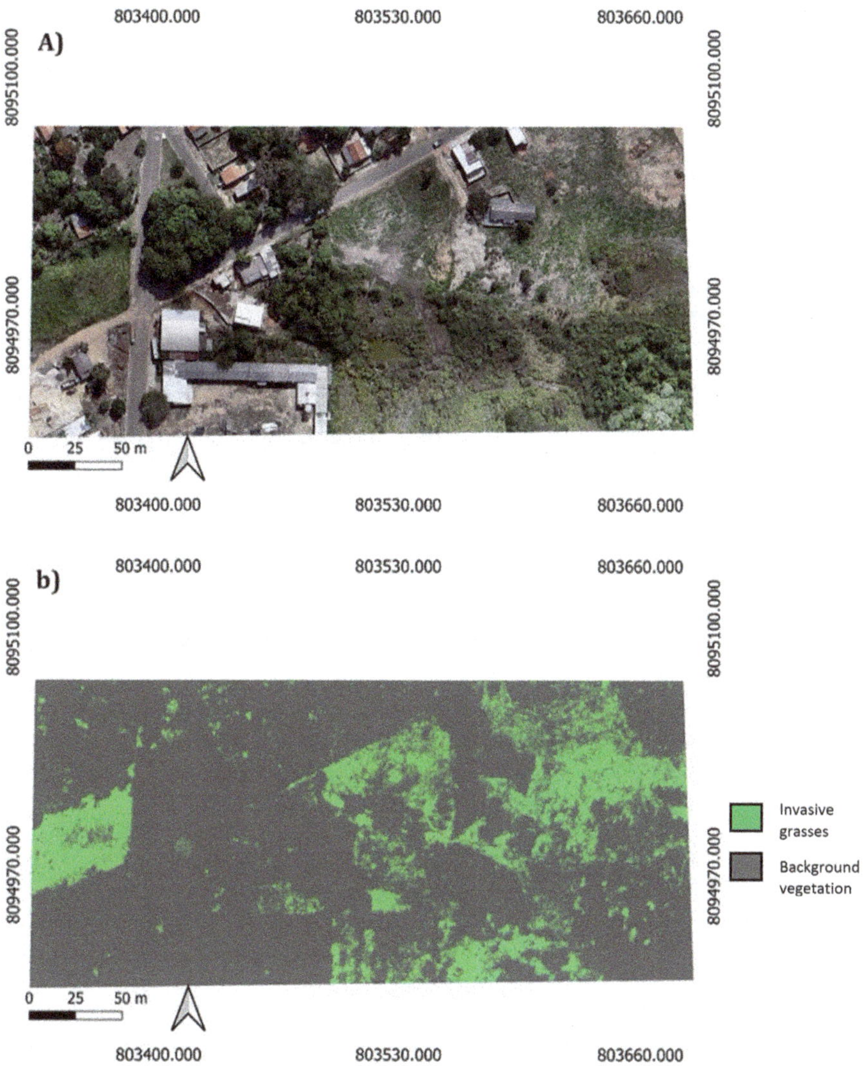

Fig. 13.5 Result of the hybrid classification between the vegetation indexes and the height ratio for the Itiquira area: the mosaic of images from the Itiquira area with the presence of studied invasive grass *Urochloa* spp., remnants of tree-like vegetation, built area, and bare soil (**a**) and the result of classification based on the sum of the two models used to identify grasses (green, **b**)

native tree species. This achievement is noteworthy because it was accomplished solely using RGB images, which are more accessible and readily available compared to multispectral sensors. In contrast, previous studies often relied on expensive hardware, such as multispectral images, to enhance performance. While our method using UAVs may have certain limitations in terms of spatial scale compared to methods employing airborne sensors, the cost-effectiveness and ease of use of

Table 13.3 Performance metrics for four model types developed for the Granville and Itiquira sites, as well as other sites

Model-based on	Object	Model	TN	FP	FN	TP	AUC	Accuracy	Kappa	Sensitivity	Specificity	TSS
Granville	Granville	Binomial GLM	1492	508	476	1524	0.807	0.754	0.508	0.762	0.746	0.508
Granville	Granville	RF	1662	338	268	1732	0.900	0.848	0.697	0.866	0.831	0.697
Granville	Granville	SVM	1495	505	476	1524	0.799	0.754	0.509	0.762	0.747	0.509
Granville	Granville	GBM	1662	338	268	1732	0.911	0.848	0.697	0.866	0.831	0.697
Itiquira	Itiquira	Binomial GLM	1163	837	836	1164	0.600	0.582	0.163	0.582	0.581	0.163
Itiquira	Itiquira	RF	1427	573	573	1427	0.755	0.713	0.427	0.713	0.713	0.427
Itiquira	Itiquira	SVM	1160	840	838	1162	0.591	0.580	0.161	0.581	0.580	0.161
Itiquira	Itiquira	GBM	1422	578	574	1426	0.781	0.712	0.424	0.713	0.711	0.424
Itiquira	Granville	Binomial GLM	1387	613	471	1529	0.791	0.729	0.458	0.764	0.693	0.458
Itiquira	Granville	RF	952	1048	1049	951	0.508	0.476	−0.048	0.475	0.476	−0.048
Itiquira	Granville	SVM	1384	616	471	1529	0.769	0.728	0.456	0.764	0.692	0.456
Itiquira	Granville	GBM	1068	932	730	1270	0.587	0.584	0.169	0.635	0.534	0.169
Granville	Itiquira	Binomial GLM	1046	954	953	1047	0.557	0.523	0.046	0.523	0.523	0.046
Granville	Itiquira	RF	1059	941	930	1070	0.552	0.532	0.064	0.535	0.529	0.064
Granville	Itiquira	SVM	1124	876	876	1124	0.580	0.562	0.124	0.562	0.562	0.124
Granville	Itiquira	GBM	1210	790	1145	855	0.514	0.516	0.032	0.427	0.605	0.032

Abbreviations: *TN* true negative, *FP* false positive, *FN* false negative, *TP* true positive, *AUC* area under the receiver-operator curve, *TSS* true skill statistic

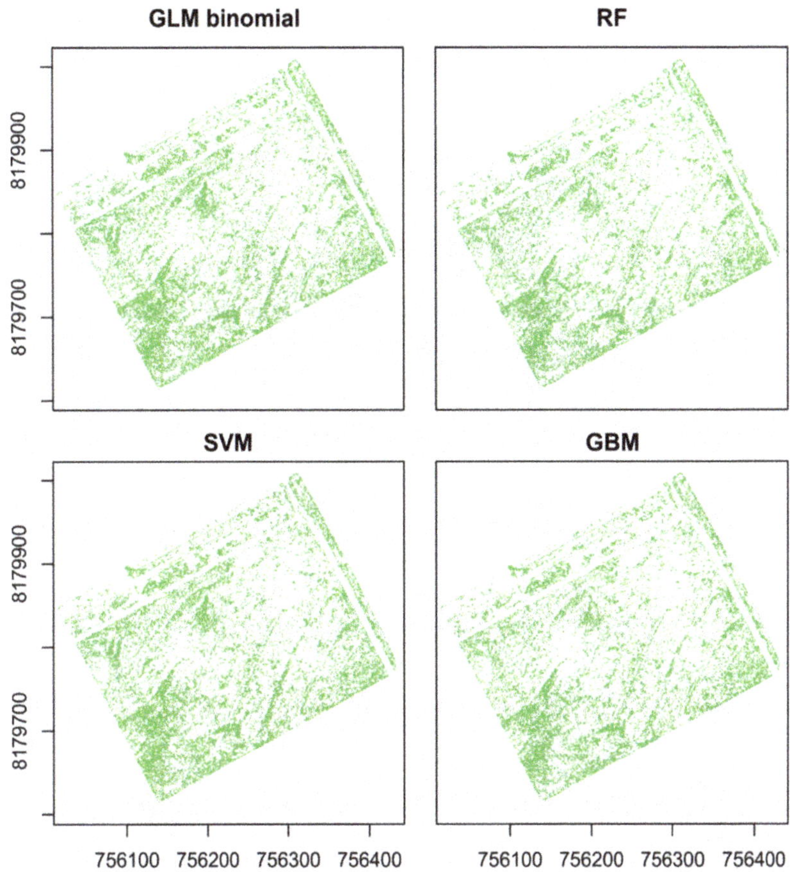

Fig. 13.6 Predictions of invasive species presence (green) in the Granville site using four methods

UAVs enable regular monitoring. On the other hand, other UAV-based research focused only on a limited number of tree species or detecting rectangular objects (Alvarez-Taboada et al., 2017; Dash et al., 2019).

Main disadvantage is lack of constant flight parameters suitable to weather conditions and studied taxta. The routine needs to be adjusted due to various factors, including the UAV camera and the image files it generates. UAV flight parameters, such as equipment speed and camera calibration (e.g., ISO and shutter speed), can result in image files containing noise and distortion, which can negatively impact the method's performance. Additionally, since we are working with a passive sensor, the flight schedule and brightness window are also important considerations.

Given the high spatial resolution of the files, with ground sample distance (GSD) of centimeters, and the jumble of information represented by the pixels, a standardized operational protocol must be adopted, and technicians must ensure that the classified images express a target characteristic of the monitored environment based

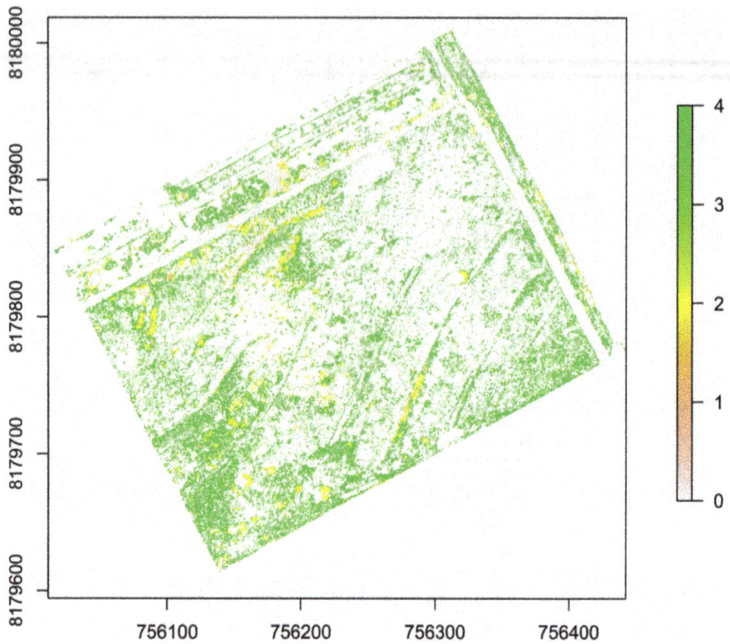

Fig. 13.7 Convergence of predictions of invasive species presence in the Granville site using four methods, expressed as the number of models predicting species presence in a specific pixel

on validations (samples) in the field. Therefore, for example, areas covered by small vegetation, such as invasive grasses, will be detected more effectively, and based on their geographical location—the occupied area—and context of their surrounding environment, such as residences, public areas, industries, forests, and so on, the urban manager will be able to determine a management action, preventing environmental damage and pollution.

4.2 *Expert-Based Assessment*

QGIS, a geographic information system renowned for its speed and efficiency in processing environmental data, has emerged as a prominent tool. In this study, we employed the QGIS methodology, combining intercropping vegetation indices and digital height models, which proved to be remarkably effective in detecting plants of diverse sizes. This approach can serve as a viable alternative for target detection, particularly in urban areas where grasses proliferate, especially in vacant spaces like abandoned lands. Similar approach produced a sound results in the works of Pessi et al. (2020), who conducted a study in an area with invasive grasses and fragments of remaining vegetation and used the DSM-DEM ratio to separate grassy vegetation from fragments of tree vegetation. The processing

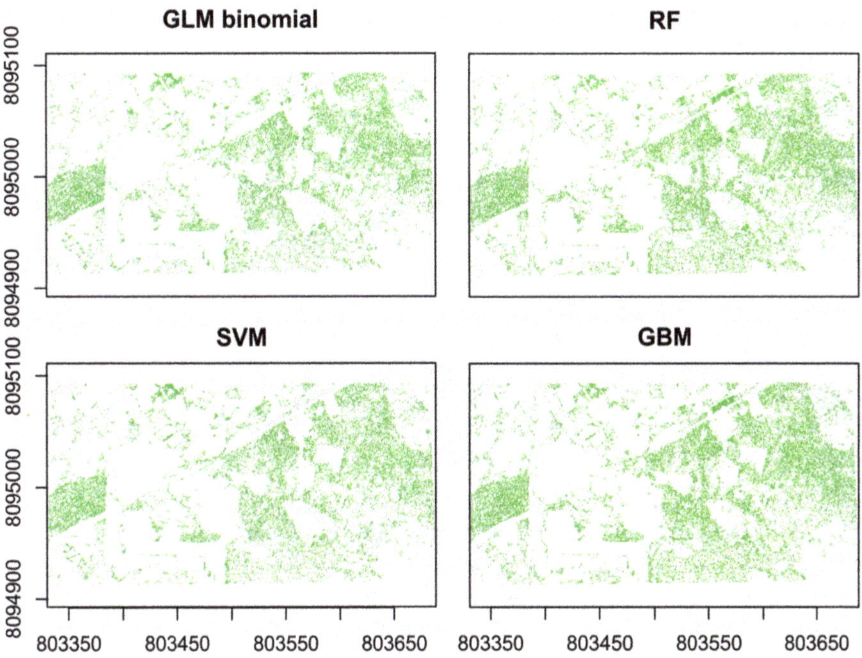

Fig. 13.8 Predictions of invasive species presence (green) in the Itiquira site using four methods

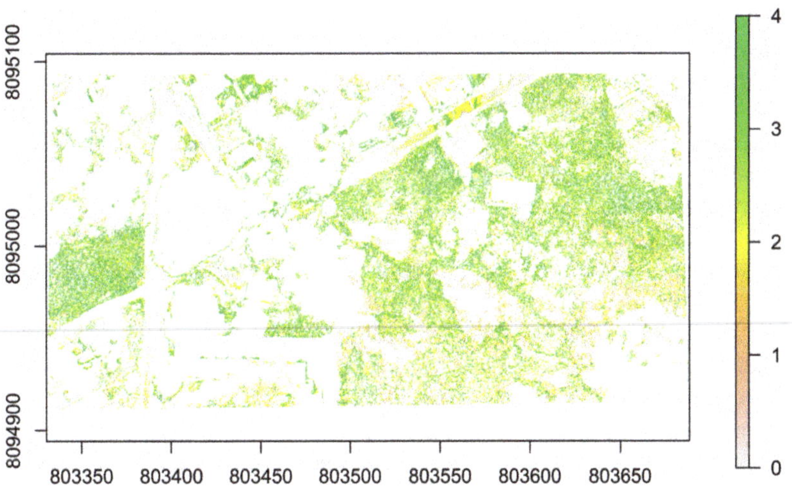

Fig. 13.9 Convergence of invasive species presence predictions in the Itiquira site using four methods, expressed as the number of models predicting species presence in a given pixel

sequence used in the SIG QGIS can be automated, resulting in greater data collection agility. Already available tools built into QGIS, and ability to expand them using a Phyton language, can help with this automation. However, for urban managers with little experience with technologies, non-automated execution processes, such as the one proposed here represent a significant gain in time and financial resources, because mapping critical environments (e.g. in risk of endemic diseases or epidemics, loss of native biodiversity, or fires) for urban planning, data from a low-cost drone equipped with a conventional camera sensitive to the visible range of the electromagnetic spectrum and a free GIS, such as QGIS, can be easily mapped for urban planning. Therefore, this approach can be easily adapted for other invasive species and cities.

4.3 Machine Learning Approach

One of the most popular tools for programming is R, an open-source computer language for statistical and graphics computing, that allows a user with no experience in R to run their data and generate their results by using a workflow already prepared by another user. R has grown in popularity and is now widely used in research (Loewen et al., 2014; Mizumoto & Plonsky, 2016) because it is a software that delivers good processing and automation performance with pre-established routines by any user and is available as open-source software to the scientific community. R has recently "become the most widely used analytical software for all academic articles," according to (Lai et al. 2019). Another important aspect to note is that journals and research funding agencies are increasingly requiring that data be made publicly available after publication (Mizumoto & Plonsky, 2016). Furthermore, open access to R codes facilitates data analysis reproduction and replication, allowing other researchers from anywhere in the world to use the same script for their data analysis. Because R is a command-line program, this is possible. To replicate the analysis and obtain the same results as the original study, one must simply run the code in R. This R feature also enables researchers to experiment with and learn from different analytical approaches based on the same data set (Mizumoto & Plonsky, 2016).

The models developed in this study using R have been site- and input data-specific. Therefore, the transferability of the approach will require adjustments for particular conditions (inspection of vacant lots, parks), and different image formats (different spectral bands, spatial resolution, etc.) to receive the satisfying level of accuracy in terms of false positives and false negatives ratio. On the other hand, site-specificity can be helpful to account for the context-dependence of biological invasions (Sapsford et al., 2020), maximizing the accuracy of detection in particular conditions.

Our models were developed to achieve a balance in the ratio of false positives to false negatives. As a result, we assume a similar cost for false alarms and omissions. This strategy can be implemented in areas with high invasive species coverage, but it may not be sufficient for early detection. In such cases, the models may be insufficiently sensitive, and we recommend the use of models that increase sensitivity. This will result in a higher rate of false alarms but a lower rate of omissions. For sites with a low prevalence of invasive species, we can modify our approach by using techniques developed for unbalanced samples, such as SMOTE (Chawla et al., 2002).

5 Conclusions

Our study provided a new routine for detecting invasive species using low-cost drones in urban ecosystems. We developed two frameworks for data processing, model development, and validation that can be used at any location in the world. One of them employs QGIS, a graphical user interface-based geospatial information system, and has a lower threshold of knowledge required to work, while the second requires basic knowledge of machine learning methods and using R. Both approaches are based mostly on freeware and open source software, to maximize accessibility and transferability of approach. We achieved our goal by employing only the most basic predictors: red, green, and blue bands, a simple vegetation index based on them, and a digital terrain model derivative. These predictors can be obtained using a low-cost device, making them accessible to a wide range of nature conservation and urban planning units, as well as researchers worldwide. Furthermore, the use of these solutions can provide a reasonable estimate of invasive species distributions, which can aid in the management of invasive species in cities. With a low financial investment of around US $ 2000, a city administration can set up a basic workstation, which includes an image processing computer, UAV, and generate technology products to support the decision-making process for environmental management of invasive species. However, depending on target species characteristics, the approach can be extended to other utilities, e.g. expansive native species or threatened species.

Acknowledgments The authors would like to express their gratitude to the Graduate Program in Environmental Technologies (PPGTA) of the Faculty of Engineering, Architecture, and Urbanism (FAENG) of the Federal University of Mato Grosso do Sul (UFMS). They also thank the Graduate Program in Environmental Management and Technology (PGGTA) of the Federal University of Rondonópolis, Mato Grosso.

This work was supported by the Coordination for the Improvement of Higher Education Personnel—Brazil (CAPES)—Financing Code 001. Finally, the authors would also like to thank CAPES for the PhD fellowship granted to Dhonatan Diego Pessi, process number 88887.494036/2020-00, and CNPq for their support (Process number 441975/2018-6 and 315170/2018-2). MKD acknowledges partial financial support from the Institute of Dendrology, Polish Academy of Sciences.

References

Accatino, F., De Michele, C., Vezzoli, R., Donzelli, D., & Scholes, R. J. (2010). Tree–grass coexistence in savanna: Interactions of rain and fire. *Journal of Theoretical Biology, 267*, 235–242.

Alexander, N., & Jenkins, L. (2015). *An application of aerial drones in zoning and urban land use* (100 pp). Ryerson University.

Alvarez-Taboada, F., Paredes, C., & Julián-Pelaz, J. (2017). Mapping of the invasive species *Hakea sericea* using unmanned aerial vehicle (UAV) and WorldView-2 imagery and an object-oriented approach. *Remote Sensing, 9*, 913. https://doi.org/10.3390/rs9090913

Alvey, A. A. (2006). Promoting and preserving biodiversity in the urban forest. *Urban Forestry & Urban Greening, 5*, 195–201. https://doi.org/10.1016/j.ufug.2006.09.003

Angeoletto, F., et al. (2016). Tipología socio-ambiental de las ciudades medias de Brasil: aportes para un desarrollo urbano sostenible. *Urbe. Revista Brasileira de Gestão Urbana, 8*, 272–287.

Angeoletto, F., et al. (2017). The grass is always greener on the other side of the fence: The flora in urban backyards of different social classes. *Ambiente & Sociedade, 20*, 1–20.

Angeoletto, F., et al. (2022). Will Brazilian city dwellers actively engage in urban conservation? A case study with the charismatic neotropical Blue-and-yellow macaw (*Ara ararauna*). *Birds, 3*(2), 234–244.

Angeoletto, F. H. S., Richard, E., Santos, J., Guedes, N., Oliveira, S. L., Albertín, R. M., et al. (2023). When we think of urban afforestation in Brazil, what comes to mind is a mouth that has lost its teeth. *Terr@ Plural, 17*, e2321654.

Bakogiannis, E., Kyriakidis, C., & Zafeiris, V. (2019). Using unmanned aerial vehicles (UAVS) to analyze the urban environment. *European Journal of Engineering and Formal Sciences, 3*(2), 20–28.

Baptistella, J. L. C., De Andrade, S. A. L., Favarin, J. L., & Mazzafera, P. (2020). *Urochloa* in tropical agroecosystems. *Front Sustain Food Systems, 4*. https://doi.org/10.3389/fsufs.2020.00119

Bartz, R., & Kowarik, I. (2019). Assessing the environmental impacts of invasive alien plants: A review of assessment approaches. *NeoBiota, 43*, 69–99.

Beerling, D. J., & Osborne, C. P. (2006). The origin of the savanna biome. *Global Change Biology, 12*, 2023–2031.

Bolognez, A. K. A. C., Schilichting, A. F., Da Silva, N. M., Paranhos Filho, A. C., Haupenthal, M. R., Pessi, D. D., Barbosa, D. S., Branco, K. R. J. J. C., & Miranda Júnior, P. L. (2021). Aeronaves Remotamente Pilotadas – RPA. In P. Filho et al. (Orgs.), *Geotecnologias para Aplicações Ambientais* (pp. 351–972). Editora Uniedusul, Maringá, PR

Breiman, L. (2001). Random forests. *Machine Learning, 45*, 5–32. https://doi.org/10.1023/A:1010933404324

Caizergues, A. E., Santangelo, J. S., Ness, R. W., Angeoletto, F., Anstett, D. N., Anstett, J., ... & Johnson, M. T. (2024). Does urbanisation lead to parallel demographic shifts across the world in a cosmopolitan plant?. *Molecular Ecology, 33*(7), e17311.

Caramaschi, G. M. C. L. (2015). *Crescimento inicial das gramíneas Paspalum atratum(Swallen) e Urochloa brizantha (Hochst. Ex A. Rich) Stapf. e seus efeitos na rebrota de espécies lenhosas do do Cerrado*. xi, 104 f., il. Tese (Doutorado em Botânica)—Universidade de Brasília.

Chawla, N. V., Bowyer, K. W., Hall, L. O., & Kegelmeyer, W. P. (2002). Smote: Synthetic minority over-sampling technique. *Journal of Artificial Intelligence Research, 16*, 321–357. https://doi.org/10.1613/jair.953

Chen, Y., Shioi, H., Montesinos, C. F., & Koh, L. P. (2014). Active detection via adaptive submodularity. In *International conference on machine learning* (pp. 55–63).

Clark, J. R., Matheny, N. P., Cross, G., & Wake, V. (1997). A model of urban forest sustainability. *Journal of Arboriculture, 23*, 17–30.

Colomina, I., & Molina, P. (2014). Unmanned aerial systems for photogrammetry and remote sensing: A review. *ISPRS Journal of Photogrammetry and Remote Sensing, 92*, 79–97.

da Silva, N. M., Pessi, D. D., Mioto, C. L., Paranhos Filho, A. C., Barbosa, D. S., Chaves, J. R., Miranda Junior, P. L., & Guarlotti, M. (2021). *Guia operacional básico (GOB) – Aeronaves Remotamente Pilotadas* (p. 58). Editora Uniedusul, Maringá, PR.

Dash, J. P., Watt, M. S., Paul, T. S. H., Morgenroth, J., & Pearse, G. D. (2019). Early detection of invasive exotic trees using UAV and manned aircraft multispectral and LiDAR data. *Remote Sensing, 11*, 1812. https://doi.org/10.3390/rs11151812

Dyderski, M. K., & Jagodziński, A. M. (2019). Context-dependence of urban Forest vegetation invasion level and alien species' ecological success. *Forests, 10*, 26. https://doi.org/10.3390/f10010026

Dyderski, M. K., Gdula, A. K., & Jagodziński, A. M. (2015). "The rich get richer" concept in riparian woody species – A case study of the Warta River Valley (Poznań, Poland). *Urban Forestry & Urban Greening, 14*, 107–114. https://doi.org/10.1016/j.ufug.2014.12.003

FATMA (Fundação do Meio Ambiente de Santa Catarina). (2016). *Lista comentada de espécies exóticas invasoras no estado de Santa Catarina: espécies que ameaçam a diversidade biológica*. Governo de Santa Catarina.

Foxcroft, L. C., Richardson, D. M., Rejmánek, M., & Pyšek, P. (2010). Alien plant invasions in tropical and sub-tropical savannas: Patterns, processes and prospects. *Biological Invasions, 12*, 3913–3933. https://doi.org/10.1007/s10530-010-9823-7

Gallacher, D. (2016). Drone applications for environmental Management in Urban Spaces: A review. *International Journal of Sustainable Land Use and Urban Planning [IJSLUP], 3*(4), 1–14.

Gilbert, O. L. (1989). *The ecology of urban habitats*. Chapman and Hall.

Goddard, M. A., et al. (2021). A global horizon scan of the future impacts of robotics and autonomous systems on urban ecosystems. *Nature Ecology & Evolution, 5*(2), 219–230.

Horus. (2015). *I3N-Brasil. Base de dados nacional de espécies exóticas invasora (National database of exotic invasive species)*. Florianópolis – SC, Brazil: I3N Brasil, Instituto Hórus de Desenvolvimento e Conservação Ambiental. http://i3n.institutohorus.org.br

IBGE. Instituto Brasileiro de Geografia e Estatística. (2020). *Malhas Territoriais*. Disponível em: https://www.ibge.gov.br/geociencias/organizacao-do-territorio/malhas-territoriais/15774-malhas.html?=&t=downloads. Acesso em: 28 de fevereiro de 2021.

Keeley, J. E., & Rundel, P. W. (2005). Fire and the Miocene expansion of C4 grasslands. *Ecology Letters, 8*, 683–690.

Knapp, S., Kühn, I., Stolle, J., & Klotz, S. (2010). Changes in the functional composition of a central European urban flora over three centuries. *Perspectives in Plant Ecology, Evolution and Systematics, 12*, 235–244. https://doi.org/10.1016/j.ppees.2009.11.001

Kowarik, I. (1995). On the role of alien species in urban flora and vegetation. In P. Pyšek, K. Prach, M. Rejmánek, & M. Wade (Eds.), *Plant invasions: General aspects and special problems* (pp. 85–103). SPB Academic Publishing.

Kowarik, I. (2011). Novel urban ecosystems, biodiversity, and conservation. *Environmental Pollution, 159*, 1974–1983. https://doi.org/10.1016/j.envpol.2011.02.022

Kowarik, I., Hiller, A., Planchuelo, G., Seitz, B., Von Der Lippe, M., & Buchholz, S. (2019). Emerging urban forests: Opportunities for promoting the wild side of the urban Green infrastructure. *Sustainability, 11*, 6318. https://doi.org/10.3390/su11226318

Kuhn, M. (2008). Building predictive models in R using the caret package. *Journal of Statistical Software, 28*, 1–26. https://doi.org/10.18637/jss.v028.i05

Kühn, I., Brandl, R., & Klotz, S. (2004). The flora of German cities is naturally species rich. *Evolutionary Ecology Research, 6*, 749–764.

Lajo, A. A. M. (2021). Geoprocessamento em Áreas Urbanas. In P. Filho et al. (Org.). *Geotecnologias para aplicações Ambientais* (394 p). Editora UNIEDSUL.

Lai J, Lortie CJ, Muenchen RA, et al (2019) Evaluating the popularity of R in ecology. Ecosphere 10:e02567. https://doi.org/10.1002/ecs2.2567

Laliberte, A. S., Goforth, M. A., Steele, C. M., & Rango, A. (2011). Multispectral remote sensing from unmanned aircraft: Image processing workflows and applications for rangeland environments. *Remote Sensing, 3*, 2529–2551.

Lambert, M. R., & Donihue, C. M. (2020). Urban biodiversity management using evolutionary tools. *Nature Ecology & Evolution, 4*(7), 903–910.

Lima, G. F., Correa Santos, J. W. M., Albertin, R. M., Martínez-Miranzo, B., Souza, F. L., & Angeoletto, F. (2023). Backyards are a way to promote environmental justice and biodiversity conservation in Brazilian cities. *Diversity, 15*(7), 815.

Llano, X. C. (2019). AcATaMa – QGIS plugin for Accuracy Assessment of Thematic Maps, version XX.XX. Disponível em: https://plugins.qgis.org/plugins/AcATaMa/. Acesso em: 03 de junho de 2021.

Loewen, S., Lavolette, E., Spino, L., Papi, M., Schmidtke, J., Sterling, S., & Wolff, D. (2014). Statistical literacy among applied linguists and second language acquisition researchers. *TESOL Quarterly, 48*(2), 360–388. https://doi.org/10.1002/tesq.128

Mao, W., Wang, Y., & Wang, Y. (2003). Real-time detection of between-row weeds using machine vision. In *2003 ASAE annual meeting* (p. 1). American Society of Agricultural and Biological Engineers.

Mckinney, M. L. (2002). Urbanization, biodiversity, and conservation. *Bioscience, 52*, 883–890. https://doi.org/10.1641/0006-3568(2002)052[0883:UBAC]2.0.CO;2

Mckinney, M. L. (2006). Urbanization as a major cause of biotic homogenization. *Biological Conservation, 127*, 247–260. https://doi.org/10.1016/j.biocon.2005.09.005

Meyer, G. E., & Camargo Neto, J. (2008). Verification of color vegetation indices for automated crop imaging applications. *Computer and Eletronics in Agriculture, 32*(2), 282–293. https://doi.org/10.1016/j.compag.2008.03.009

Meyer, D., Dimitriadou, E., & Hornik, K., et al. (2018). e1071: Misc Functions of the Department of Statistics, Probability Theory Group (Formerly: E1071), TU Wien. Version 1.7-0. Disponível em: https://CRAN.R-project.org/package=e1071. Acesso em 03 de junho de 2021.

Miller, J. R., & Hobbs, R. J. (2002). Conservation where people live and work. *Conservation Biology, 16*, 330–337. https://doi.org/10.1046/j.1523-1739.2002.00420.x

Mizumoto, A., & Plonsky, L. (2016). R as a lingua Franca: Advantages of using R for quantitative research in applied linguistics. *Applied Linguistics, 37*(2), 284–291. https://doi.org/10.1093/applin/amv025

Moroń, D., Beim, M., Gudowska, A., Angeoletto, F., Celary, W., Cwajna, A., ... & Tryjanowski, P. (2024). Evaluating tramway infrastructure on biodiversity and ecosystem services. *Scientific Reports, 14*(1), 9394.

Mostert, E., et al. (2018). A multi-criterion approach for prioritizing areas in urban ecosystems for active restoration following invasive plant control. *Environmental Management, 62*, 1150–1167.

Paranhos Filho, A. C., Catalani, T. G. T., Pessi, D. D., & Mioto, C. L. (2021). Modelagem do Terreno. In P. Filho et al. (Orgs.), *Geotecnologias para aplicações ambientais* (pp. 235–244). Editora Uniedusul, Maringá, PR.

Paź-Dyderska, S., Ladach-Zajdler, A., Jagodziński, A. M., & Dyderski, M. K. (2020). Landscape and parental tree availability drive spread of *Ailanthus altissima* in the urban ecosystem of Poznań, Poland. *Urban Forestry & Urban Greening, 56*, 126868. https://doi.org/10.1016/j.ufug.2020.126868

Pessi, D. D., José, J. V., Mioto, C. L., & Silva, N. M. (2020). Aeronave remotamente pilotada de baixo custo no estudo de plantas invasoras em áreas de cerrado. *Nativa, Sinop, 8*(1), 66–80.

Pivello, V. R., Carvalho, V. M. C., Lopes, P. F., Peccinini, A. A., & Rosso, S. (1999). Abundance and distribution of native and alien grasses in a "Cerrado" (Brazilian savanna) biological reserve. *Biotropica, 31*, 71. https://doi.org/10.2307/2663960

Possoch, M., Bieker, S., Hoffmeister, D., Bolten, A., Schellber, J., & Bareth, G. (2016). Multi-temporal crop surface models combined with the Rgb vegetation index from UAV-based images for forage Momitoring in grassland. In *The international archives of the photogrammetry. Remote sensing and spatial information sciences, volume XLI-B1, 2016 XXIII ISPRS Congress*. https://www.int-arch-photogramm-remote-sens-spatial-inf-sci.net/XLI-B1/991/2016/isprs-archives-XLI-B1-991-2016.pdf. Accessed 5 May 2021.

Pyšek, P. (2016). Alien and native species in central European urban floras: A quantitative comparison. *Journal of Biogeography, 25*, 155–163. https://doi.org/10.1046/j.1365-2699.1998.251177.x

QGIS Development Team. (2021). *Versão 3.16. 'Hann.over'. Geographic information system.* Open Source Geospatial Foundation Project. Disponível em: https://www.qgis.org/pt_BR/site/forusers/download.html. Acesso em: 03 de junho de 2021.

R Core Team (2024) R: A language and environment for statistical computing.

Ramana, M. V., Aditya, S. V., & Kothari, M. (2016). Motion planning for a fixed-wing UAV in urban environments. *IFAC-PapersOnLine, 49*(1), 419–424.

Ridgeway, G. (2017). *GBM: Generalized boosted regression models. Version 2.1.3.* Disponível em: https://CRAN.R-project.org/package=gbm. Acesso em: 03 de junho de 2021.

Riet-Correa, B., Castro, M. B., Lemos, R. A., Riet-Correa, G., Mustafa, V., & Rietcorrea, F. (2011). *Brachiaria* spp. Poisoning of ruminants in Brazil. *Pesquisa Veterinária Brasileira, 31*(3), 183–192.

Rumble, H., et al. (2019). Understanding and applying ecological principles in cities. In *Planning cities with nature: Theories, strategies and methods* (pp. 217–234).

Sandino, J., et al. (2020). UAV framework for autonomous onboard navigation and people/object detection in cluttered in door environments. *Remote Sensing, 12*, 3386.

Santangelo, J., et al. (2022). Global urban environmental change drives adaptation in white clover. *Science, 375*(6586), 1275–1281.

Sapsford, S. J., Brandt, A. J., Davis, K. T., Peralta, G., Dickie, I. A., Gibson, R. D., Green, J. L., Hulme, P. E., Nuñez, M. A., Orwin, K. H., Pauchard, A., Wardle, D. A., & Peltzer, D. A. (2020). Towards a framework for understanding the context dependence of impacts of non-native tree species. *Functional Ecology, 34*, 944–955. https://doi.org/10.1111/1365-2435.13544

Shaffer, H. B. (2018). Urban biodiversity arks. *Nature Sustainability, 1*, 725–727.

Soanes, K., & Lentini, P. E. (2019). When cities are the last chance for saving species. *Frontiers in Ecology and Environment, 17*, 225–231.

Soanes, K., et al. (2018). Correcting common misconceptions to inspire conservation action in urban environments. *Conservation Biology, 33*, 300–306.

Souza, F. L., Fenoglio, M. S., & Angeoletto, F. (2023). To be a Brazilian City dweller, sometimes we must learn to say enough! *Sustainability, 15*(4), 3699.

Szlavecz, K., Warren, P., & Pickett, S. (2011). Biodiversity on the urban landscape. In R. P. Cincotta & L. J. Gorenflo (Eds.), *Human population* (pp. 75–101). Springer.

Themistocleous, K. (2019). DEM modeling using RGB-based vegetation indices from UAV images. In *Seventh international conference on remote sensing and geoinformation of the environment—RSCy*.

UEMRI – Urban Environmental Management Research Institute. (2008). *Introduction: Urban environmental management*. Disponível em: www.gdrc.org/uem/doc-intro.html. Acesso em: 04 de fevereiro de 2021.

Vendruscolo, D. G. S., Drescher, R., Souza, H. S., Moura, J. P. V. M., Mamoré, F. M. D., & Siqueira, T. A. S. (2015). Estimativa da altura de eucalipto por meio de regressão não linear e redes neurais artificiais. *Revista Brasileira de Biometria São Paulo, 33*(4), 556–569.

Wasser, L. (2018). *earthlab/earth-analytics-r-course: earth analytics course in the R programming language*. Available in: https://www.earthdatascience.org/courses/earth-analytics/lidar-raster-data-r/lidar-chm-dem-dsm/, https://zenodo.org/record/1326873. Accessed 27 Feb 2021.

Williams, N. S. G., Hahs, A. K., & Vesk, P. A. (2015). Urbanisation, plant traits and the composition of urban floras. *Perspectives in Plant Ecology, Evolution and Systematics, 17*, 78–86. https://doi.org/10.1016/j.ppees.2014.10.002

Chapter 14
Evaluation of Urban Green Areas: A Methodological Proposition

Dulcina de Aquino Hernandez de Oliveira Queiroz, Silvia Méri Carvalho, Lindon Fonseca Matias, and César Ángel Peña Salmón

Abstract Currently, the attention to a green agenda is becoming more and more visible in the search for sustainability in cities. The use of benchmarks, when based on mere arithmetic operations, which distribute an area per inhabitant, masks the real situation of urban green areas and allows methodological differences from one place to another, thus preventing a comparative analysis. The use of conventional standards would allow cities to share goals and compare results related to the quality and distribution of green areas, providing institutions with a methodology that would aid in the diagnosis phase of urban planning. The proposed methodology is a quali-quantitative index that evaluates the Quality of Urban Green Areas (IQAVU, Brazilian acronym for Índice de Qualidade das Áreas Verdes Urbanas), utilizing measures for indicators (sub-indexes) representative of Social, Environmental and Aesthetic functions, all of equal importance. The index is intended to be applicable to any urban area, regardless of demographic, geographic or political reality. In each indicator, fundamental parameters for the characterization of an urban green area will be considered. Therefore, an urban green area deficient in at least one of these indicators cannot be considered functional. An application of a case study is presented in the city of Ponta Grossa, Southern Brazil.

Keywords Urban planning · Indexes · Indicators · Public spaces

---

D. de Aquino Hernandez de Oliveira Queiroz (✉) S. M. Carvalho
Universidade Estadual de Ponta Grossa, Ponta Grossa, PR, Brazil
e-mail: silviameri@uepg.br

L. F. Matias
Universidade Estadual de Campinas, UNICAMP, Campinas, SP, Brazil

C. Á. P. Salmón
UABC-Universidad Autónoma de Baja California, Mexicali, Mexico

F. Angeoletto et al. (eds.), *Ecology of Tropical Cities*,
https://doi.org/10.1007/978-3-031-48012-6_14

1 Introduction

Well-planned cities offer good traffic conditions, efficient public services and suitable infrastructure, in addition to physical amenities, among which urban green areas might be included. There is no doubt that the presence of green areas in urban environments is beneficial for the improvement of the population's quality of life, and for the biodiversity conservation in urban ecosystems (Rumble et al., 2019; Angeoletto et al., 2022; Santangelo et al., 2022), mainly when these spaces are strategically thought and planned. Therefore, an efficient plan optimizes the levels of urban sustainability and quality of life (Angeoletto et al., 2023; Souza et al., 2023).

To achieve this aim, it seems relevant to emphasize how important it is to observe realistically the conditions of the existing green areas in a certain urban space through a diagnostic phase, which should be the starting point to guide mitigation and intervention actions to be developed.

Regarding green areas, in the diagnostic phase, the use of isolated indexes divided by the number of inhabitants is common, among these are the green area index per inhabitant (IAV/hab, Brazilian acronym for Índice de Área Verde por Habitante), the vicinity park green area index (IAVPV, Brazilian acronym for Índice de Área Verde para Parques de Vizinhança), the usable green area index (IAVU, Brazilian acronym for Índice de Áreas Verdes Utilizáveis), and the vegetation cover index (ICV, Brazilian acronym for Índice de Cobertura Vegetal), among others (Cavalheiro & Del Picchia, 1992; Guzzo, 1999). Although this is a widely used methodology in the scientific community, it shows reduced efficiency, which is sometimes even convenient.

Such indexes when applied in isolation are complemented with reference value resources, which are available in the literature and are used to classify the quality of urban green areas. However, these reference values are not standardized either, which leads to different positions as a function of location in different continents, countries or regions.

It seems also relevant to highlight the importance that the green area planning theme has achieved in urban spaces with time, particularly, regarding the environmental perspective (Lima et al., 2023). It has gained space in global forums on themes related to the sustainability of urban spaces from a global perspective. Such attempts reveal the pursuit of analysis, comparison, creation of indicators and global classifications, whose efficacy depends on the creation of a single conventionalized standard.

In such context, we introduce a methodology developed in the academic environment, which proposes and applies a compound index to evaluate urban green areas, utilizing not only environment-related indicators, but also social and aesthetic parameters. This methodology provides the users with a concrete reading of the current reality through the diagnosis of the conditions of these areas aiming to facilitate the urban planning process.

2 Concepts and Reference Indexes

The literature on urban green areas introduces some concepts and most of the applied studies include some kind of calculation or propose indexes to express their spatial distribution. The term covers several approaches, observing diverse methodologies for the classification and mapping of the urban environment and later comparison with other indexes from different municipalities. Knowing this theme helps to understand the role of these areas in the urban space and how their different functions contribute to the improvement of the population's quality of life.

Cavalheiro and Del Picchia (1992) stated that green areas refer to free spaces and that the use of this term is better employed for having a broader meaning, even including superficial water, for example. Milano (1992) emphasized that every green area constitutes a free area, even if its function and nature might be limited, and that adaptations of the concept consider the existence of a significant vegetable cover in these areas. Mazzei et al. (2007), however, pointed out that green areas and free areas cannot be understood as synonyms.

Cavalheiro et al. (1999) and Guzzo (2006) considered that green areas are related to the percentage of vegetation cover (70%) and unsealed soils. Considering the term soil permeability in urban areas is a relevant factor, since it is mainly related to natural spaces, parks, gardens and squares that are usually free from buildings and are associated to environmental quality, since the vegetal cover is a fundamental element in the composition of green areas.

Di Fidio (1990) classified green spaces according to their access such as (i) private and semi-public urban green spaces, which include home gardens and urban gardens; (ii) public urban green spaces such as squares and urban parks; and (iii) suburban green spaces, like green belts.

Among the conceptual proposals, Bargos (2015, p. 29) understands that:

> […] the concept of urban green area should consider that this is a category of urban free space containing trees and shrubs (including trees on the streets, provided that they reach an influence radius that allows them to provide the functions of a green area), with soil free from buildings or soil sealing materials (at least 70% of its area), public access or not, and that can minimally provide ecological functions (increase in thermal comfort, air and sound pollution control, rainwater drainage, and fauna shelter), aesthetic functions (visual and ornamental improvement of the environment and diversification of the built landscape), and leisure functions (entertainment).

Guzzo (1999) lists the main functions of green areas as ecological, aesthetic and social. The ecological function contributes as a way of minimizing the negative effects of the city through natural elements; the aesthetical function is associated to the integration of green spaces with built spaces and population circulation spaces; while the social function would be related to the use of these spaces for the population's leisure activities. Vieira (2004) adds an educational function that is the use of these spaces for the development of teaching and learning activities; and the psychological function that is the use of these spaces by the population with anti-stress activities.

Although some studies present theoretical and empirical similarities, a consensus on the concept of green areas has not been achieved yet. The same can be said of the different indexes found in the literature. The lack of agreement regarding the terminology and the different methodologies used to obtain the Urban Green Area Index (IAVU, Brazilian acronym for Índice de Área Verde Urbana), with its different meanings, might lead to inaccurate interpretations and an incorrect political use of this data. As a consequence, such indexes, which have been developed to help the study and monitoring of the urban vegetation, are presented in different ways, portraying different places and purposes.

Comparative studies developed including indexes from different urban areas, both national and international, most times adopt distinct methodological principles and also conceptual, which do not favor the simple comparison of results or the replication of the same indexes in areas with very different characteristics (Bargos & Matias, 2012).

The IAVU is calculated from different parameters of the area and the adoption of a certain area for specialization is defined by the researcher to meet certain interests, and can be expressed considering the green area as a whole or focusing on more specific sub-spaces (zones, neighborhoods, sectors, etc.) divided by the total number of inhabitants considered. The trend is to achieve greater accuracy in the data analysis, which implies less generalization, to the extent in which the data is distributed in smaller units of area, making the observation of the presence or absence of green areas more objective and, consequently, the observation of their benefits more realistic. In addition, it is very common to find the calculation that expresses the relation between urban green areas and the number of inhabitants in a certain place. However, it is not always possible to confirm the existence of a direct relation of use of such area, since it is necessary to consider the spatial distribution of such areas and that of the population in the relevant urban space (Bargos, 2015).

Perin (2019) recorded 80 applied studies in the last two decades, in cities of different Brazilian regions. The most cited index is the Green Area Index (IAV, Brazilian acronym for Índice de Área Verde), even if this is the most used index, it is also the one that presents greatest disagreement in relation to the calculation methodology, based on the division of the green areas found by the existing urban population. Such disagreement generates problems when a simple comparison between cities is carried out, and it is a mistake to use it without considering the parameters that produced such index. Rondino (2005) observed that the index should initially be calculated as a function of the total number of existing areas and, later on, recalculated showing how many of these areas are being really used, after evaluating their conditions of use and conservation.

The discussion of the value obtained by using the IAV has provoked many debates on the urban vegetation issue in the municipalities, such as the fact that a reduction in the IAV value does not mean necessarily that the green areas have reduced, but rather that the population might have increased and, as a result, the index high value does not mean that the whole population has access to these areas, since they might be concentrated in a particular area of the city or even in one large fragment. For these reasons, the research must be spatialized considering geographical units as census sectors or neighborhoods to evidence the actual distribution of the green areas throughout the city.

According to Nucci (2001), this index represents the number of leisure areas available to be used by the community. This factor is what differentiates the vegetation cover as a whole that influences the city ecological conditions, from those green areas located in the urban zone and that, mandatorily, offer leisure options. When using this concept as a foundation for the calculation, the quantitative aspect of green areas is considered first and, later on, the qualitative aspect, since even if a green area has been spotted, it is important to qualify its conditions of use.

Although the green area indexes per inhabitant are widely used as indicators in the determination of the environmental quality of urban areas, alone they are not enough to reach this objective. In other words, in addition to the number of green areas, other factors, which are not less important, must also be considered, such as quality, distribution and infrastructure of the green areas.

Colet and Soares (2013) presented a classification of public green areas as squares, gardens, cemeteries, parks, and public woods that show a maximum of 30% soil sealing and where the vegetation cover predominates. Such vegetation cover includes grass cover and treetops, which provide the community with wellbeing. Private green areas were classified as pieces of land without any building and with an area over approximately 50 m^2, with limited use by their owners, and being covered by the predominant vegetation, which also include grass cover and treetops. These definitions have been used as parameters for the elaboration of diagnosis in urban areas considering both public and private green areas.

A calculation of the green area index for parks, considers two main indices: vicinity park green area index (IAVPV) and neighborhood park green area index (IAVPB, Brazilian acronym for Índice de Área Verde para Parques de Bairro), both divided by the urban population. Urban parks have ecological importance, since they present a dimension of environmental conservation. However, the requirements of the conservation dimension are not always met, since many times what is privileged in parks is the landscaping (Alencar, 2017). Urban parks contribute to reduce the damage resulting from the space disorganized occupation, which has consequences to the economic, social and cultural development of a city, in addition to impacts on the remaining vegetation and the natural resources that provide balance to the urban ecosystem (Toledo & Santos, 2012).

The use of squares in the index calculation appears as total green area index (IAVT, Brazilian acronym for Índice de Áreas Verdes Total), usable green area index (IAVU), treated green area index (IAVT, Brazilian acronym for Índice de Áreas Verdes Tratadas), and green area index per neighborhood (IAVB, Brazilian acronym for Índice de Áreas Verdes por Bairro), divided by the number of inhabitants. Some variables, according to Souza et al. (2014), might be used for the calculation of squares as (i) treetop diameter; (ii) total height and first branch height; and (iii) circumference at 1.30 m from the soil.

Lima et al. (1994) sought to define the several terms employed to classify the urban space categories, considering free spaces as the best term, for its opposition to the meaning of built space in urban areas. The square as a green area has the main function of entertainment. A square might not be a green area when it does not have vegetation cover or its ground has been sealed. Gomes and Soares (2003) reported the evolution of squares as green areas in Brazil and their relation with the valorization of green spaces in cities. They emphasized that the insertion of urban green areas in Brazilian cities occurred simultaneously to the evolution of the square functions. In the past, they used to be large open spaces without the presence of vegetation cover, and were mostly used as a place where people would gather. More recently, however, they started to be improved in the cities and began to take on to have the characteristics of a garden. They became more enjoyable both aesthetically and functionally, therefore, garden-squares became a fundamental mark of the improvement and valorization of gardening in cities, mainly in public places.

The green area index of vegetable massifs (IAVMV, Brazilian acronym for Índices de Áreas Verdes por Maciços Vegetais) is rarely found in the Brazilian literature. Vegetable massifs are considered as parks, squares and woods where there is predominance of trees, with areas over 10,000 m^2 (Otani, 2019). This index might complement the IAV, since vegetable massifs are larger areas usually away from urban green areas.

The public green area index (IAVP, Brazilian acronym for Índice de Áreas Verdes Públicas) has been constantly used, disregarding private green areas. The IAVP, according to Resende et al. (2009) is a socioenvironmental indicator resulting from the quantification of the existing treetop area mean value divided by the number of inhabitants, which expresses the relation between the vegetation cover area in public spaces with the populational density of each neighborhood in the city.

3 Green Area Evaluation Methodologies and Focus

To evaluate the quality of urban green areas, this chapter presents the composition of sub-indexes in a compound index, characterized by indicators related to the environmental, social and aesthetics functions.

The research methods are a set of organized procedures that enable the analysis and categorization of segments of the reality that were defined as problems. However, since there is no single method to develop research works, but rather a variety of methods, a methodology must be chosen as the reasoning that guides the use of methods, regarding efficacy and strengths to produce relevant knowledge (Aguilera Hintelholher, 2013).

The methodologies employed in the analysis of green areas are several and depend basically on three factors: 1. Objectives, which represent the set of aims to achieve; 2. The concept of green area, in which the values ascribed to these places are found, and 3. The disciplines from which the studies are elaborated and use their own methods and techniques that might lead to different meanings.

The variables used in green area studies determine the focus of analysis, which might be quantitative or qualitative. For example, the variable square meters of green area per inhabitants or the number of parks smaller than 2 ha, correspond to a quantitative focus, while the variables accessibility to users with special needs or the conditions of amenities and facilities, correspond to a qualitative focus (Taylor & Houchuli, 2017).

The review of magazine articles about green areas put forward by Taylor and Houchuli (2017) pointed out that within social, natural and human sciences, most of the research is multidisciplinary and involves the disciplines: urban planning, biology, medicine and health, agriculture and environmental sciences, cognitive sciences, social sciences, earth sciences, education and history, and archeology.

Therefore, studies that contribute to the analyses of green areas from a multidisciplinary perspective include several dimensions such as social, economic, biological, and planning, as well as the use of different methods and analyses when dealing with quantitative and qualitative variables (Diaz & Sanchez, 2014; Gonzalez-Duque & Panagopoulos, 2013; Baycan-Levent et al., 2009).

Regarding the multidimensional evaluation in the comparative study of urban green spaces in European cities (Baycan-Levent et al., 2009), a broad conceptual structure of green areas was established in which five values were devised and applied: Ecological value (genetic diversity, life support value), Economic value (market value, direct and indirect use), Social value (entertainment, aesthetic, historical-cultural, therapeutic and social interaction), Planning value (instrumental and structural) and Multidimensional value (scientific and public policy). Based on that, a multicriteria analysis was carried out for each of the values listed utilizing quantitative methods from the monetary assessments and qualitative methods with non-monetary valuation.

That study concluded that the nature of urban green areas contains a multidimensional set of particular features that require a methodological perspective including several criteria for its evaluation with quantitative and qualitative methods.

Another example of multidimensional research was developed in the Metropolitan Region of Santiago do Chile (Arellano et al., 2016), to evaluate the quality of urban green areas through the creation of an urban green area quality index by measuring four factors: Ecological, Social, Physical infrastructure and Management. With a quantitative focus, it analyzed the relation between square meters and green areas per inhabitant, while the qualitative focus assessed the quality of the urban furniture elements, safety and accessibility, among others, which determined how the green areas were used by the population. This study privileged the creation of specific knowledge and transversal abilities, through applied research interdisciplinary work.

On the other hand, there are also studies carried out from a single perspective, which are concerned with the analysis of a single value in the green areas, in the light of one discipline, whose results focus one direction only.

The ecological perspective is within the scope of the Natural Sciences, according to the Environment European Agency (Abdul-Malak & Erhard, 2015). The ecological evaluation consists in monitoring the current conditions and the changes of ecological resources and understanding the structure and function of ecosystems to develop better management options. Such assessment includes foreseeing human impact on the environment and evaluating the ecological consequences of the management actions for those responsible for the decision making. In this sense, there are studies that set indicators of the green infrastructure ecosystem services to evaluate their benefits and contribution to the sustainability of the place (Tudorie et al., 2019); studies that propose the characterization of ecosystems from the planning and soil use management discipline, using a hierarchical scale linking the biotic and abiotic components of the land systems and their relation with the existing water systems (Bourgeron et al., 2001); or those whose purpose is to reinforce the conservation of natural spaces upon the evaluation of ecological qualities and benefits of the landscape patterns of the urban green spaces (Yuhong et al., 2014).

Within the Social Sciences field, there are studies focusing on the way urban communities use their urban green spaces as part of their lifestyle, which analyze how the users interact with green spaces and whether they examine their preferences, identifying positive factors that contribute to their use and/or negative factors that discourage their use. These generally imply the consideration of three interaction areas: the location of the site that allows it access and use, the quality of the experience defined by the satisfaction that the users obtain from the site and the functional aspects defined by the users' activities (Coles & Caserio, 2015).

This group gathers the studies that analyze the influence of green spaces in the residents' socialization behavior in a certain neighborhood (Aram et al., 2019), or the patterns of use of green areas in relation to the spatial interaction between different social groups (Krellenberg et al., 2014), or even the contribution of urban green areas to the urban quality of life and social sustainability (Cengiz et al., 2018).

In the Human Sciences area, research on green areas is found under the aesthetic perspective that assesses the visual quality of spaces, either using objective methods that evaluate the aesthetic quality of the intrinsic properties of the object or upon subjective methods that are based on the users' or viewers' perceptions (Terry, 2001). This group includes studies on the influence of perceived characteristics of urban green areas, on the wellbeing variables: visitors' level of satisfaction and the self-informed quality of life (Kothencz et al., 2017), studies that investigate how urban planning strategies aiming at eco-aesthetics factors might help the socio-ecological resilience of cities (Haruna et al., 2018), and studies on how the urban gardening can contribute to the attractive perception of urban areas (Lindemann-Matthies & Brieger, 2016).

Summarizing, studies on green areas and their methodologies can be classified as unidisciplinary, whose analysis focuses on a single dimension of knowledge, or multidisciplinary, which integrates several dimensions. They can also be quantitative, since they describe and analyze green areas objectively, or qualitative for evaluating them, usually, from subjective criteria, however, they can be combined in mixed studies (Fig. 14.1).

Since the disciplines have different aims and utilize different methodologies, this might result in different meanings. According to Taylor and Houchuli (2017), to guarantee that the political decision making regarding green areas be relevant, operational definitions must be provided that can be interpreted by all sectors. Operational definition here means the term that can be qualified and quantified, which might increase the likelihood of it being understood in multiple disciplines and research contexts.

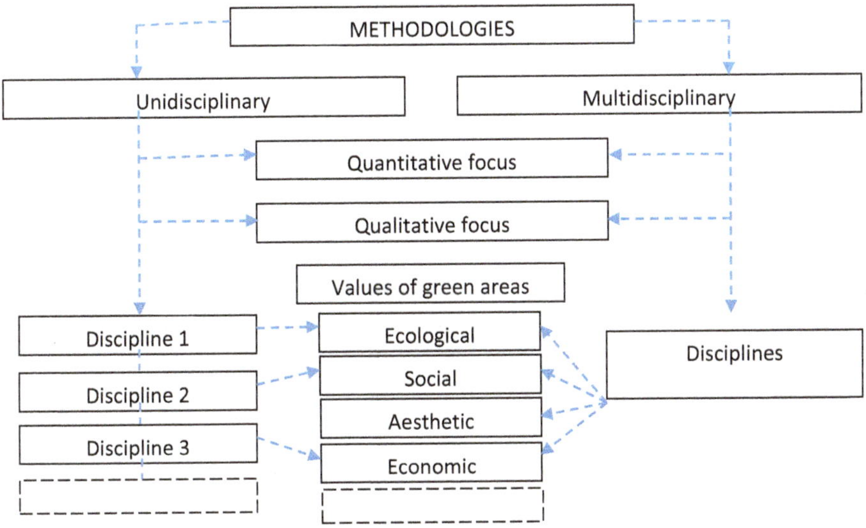

Fig. 14.1 Classification of methodologies based on disciplines and focus

Chart 14.1 Matrix of indicators included in the IQAVU

Compound index	Functions	Indicators	Data analyzed
URBAN GREEN AREA QUALITY INDEX (IQAVU)	**Environmental Function (EF)**	Tree cover index (TCI)	Treetop area/total urban area
		Presence of native species (NS)	% native species
		Fragmentation (FRAG)	Isolated/Linear/ connected
	Social Function (SF)	Distribution Fairness (DF)	Jain's Index
		Categories of green areas (C)	Audience/potentially collective/private
		Presence of leisure equipment (LE)	Presence/Absence
	Aesthetic Function (AF)	Urban furniture (UF)	Conservation, Comfort
		Pavement (PAV)	Conservation, Safety
		Water element (WE)	Presence/Absence

4 Methodological Proposal to Evaluate Urban Green Areas: IQAVU

Due to the global nature of the proposal presented in this paper, which should be applicable to any georeferenced urban unit, indicators were considered (described in Chart 14.1) with more general characteristics.

4.1 Functions of Green areas in Urban Areas

Authors such as Alvarez (2004) and Nucci (2008) understand that the performance of the environmental function by green areas is directly related to the presence of vegetation cover. Depending on their characteristics, vegetation covers act through interception, reflection, absorption and transmission of direct or reflected radiation, maintenance of high absorption rates, contribution to temperature reduction and decrease in thermal amplitudes, maintenance of air humidity, therefore, providing populations with more thermal and lighting comfort and keeping the local climate.

They also work as a physical protection and root stabilizer, and intercept rain drops through the leaves, preventing the destruction of the soil superficial horizon. When native vegetation is considered, the organic matter produced promotes the formation of suitable structures to the soil.

As pointed out by Almeida (2015), the vegetation cover can be analyzed from the soil perspective, but also from the air volume it occupies, mainly, trees that increase the shading degree and the purifying action in the atmosphere and reducing the effect of the sound pollution, which is a characteristic of the urban environment.

It seems also relevant to emphasize that the environmental function of urban green areas is more efficient when provided by native species, mainly due to their positive action on hydrological ecosystems, nutrient cycle and soil erosion, in the vegetable productivity, trophic chains and function of the species, in the biomass distribution and accumulation, among others (Marchante & Marchante, 2007).

In addition to the environmental function, green areas also have a relevant social function. When the green areas are managed by the government, the social component performance is reinforced by the democratization they promote, since they provide every citizen, regardless of their social class or other forms of differentiation, with the availability of leisure spaces and natural value.

Therefore, in the evaluation of this function, the pertinence of the use of the Jain's fairness index is highlighted. This index translates numerically the degree of fairness in the distribution of green areas throughout the whole urban area (Jain et al., 1984), the presence and absence of equipment providing the population with leisure and the favoring of public management in these spaces.

Despite their complexity, urban green areas play a relevant aesthetic role because they promote beautification of their surroundings and contemplation. Although thinking and planning green areas that are aesthetically functional might belong to a more subjective field, some guidelines related to it were considered in this proposal.

We considered that the design elements for open spaces, as pointed out by Salmon (Peña Salmon, 1990): pavement, urban furniture and the presence of water are relevant to provide greater aesthetic contribution by the green areas.

Pavements provide a surface for circulation and also play other intangible roles that help the development of activities in free spaces such as promoting a relaxation sensation, indicating danger, reducing the scale of spaces and reinforcing the place identity.

Urban furniture are the objects that support the proper development of urban activities such as benches, garbage bins, telephone booths, fences, sculptures, fountains, bridges, public clocks, ashtrays, drinkers, pachymeters, information and advertisement boards, displays, bicycle racks, kiosks, stands, pavilions, mail boxes, bus stops or taxi ranks, utility cabinets (electrical power, telephone lines, etc.) shelters, awnings, windbreaks, roofs, platforms, shop windows, public and chemical toilets, and other similar elements. Such elements are characterized for being relatively small when compared to their surroundings, but they are also an important part of the urban landscape (Peña Salmon, 1990; Corona, 2001; Pacheco, 2010).

Finally, the inclusion of water bodies as aesthetic elements can be considered as a strategy to improve thermal and visual comfort. They are strongly attractive elements that invite to leisure and contemplation and promote human contact with nature, either through water mirrors, fountains, lakes, and rivers, among others. They improve considerably the environmental quality of the space and therefore, they become a hierarchically superior element (Minda, 2009; Hannes, 2016).

4.2 Methodology for the Calculation of IQAVU

The IQAVU is the proposal of a compound index, inspired in the widely known Human Development Index (HDI), whose application is so wide that a hierarchy was created between 104 countries, measuring health, education and education conditions, and the performance of a country, city or region (Nahas, 2001). Therefore, the IQAVU is represented by the following formula:

$$IQAVU = \sqrt[3]{EF \times SF \times AF} \tag{14.1}$$

Where:

EF = environmental function
SF = social function
AF = aesthetic function

The value of each function that is part of the main index is presented in the following formulas:

Environmental Function

$$EF = \sqrt[3]{TCI \times IFrag \times NSI} \tag{14.2}$$

Where:

TCI = Tree cover index

$$TC = \frac{Treetoparea}{Urbanarea} \tag{14.3}$$

IFrag = Fragmentation index (Isolate = 0; linear = 0,5; connected = 1)

NSI = Native species index

$$NSI = \frac{\text{QNat} \times 100}{\text{total of species}} \tag{14.4}$$

Social Function

$$SF = \sqrt[3]{JI \times GAC \times LE} \tag{14.5}$$

Where:

JI = Jain's index
GAC = Green area category (public)
LE = Leisure elements

Aesthetic Function

$$AF = \sqrt[3]{UFI \times PAVI \times WEI} \tag{14.6}$$

Where:

UFI = Urban furniture(aesthetic) index

$$UFI = \frac{UF - 1}{4 - 1} = \frac{UF - 1}{3} \tag{14.7}$$

PAVI = Pavement (aesthetic) index

$$PAVI = \frac{PAV - 1}{4 - 1} \tag{14.8}$$

WEI = Water element index (Obtained from a normalization operation)

Taking the above into consideration, in addition to using the same calculation formula, this proposal also considers the same classification scale as the HDI, as shown in Table 14.1.

The calculation formula used in all functions was the geometric mean, so that no value obtained could influence another value. It also enabled the observation of the

Table 14.1 IQAVU different classification scales

Class	Score
Very high	0.80–100
High	0.70–0.79
Medium	0.55–0.69
Low	Below 0.55

Source: Adapted from the HDI classification
Org: Authors

index dimensions that might suffer interventions in the course of an efficient urban planning that is concerned with the valorization of such spaces. We also sought to make it the most representative possible regarding its main functions according to the literature in the area, and at the same time with a general approach that would make it suitable to be applied to any urban space.

5 Applying the Index to the Municipality of Ponta Grossa, Southern Brazil

The application considered the urban area of Ponta Grossa, a city located in the South of Brazil, coordinates 25°09′S and 50°16′W, specifically on the Second Plateau of the State of Parana, in the Campos Gerais region (Fig. 14.2). In the Brazilian context, Ponta Grossa is a medium-sized city recognized as an important road and rail junction in the Campos Gerais (Sahr, 2001).

The IQAVU analysis in Ponta Grossa included 40 green areas distributed throughout 15 neighborhoods and the central area of the city, whose distribution configured 23 squares, 8 parks, 4 clubs and 5 others, which represented 19,928,000 m^2 (19,928 km^2), corresponding to approximately 0.5% of the total urban area, resulting in 7.31 m^2 per inhabitant.

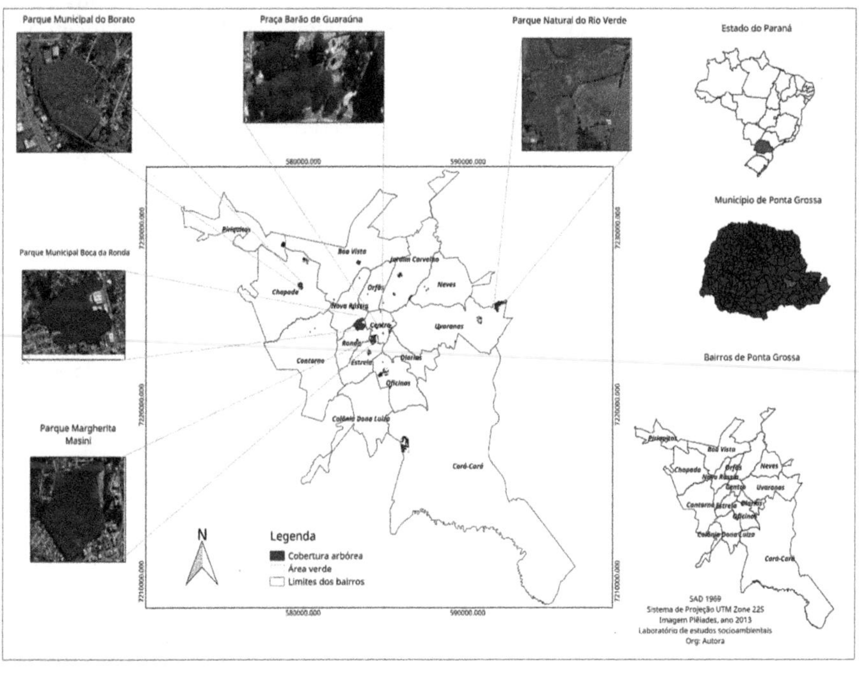

Fig. 14.2 Map locating urban green areas

Numerically, the square category is the most representative, for this reason it becomes quite relevant. For many residents, they are the only a source of public leisure and a place for being in contact with nature (PDMPG, 2006).

As spatialized in Fig. 14.2, the city center holds the higher concentration of squares and parks, clubs and other types of green areas, which are also larger spaces when compared to those found in the neighborhoods. As for the neighborhoods, the *Rio Verde* natural park in *Uvaranas*, the *Boca da Ronda* municipal park in *Ronda* neighborhood, and the *Margherita Masini* park in *Estrela* neighborhood outstand for their size.

The results obtained from the IQAVU application are presented in Chart 14.2, enabling an analysis of the quality profile of the green areas in Ponta Grossa.

When analyzing, the TCI—tree cover index verified is the least contributing value to the environmental function of green areas, with a 0.04 value. The poor performance of this indicator is mainly due to the low representativeness of the treetop coverage proportion in relation to the total urban area.

To classify the indicator percentage of native species, the value presented by Santos Eurich (2014) was considered. In a quantitative study on the afforestation of squares in the municipality of Ponta Grossa, the author reported the existence of 55% exotic species and 44.43% native species. It seems relevant to highlight the diagnoses that have been presented since the municipality 1992 Master Plan was launched and that have recorded the degradation and suppression of native forests in the city.

When evaluating three cover patches for the indicator fragmentation, the most found form was that corresponding to the isolated-aggregated. This classification is due to the fact that the largest group of green areas is formed by squares, which are predominantly located in strongly built places with streets, and sealed surfaces, whose tree coverage became larger units surrounded by a built matrix.

Chart 14.2 Results obtained by applying the IQAVU to the urban area of Ponta Grossa

Functions	Indicators	
Environmental function	TCI	0.04
	FRAGI	0.50
	NSI	0.44
	EFI	**0.21**
Social function	JainI	0.32
	Public catI	0.82
	LEI	0.70
	SFI	**0.57**
Aesthetic function	UFI	0.60
	PAVI	0.70
	waterI	0.22
	AFI	**0.45**
IQAVU		**0.38**

Source: Queiroz (2018)

Four patches (municipal parks) were also verified, which fit the connected class of a branched subtype. They correspond to remaining forests that were established before the urbanization, but for presenting wide vegetation cover, they provide habitat to several species. However, these places are usually strongly under pressure, mainly those close to valley floors such as the parks *Margherita Masini* and *Rio Verde*, with the presence of irregular occupations/invasions.

In the city of Ponta Grossa, the *Jain* index result was 0.32, revealing that the distribution of green areas (m^2) per inhabitant is unfair. The absence of green areas in 2 neighborhoods (Olarias and Colonia Dona Luiza), for example, contributes to this scenery, as well as the concentration of green areas downtown and in closer neighborhoods.

The best performance of this function is due to the indicator of categories of green areas, in which public green areas represent 82% of these elements, totalling around 33 public green areas.

Six private use green areas were found, The *Ponta Lagoa* Club outstands among them. It is a space that integrates a lake that allows fishing, with a wood and several gardens, whose use is limited to members only. Another private property is the *Recanto Monteiro* park, with a complex of swimming pools that are mostly used during summer and the members of the community can use, but need to pay an entrance fee.

Also, in the category of private green areas, a considerable patch of forest used as a wood inside a private condominium is included, it has a swimming pool and a track for walking or jogging, all these amenities are for exclusive use of the residents.

Regarding the indicator presence of leisure and sports equipment, it was found in around 70% of the green areas. These figures are justified by the concentration of playgrounds in squares such as the *Barão de Rio Branco* square in the city center, and the sports and physical activity infrastructure observed in most public parks and private clubs for occupying larger areas.

The indexes characterizing the social function in the urban area of Ponta Grossa totalled 0.57, showing a performance over that of the environmental function, which was 0.27.

As regards the aesthetic function, it seems relevant to mention that the indicator that achieved the best classification was the urban furniture index. It obtained 2.8 points, the average between the 114 categories of items analyzed, where over 580 elements of urban furniture are allocated, some examples of such items are: lamp posts, benches, tables, barbecues, playgrounds, gymnastics equipment, multisport courts, toilets, monuments, bus stops, taxi ranks, newsstands, and football pitches.

The best classification of urban furniture, pavement and water elements were found in private green areas. Some of these are: *Ponta Lagoa* Club, with 2 olympic swimming pools, 1 child pool, 3 event rooms, 9 barbecues, 5 football pitches, 13 multi sports courts, 9 kiosks, 1 playground, 1 lake for sport fishing, 1 game room and 20 lamp posts, ranked 3 or 4, as in the examples highlighted in Fig. 14.3.

Another outstanding private area is the *Recanto do Monteiro* park, a large green area with swimming pools, water slides, barbecue, restaurant, among others, which is open to the community, upon payment, in the summer months.

Fig. 14.3 Images of the *Ponta Lagoa* Club structure. (Source: Queiroz, 2018)

The worst classification for this function was obtained in areas that most contribute to the environmental function, due to the expressive tree coverage patch. The *Margherita Masini* and *Rio Verde* parks are among them, in a recent past both parks have already been relevant leisure options to the population, however, currently they have been spoiled by vandals and present strong signs of abandonment and wearing of their infrastructure, despite several attempts by the municipal environment secretariat to renovate these places.

The public green areas that positively outstand in the composition of their urban furniture and pavement are the *Monteiro Lobato* Park in *Jardim Carvalho*, with 6 concrete benches, 52 lamp posts, 12 garbage bins, 7 concrete tables, a playground, 7 gymnastics devices, and 6 football pitches, ranked between 2 and 3. The Environmental Park, downtown, presents similar characteristics, with several items such as a playground, two multi sports courts, two toilets, eight plastic garbage bins, and 7 gymnastics devices. In this park, there is also a police unit, a significant element to guarantee the safety of the users of that space.

Fig. 14.4 Benches in Ponta Grossa squares. (Source: Queiroz, 2018)

Fig. 14.5 Urban furniture with signs of degradation. (Source: Queiroz, 2018)

Squares, in general, present a standard in the composition of their furniture and design. Their benches are mainly made of wood, similar to Fig. 14.4, and except for the *Barão do Rio Branco* square, some of the furniture pieces present signs of degradation as shown in Fig. 14.5.

Regarding pavements, 15 different types were analyzed in relation to their safety and conservation, and were ranked 2.8 (Regular), the most frequent among the floor coverings were stones, concrete and tar, and those presenting more conservation problems were those made of stones, which are aesthetically more suitable, mainly

Fig. 14.6 Stone sidewalk with signs of degradation. (Source: Queiroz, 2018)

in squares, however, present maintenance problems, with some loose pieces and others missing (Fig. 14.6).

The presence of the water element was registered in 22% of the green areas in the municipality of Ponta Grossa, and its greatest occurrence is observed in private green areas, with the inclusion of swimming pools and lakes. In public squares, this element appears in the form of fountains, and the presence of water mirrors in the *Rio Verde* Municipal Park.

Taking that into consideration, the aesthetic function classification resulted in 0.45 contribution to the IQAVU.

6 Conclusions

The methodology presented was seen to be an efficient response to the challenges of diagnosing in the planning of urban green areas. From a quali-quantitive framework, the qualification of green areas is more realistic and complete since it covers the main dimensions that characterize green areas and, among those, the ones that need to be prioritized for intervention and public policies. The calculation using the geometrical mean, similar to what happens with the HDI, ensures that the performance of one of the functions does not hide the performance in another function.

The application of this methodology in the municipality of Ponta Grossa, revealed that the index lowest values are in the environmental and aesthetic functions, indicating that implementing new types of green areas such as parks, squares, among others, in empty spaces, along with the planting of trees and improvements in the quality and conservation of the urban furniture and safety of the flooring might affect positively the value and category of the IQAVU in the city.

References

Abdul-Malak, D., & Erhard, M. (2015). *European ecosystem assessment — Concept, data, and implementation*. European Environment Agency. https://doi.org/10.2800/629258

Aguilera Hintelholher, R. M. (2013). Identidad y diferenciación entre Método y Metodología. *Estudios políticos, 9*(28), 81–103. https://doi.org/10.1016/S0185-1616(13)71440-9

Alencar, C. d. J. (2017). *Áreas Verdes e Risco de Contaminação Parasitológica Parques Urbanos de Aracaju – SE*. Dissertação de Mestrado. Universidade Federal de Sergipe.

Almeida, C. (2015). *O Verde na cidade. Reflexão inspirada na cidade de Coimbra (Portugal)*. Cadernos de Geografia n°34. Faculdade de Letras da Universidade de Coimbra, 03–09.

Alvarez, I. A. (2004). *Qualidade do verde urbano: uma proposta de índice de avaliação. Tese (Doutorado em Agronomia) – Escola superior de Agricultura Luiz de Queiroz. Universidade de São Paulo.*

Angeoletto, F., Tryjanowski, P., Santos, J., Martinez-Miranzo, B., Leandro, D., Bohrer, J., et al. (2022). Will Brazilian City dwellers actively engage in urban conservation? A case study with the charismatic Neotropical blue-and-yellow macaw (Ara ararauna). *Birds, 3*(2), 234–244.

Angeoletto, F. H. S., Richard, E., Santos, J., Guedes, N., Oliveira, S. L., Albertín, R. M., et al. (2023). When we think of urban afforestation in Brazil, what comes to mind is a mouth that has lost its teeth. *Terr@Plural, 17*, e2321654.

Aram, F., Solgi, E., & Holden, G. (2019). The role of green spaces in increasing social interactions in neighborhoods with periodic markets. *Habitat International, 84*, 24–32. https://doi.org/10.1016/j.habitatint.2018.12.004

Arellano, E., Ovalle, J., Acuña, S., Goldschmidt, G., Orozco, J., & Quiroga, D. (2016). Propuesta para la evaluación de calidad de áreas verdes para la Región Metropolitana. In Centro de Políticas Públicas UC, *UC Propone: Iniciativas de Investigaciones en Pregrado* (1a Edicion ed., págs. 18–39). Centro de Politicas Publicas UC. Obtenido de https://www.researchgate.net/publication/308632048

Bargos, D. C. (2015). *Cartografia das áreas verdes urbanas: efetividade e representação da influência no espaço urbano em Paulínia (SP)*. Tese. Universidade Estadual de Campinas.

Bargos, D. C., & Matias, L. F. (2012). Áreas verdes urbanas: um estudo de revisão e proposta conceitual. *Piracicaba: RSBAU, 6–3*, 172–188.

Baycan-Levent, T., Vreeker, R., & Nijkamp, P. (2009). Multidimensional evaluation of urban green spaces : A comparative study on European cities. *European Urban and Regional Studies, 16*(2), 193–213. https://doi.org/10.1177/0969776408101683

Bourgeron, P., Humphries, H., & Jensen, M. (2001). Ecosystem characterization and ecological assessments. In E. M. Jensen & P. Bourgeron (Eds.), *A guidebook for integrated ecological assessment* (pp. 40–54). Springer. https://doi.org/10.1007/978-1-4419-8620-7_4

Cavalheiro, F., & Del Picchia, P. C. D. (1992). *Áreas Verdes: conceitos, objetivos e diretrizes para o planejamento* (pp. 29–35). Congresso Brasileiro Sobre Arborização Urbana.

Cavalheiro, F, Nucci J. C., Guzzo. P, & Rocha. T. (1999). *Proposição de Terminologia para o Verde Urbano*. Boletim Informativo da Sociedade Brasileira de Arborização Urbana. Rio de Janeiro, jul/ago/set.

Cengiz, C., Cengiz, B., & Keçecioğlu Dağlı, P. (2018). The role of urban green areas on social sustainability: Bartın case study. Conference: V. International multidisciplinary congress of Eurasia (págs. 136–144). Barcelona: https://www.researchgate.net/publication/328828693

Coles, R., & Caserio, M. (2015). *Social criteria for the evaluation and development of urban green spaces*. Urban Green Environment URGE. https://doi.org/10.13140/RG.2.1.4650.2488

Colet, K. M., & Soares, A. K. (2013). Diagnóstico e índices de qualidade ambiental da bacia do Córrego do Barbado, Cuiabá-MT. *Engenharia Ambiental, Espírito Santo do Pinhal, 10-2*, 22–40.

Corona, M. A. (2001). *Las areas verdes en el contexto urbano: estudio de caso ciudad de Guadalajara*. Tesis de maestria en Ciencias en medio ambiente desarrollo integrado. Mexico. D.F. 133 p.

Di Fidio, M. (1990). *Architettura del paesaggio*. Pirola.

Diaz, A., & Sanchez, R. (2014). La importancia de un enfoque multidimensional en la planeación de los parques: Ensenada. En L. Ojeda, & I. Espejel, *Cuando las areas verdes se transforman en paisajes, La vision de Baja California* (1a ed., págs. 149–174). El Colegio de la Frontera Norte.

Gomes, M. A. S. & Soares, B. R. (2003). *A vegetação nos centros urbanos: considerações sobre os espaços verdes em cidades médias brasileiras*. Estudos Geográficos: Revista Eletrônica de Geografia. Rio Claro, Disponível em: http://www.periodicos.rc.biblioteca.unesp.br/index.php/estgeo/article/view/270

Gonzalez-Duque, J. A., & Panagopoulos, T. (2013). Evaluation of the urban green infrastructure using landscape modules, GIS and a population survey: Linking environmental with social aspects in studying and managing urban forests. *Journal of Spatial and Organizational Dynamics, 1*(2), 82–95.

Guzzo, P. (1999). *Estudo dos espaços livres de uso público da cidade de Ribeirão Preto/SP, com detalhamento da cobertura vegetal e áreas verdes públicas de dois setores urbanos*. Dissertação de Mestrado. Universidade Estadual Paulista Júlio de Mesquita Filho.

Guzzo, P. (2006). *Cadastro Municipal de Espaços Livres Urbanos de Ribeirão Preto (SP): Acesso Público, Índices e Base para Novos Instrumentos e Mecanismos de Gestão*. Piracicaba: RSBAU, 1–1.

Hannes, E. (2016). Desenho ambiental e forma urbana o caso do bairro Riverside. *São Paulo, 196*, 07.

Haruna, A. I., Oppong, R. A., & Marful, A. B. (2018). Exploring eco-aesthetics for urban green infrastructure development and building resilient cities: A theoretical overview. *Cogent Social Sciences, 4*(1), 1–18. https://doi.org/10.1080/23311886.2018.1478492

Jain, R., Chiu, D. M., & Hawe, W. (1984). *A quantitative measure of fairness and discrimination for resource allocation in shared computer systems*. DEC Research Report TR-301.

Kothencz, G., Kolcsár, R., & Cabrera-Barona, P. (2017). Urban green space perception and its contribution to Well-being. *Interantional Journal of Environmental Research Public Health, 14*(7), 766. https://doi.org/10.3390/ijerph14070766

Krellenberg, K., Welz, J., & Reyes-Päcke, S. (2014). Urban green areas and their potential for social interaction—A case study of a socio-economically mixed neighbourhood in Santiago de Chile. *Habitat International, 44*, 11–21. https://doi.org/10.1016/j.habitatint.2014.04.004

Lima, A. M. L. P., Caavalheiro, F, Nucci, J. C., Sousa, M. A., Fialho, N. O., & Del-Picchia, P. C. D. (1994). Problemas de utilização na conceituação de termos como espaços livres, áreas verdes e correlatos. In *2° Congresso Brasileiro Sobre Arborização Urbana*, São Luís, pp. 539–550.

Lima, G. F., Correa Santos, J. W. M., Albertin, R. M., Martínez-Miranzo, B., Souza, F. L., & Angeoletto, F. (2023). Backyards are a way to promote environmental justice and biodiversity conservation in Brazilian cities. *Diversity, 15*(7), 815.

Lindemann-Matthies, P., & Brieger, H. (2016). Does urban gardening increase aesthetic quality of urban areas? A case study from Germany. *Urban Forestry & Urban Greening, 17*, 33–41. https://doi.org/10.1016/j.ufug.2016.03.010

Marchante, E., & Marchante, H. F.. (2007). As exóticas e invasoras. In J. Sande Silva (Coord. ed.), *Do freixo à bétula, as outras espécies da floresta Portuguesa*. Colecção Árvores e Florestas de Portugal, 5, pp. 179–198.

Mazzei, K., Colesanti, M. T. M., & Santos, D. G. (2007). Áreas verdes urbanas, espaços livres para o lazer. *Uberlândia: Sociedade & Natureza, 19-1*, 33–43.

Milano, M. S. (1992). *A cidade, os espaços abertos e a vegetação*. Congresso brasileiro sobre arborização urbana, Vitória, 03–14.

Minda, Jorge Eduardo Calderon. (2009). Os espaços livres públicos e o contexto local: o caso da praça principal de Ptalito huia- Colômbia. Dissertação de mestrado. Faculdade de arquitetura e urbanismo da Unibversidade de Brasília.

Nahas, M. I. P. (2001). Metodogia de construção de índices e indicadores sociais, como Instrumentos balizadores da gestão municipal da qualidade de vida urbana: uma síntese da experiência de Belo Horizonte. In D. Hogan, et al. (org.), *Migração e ambiente nas aglomerações urbanas*. Núcleo de Estudos de População/UNICAMP.

Nucci, J. C. (2001). *Qualidade ambiental & adensamento urbano: um estudo de Ecologia e Planejamento da Paisagem aplicado ao distrito de Santa Cecília (MSP)*. Humanitas/FFLCH-USP.

Nucci, J. C. (2008). *Qualidade Ambiental e adensamento urbano: um estudo de ecologia e planejamento da paisagem aplicada ao Distrito de Santa Cecília (MSP). 2 ed. Curitiba: Ediçãodo autor.151p.*

Otani, C. T. (2019). *Influência da presença da vegetação na temperatura e na umidade do ar: estudo em bairros da cidade de Cascavel – PR*. Dissertação de Mestrado. Universidade Estadual do Oeste do Paraná.

Pacheco, C. P. N. (2010) *Equipamentos Nos Espaços verdes para a 3ªidade-O caso Quinta das Conchas*. Dissertação de Mestrado. Faculdade de Arquitetura da Universidade Técnica de Lisboa.

PDMPG, Plano Diretor Municipal de Ponta Grossa. (2006). Disponível em: www.pontagrossa.pr.gov.br. Acessado a 15 de Abril de 2018.

Peña Salmon, C. P. (1990). *Usos, Funciones y caracteristicas de las Plantas en el deseño del paisaje: Funciones ecológicas*. ed. elustrada. Universidad de Baja California.

Perin, J. V. (2019). *Análise crítica dos índices de áreas verdes urbanas*. Trabaldo de Concusão de Curso. Universidade Estadual de Campinas.

Queiroz, D. H. O. (2018). *Avaliação da qualidade das áreas verdes: uma abordagem com base nas suas funções*. Tese. Universidade Estadual de Ponta Grossa.

Resende, W. X., Souza, T. R., & Souza, R. M. (2009). *Índices de Áreas Verdes Públicas: Uma avaliação fitogeográfica da qualidade Ambiental em Aracajú*. Simpósio Nacional de Geografia Física Aplicada, Viçosa.

Rondino, E. (2005). *Áreas verdes como redestinação de áreas degradadas pela mineração: estudo de casos nos municípios de Ribeirão Preto, Itu e Campinas, Estado de São Paulo*. Dissertação de Mestrado. Escola Superior de Agricultura Luiz De Queiróz.

Rumble, H., Angeoletto, F., Connop, S., Goddard, M. A., & Nash, C. (2019). Understanding and applying ecological principles in cities. In *Planning cities with nature* (pp. 217–234). Springer.

Sahr, C. L. (2001). Estrutura interna e dinâmica social na cidade de Ponta Grossa. In: C. De H. M. Ditzel, & Löwen Sahr, C. L. (Orgs). *Espaço e cultura: Ponta Grossa e os Campos Gerais*. UEPG.

Santangelo, J. S., Ness, R. W., Cohan, B., Fitzpatrick, C. R., Innes, S. G., Koch, S., et al. (2022). Global urban environmental change drives adaptation in white clover. *Science, 375*(6586), 1275–1281.

Santos Eurich, Z. R. S. (2014). *As praças de Ponta Grossa: Arborização urbana, infraestrutura e distribuição espacial. Dissertação de mestrado*. Programa de pós-Graduação em Geografia. Universidade Estadual de Ponta Grossa.

Souza, O. D. P. S., Souza, P. T. S., Freitas, A. D. D., Paraense, V. C., & Souza, D. V. (2014). Indicadores de área verde e cobertura arbórea para as praças do município de Altamira. *Enciclopédia Biosfera: Centro Científico Conhecer, 18-10*, 1955–1962.

Souza, F. L., Fenoglio, M. S., & Angeoletto, F. (2023). To be a Brazilian City dweller, sometimes we must learn to say enough! *Sustainability, 15*(4), 3699.

Taylor, L., & Houchuli, D. F. (2017). Defining greenspace: Multiple uses across multiple disciplines. *Landscape and Urban Planning, 158*, 25–38.

Terry, D. C. (2001). Whiter scenic beauty? Visual landscape quality assessment in the 21st century. *Landscape and Urban Planning, 54*, 267–281.

Toledo, F. S., & Santos, D. G. (2012). Espaço Livre de Construção—Um passeio pelos Parques Urbanos. *Revista da Sociedade Brasileira de Arborização Urbana, 7-2*, 10–23.

Tudorie, C., Gielen, E., Valles-Planells, M., & Galiana, F. (2019). Urban green indicators: A tool to estimate the sustainability of our cities. *International Journal of Design & Nature and Ecodynamics, 14*, 19–29. https://doi.org/10.2495/DNE-V14-N1-19-29

Vieira, P. B. H. (2004). *Uma visão geográfica das áreas verdes de Florianópolis, SC: estudo de caso do Parque Ecológico do Córrego Grande (PECG)*. Trabalho de Conclusão de Curso. Universidade Federal de Santa Catarina.

Yuhong, T., Jim, C., & Haiqing, W. (2014). Assessing the landscape and ecological quality of urban green spaces in a compact city. *Landscape and Urban Planning, 121*, 97–108. https://doi.org/10.1016/j.landurbplan.2013.10.001

Chapter 15
Social-Ecological Dynamics of Green and Blue Infrastructure in the Indian Megacity of Bengaluru

Pramila Thapa, Arvind Lakshmisha, Dhanya Bhaskar, Harini Nagendra, Tobias Plieninger, and Andreas Thiel

Abstract With rapid urbanization currently observed and forecasted, especially in Asia, green and blue infrastructure (GBI) has emerged as an important component of urban spaces, providing multiple ecosystem services, and augmenting human well-being. GBI promotes a better communication of ecosystem services and biodiversity conservation as well as a better prioritization, integration, and structuring of implementation strategies. A focus on GBI is particularly useful in a country like India that needs holistic approaches including multiple stakeholders and sectors to facilitate inclusive and sustainable urban planning.

P. Thapa
University of Kassel, Witzenhausen, Germany

Leuphana University of Lueneburg, Lueneburg, Germany
e-mail: pramila.thapa@leuphana.de

A. Lakshmisha
University of Kassel, Witzenhausen, Germany

Azim Premji University, Bengaluru, India
e-mail: arvind.lakshmisha@apu.edu.in

D. Bhaskar (✉)
Indian Institute of Forest Management, Bhopal, India
e-mail: dhanyab@iifmbhopal.edu.in

H. Nagendra
Azim Premji University, Bengaluru, India
e-mail: harini.nagendra@apu.edu.in

T. Plieninger
University of Kassel, Witzenhausen, Germany

University of Göttingen, Göttingen, Germany
e-mail: plieninger@uni-kassel.de

A. Thiel
University of Kassel, Witzenhausen, Germany
e-mail: thiel@uni-kassel.de

F. Angeoletto et al. (eds.), *Ecology of Tropical Cities*,
https://doi.org/10.1007/978-3-031-48012-6_15

This book chapter developed a synthesis view that interprets outcomes in the governance and management of three GBI elements: wooded groves, a protected area, and lakes under conditions of rapid urbanization in India's fastest growing metropolis through a set of social-ecological indicators. For these elements of local GBI in greater Bengaluru region, a snapshot of biophysical, ecological, and institutional characteristics is provided, together with an indication of the direction they have developed over time. Subsequent cross-comparative analyses identified individual and overarching trends in urbanization driven transformations of GBI. The selected cases show that urbanization leads to encroachment of GBIs, increasing the number of actor groups and changes in ecosystem services derived by people. We conclude that GBIs are often ignored by policy and planning. We advocate the need for greater attention of policymakers to consider these important elements of urban fabric especially for their contributions to societal well-being.

Keywords Biophysical characteristics · Ecosystem services · Governance · Lakes · Protected areas · Urbanization · Wooded groves

1 Introduction

Urbanization is a key demographic trend in the Global South. It is not just a socio-economic phenomenon but also a process of ecological transformation by humans, and by that a key driver of global environmental change (D'Souza & Nagendra, 2011; Grimm et al., 2008; Huang et al., 2010; Souza et al., 2023). As the size and number of cities grow globally, the intensity and extent of urban impacts on ecosystems continue to increase. Expanding cities tend to appropriate a high share of ecosystems in terms of resource inputs and waste sinks driving large-scale modifications of ecosystems corresponding to ongoing transformations.

With urban land use strongly limiting the availability of places where people can access nature and its services (D'Souza & Nagendra, 2011), creating and maintaining green and blue infrastructure (GBI) is increasingly recognized as a pathway for sustainable urban planning and development. GBI is an interconnected network of natural or semi-natural designed landscape components including trees/shrubs/grass (i.e., green spaces), water bodies (i.e., blue spaces), built-up spaces (i.e., grey spaces) or open spaces (Ghofrani et al., 2017). GBI helps build resilience of cities against the impact of climate change and natural hazards, reduces urban heat island effects, improves water retention capacity, and enhances local biodiversity (Angeoletto et al., 2023; Oijstaeijen et al., 2020). Functional connections of these spaces extending to urban peripheries enable them to provide a broad range of ecosystem services to residents (Andersson et al., 2019; Lima et al., 2023). They support recreational, educational, and cultural services that improve human health and well-being (Kabisch et al., 2016). GBI in urban areas also help people build an identity attached to their landscape enabling conservation action (Dushkova et al., 2021). These ecosystem services are being revalorized widely in the context of the COVID-19 pandemic. Coinciding with the pandemic-induced lockdown phase, increased interest in going out and spending time in nature has been documented from several cities (Kleinschroth & Kowarik, 2020; Ugolini et al., 2020).

While multiple services of GBI in urban areas are being appreciated, there is concomitantly a growing recognition of the rebound effects of urban growth on the social-ecological integrity of GBI. Research has documented challenges related to accessibility and distribution of GBI (Andersson et al., 2019). In the Global South, transformations due to urbanization are often rapid and ill-planned. Cities in economically weaker countries have been relatively little studied in comparison to those in Europe and North America, especially in terms of urban development pathways and institutional arrangements such as property rights, and governance (Thapa et al., 2021). The vital role that GBI elements play in enhancing urban ecological resilience and their centrality to the livelihoods of poor and marginalized urban residents call for a markedly different outlook in planning and creating GBI in the context of the Global South. This chapter attempts to synthesize context-specific knowledge on the impacts of urbanization on GBI in a city of the Global South, characterized by staggering rates of urban expansion, spatial heterogeneity and mixed land use, rapid population growth and weak institutional structures.

In India, the nature of urbanization processes has changed drastically since the neo-liberal reforms of 1991, with more space created for private enterprises and real estate (Narain & Singh, 2017; Vji & Narain, 2016). Several Indian cities have been on a trajectory of steady growth, sustained by the boom in real estate and the rapid growth of outsourcing and other services, a trend that is most pronounced in Bengaluru as the Information Technology (IT) capital of India. We use the case of Bengaluru to trace the changing nature of GBI, its uses and governance in a rapidly urbanizing context. This chapter compiles cases on three different elements of GBI that present changes triggered by urban growth in Bengaluru. The chapter is focused on transformations of GBI elements that have taken place over the past three to five decades.

The next section of the chapter describes the study area—the Bengaluru Metropolitan Region (BMR), with a justification to the choice of the GBI cases for analysis. This is followed by a detailed description of the selected cases in terms of their biophysical and ecosystem service changes, and institutional and governance transitions. The discussion section draws on the commonalities and differences between the cases to unravel the patterns and trends of social-ecological transformations of GBI in Bengaluru, followed by the concluding section that foregrounds the implications of the study for incorporating GBI in urban planning processes in the Global South.

2 Study Area—Bengaluru Metropolitan Region (BMR)

Bengaluru is one of the fastest growing cities and, with a population of 12.7 million (World Population Review, 2021), the third most-populous city in India. The city was once known as the 'garden city'—due to the number of parks and green spaces—and as the 'city of 1000 lakes'—due to the large number of interconnected lakes that provided water for drinking and irrigation (Mundoli et al., 2015). Most of Bengaluru's spatial expansion and population growth in the pre-1990 period is attributed to industrial development. However, the IT revolution of the 1990s actively facilitated by the central and state government policies is widely touted as the key driving factor behind the ongoing urbanization boom. Much of the current spatial expansion is taking place towards the north and south, owing to the emergence of IT hubs in these areas (Patil et al., 2018). An analysis of land-use changes in the region indicates a growth of 584% in built-up areas since the 1970s. This has been associated with a decline of 66% in vegetation and 74% in water bodies (Ramachandra et al., 2012). The Bengaluru Metropolitan Region Development Authority was created in 1985 for the coordinated planning and development of expanding metropolitan areas. The BMR (location map in Fig. 15.1) encompasses an area of 8005 km^2 covering three administrative districts—Bengaluru Urban, Bengaluru Rural, and Ramanagara district. The astounding transformations of Urban Bengaluru have spilled over to Bengaluru Rural and Ramanagara where they led to a mosaic of rural and urban features that correspondingly combine urban and rural characteristics. Rapid settlement and infrastructure development have transformed agricultural landscapes and ecological commons in these districts (Patil et al., 2018) and led to the dramatic losses of water bodies, often through illegal encroachment, construction, and draining of lakes (Enqvist et al., 2016). Dumping of debris and release of urban and industrial sewage into lakes are well-documented (Mundoli et al., 2015; Nagendra & Ostrom, 2014).

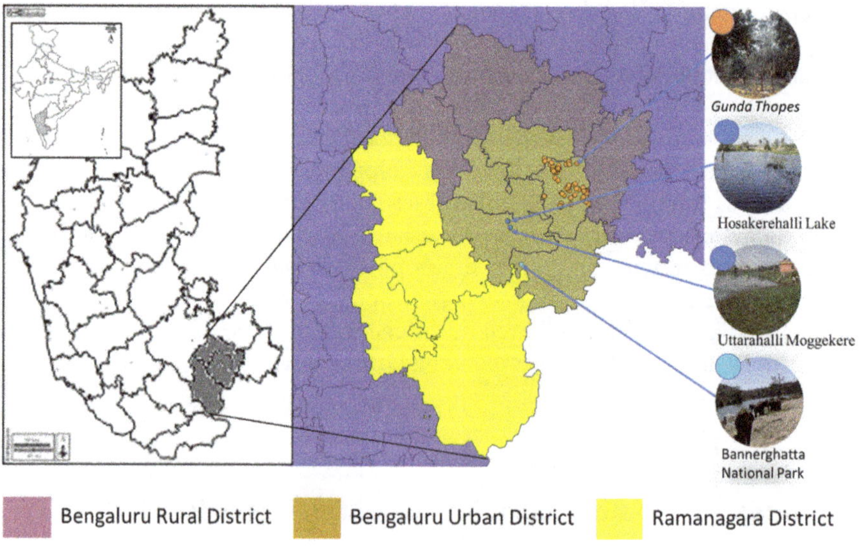

Fig. 15.1 Map of Karnataka State of India (left) with highlighted Greater Bengaluru Metropolitan Region (right) covering three administrative districts and four green and blue infrastructure cases included in this chapter. Many wooded groves are scattered in the Bengaluru East Taluk of Bengaluru Urban District. (Photos taken by Pramila Thapa in February 2020)

We selected cases of both green (wooded groves and protected area) and blue (lakes) infrastructure elements that are commonly seen in BMR. We also considered spatial location with respect to the urban centre to ensure adequate representation of GBI from both core and peripheral regions of BMR. Such a selection of cases further allowed us to provide a better outlook on the dynamics that unfold in urbanization processes in Greater Bengaluru. Further, the chosen GBI elements have been in use since centuries and continue to be so, although their services have been repurposed to suit urbanization-driven changes in demand. To understand the impacts of urban transformation, the foci on not just biophysical and ecological elements, their services, and disservices, but also on changes in actors, property rights, and local institutions are highlighted in this chapter. Hence the cases were selected to cover a range of governance regimes from predominantly community-managed to mostly state-managed. Availability of reliable data on multiple social-ecological aspects was another criterion for selecting the cases for analysis. Information from secondary sources, mostly peer-reviewed articles and reports for wooded groves and protected areas are used. The lake cases used primary data collected during 2018–2019 through focus-group discussions and semi-structured interviews with various actors involved in lake management identified using purposive followed by snowball sampling, along with existing literature on the topic.

3 Wooded Groves

Wooded groves or *gunda thopes* (in local terms) are traditional orchards planted and maintained by local communities as commons to meet the needs for fruits, fodder, timber, and other natural resources. Wooded groves, mostly located in peri-urban areas of Bengaluru, contain locally preferred native species such as mango (*Mangifera indica*), jamun (*Syzygium cumini*), jackfruit (*Artocarpus heterophyllus*), tamarind (*Tamarindus indica*), banyan (*Ficus benghalensis*), and sacred fig (*Ficus religiosa*) that are grown and managed for multiple products and at the same time admired for their religious, social, or recreational values. Native grass undergrowth in wooded groves traditionally provided fodder for livestock in the villages (Mundoli et al., 2017a).

Already in 1894, several wooded groves were scattered widely throughout the urban and peri-urban landscapes of Bengaluru (Rice, 1897) and accessed by residents for traditional uses. Still today, they are often located close to lakes and therefore function as catchment areas. They also promote native biodiversity (Gadgil, 2004). The scale, type, and context of uses of wooded groves as village commons have changed over time. However, there is scant data available on the history of the groves, and on changing patterns of their uses, quality, and resilience (Mundoli et al., 2017a).

3.1 Changes in Biophysical Characteristics

Traditionally, wooded groves (Fig. 15.2) were rich in native trees. Urbanization has resulted in alienation of the communities from wooded groves, with substitutes to many of the tree products now being purchased rather than harvested from wooded groves. The alienation also resulted in lack of interest among villagers in managing the wooded groves. Some of the native trees have been replaced with ornamental and non-native trees, sometimes with negative repercussions on biodiversity. In addition, construction of grey infrastructure such as roads and buildings replacing the grove spaces increased with urbanization. The state also intervened in the use of grove spaces by clearing wooded groves for construction of government buildings such as schools, hospitals, or bus parking lots. Resulting construction debris and garbage were often dumped in the space of the groves. Wooded groves are often misused for open defecation by residents of nearby squatter settlements. Some wooded groves were (partially) converted into farmlands and plantations (e.g., of acacia) by locals, particularly by socially disadvantaged groups who had been displaced elsewhere (Mundoli et al., 2017a).

3.2 Changes in Ecosystem Services

Urbanization triggered substantial shifts in the supply of, demand for, and access to provisioning, regulating, and cultural services from the wooded groves (Mundoli et al., 2017a; Mundoli & Nagendra, 2021). As for provisioning services, people continue to eat fresh or pickled fruits harvested from wooded groves, but they do not sell the fruits to the markets anymore. In earlier times, people used to harvest seasonal fruits and sell them in (nearby) markets. Some people from the disadvantaged/landless groups used to utilize the spaces between the trees of the groves for cultivating millets and vegetables. These days, cultivation is rather considered as

Fig. 15.2 A traditional wooded grove with relatively less human influence (left, source: Pramila Thapa) and a wooded grove with land use changes indicating substantial increase in built-up areas (right, source: Raghvendra S Vanjari)

encroachment, however illegal cultivation of commercial acacia *(Acacia auriculiformis)* is a new practice. The practice of using wood from wooded groves to renovate houses, temples, or as fuelwood for cremation or preparing communal meals has disappeared, except for firewood harvested and used by a few (disadvantaged) local and migrant households (Mundoli et al., 2017a).

Wooded groves continue to provide vital regulating services such as habitat support for biodiversity along with climate, air, and soil regulation. These are still supported by the wooded groves despite urbanization, although the extent of appreciating such services has decreased. With urbanization, conversion of the wooded groves to manicured parks and more preference of non-native and aesthetic trees in those parks lead to a loss of native trees resulting in degrading biodiversity (Mundoli et al., 2017a).

Cultural ecosystem services are the key features of the wooded groves. The wooded groves used to be sites for community feasts and gatherings during festivals and village meetings, or even offered temporary shelters for nomadic communities when they were passing by the village. These activities are not common anymore. However, worshiping family deities, in shrines located within the wooded groves and using the leaves from sacred trees such as mangoes for decoration during festive events are still in practice. A very significant change is the growing perception of wooded groves as recreational and aesthetic spaces, especially in villages where urbanization has triggered a boom in real estate and housing projects. Wooded groves are converted into recreational parks, clearing off the native grasses and beautifying with manicured lawns, ornamental shrubs, and paved tracks for jogging (Thapa et al., 2023). Such parks are also fenced, and entry times are restricted, curtailing access to livestock graziers and other marginalized communities (Mundoli et al., 2017a).

To sum up, provisional services and religious services (being part of wider cultural ecosystem services) of the wooded groves are being less realized and less valued with urbanization, perhaps because the society is being more individualistic. Yet, some of the cultural services such as recreation and aesthetics are valued more than in earlier times—mostly by economically sound communities. Such priority might marginalize traditional land users such as livestock grazers.

3.3 Changes in Governance and Institutions

Mundoli et al. (2017a) classified wooded groves into three types based on formal and informal ownership and use: (a) government wooded groves, which were planted by government authorities or have not been claimed by anyone since immemorial time, and whose produce was auctioned by the state; (b) charitable or public wooded groves, planted by individuals on private or government land that were accessible to all members of the community; (c) private wooded groves which were planted on government land by private individuals from which the government may or may not receive revenue or a share of the produce. Most of the decisions regarding the management were handled by a local village leadership, which was responsible for monitoring and enforcement. Communities rendered voluntary support as per the instructions of the village leadership. With state overtaking of the commons, and abolition of traditional village offices in 1960s, there were changes in governance and management of the commons. Different sectoral administrations at different levels became responsible for grove management, including local administration (village and city corporation ward offices), the revenue department which officially became the owner of the commons, and the forest department which is responsible for planting and protection of trees. Sometimes the management of groves is also transferred temporarily from one department to another. For example, one of the village administrations seems to have handed over their grove to the forest department, the latter having planted saplings as part of its afforestation programs. While an array of government departments is now responsible for the management of wooded groves, the role of the local community has diminished greatly. There is no transparency in the division of powers among various government bodies like the City Corporation and the forest department. With the state custody, some of the groves are fenced and locals must take permission to use the ecosystem services from the groves, which is bureaucratic and takes a long time, and is therefore discouraging. Local community members believe the grove management has become politicized and often express preference for the traditional village-based institutional system (Mundoli et al., 2017a).

4 Bannerghatta National Park

Located only about 20 km away from the city centre, Bannerghatta National Park (BNP) (Fig. 15.3) is the largest natural ecosystem in the larger Bengaluru metropolitan landscape. BNP is an uneven and rocky landscape comprising dry deciduous and scrub forests characteristic of the Eastern Ghats. The park acts as a vital ecological corridor closely linked to other protected areas like the Cauvery Wildlife

Fig. 15.3 Bannerghatta National Park with less human influence (*left*) and more land use changes (*right*), indicating substantial increase in built-up areas. (Sources of both photos: Dhanya Bhaskar)

Sanctuary in the south and Bandipur Tiger Reserve to the northeast. Most importantly, BNP forms a part of Mysuru Elephant Reserve—the largest protected habitat for the Asian Elephant in South-east Asia (Varma et al., 2009). The park harbours over 128 types of tree species, predominantly woody and scrub vegetation of *Acacia chundra*, *Anogeissus latiflora*, *Wrightia tinctoria*, etc. and medicinally important plants such as *Shorea roxburghii* and *Cassia fistula*. Plantations of Eucalyptus can be observed in and around the park because of reforestation programs.

BNP has over 112 villages in its fringes and five within the park with communities traditionally dependent on the forest for agriculture, livestock grazing, and diverse produce such as wild fruits, meat, manure, firewood, lac, and honey (Gopalakrishna et al., 2015; Varma et al., 2009). A large population of marginalised communities reside in the forest peripheries such as the nomadic *Hakki Pikki* tribe and the forest dwelling *Iruliga* tribe who were resettled through a tribal welfare scheme (Singh, 2008; The Alemaari Trails, 2021).

4.1 Changes in Biophysical Characteristics

Scrub grasslands with thorny trees (predominantly *Banni* or *Prosopis cineraria*, which lends the name *Bannerghatta* or the mountain in the village of *Banni* tree to the protected area) once covered the landscape, as oral histories and archival documents reveal. The region supported thriving agropastoral livelihoods, while the undulating and rocky terrain limited agrarian expansion in certain areas where patches of deciduous forests occurred (Jayaprakash & Hickey, 2019).

With the advent of British rule, Bengaluru was chosen as a cantonment and commercial forestry of hardwood species such as teak (*Tectona grandis*) to cater to the rising wood demands and Karachi (*Hardwickia binata*) for charcoal production, was initiated in the Bannerghatta forests. Other commercially valuable species including sandalwood (*Santalum album*), dindiga (*Conocarpus latifolia*) and jalari (*Vatica laccifera*) were also prioritised for produces such as aromatic wood, gum, and lac respectively. In fact, Bangalore was selected by the colonial rulers for cantonment considering the availability of cheap firewood from the surrounding forests (Jayaprakash & Hickey, 2019). With forests coming under the State control, access and harvest of produces became a challenge to the locals. A decline in the native scrub forest area and significant increase in areas under tree plantations ensued (Adhikari et al., 2014).

Following the designation of the areas as National Park under Wildlife Protection Act (WPA), 1972 a stable vegetation cover could be maintained within the BNP, but the surrounding areas suffered large scale changes. BNP's peripheries have seen tremendous land conversions for industrial, residential and tourism development that have eroded the ecological corridors. Proximity to three important development corridors, and the information technology hub of Electronic city located less than 10 km radially from the park boundaries, triggered a cascade of land use and socio-cultural changes (Jayaprakash & Hickey, 2019). While extensive grazing, illegal sand and stone mining and unsustainable timber harvesting affected the forest surroundings, the proximity to the urban centre triggered extensive cultivation of commercial crops like coconut, teak, mango, and eucalyptus replacing subsistence produce like finger millet in agricultural fields in peripheries (Adhikari et al., 2017).

A comparative study of land use change between 1973 and 2007 shows conversion of BNP from forest to non-forest area and the expansion of large-scale Eucalyptus and teak plantations that changed the ecology of the park (Adhikari et al., 2017). Land use change analysis of the BNP with a 5 km buffer shows a marked increase in urban land use from 0.4% in 1973 to 4.5% in 2015. Area under deciduous forests declined from 52.7% in 1973 to 32.5% in 2015, while area under mining has seen a jump from 0.1% to 1.3% during the same period (Ramachandra & Setturu, 2019). Due to the availability of suitable coarse granites and complex gneiss rock formations intense mining can be observed around BNP, especially in villages close to the critical elephant corridors (Varma et al., 2009). Land use projections indicate continued decline of forests owing to further expansion in urban area to 9.62% in 2027 (Ramachandra & Setturu, 2019). Forest fires both due to natural and anthropogenic factors and the spread of invasive species also pose threats for the conservation and management of BNP (Varma et al., 2009). Landuse changes and degradation of ecological corridors have also fuelled growing human-wildlife conflict around BNP leading to crop depredation and loss of human life and property. The wild Asian elephant (*Elephas maximus*) is reported by local community as the most serious threat, followed by the wild boar (*Sus scrofa)* (Prasad et al., 2016). Mapping of damage shows that crop depredation was not uniformly spread but was relatively concentrated along the migratory routes of elephants (Bantalpad et al., 2017). Human wildlife conflicts also result in a shift in cropping patterns from agriculture to tree plantations to avoid the high frequency of elephant raids on agricultural areas and fruit orchards (Adhikari et al., 2014).

4.2 Changes in Ecosystem Services

The savanna landscape of BNP has a long history of offering provisional services supporting pastoralism, agriculture and forest produce collection (Kamat, 1980). Collection of bamboo for basket weaving by locals has also been documented (Varma et al., 2009). However, with the designation of protected area, grazing and forest produce collection has been legally curtailed, though a tacit understanding operates between local forest officials and villagers by which livestock grazing is allowed and villagers return the favour by helping to control forest fires on public land (Jayaprakash & Hickey, 2019). The water from BNP drains into river Arkavathi through tributaries such as Suvarnamukhi, Rayathmala holé and Anthragange and joins river Cauvery, promoting agriculture, livestock grazing, silkworm rearing and fodder collection along its course. Illegal quarrying of granite from rocky outcrops and mining of sand from the riverbeds for urban construction has become rampant in many peripheral areas of the park (Adhikari et al., 2014).

As for regulating services, along with providing a critical habitat for diverse flora and fauna, BNP's role in regulating Bengaluru's microclimate and ground water (Gopalakrishna et al., 2015) and aiding carbon sequestration is well recognized. Comparison of agroecology between northern and southern peripheries of Bengaluru found the southern side to be more agriculturally vibrant, facilitated by the flow of watershed services and pollination services from Bannerghatta forests (Patil et al., 2018).

BNP houses important cultural elements such as megalithic burial sites (Gopalakrishna et al., 2015). The Bannerghatta Biological Park with its safari units, butterfly park and animal rescue centre covering an area of 731.88 ha is an important tourist attraction. It aims at promoting awareness of ex-situ conservation and focuses on environmental education for school children and provides research opportunities.

A discernible trend in terms of ecosystem service changes is the declining salience of provisioning roles of the forests, while cultural services such as recreation and some regulating services (e.g., biodiversity protection) are better recognized and conserved. This also leads to serious equity concerns as provisioning services are crucial for local agropastoral livelihoods that are already pushed to margins in the globalized city, whereas recreational services are mostly captured bythe wealthy residents. Trade-offs between and within provisioning services and regulating services are also growing; for instance, the increasing human wildlife conflicts mars the profitability of farming in peripheries that benefit from pollination and water flow from forests.

4.3 Changes in Governance and Institutions

From a community-managed scrub vegetation landscape, Bannerghatta became a state forest during the colonial rule and continues to be state-controlled, but the extent of protection has changed over time. BNP was carved out of state reserve forests in 1974 under the Wildlife Protection Act of 1972, according to the highest level of protection to the area. The park comprises 12 reserve forests within its range (Patil & Raj, 2019). Part of the national park was designated as a zoo in 1977 and later renamed Bannerghatta Biological Park in 2002 under the administration of the Zoo Authority of Karnataka. The initial park area of 102.7 km^2 was expanded in 2011 with the addition of reserved forests of 157.7 km^2 area from Ramanagara district along with a buffer zone of 5 km^2, bringing a total area of 260 km^2 under protection.

Despite the strictly protected status of the forests, there are institutional complexities that render governance of BNP extremely challenging. The park is managed by Karnataka Forest Department, while its northern side and parts of the southern boundary come under the Bangalore Development Authority and Bangalore Metropolitan Regional Development Authority. It is geographically distributed across two districts—the eastern side belongs to Anekal Taluk of Bengaluru urban district and the Western one to Kanakapura Taluk of Ramanagara district, each with its own Local Planning Authority (Patil & Raj, 2019). Poor coordination between the multiple planning agencies often leads to complex and conflicting management decisions. The peri-urban location of BNP augments the connections with liberalization policies at national level which have created new pressures on local institutions. Socio-cultural relations mediated by caste have traditionally and continue to play a major role in determining property rights, resource access and usage rights, and significantly affect institutional functioning (Jayaprakash & Hickey, 2019).

The failure of local institutions in protecting the integrity of the park in the context of growing demand for urban developmental needs is quite evident. BNP was demarcated with an Eco-sensitive zone (ESZ) of 268.9 km^2 with radii varying between 100 m to 1 km from the park boundary under a central Government notification. The ESZ regulates land-use activities and acts as a transition zone between the protected area and the highly human modified landscapes. Unfortunately, a whopping 30% of this ESZ was later denotified by the Karnataka State Government in 2018 to make way for construction projects, highways, and quarrying (The Hindu, 2020). A proposal to develop a satellite town ring road through the ESZ of BNP is also under active consideration. The Revised Master plan for Bengaluru 2031 (BDA, 2017) claims to protect the ESZ while retaining the approved developmental projects, a proposition that seems highly impractical given the intense pressure on the landscape. It is also to be noted that the land title claims of traditional forest dwelling communities such as *Iruligas* under the Forest Rights Act of 2006 are not being recognized citing conservation needs, while forest land diversion for developmental projects continues unencumbered, raising questions of social justice.

5 Lakes

Lakes are perhaps the best-studied GBI elements in Bengaluru. We illustrate the impacts of urbanization on these commons in general and the changes in two urban lakes. Lakes in Bangalore form an interconnected network mainly through channels adapted according to the spatial gradient of the city and form an essential part of the city's landscape. We first provide an account of changes in characteristics of lakes, ecosystem services provided and governance and institutions that hold for all lakes in general terms. Subsequently, we use two specific examples to illustrate these changes in relation to urban lakes. *Uttarahalli Moggekere*[1] lake is considered one of the well-managed lakes in Bengaluru, with sharing of management responsibilities between the government and citizens led by a third-party non-governmental organisation, United Way Bengaluru. The lake was created in 1869 as a source of water

[1] Kere in the local language (Kannada) means lake, hence moggekere can be roughly translated as "flower bud lake"

for irrigation (EMPRI, 2018). Currently, it is in densely populated localities of the city. The second lake under consideration is *Hosakerehalli lake*, located in one of the fastest urbanizing locations of Bengaluru, with limited horticultural activities still practiced on one side of the lakebed. This lake was created in 1903 (EMPRI, 2018), mainly for irrigation.

5.1 Changes in Biophysical Characteristics

There have been tremendous changes in the biophysical characteristics of lakes and their surroundings in the city. Urbanization has resulted in conversion of these lakes into real estate that have been encroached (converted into residential and commercial uses) and polluted (release of domestic sewage into the lake and dumping grounds for garbage) leading to their degradation and damage to channels and interconnections between the lakes (Mundoli et al., 2015). The fencing of the lake area keeps off people from using the lake. It promotes the overgrowth of water hyacinth and other weeds and converts them into spaces prone to anti-social activities, thus alienating the local communities from using them.

5.2 Changes in Ecosystem Services

Lakes in Bengaluru traditionally served the community by providing multiple services. In earlier times, lake (water) was commonly used for drinking (for both human and livestock), washing clothes and utensils, and bathing of human and livestock, in addition to irrigation, fishing, and the provision of grazing material, fuelwood and wild plants, as well as of vegetables used for cooking. However, today, the use of the lake water is limited mainly to irrigation and fishing. Often the spaces around lakes are used for grazing livestock and collecting medicinal herbs and fuelwood. Moreover, silt from the lakes is used to fertilize farmlands (D'Souza & Nagendra, 2011; Mundoli et al., 2015; Plieninger et al., 2022).

Urban lakes are also well-known for regulating services, such as supporting habitats for (both migratory and local) birds, reptiles and aquatic flora and fauna. Climate, water and soil regulation are other important regulatory services offered by the lakes. However, most of the regulatory services are diminishing with the decreased quality of lakes due to increased dumping of wastes/sewages in and around them, followed by the expansion of urban areas (Mundoli et al., 2018).

As for cultural services, one of the most prominent services urban lakes provide is birdwatching. In addition, they also offer space for recreation activities such as sports, and jogging. For some urban dwellers, the lake spaces are a location for social activities like meeting friends.

5.3 *Changes in Governance and Institutions*

In the governance of the urban lakes of Bengaluru, there has been a clear change from a community-managed system to a more centralized dependence on the state. The state in 1961 officially halted community participation in lake management by the passing of the Karnataka Village offices abolition act leading to abolition of numerous traditional offices that were responsible for the upkeep of the lakes and other common pool resources in the state (Nagendra, 2016). This was further accentuated when the state took control of the lakes in 1964 with the passing of the Karnataka Land Revenues Act, which declared all common property resources as state property. Lakes were now placed under the ambit of numerous state agencies based on location and use; thus, responsibilities were divided between the state authorities (Nagendra, 2016). For example, fishing in the lakes came under the purview of the fisheries department, whereas water allocation and use was the responsibility of the village panchayat in collaboration with the irrigation department. The Pollution Control Board is responsible for monitoring pollution levels, the Bangalore Urban Water Supply and Sewerage Board for preventing the discharge of wastewater in the lake. Each of these state actors had a distinct policy on lakes, with the forest department taking a conservation approach, whereas the others focused more on their physical and infrastructural development. These changes in formal regulations have led to a change in the actor constellation from traditional actors to more state-based actors. The state as the custodian of the lakes plays an important role in lake management, whereas the role of the residents has been reduced to that of users and managers with no/limited powers of enforcement. In the following, for illustrative purposes, we briefly specify these broad changes for two urban lakes.

Illustrative Case 1: Uttarahalli Moggereke
Uttarahalli Moggekerewas once a major source of irrigation but is currently seen as a social space. There has been limited encroachment along the lake, as it was fenced in 1993–1994 by the forest department. There have been numerous proposals by various committees examining the feasibility of preserving the lake or converting the area into parks or open spaces (Lakshman Rao Committee, 1986). Land-use change around the lake has led to the reduction in catchment area with increasing concretization (Fig. 15.4). The lake was under the governance of the forest

Fig. 15.4 Urbanization around Uttarahalli lake 2000 (left) and 2021 (right). (Source: Google Earth)

department and thus protected from the impacts of urbanization, as the objective of the department was conservation. The lake custody was transferred to the city municipality in 2000s, and was restored in 2010 (resulting in desilting, pitching of embankments, the construction of walking paths and a play area for children). After the rejuvenation in 2010, the state custodian approached a non-governmental organisation, United Way Bengaluru as a partner to raise funds and maintain the lake. United Way Bengaluru revitalized local institutions by constituting a community-based organization comprising residents responsible for monitoring the activities and the performance of day-to-day activities of management. The tripartite agreement between the three actors clearly spells out the various responsibilities of each actor and indicates the penalties for not fulfilling their roles. United Way Bengaluru has managed the lake using an ecosystem-based approach in collaboration with the local community association. Since the lake restoration in 2010, it provides both regulating (micro-climate, groundwater) and cultural services to the local community, as well as a space for recreational activities (jogging, walking). The local community association in collaboration with United Way Bengaluru aim to make the lake a hotspot for birdwatching and nature lovers. The community regularly organizes '*lakeathons*' with various activities around the lake, in addition to the celebration of various festivities such as *deepothsava* (festival of lights), bringing the community closer together.

Illustrative Case 2: Hosakerehalli Lake

Hosakerehalli lake was a major source of irrigation and grazing lands. With urbanization, there has been a change in the land use with urban settlements being developed all around the lake. It is estimated that three percent of the lake area is encroached by houses and factories; its outlet has been blocked by the road built under the Bengaluru-Mysore Infrastructure corridor which is constructed on the lake (Management Lakes 2018) (Fig. 15.5).

The lake falls under the jurisdiction of the Bangalore Development Agency (BDA), which is undertaking restoration activities (in 2018, since 2020 the lake has been transferred to the city municipality). Since then, various government agencies are responsible for different aspects of the lake, each of them undertaking its

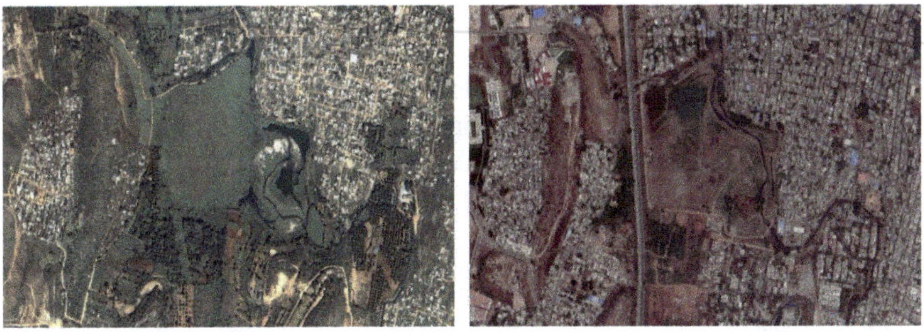

Fig. 15.5 Urbanization around Hosakerehalli Lake 2000 (left) and 2020 (right). (Source: Google earth)

activities without consultation with the others and depending on the funds available to them, which leads to delays and confrontations with not just the various government agencies but also with the community. Social and economic differences and different perceptions of the lake among the residents divide them in their approaches to lake management, leading to a lack of social cohesion. The lake is currently covered by weeds and used as a dumping ground for garbage and construction debris. It is a clear example of mismanagement. However, it has been under restoration since a few years. Yet, due to the lack of interactions between various government agencies and with the community, it is still used for dumping of debris, and the unregulated inflow of urban sewage turns it into a space for mosquito breeding. The bad odours emanating from it create nuisance to the residents.

6 Discussion

In this section, we attempt to synthesize the trajectories of urbanization-induced social-ecological transformations of the different GBI cases analyzed, to unveil the similarities and differences in patterns and trends (Table 15.1).

As Table 15.1 indicates, the direction and nature of changes in social-ecological characteristics of the GBI cases vary. However, certain overarching trends are discernible. Biophysical changes due to sealing of surfaces by (expanding) built-up area, ecological degradation in terms of diminishing quality of water and soil resources, as well as decreasing richness and diversity of indigenous species in GBI are quite pronounced. Even in the peri-urban GBI cases, the tremendous impacts of urban stressors are very clear—most of these ecosystems have degraded as sinks of urban waste (as in the case of wooded groves used for dumping of construction debris) or have been transformed to produce goods and services consumed in the urban core (as in the case of the shift in farming practices from traditional food crops to horticultural or timber crops around BNP).

Provisioning services from most GBI elements have been invariably on the decline, owing to limited access for the community due to fencing around lakes, groves, and protected areas after state interventions in their management, compounded by increasing preferences for substitutes of goods and services provided by GBI driven by socio-cultural changes accompanying urbanization. Blue infrastructure still supports provisioning services such as water for irrigation, as compared to green infrastructure, which is increasingly being viewed as a source of recreation or aesthetic value, with restrictions imposed on active community use. In protected areas, state intervention has amplified regulating services including carbon sequestration and habitat services. Some other urban lakes that have been restored with active community engagement such as Jakkur lake, and Puttenahalli lake have also flourished into excellent habitats for mammals, insects and other flora and fauna (Gadgil, 2004). There has been a visible disappearance of livelihood linkages with the ecosystem, making it difficult to assess the direct use values. With the

Table 15.1 Social-ecological transformations of selected GBI cases

Feature	Wooded groves	Protected area	Lakes
(a) Biophysical characteristics			
Spatial extent	Shrinking	Expanding	Shrinking
Boundaries	From open to fenced	From open in precolonial times to fenced during British rule and strictly fenced following protected Area declaration in 1974	Open to fenced
Land use	Increasing built up area within and around	Increasing built up area outside protected area	Increasing built-up areas in the surroundings
Quality of ecosystem	Degrading owing to decline of native trees and biodiversity	Degrading due to exotic tree plantations and vanishing ecological corridors with other protected forests	Context specific, as illustrated by the two cases
Proximate stressors	Misuse of the space for waste dumping and defecation	Human-wildlife conflicts, waste dumping	Space for dumping waste, inflow of sewage, health of the residents
(b) Ecosystem services			
Provisioning	Dwindling	Dwindling	Dwindling
Regulating	Important, but less appreciated	Habitat services better appreciated	Important but starting to be appreciated
Cultural	Religious values are declining while aesthetic and recreational values are better appreciated	Aesthetic, recreational, and educational services better appreciated	Recreational, aesthetic, and social spaces used for community building
(c) Governance and institutions			
Actors	Number of actor groups increasing, migrant communities also involved in addition to local communities and state actors	Number of actor groups increasing, Involves forest department, urban planning and development bodies, conservation and civil society organizations, academia	Increasing number of actors, mainly state agencies, but since recently there have been local community-based organizations and non-governmental organizations
Property rights	Earlier owned by village institutions or private individuals, now mostly owned by forest department, revenue department or local village administration	Forest department, usufruct rights of communities denied after protected area declaration	Earlier in custody of the village community, now in the custody of various state agencies. There is a shift of certain rights to the local communities in specific lakes
Governance regime	From community managed to state managed and community managed	State managed after the protected area declaration	Community managed to state management to shared practices in certain cases

availability of ample informal urban job opportunities, dependence on GBI for livelihood sustenance has diminished considerably.

There has been a change from community governance of GBI to state management since state acquisition of the commons in 1964. This has led to an alienation of the local communities dependent on these commons mainly due to fencing of the boundaries and inappropriate management by the state, and due to lack of information, awareness, and resources with state actors. Even the expanding area of the strictly protected forests of BNP, which appears to be an ecologically beneficial step, is a result of a tightening state control over forest commons, excluding traditional forest users with enormous social costs. The dilution of the buffer ESZ zones around BNP also indicates prioritization of development demands over community needs in GBI management. Segregation of property rights between multiple state actors, with limited roles for the community, has led to a complex interplay between managing actors, leading to confusion among the community regarding responsibilities and powers. However, in the case of some urban lakes, the communities have claimed certain rights through participatory legitimation—mostly limited to rights to access and monitoring.

The changing perceptions and degree of awareness among the community is also an important aspect to be addressed for sustainable management of urban GBI. Most GBI elements are increasingly viewed as cultural spaces to meet the demands of the urban citizens (for jogging and bird watching) rather than as traditionally used community resources for irrigation, grazing etc. A transformed social demography accompanying urbanization has made collective action for ecosystem conservation challenging. This is attributed mainly to the arrival of new residents with limited understandings of local ecosystems and scant regard for the services obtained from common resources (Mundoli et al., 2017a, 2018). There is a complete lack of awareness among both the community members and the government officials concerning wooded groves, as compared to lakes and BNP where civil society and conservation organizations actively collaborate with the state to restore and protect these ecosystems (Nagendra & Ostrom, 2014). Academia and the media also have played crucial role in creating awareness and changing the state's attitude to consider diverse non-state stakeholders as legitimate partners in GBI management.

7 Conclusions

In the urbanizing context, uses and values of GBI ecosystems as urban commons tend to be ignored. Yet, as we demonstrate, a diverse array of urban GBI—wooded groves, a protected national park, and lakes—act as commons, providing important provisioning and cultural ecosystem services for a wide range of users. This facet of traditional commons is largely ignored in urban policy and planning. We therefore call for greater attention of planners and policymakers to these important elements of the urban fabric and their contributions to the well-being of communities living and working there.

Socio-economic groups who value and use the ecosystems depend on their multiple goods and services for their resilience (Unnikrishnan et al., 2016; Thapa et al.,2024). However, the developmental priorities of the city region of Bengaluru and the prevailing imaginations of a 'global city' modelled after cities of the Global North ignores not just the local ecosystems but local communities as well (Nagendra et al., 2018). In contrast, we advocate for prioritizing provisioning and cultural ecosystem services beyond recreation, such as sense of place and a motivation for a collective action in policy making. They are important to foster the resilience of the city, as well as the most vulnerable residents within the city (Nagendra, 2016). To arrive at such a greater appreciation of GBI as urban commons, management of urban GBI calls for a pro-people approach recognizing and integrating the multiple values of these ecosystems in urban planning and policy in the Global South.

Acknowledgement This research has been funded by the Deutsche Forschungsgemeinschaft (DFG, German Research Foundation)—project number 279374797. The research on Bannerghatta National Park is funded by the Department of Biotechnology, Government of India. We would like to also acknowledge Samudyatha Ramananda for assisting in literature review and Imke Horstmannshoff for formatting and proof-reading the manuscript of the book chapter.

References

Adhikari, S., Southworth, J., & Nagendra, H. (2014). Understanding forest loss and recovery: A spatiotemporal analysis of land change in and around Bannerghatta National Park, India. *Journal of Land Use Science, 10*, 4–424. https://doi.org/10.1080/1747423X.2014.920425

Adhikari, S., Fik, T., & Dwivedi, P. (2017). Proximate causes of land-use and land-cover change in Bannerghatta National Park: A spatial statistical model forests. *Forests, 8*, 342. https://doi.org/10.3390/f8090342

Andersson, E., Langemeyer, J., Borgström, S., & Mcphearson, T. (2019). Enabling green and blue infrastructure to improve contributions to human well-being and equity in urban systems. *Bio Science, 69*(7), 566–574. https://doi.org/10.1093/biosci/biz058

Angeoletto, F. H. S., Richard, E., Santos, J., Guedes, N., Oliveira, S. L., Albertín, R. M., et al. (2023). When we think of urban afforestation in Brazil, what comes to mind is a mouth that has lost its teeth. Terr@. *The Pluralist, 17*, e2321664–e2321654.

Bangalore Development Authority (BDA). (2017). Revised master plan 2031, Planning District Report. Open City Urban Data Portal. https://opencity.in/documents/bda-revised-master-plan-2031-planning-district-report

Bantalpad, M., Gayathri, A., & Krishnan, A. (2017). Seasonal crop raiding of fruit trees by Asian elephants: An insight into foraging preferences from croplands abutting Bannerghatta National Park, Bengaluru, Karnataka, India. *Asian Journal of Environment & Ecology, 2*, 1–12. https://doi.org/10.9734/AJEE/2017/31851

D'Souza, R., & Nagendra, H. (2011). Changes in public commons as a consequence of urbanization: The Agara lake in Bangalore. *Environmental Management, 47*(5), 840–850. https://doi.org/10.1007/s00267-011-9658-8

Dushkova, D., Ignatieva, M., Hughes, M., Konstantinova, A., Viacheslav, V., & Elvaria, D. (2021). Human dimensions of urban blue and green infrastructure during a pandemic. Case study of Moscow (Russia) and Perth (Australia). *Sustainability, 13*(8), 4148. https://doi.org/10.3390/su13084148

EMPRI. (2018). *Inventorization of water bodies in Bengaluru metropolitan area (BMA)* (Vol. 1). Environmental Management and Policy Research Institute (EMPRI).

Enqvist, J., Tengö, M., & Boonstra, W. J. (2016). Against the current: Rewiring rigidity trap dynamics in urban water governance through civic engagement. *Sustainability Science, 11*(6), 919–933. https://doi.org/10.1007/s11625-016-0377-1

Gadgil, M. (2004). *Karnataka state biodiversity strategy and action plan*. ENVIS Technical Report No 15, Environmental Information Systems, Centre for Ecological Sciences.

Ghofrani, Z., Sposito, V., & Faggian, R. (2017). A comprehensive review of blue-green infrastructure concepts. *International Journal of Environment And Sustainable Development, 6*(1), 15–36. https://doi.org/10.24102/IJES.V6I1.728

Gopalakrishna, S. P., Kaonga, M. L., Somashekar, R. K., Suresh, H. S., & Suresh, R. (2015). Tree diversity in the tropical dry forest of Bannerghatta National Park in eastern Ghats, Southern India. *European Journal of Ecology, 1*(2), 12–27. https://doi.org/10.1515/eje-2015-0013

Grimm, N. B., Faeth, S. H., Golubiewski, N. E., Redman, C. L., Wu, J., Bai, X., et al. (2008). Global change and the ecology of cities. *Science, 319*(5864), 756–760. https://doi.org/10.1126/science.1150195

Huang, S. L., Yeh, C. T., & Chang, L. F. (2010). The transition to an urbanizing world and the demand for natural resources. *Current Opinion in Environmental Sustainability, 2*(3), 136–143. https://doi.org/10.1016/j.cosust.2010.06.004

Jayaprakash, L. G., & Hickey, G. M. (2019). Mistaking the map for the territory: What does the history of Bannerghatta National Park, India, tell us about the study of institutions? *Society & Natural Resources, 32*(12), 1433–1450. https://doi.org/10.1080/08941920.2019.1643431

Kabisch, N., Strohbach, M., Haase, D., & Kronenberg, J. (2016). Urban green space availability in European cities. *Ecological Indicators, 70*, 586–596. https://doi.org/10.1016/j.ecolind.2016.02.029

Kamat, S. U. (1980). *A concise history of Karnataka: From pre-historic times to the present*. Archana Prakashana.

Kleinschroth, F., & Kowarik, I. (2020). COVID-19 crisis demonstrates the urgent need for urban greenspaces. *Frontiers in Ecology and the Environment, 18*(6), 318–319. https://doi.org/10.1002/fee.2230

Lima, G. F., Correa Santos, J. W. M., Albertin, R. M., Martínez-Miranzo, B., Souza, F. L., & Angeoletto, F. (2023). Backyards are a way to promote environmental justice and biodiversity conservation in Brazilian cities. *Diversity, 15*(7), 815.

Mundoli, S., & Nagendra, H. (2018). *Lakes of Bengaluru: The once living, but now endangered peri-urban commons*. http://publications.azimpremjifoundation.org/1/1/APU-181213-Working-Paper-Series-10-28.03.2018.pdf

Mundoli, S., & Nagendra, H. (2021). *Where have all our gunda thopes gone? A meeting of old friends Maranna and Lakshmamma*. Azim Premji University.

Mundoli, S., Manjunath, B., & Nagendra, H. (2015). Effects of urbanisation on the use of lakes as commons in the peri-urban interface of Bengaluru, India. *International Journal of Urban Sustainable Development, 7*(1), 89–108. https://doi.org/10.1080/19463138.2014.982124

Mundoli, S., Manjunatha, B., & Nagendra, H. (2017a). Commons that provide: The importance of Bengaluru's wooded groves for urban resilience. *International Journal of Urban Sustainable Development, 9*(2), 184–206. https://doi.org/10.1080/19463138.2016.1264404

Mundoli, S., Unnikrishnan, H., & Nagendra, H. (2017b). The "sustainable" in smart cities: Ignoring the importance of urban ecosystems. *Decision, 44*(2), 103–120. https://doi.org/10.1007/s40622-017-0152-x

Nagendra, H. (2016). *Nature in the City: Bengaluru in the past, present and future*. Oxford University Press.

Nagendra, H., & Ostrom, E. (2014). Applying the social-ecological system framework to the diagnosis of urban lake commons in Bangalore, India. *Ecology & Society, 19*(2), 67. https://doi.org/10.5751/ES-06582-190267

Nagendra, H., Bai, X., Brondizio, E. S., & Lwasa, S. (2018). The urban south and the global predicament of global sustainability. *Nature Sustainability, 1*, 341–349. https://doi.org/10.1038/s41893-018-0101-5

Narain, V., & Singh, A. K. (2017). Flowing against the current: The socio-technical mediation of water (in)security in periurban Gurgaon, India. *Geoforum, 81*, 66–75. https://doi.org/10.1016/j.geoforum.2017.02.010

Oijstaeijen, W. V., Passel, S. V., & Cools, J. (2020). Urban green infrastructure: A review on valuation toolkits from an urban planning perspective. *Journal of Environmental Management, 267*, 110603. https://doi.org/10.1016/j.jenvman.2020.110603

Patil, D. R., & Raj, M. P. (2019). Urban fringe development around Bannerghatta National Park, Bangalore. *Tekton, 6*(1), 40–57.

Patil, S., Bhaskar, D., Vanjari, R., & Purushothaman, S. (2018). Urbanisation and new agroecologies. The story of Bengaluru's peripheries. *Economic & Political Weekly, 53*(41), 71–77. https://www.epw.in/journal/2018/41/review-environment-and-development/urbanisation-and-new-agroecologies.html

Plieninger, T., Thapa, P., Bhaskar, D., Nagendra, H., Torralba, M., & Zoderer, B. M. (2022). Disentangling ecosystem services perceptions from blue infrastructure around a rapidly expanding megacity. *Landscape and Urban Planning, 222*, 104399.

Prasad, A., Jayashankar, M., & Avinash, K. (2016). Perspective survey on the impact of vertebrates in agri-horticultural ecosystems in and around selected villages adjoining the Western margins of Bannerghatta National Park. *The Bioscan, 11*(2), 1295–1298.

Ramachandra, T. V., & Setturu, B. (2019). Sustainable management of Bannerghatta National Park, India, with the insights in land cover dynamics. *FIIB Business Review, 8*(2), 118–131. https://doi.org/10.1177/2319714519828462

Ramachandra, T. V., Aithal, B. H., & Sanna, D. D. (2012). Insights to urban dynamics through landscape spatial pattern analysis. *International Journal of Applied Earth Observation and Geoinformation, 18*, 329–343. https://doi.org/10.1016/j.jag.2012.03.005

Rau, L. Committee. (1986). Report of the expert committee for preservation, restoration or otherwise of the existing tanks in Bangalore Metropolitan area.

Rice, B. L. (1897). Mysore: A gazetteer compiled for government. In *Mysore in General* (Vol. 1). Archibald Constable & Company.

Singh, V. V. (2008). *Wildlife management plan for Bannerghatta National Park 2008–2013*. Forest Department, Government of Karnataka.

Souza, F. L., Fenoglio, M. S., & Angeoletto, F. (2023). To be a Brazilian city dweller, sometimes we must learn to say enough! *Sustainability, 15*(4), 3699.

Thapa, P., Torralba, M., Buerkert, A., Dittrich, C., & Plieninger, T. (2021). Ecological and social outcomes of urbanization on regional farming systems: A global synthesis. *Ecology and Society, 26*(3), 24.

Thapa, P., Torralba, M., Bhaskar, D., Nagendra, H., & Plieninger, T. (2023). Green in grey: ecosystem services and disservices perceptions from small-scale green infrastructure along a rural-urban gradient in Bengaluru, India. *Ecosystems and People, 19*(1), 2223307.

Thapa, P., Torralba, M., Nölke, N., Chowdhury, K., Nagendra, H., & Plieninger, T. (2024). Disentangling associations of human wellbeing with green infrastructure, degree of urbanity, and social factors around an Asian megacity. *Landscape Ecology, 39*(8), 152.

The Alemaari Trails. [date unknown]. *The Hakki Pikki Tales*. https://alemaaripeople.wordpress.com/the-landscape/. Accessed on 27 May 2021.

The Hindu [author unknown]. (2020). It's official: BNP's eco-sensitive zone reduced by 100 sq.km. *The Hindu*, March 13.

Ugolini, F., Massetti, L., Calaza-Martínez, P., Cariñanos, P., Dobbs, C., Ostoic, S. K., et al. (2020). Effects of the COVID-19 pandemic on the use and perceptions of urban green space: An international exploratory study. *Urban Forestry & Urban Greening*, 126888. https://doi.org/10.1016/j.ufug.2020.126888

Unnikrishnan, H. S., Mundoli, B. M., & Nagendra, H. (2016). Down the drain: The tragedy of the disappearing urban commons of Bengaluru. *SAWAS Journal, 5*(3), 7–11.

Varma, S., Anand, V. D., Gopalakrishna, S. P., Avinash, K. G., & Nishant, M. S. (2009). *Ecology, conservation and management of the Asian elephant in Bannerghatta National Park, Southern India*. A Rocha India & Asian Nature Conservation Foundation.

Vji, S., & Narain, V. (2016). Land, water & power: The demise of common property resources in periurban Gurgaon, India. *Land Use Policy, 59*, 59–66. https://doi.org/10.1016/j.landusepol.2015.08.030

World Population Review. (2021). *Bangalore population 2021*. https://worldpopulationreview.com/world-cities/bangalore-population. Accessed 28 May 2021.

Chapter 16
Environmental Quality Indicators in Brazilian Squares as a Conservation Strategy for Urban Biodiversity

Silvia Méri Carvalho, Zingara Rocio dos Santos Eurich, Rosa Imelda Rojas Caldelas, Rogério Bobrowski, Daniela Biondi, Eliana Cardoso-Leite, and Marcos Clair Bovo

Abstract Well-planned squares are important spaces that can ensure quality of life and contribute to the urban biodiversity conservation. These places have been socially and economically important since their creation, and more recently their environmental contribution and aesthetic importance have been highlighted. Vegetation in squares is an important indicator of environmental quality. Therefore, data acquisition and monitoring are necessary for decision-making aimed at improving urban environments. The tree-shrub composition in squares should be evaluated along with spatial and coverage indices traditionally used, since these indices in isolation do not provide sufficient information for proper planning. Other information, such as estimated permeable area and proportion of local native species, also help to guide planning decisions in the management process to create better spaces, allocate resources for maintenance, and improve public use. The planned tree-shrub composition in squares occurs together with the development of other functions, aesthetic and functional, which can be considered indicators of the quality of these open spaces. This chapter proposes: (a) to present an understanding of squares from the perspectives of geography, landscape architecture, forestry, and biological

S. M. Carvalho (✉) · Z. R. dos Santos Eurich
Universidade Estadual de Ponta Grossa, Ponta Grossa, Brazil
e-mail: silviameri@uepg.br

R. I. R. Caldelas
Universidad Autónoma Baja California, Mexicali, Mexico

R. Bobrowski
Univeridade Estadual do Centro Oeste, Irati, Brazil

D. Biondi
Universidade Federal do Paraná, Curitiba, Brazil

E. Cardoso-Leite
Universidade Federal de São Carlos, Sorocaba, Brazil

M. C. Bovo
Universidade Estadual do Paraná, Campo Mourão, Brazil

F. Angeoletto et al. (eds.), *Ecology of Tropical Cities*,
https://doi.org/10.1007/978-3-031-48012-6_16

sciences; (b) to highlight the process of insertion and recognition of the role of vegetation in que quality of these public spaces, mainly in Brazil; (c) to propose a set of environmental indicators for the characterization and assessment of vegetation in squares to contribute to the urban biodiversity conservation.

Keywords Open spaces · Environmental quality · Vegetation · Green area · Public space

1 Introduction

Urban green spaces have become a relevant factor to be considered in the urbanization process. Their aesthetic, landscaping, and environmental benefits contribute to improve the population's quality of life in the urban scenery (Angeoletto et al., 2018; Lima et al., 2023; Santangelo et al., 2022; Moroń et al., 2024). Urban green might be represented by elements of local influence that cover small areas such as squares, and when connected to other structures preserve functions and values of the natural ecosystems and provide the human population with associated benefits (Rumble et al., 2019; Angeoletto et al., 2022, 2023).

It seems relevant to emphasize that it is not enough to create green areas in a city, it is also necessary to be aware of their quality, considering social, ecological, and environmental coexistence (Angeoletto et al., 2017; Fellowes et al., 2020; Souza et al., 2023). Current trends in international organizations such as the United Nations regarding planning, management, and design of urban green public spaces include issues related to the Sustainable Development Goals (United Nations, 2015). For example, goal 11 raises the issue of city subsistence regarding environmental quality and accessibility to open public spaces and communities, while goal 15 refers to the sustainable management of forests, as well as halt and reverse of land degradation and loss of biodiversity.

The New Urban Agenda (UN-HABITAT, 2020) highlights the multiple functions of urban open spaces inside cities and their relationship with the environment. It also demands the protection, conservation, and restoration of ecosystems, in which biodiversity protection becomes a core strategy for sustainable urban development. This strategy requires, on the one hand, the development of technical studies to support assessment and monitoring; and, on the other hand, the integration of different disciplines and sector involved in the territory organization and management of natural resources. Environmental sustainability is approached through biological diversity, urban green areas reducing heat island effects, linking urban, peri-urban, agricultural, and forestry areas, thus creating green corridors, green infrastructure, urban gardens, and promoting green building.

At the same time, biodiversity has been integrated into the Climate Change Agenda aiming at the promotion of sustainable management and conservation of forests (United Nations Climate Change Conference, 2021). In this perspective, vegetation is fundamental for the conservation and restoration of ecosystems, afforestation, biodiversity protection, natural risks associated with loss of vegetation due to changes in soil use, and the need to reduce the impact of urban expansion, by creating and strengthening urban green space systems.

Squares can be considered a type of green infrastructure. In this chapter they are discussed as essentially public spaces free of buildings, which provide their users with leisure and community coexistence regardless of their size (Cavalheiro et al., 1999; De Angelis, 2000; Robba & Macedo, 2002). Squares are approached from the perspective of geography, landscape architecture, forestry engineering, and biology considering vegetation aspects as indicators of environmental quality. The presence, management, and monitoring of vegetable resources are essential to improve urban ecosystems. We envisage possibilities of promoting a relationship between the landscaping of squares with the native vegetation in the region, thus, restoring and/or reconnecting, at least partly, the native vegetation and privileging the local biodiversity.

This chapter has three main objectives. First, it aims to present an understanding of squares from the perspectives of geography, landscape architecture, forestry engineering, and biology. Second, it highlights the process of insertion and recognition of the role of vegetation in que quality of these public spaces, mainly in Brazil. Third, it seeks to propose a set of environmental indicators for the characterization and assessment of vegetation in squares to contribute to the urban biodiversity conservation.

2 Squares: Conceptual Aspects and Functionalities

Squares can be seen under different approaches depending on the area of knowledge and the perspective of study applied. Then, a deep analysis of these spaces requires a multidisciplinary effort.

In a geographic perspective, squares are understood as urban open spaces, with a public and community character, where social relationships develop along with the interaction society/nature, even if in many cases it is artificialized (Cavalheiro et al., 1999; De Angelis, 2000; Lima et al., 1994). Despite the existence of several concepts, there is one common point between them, squares are first of all places of coexistence and relationships, which currently represent important spaces for the urban population's leisure (Cavalheiro et al., 1999; De Angelis, 2000; Lima et al., 1994; Sousa & Oliveira, 2010).

In the landscape architecture, these spaces are approached from the landscaping perspective, focusing on people, with diversity, accessibility, inclusion, and equality criteria. This presents a challenge in the conditioning of urban spaces for habitability, accompanied by a healthy environment, drawing attention to the environmental services rendered by their biodiversity, and welcoming the people's culture and artistic expression (Gehl, 2017). The professional field with greater visibility in this approach is the design of urban open green spaces grouped into three basic types: streets, squares, and parks (Schjetnan et al., 2008).

In forestry engineering, squares are considered urban microecosystems or spaces of socio-cultural-environmental interrelations. In such spaces, different types of vegetation cover such as native forests, take diversified functions, provide green structure, and offer a range of ecosystem services that help to keep and improve urban life quality.

In the biology perspective, squares represent forested spaces, and their vegetation elements might provide some animals such as birds (Moura et al., 2018) and insects (IPBES, 2016) with "stepping stones" or "landing points". Such features contribute to the maintenance of plant and animal populations (Moura et al., 2018; Santos et al., 2013; Souza et al., 2011) and support urban biodiversity.

Although such concepts were presented by different authors from different areas, functionality is always included. According to Sitte (1992), squares are the public life spaces, able to recover the urban life social dimension. Robba and Macedo (2002) observed that squares are open spaces, mainly for leisure and the population coexistence, and they must be public. In the same view, Demattê (1997) explained that the main function of squares is to incentivize community life and reflect the costumes, beliefs, and other cultural aspects of their users' lives, and they might originate from planning or not. According to Lima (2006, p. 28) the square is a "permanence element, since it is one of the urban facts that persists longer in the urban tissue resisting to transformations, it is a determiner of the city image".

When considering each area's approach, different perspectives were verified. While forestry engineering and biology focus on the presence of vegetation, in geography and architecture, social relationships and the relation between society and the natural environment in squares are emphasized. Therefore, the common point between the knowledge areas considered is the environmental function of squares within the urban space.

Focusing on this relevant function, the common approach developed to assess urban green areas considers its vegetation cover, which resulted in the definition of an urban index establishing the m2 of green area /inhabitant relation. However, green areas are key elements for the performance of social, ecological, environmental, and aesthetic functions, which represent more than a cold number that might not consider the biogeographic, historical, and urban contextual conditions that laid the foundations of their origin, dynamics, and role in cities.

Therefore, facing the problem of public open spaces, that is "squares", their distinct functions and specifically, biodiversity has required the joint work of different disciplines. Their approaches were considered to build a proposal to set a group of environmental quality indicators to assess public urban green areas, and to guide planning, design, and management at site level. The core interdisciplinary topic is vegetation in urban public green spaces, considering the standpoint of geography, landscape architecture, ecology, and forestry management (see Table 16.1).

The disciplines listed above share a common framework that includes geographical and ecological knowledge, different scales of territorial intervention, use and management of vegetation, spatial planning, and geographic information systems. Each discipline contributes to the characterization of the historical context of the squares and their evolution in cities, the significant role of vegetation in the planning

Table 16.1 Approaches contributing to planning and assessing urban green areas

	Disciplinary contributions	Shared knowledge and skills
Geography	(a) Characterization of geophysical, social, and economic processes in the territory (b) Preparation of cartography and processing of satellite images (c) Management of databases through geographic information systems	(1) Knowledge regarding biodiversity, natural ecosystems, and ecosystem services
Landscape architecture	(a) Spatial planning and territorial management of the multiple use of natural resources for development and conservation (landscape units) (b) Planning and management of urban green space systems (c) Design and construction of public green spaces; streets, squares, parks, and private spaces (d) Visual and aesthetic perception of the landscape by the public	(2) Knowledge dealing with the use and management of vegetation; regional, urban, and public green spaces (3) Resource planning for conservation and development purposes
Biology	(a) Knowledge of living beings (b) Biological diversity (c) Ecology and nature conservation (d) Ecosystem services (e) Broad knowledge of plants and biological collections (f) Genetics of the species	(4) Ecological and spatial planning, management plans and programs concerning conservation and restoration of ecosystems in an urban-rural-regional continuum
Forest engineering	(a) Conservation and sustainable management of forest resources for their use in productive and conservation activities (b) Planning and management of forestry (c) Formulation of programs for the prevention and mitigation of forest fires, pests, and diseases (d) Management of conservation units for sustainable use of wildlife	(5) Use of geographic information systems

of urban green spaces, the environmental services it provides, and the proper management they require.

Multidisciplinary study and analyses enable the presentation of attributes of each area, which together can produce a response regarding the improvement of planning and management of these spaces. This consideration justifies the importance of gathering such areas attempting to improve the quality of urban environments, and consequently, the population's quality of life.

3 Vegetation Role in Brazilian Squares

In Brazil, the vegetation in public areas, as well as the Brazilian landscaping history, have received different influences since its discovery, throughout the colonization period, several political periods, and economic phases. According to Macedo (1999), in the colonial period in Brazil (1500 to 1822), the vegetation was confined in backyards and mansion patios of rich people's properties.

Only in the XVIII century, botanical gardens and public gardens were built in several capitals of the country. However, they were not open to the public, and were only used for research on their vegetation. This model remained up to the XIX century, accompanying the formal landscape called eclectic, which characterized the abandonment of colonial patterns. With the appearance of the modern landscaping line, in the XIX century, botanical gardens became public, and in 1870 large squares were created in Rio de Janeiro with several connotations, which sometimes did not prioritize vegetation, but always praised the native flora. Only in the XX century, with the urban expansion, the contemporary project line appeared, in which the vegetation in public areas started to be treated with an ecological view and organized open spaces were created aiming at the preservation of the urban biodiversity.

The use of native vegetation and the characteristic design of Brazilian squares became world-famous with the works by the landscape architect Roberto Burle Max in the XX century. At that time, he created the Brazilian modern garden as a work of art molded by vegetation, which was the main plastic element in the composition, along with water elements, walls, stones, edifications, and sculptures (Carneiro, 2014). During his stay in Recife-Pernambuco (northeastern Brazil), the landscape architect explored the northeastern regional flora to create several theme gardens such as the Casa Forte Square (1934) divided into three parts: one with species of the Brazilian flora, including that of the Atlantic Forest; another with species from Amazonia; and the third, with exotic plants; the Euclides da Cunha Square, with plants from the Brazilian *caatinga*, and the Artur Oscar Square, with typical sandbank plants, due to its proximity to river and ocean waters. The gardens in these three squares outstand in his production due to the themes focusing on the Recife native vegetation (Carneiro, 2014). This connotation adopted by Burle Max, in which the vegetation was more expressive in squares, was disseminated in other projects he implemented all over the country.

Robba and Macedo (2010) pointed out that in the late XX century, some plurality was observed expressed in square projects that sometimes presented stricter design and other times showed the lightness of simple drawings, providing several solutions that explored scenic, symbolic, and sculpture possibilities of the elements built and the vegetation.

In addition to following project lines that characterized a chronological and evolutionary order in the Brazilian landscaping, the square vegetation was increasingly assigned its main role in the aesthetic or visual, ecological, and social aspects.

In such context, Viezzer et al. (2018) studied the vegetation found in the landscaping of squares in Curitiba (southern Brazil) and observed that, in general, there is a large number of species (141) and plant families (58), when studies on flora survey and inventory in urban areas are considered. Although a historical concern of the municipality with urban afforestation was observed, most species were exotic trees in relation to the region original ecosystem. The most common tree found in those squares was the floss silk tree (*Ceiba speciosa* (A. St.-Hill.) Ravenna), followed by species that present great symbology to the region such as the golden trumpet tree (*Handroanthus chrysotrichus* (Mart. ex A. DC.) Mattos), and Brazilian pine (*Araucaria angustifolia* (Bertol.) Kuntze). Azalea (*Rhododendron simsii* Planch.) is notably the shrub species that presents the highest number of individuals. The same study also found squares with up to 40 different species and other with very low diversity, that is, only one species. Those authors pointed out the importance of considering the number of individuals, the diversity of species, and their characteristics, so that the ecological benefits provided by squares can be guaranteed.

The vegetable component, mainly in squares, green area typology that, according to Biondi (2015), can be defined as environments that present several kinds of vegetation, including tree species, demonstrates the existence of a landscaping treatment to meet social, aesthetic, and/or ecological requirements.

As previously suggested, the presence of vegetation in a green area implies several aspects. However, in this chapter we will focus on the urban biodiversity conservation. To achieve such aim, the vegetation must be planned and evaluated through monitoring, which include indicators, parameters, and values to quantify the environmental quality of a space.

As already mentioned, the presence of vegetation in a green area implies several aspects, but in this chapter, we focus on the preservation of urban biodiversity. To achieve this purpose, the vegetation must be planned and evaluated through monitoring, which must include indicators, parameters, and values to quantify its environmental quality.

Although the size of squares is not standardized, these green infrastructures perform several ecosystem services to the urban population, regardless of their size. However, as regards their offer of ecosystem services, there are no systematized and widely accepted procedures to compare the suitability of these spaces, either when the objective is to analyze the neighborhood or different cities during development processes and policies for local and regional urban planning.

Proposals of indicators and analysis parameters have been presented by several Brazilian authors (De Angelis, 2000; De Angelis Neto, 2000; Lima Neto & Souza, 2009; Schallenberger et al., 2010; Fermino et al., 2012; Viezzer, 2015; Oliveira et al., 2019; Bobrowski et al., 2020), aiming to characterize the quality of squares regarding their environmental, structural, social, and cultural factors, but focused on different environmental indicators.

To highlight the environmental relevance of squares as green infrastructures available to the improvement of the urban quality of life, through different ecosystem services, this chapter presents indicators and parameters in a systematized way. These indicators were adapted from different sources that had varied research aims, not necessarily related to the assessment of vegetation in squares, but which still apply to the theme (Chart 16.1). Some of them (Risks of fall and damage, Damage

ENVIRONMENTAL INDICATORS	Class 1	Class 2	Class 3	Class 4	Class 5
Tree density	< 1.25	≥ 1.25 and < 2.50	≥ 2.50 and < 3.75	≥ 3.75 and < 5.00	≥ 5.00
Tree shade	< 0.2	≥ 0.2 and < 0.4	≥ 0.4 and < 0.6	≥ 0.6 and < 0.8	≥ 0.8
Proportion of native species	< 0.2	≥ 0.2 and < 0.4	≥ 0.4 and < 0.6	≥ 0.6 and < 0.8	≥ 0.8
Proportion of native tree species	< 0.2	≥ 0.2 and < 0.4	≥ 0.4 and < 0.6	≥ 0.6 and < 0.8	≥ 0.8
Diversity of tree species	< 1.25	≥ 1.25 and < 2.50	≥ 2.50 and < 3.75	≥ 3.75 and < 5.00	≥ 5.00
Variety of taxon (% trees) [(a)]	S>10+G>20+F>30	S>10	G>20	F>30	S<10+G<20+F<30
Adaptability of species	< 0.2	≥ 0.2 and < 0.4	≥ 0.4 and < 0.6	≥ 0.6 and < 0.8	≥ 0.8
Risk of fall and damage[(b)]	≥ 0.8	≥ 0.6 and < 0.8	≥ 0.4 and < 0.6	≥ 0.2 and < 0.4	< 0.2
Damage caused to structures[(c)]	≥ 0.8	≥ 0.6 and < 0.8	≥ 0.4 and < 0.6	≥ 0.2 and < 0.4	< 0.2
Soil permeability	< 0.2	≥ 0.2 and < 0.4	≥ 0.4 and < 0.6	≥ 0.6 and < 0.8	≥ 0.8
Visual variety of plant composition[(d)]	G or G+F	T	T+F+G	T+S+G	T+S+F+G
Tree diameter distribution [(c)]	NT	NJ	EC	IJ	UN
Horizontal structure	1 Stratum		2 Strata		> 2 Strata
Longevity of most species (>50%)	Low		Medium		High
Size of most species (>50%)	Small		Medium		Large
Growth of most species (>50%)	Fast		Medium		Slow
Functions in the local ecosystem	1 Function		2 Functions		> 2 Functions

[(a)] Verifying whether there is predominance of one or more species (S>10%), genus (G>20%) or family (F>30%) in the composition.
[(b)] Proportion of trees in a condition of high risk, according to the ISA protocol decision matrix (2013)
[(c)] Proportion of trees causing damages to the square structure.
[(d)]Verifying whether there is grass (G), forage (F), shrubs (S) and/or trees (T) in the composition.
[(e)] Verifying characteristics: no-tree class (NT), normal J (JN), equalization between classes (EC), inverted J (JI), unimodal (UN)
[(f)] Verifying condition: low, pioneer species; medium, secondary species; high, climax species.

Chart 16.1 Environmental indicators for the qualification of squares, with their evaluation parameters and respective value classes for comparative analysis between neighborhoods, urban zones, or cities caused to structures, soil permeability, and visual variety of plant composition) have not been proposed and/or used in other studies, thus constituting innovative indices used as tools to promote evaluation process and square management, for considering issues related to vegetable species, local safety, environment qualification, and users' perception. It seems relevant to point out that this is a proposal, that is, it has not been applied yet, and represents a collection of information from several authors that have been working on such perspective. The rationale behind this proposal is that reflection upon the several indices presented altogether might result in a different perspective for the planning and management of squares in different environments.

For each indicator, classes of values associated with each parameter are set to better promote statistical procedures for comparative analysis (descriptive statistics and experimental analysis, among others) between distinct locations or periods of evaluation, allowing a more suitable system of evaluation. The comparison between places with different numbers of squares, or squares of different sizes (common fact between neighborhoods and cities) is possible by using the parameters proposed, given the weighting per unit of area whenever needed.

The indicators presented seek to evidence the characteristics of squares as live and dynamic micro ecosystems within the urban environment, where the existing vegetation, mainly trees, are important elements in the offer of ecosystem services such as those measured in native forest remnants. Therefore, their soil permeability, floristic composition diversity, varied floristic structure, and specific microclimate conditions more suitable than their non-floristic surroundings, make square environments that help to improve the urban quality of life.

The tree density index (TDI) and tree shade index (TSI) indicators show the value of the number of trees and canopy cover in a certain area of the urban space (Chart 16.2), which reveals their capability of generating ecosystem services. The advantage of using such indicators is that they can be applied to direct comparison of places with different sizes, since they weight the results per specific area unit. These indices are determined using the equations proposed by Lima Neto and Souza (2009):

The proportion of native species (PNE), proportion of native trees (PNT) and diversity of tree species (DIV) indices were proposed by Bobrowski, Aguiar and Cuchi (2020) to characterize, in a weighted way, the floristic composition of squares

Tree Density Index (TDI)	Tree Shade Index (TSI)
$TDI = \frac{N}{A} \times 100$	$TSI = \frac{TCA}{A} \times 100$
TDI: tree density (N/m^2) N: total number of trees (N) A: total area (m^2)	TSI: tree shade (m^2/m^2) TCA: total tree crown area (m^2) A: total area (m^2)

Chart 16.2 Equations to determine the tree density and the tree shade indices in squares. (Source: Lima Neto & Souza, 2009)

Proportion of native species (PNE)	Proportion of native trees (PNT)	Diversity of tree species (DIV)
$PNE = \frac{NNS}{NS} \times 1000$	$PAN = \frac{NNT}{NT} \times 1000$	$DIV = \frac{NS}{A} \times 1000$
NNS: number of native species NS: total number of species	NNT: number of native species trees NT: total number of trees	NS: total number of species A: total area

Chart 16.3 Equations used to determine the proportion of species and native trees, and diversity of species in the composition of squares. (Source: Bobrowski et al., 2020)

(Chart 16.3). When analyzing the floristic composition diversity, the richness of native species is assessed, but the number of trees of these species should be of chief interest, since diversity values might be hidden in a composition with exotic species or invasive exotic ones. The equations presented prevent problems with divergent size of squares in comparative analyses, which might lead to error in the estimates, mainly when diversity indices are used without procedures for rarefaction of values (Magurran, 2011).

The selection of species able to tolerate urban conditions is an essential requirement for a healthy green structure, but it might lead to the globalization of few species and genus (Fini et al., 2017). For this reason, Santamour Junior (1990), recommends that in the urban area no more than 10% of any species should be used, no more than 20% of any genus, and not over 30% of any family, to maximize the protection against outbreaks of pests. Such general recommendation has been widely used in studies on the urban vegetation diversity composition and originated the taxon variety index shown in Chart 16.1.

There is still a lack of proper diversity limits for the composition of tree diversity in squares or for the proportion of taxon, mainly when the adaptability of species to urban stress conditions must be considered (Raupp et al., 2006). Therefore, the limits recommended by Santamour Junior (1990), even if arbitrary, serve to prevent problems with floristic simplification that compromises the visual variety and sustainability of the floristic composition over time. The determination of the proportion of taxon is carried out simply by the relation between the number of individuals of certain species and the total number of trees; the number of individuals of certain genus and the total number of trees; and the number of individuals of certain plant family and the total number of trees.

The indicators of species' adaptability, risk of fall and damage (RFD), and damage caused to structures (DCS) measure aspects of local safety, since trees that are not properly managed or are not well adapted to the place might present structural defects and favor risky situations, affecting the existing hardscape (access, pathways, benches, lamp posts, fountains, among others) and users.

The species' adaptability to a certain place, among other possibilities of analysis, including physiological evaluations, might be determined by the observation of the general condition of trees and determination of the species performance index (Bobrowski et al., 2016). The general condition of the trees can be assessed by observing their general aspect and placing them in one of the five classes:

1. Very bad: when over 50% of the trees are dead or show expressive damage/problems in trunk and crown (dieback, topping, rotten branches and twigs, broken branches).
2. Bad: when between 25–50% of the trees are dead or show expressive problems in trunk and crown (dieback, topping, rotten branches and twigs, broken branches).
3. Regular: When less than 25% of the trees are dead or show expressive damage/problems in trunk and crown (dieback, topping, rotten branches and twigs, broken branches).
4. Satisfactory: when the trees present irrelevant or small damage/problems, which are manageable by applying different pruning techniques or cabling/bracing branches and trunk.
5. Good: when the trees do not show damage or structural problems and there are no dead trees.

From the complete evaluation of the general condition of all trees, the species performance index (SPI) can be determined using the following equation (Bobrowski et al., 2016):

$$SPI = \frac{\left(\dfrac{naB + naS}{n}\right) \times 100}{\left(\dfrac{NaB + NaS}{N}\right) \times 100}$$

Where: naB = number of trees of that species "and" that are in the Good class; naS = number of trees of that species "and" that are in the Satisfactory class; n = number of trees of that species; NaB = total number of sampled trees in the Good class; NaS = total number of sampled trees in the Good class; N = total number of sampled trees in the study.

To classify the information regarding the species adaptability (Chart 16.1) the proportion of species presenting SPI higher or equal to 1.0 is analyzed. This SPI reference value indicates the species success or better adaptation to the place, when compared to the other species, evidenced by the better general condition of trees in the floristic composition analyzed.

The tree risk of fall and damage (RFD) and damage caused to structures (DCS) indicators are determined in proportional terms between the number of damaged trees, trees with problems or causing damage in relation to the total number of trees in the square (Chart 16.1). However, the analysis of tree fall, and damage risk must be carried out following a specific protocol for this detailed activity, by visual analysis, and employing the guidance protocol by the International Society of Arboriculture (ISA, 2013).

The soil permeability indicator (SPE) is related to the capability of these spaces to contribute to the reduction in surface runoff, keeping the infiltration of rainwater in the soil, which is quite important in cities of tropical regions, where heavy rains result in flooding. This indicator can be determined by the relation between the square unsealed area and the total area.

The visual variety of plant composition indicator characterizes general aesthetic aspects of the floristic composition, since it values de variety of types of vegetation and their interaction. It aims at promoting an attractive environment to the users, and at the same time, diversifying the type of vegetation able to retain pollutants and reduce noise pressure levels. It is always pleasant to find a wide variety of plants in the vegetable composition of squares, not only from the aesthetic standpoint, but also to generate an environment with several ecological opportunities and functions. Viezzer et al. (2018) studied the vegetation of squares in Curitiba-Parana-Brazil and found mainly tree species (60.3%) in their composition, followed by shrubs (27.7%), herbal plants (10.6%) and climbing plants (1.4%). The determination of this indicator, in its different parameters, is carried out by observing the variety of vegetation found in the floristic survey of each square.

The tree diameter distribution indicator is determined from the analysis of the distribution of the DBH classes of the trees, in at least five distribution classes with a 0.05 m amplitude each, after the floristic survey of each square. The way the DBH distribution curve is characterized helps to better manage these green areas. After the tree size distribution in the classes is established, it enables the planning of new planting actions and substitution of old trees or those with higher risk associated to their presence in a square (Bobrowski et al., 2020).

The remaining indicators are related to the dimension of the arboreal elements, their size, growth, and longevity, which are factors that regulate the offer of ecosystem services and the dynamics over time. The determination of such indicators, and the setting of parameters for each square occurs through the observation of the floristic composition structure, via floristic survey, identification of species and by seeking specific-species information in the specialized literature describing the biological and ecological characteristics of the sampled species.

When considering only the tree canopy cover (horizontal structure) of a square, the existence of other species located in lower strata, for example, might be ignored. However, those species can also generate several ecosystem services that are directly linked to the human wellbeing. Although Yang et al. (2009) considered the horizontal structure important to the analysis of a green area, it can be an extremely simplistic approach, because it might neglect some functions of the vegetation. Another aspect to be taken into consideration is that it might not represent what people really see around them at the ground level, missing the details of a three-dimensional landscape.

Most Brazilian cities do not present suitable infrastructure to accommodate large trees along sidewalks, however, squares can compensate the lack of street-side space. This justifies the fact that cities need large trees, since they can be eight times more beneficial than medium and small trees, even when they become older (USDA, 2014).

Therefore, the use of large trees might multiply the final results expected from ecosystem services promoted by the urban forest. Larger and older trees, with greater crown area and leaf surface store more carbon and offer more economic benefits as well as other ecosystem services than smaller trees (Cavender & Donnely, 2019). The tree size indicator might also influence other indicators such as the tree diameter distribution, tree density, species longevity and tree shade.

The longevity of species indicator is correlated to the permanence of the tree in the square for as long as possible, implying greater local ecosystem benefits. Cavender and Donnely (2019) suggested that urban environments should prioritize tree protection, ameliorating the selection and diversity of species, improving planning, patterns, and care with the trees. In addition, older trees should be privileged as well as the promotion of the longevity of trees. A suitable distribution of age or longevity of trees allows better planning of the vegetation senescence aiming to the removal and sequential substitution of dead or dying trees (Calaza et al., 2018).

Regarding environmental functions provided by squares, each local can be characterized according to the variety of functions through observation and sampling such as microclimate regulation (Viezzer et al., 2015) and fauna maintenance (Pacheco & Vasconcelos, 2007), among other services discussed below.

4 Square Ecosystem Services

Squares represent important natural aspects such as totally or partially permeable soil and the presence of vegetation cover. Considering their similarity with urban parks and municipal natural parks (Dobbs et al., 2011; Gaudereto et al., 2019), squares contribute to the ecosystem service supply (Mea, 2005). Thus, the measurement and monitoring of ecosystem services (ES) might help urban planning professionals and public managers to assess the quality of these open spaces.

Ecosystem services are the goods and services that nature offers to society (Mea, 2005; Balvanera et al., 2011). They are usually divided into supply services (food, fiber, water), support services (soil preservation, primary yield), regulation services (climate and microclimate, hydric resources, disease control) and cultural services (leisure, recreation, contemplation, spirituality), (Mea, 2005).ES can be noticed or evaluated in squares by using indicators, in which each indicator or group of indicators is associated to the supply of a certain ecosystem service.

Soil permeability (Chart 16.1) is related to the presence of vegetation and the whole piece of land occupied by vegetable species providing two relevant regulation ecosystem services, since they protect the soil against erosion processes and regulate the hydrological cycle by allowing water infiltration and preventing floods (Dobbs et al., 2011; Gaudereto et al., 2019).

The **Shade** (Chart 16.1) provided by treetops can reduce temperature and regulate the local microclimate (Alam et al., 2016; Rocha et al., 2015). Studies have shown (Martini et al., 2018) that the existence of a tree cover tends to promote significant temperature reduction. This service is extremely important, mainly in large cities, where the effects of global climate changes are added to thermal inversion phenomena, or even in small and medium-sized cities, where climate change effects are already noticeable. A study carried out in Sorocaba, interior of São Paulo State

(Barbosa et al., 2020) found up to 2 ° C lower temperature inside a group of trees, when compared to the open area are without vegetation. Frota Junior et al. (2018), studying squares of relatively small size (between 0.44 and 2.02 hectares) in Fortaleza/CE Brazil, recorded a significant decrease in soil temperature under trees in squares and under direct solar radiation, thus demonstrating that even small squares can contribute to improving the microclimate in cities. Barros (2012) reported that "in areas of urban space with low percentages of vegetated area (less than 20%), the fragmentation of this vegetation acted more decisively in the thermal environments, that is, it did not reduce the temperature. On the other hand, less fragmented vegetation resulted in reduced air temperature (Barros, 2012).

Arboreal density (Chart 16.1) is related to two regulation ecosystem services, soil protection (Queiroz, 2018) and sound regulation (Lundh, 2017). Therefore, the denser the arboreal cover is, the more efficient these services tend to be. Regarding soil protection, when a square with tree density closer to 1 and another over 5 are compared, certainly the soil in the area with higher tree density is better protected against rainfall and wind effects. Another important effect is the protection against noise, since a tree barrier might reduce noise in up to 20 decibels, as shown by (Barbosa et al., 2020) in a study developed in Sorocaba/SP.

Tree diametric distribution (Chart 16.1) is a measurement of the arboreal biomass and is related to the global climate regulation service (Franco et al., 2013), since the larger the size/diameter of trees and their density are, the higher the carbon sequestration from the atmosphere is, which contributes to the measures employed to tackle climate changes (Franco et al., 2013).

Presence (proportion) of native species (Chart 16.1) and **diversity of arboreal species** (Chart 16.1) refer to the supply of several ecosystem services. **Native species** might produce fruit (provision service), which can feed people who use the squares as well as the local fauna. Species such as cashew (*Anacardium occidentalle*), oiti (*Licania tomentosa*), and jenipapo (*Genipa americana*) are highly consumed in Brazil and can be found in squares (Santos et al., 2013; Souza et al., 2011) and in street-side trees. However, fruit producing exotic species should also be considered, since the "urban ecosystem" is already highly modified and some exotic species can also supply this important provision service such as mango (*Mangifera indica*), avocado (*Persea americana*), and blackberry (*Morus nigra*).

Native vegetable species (Chart 16.1) might be a source of food or shelter to fauna species (birds, small rodents, bats) and these animals might act as biological control agents (Gaudereto et al., 2019) against pests in the urban areas or even in farm crops around the cities (regulation service). This service tends to be more relevant when the diversity of tree species as well as shrubs and herbs is higher. Native vegetable species also provide birds with landing points (Moura et al., 2018). Moreover, the sound produced by the birds can be considered a cultural ecosystem service (Chan et al., 2012), related to leisure promotion—birdwatching—and human wellbeing (Lundh, 2017; Mace et al., 2012), since exposure to the sound produced by the birds promotes a feeling of wellbeing and belonging (Lundh, 2017, Mace et al., 2012).

Species diversity (Chart 16.1), mainly as regards native species, tends to increase plant relationship with insects and other animals that promote plant pollination. Bee species that relate to native vegetable species (which might be found in squares) provide farm species with an important pollination service (regulation service) (IPBES, 2016).

In addition to all these services, **native species** can also produce flowers and/or eye-catching fruit, which promotes a cultural service (Chan et al., 2012) "beauty" or "an aesthetic sense" (Eurich, 2014; Lundh, 2017; Queiroz, 2018). This service is highly appreciated in Brazilian squares, since there are species that blossom and produce fruit throughout the four seasons of the year, which makes it possible to plan squares so that they present flowers and fruit of varied sizes and colors all over the year. "*Ipês*" are widely used species in Brazilian squares, due to the beauty of their flowers such as the "*ipê-amarelo*" (*Handroanthus chrysotrichus), "ipê-roxo" (Handroanthus heptaphyllus),* and the "*ipê-branco*" (*Tabebuia roseoalba*) among others, along with several species of the family *Fabaceae* such as "*chuva-de-ouro*" (*Cassia ferruginea*) or the "*falso-barbatimão*" (*Cassia leptophylla*).

However, it should be noted that for squares to effectively contribute to the provision of ecosystem services, they must be connected to other vegetated spaces, as they are generally relatively small to maintain populations of native species (birds, butterflies, including tree species) or even to increase urban thermal comfort.

Melo et al. (2021), studying birds in urban areas, reported that the size of green patches (habitat structure and street trees) favored bird richness and that the reduction in species richness was positively correlated with distance from source areas (large native forests).

Campos-Silva and Piratelli (2020) and Angeoletto et al. (2017) emphasized that urban green spaces such as squares, parks, gardens, and reserves, usually created mainly for recreational purposes, are also manageable and can represent refugees for biodiversity.

In this way, squares, especially those with dense tree cover (Chart 16.1) and preferably well distributed within the urban matrix, can function as landscape connection elements, connecting larger areas of native vegetation, generally existing in rural or peri-urban areas, functioning as biodiversity refuges, and providing other ecosystem services such as climate amelioration within the gray infrastructure.

5 Final Considerations

Squares as public spaces have appeared since the beginning of their history as an urban reference marked by human interaction. They are, therefore, a relevant historical and cultural element in the urban space and are found in several cities, mainly in Brazil.

Green areas, mainly squares, play an important role in the urban mosaic, since they constitute spaces embedded in the urban system whose ecological conditions are very close to the natural ones

We understand that the theoretical and methodological contributions proposed through the parameter indicators in the geographical science, landscape architecture, biology and forestry engineering perspectives might guide the evaluation of the availability of the existing vegetation and serve as a tool for diagnosis of squares according to the functions they perform, in addition to setting indicators directly linked to the environmental quality evaluation in these public spaces.

The applicability of such indicators might enable the continuous planning of a policy for this sector, since managers who use this information might acquire a clear view of the situation of the squares in a municipality. This will enable them to create implementation or maintenance projects that meet the community interests and those of the public power. This proposal can be applied in places with different numbers and different sizes of squares, since the indicators proposed consider the weighting per unit of area whenever possible.

Thus, these open spaces in the city can be seen not only as isolated green fragments or patches in the urban design, but also as spaces where the citizens can spend their leisure time or simply enjoy the environment and be in contact with nature. This can only be guaranteed through the environmental sustainability of urban spaces that are properly planned.

References

Alam, M., Dupras, J., & Messier, C. (2016). A framework towards a composite indicator for urban ecosystem services. *Ecological Indicators, 60*, 38–44, Elsevier BV. https://doi.org/10.1016/j.ecolind.2015.05.035

Angeoletto, F., Sanz, J. P. R., Albertin, R., & Silva, F. F. D. (2017). The grass is always greener on the other side of the fence: The flora in urban backyards of different social classes. *Ambiente & Sociedade, 20*, 01–20.

Angeoletto, F., Fellowes, M. D., & Santos, J. W. (2018). Counting Brazil's urban trees will help make Brazil's urban trees count. *Journal of Forestry, 116*(5), 489–490.

Angeoletto, F., Tryjanowski, P., Santos, J., Martinez-Miranzo, B., Leandro, D., Bohrer, J., et al. (2022). Will Brazilian City dwellers actively engage in urban conservation? A case study with the charismatic Neotropical blue-and-yellow macaw (Ara ararauna). *Birds, 3*(2), 234–244.

Angeoletto, F. H. S., Richard, E., Santos, J., Guedes, N., Oliveira, S. L., Albertín, R. M., et al. (2023). When we think of urban afforestation in Brazil, what comes to mind is a mouth that has lost its teeth. Terr@. *The Pluralist, 17*, e2321664–e2321654.

Balvanera, P., Castillo, A., Chavero, E. L., Caballero, K., Quijas, S., Flores, A., & Mass, M. (2011). Marcos conceptuales interdisciplinarios para el estudio de los servicios ecosistémicos en América Latina. El valor ecológico, social y económico de los servicios ecosistémicos. Conceptos, herramientas y estudio de casos. In P. Laterra, E. Jobbágy, & J. Paruelo (Eds.), *Valoración de Servicios Ecosistémicos. Conceptos, Herramientas y Aplicaciones para el Ordenamiento Territorial* (pp. 39–67). Ediciones INTA. Capítulo 2.

Barbosa, V. L. M., Giacon, V. P., & Cardoso-Leite, E. (2020). *Protocolo Para Avaliação de Serviços Ecossistêmicos Oriundos de Áreas Verdes Urbanas*. Relatório de Iniciação Científica, CNPQ-UFSCAR campus Sorocaba.

Barros, M. P. (2012). *Dimensão fractal e ilhas de calor urbanas: uma abordagem sistêmica sobre as implicações entre a fragmentação das áreas verdes e o ambiente térmico do espaço urbano*. Tese (Programa de Pós-Graduação em Física Ambiental), Universidade Federal de Mato Grosso.

Biondi, D. (2015). Floresta urbana: conceitos e terminologias. In ______. *Floresta Urbana* (pp. 11–27). A Autora.

Bobrowski, R., Ferreira, R. L. C., & Biondi, D. (2016). Descrição fitossociológica da arborização de ruas por meio de diferentes formas de expressão da dominância e da densidade. *Ciência Florestal, 26*, 1167–1178. https://doi.org/10.5902/198050987550

Bobrowski, R., Aguiar, J. T., & Cuchi, T. (2020). How to qualify the vegetation in public squares to help the management of urban ecosystem services? *Ciência e Natura, 42*(e80), 1–14.

Calaza, P., Cariñanos, P., Escobedo, F. J., Schwab, J., & Tovar, G. (2018). Building green infrastructure and urban landscapes. *Unasylva, 69*(1), 11–21.

Campos-Silva, L. A., & Piratelli, A. J. (2020). Vegetation structure drives taxonomic diversity and functional traits of birds in urban private native forest fragments. *Urban Ecosystems*. https://doi.org/10.1007/s11252-020-01045-8

Carneiro, A. R. S. (2014). Burle Marx e os jardins do Recife. *Revista Espaço Acadêmico, 156*, 45–59.

Cavalheiro, F., Nucci, J. C., Guzzo, P., & Rocha, Y. T. (1999). Proposição de terminologia para o verde urbano. *Boletim Informativo. Sociedade Brasileira de Arborização Urbana.*, 1–3.

Cavender, N., & Donnelly, G. (2019). Intersecting urban forestry and botanical gardens to address big challenges for healthier trees, people, and cities. *Plants, People, Planet, 1*, 315–322.

Chan, K. M. A., Satterfield, T., & Goldstein, J. C. (2012). Rethinking ecosystem services to better address and navigate cultural values. *Ecological Economics, 74*, 8–18.

De Angelis, B. L. D. (2000). *A praça no contexto das cidades: o caso de Maringá PR. 2000, 367 f.* Tese de Doutorado (Geografia)—Faculdade de Filosofia, Letras e Ciência Humanas, Universidade de São Paulo.

De Angelis, B. L. D., & De Angelis Neto, G. (2000). A vegetação e as praças na cidade de Maringá/PR. *Revista Acta Scientiarum, 1455*, 1461–1422. https://doi.org/10.4025/actascitechnol.v22i0.3104

de Queiroz, D. A. H. O. (2018). *Avaliação da qualidade das áreas verdes urbanas: Uma abordagem com base nas suas funções*. Tese (Doutorado em Geografia) - Universidade Estadual de Ponta Grossa.

Demattê, M. E. S. P. (1997). *Princípios de paisagismo*. FUNEP.

Dobbs, C., Escobedo, F. J., & Zipperer, W. C. (2011). A framework for developing urban forest ecosystem services and goods indicators. *Landscape and Urban Planning, [s.l.], 99*(3-4), 196–206. Elsevier BV.

Eurich, Z.R.S. (2014). As praças da cidade de Ponta Grossa—PR: Arborização Urbana, infraestrutura e distribuição espacial. Dissertação (Mestrado em gestão do território—Geografia)—Universidade Estadual de Ponta Grossa. Ponta Grossa.

Fellowes, M. D., Acquaah-Harrison, K., Angeoletto, F., Santos, J. W., da Silva Leandro, D., Rocha, E. A., et al. (2020). Map-a-mole: Greenspace area influences the presence and abundance of the European mole Talpa europaea in urban habitats. *Animals, 10*(6), 1097.

Fermino, R. C., Reis, R. S., & Cassou, A. C. (2012). Fatores individuais e ambientais associados ao uso de parques e praças por adultos de Curitiba-PR. *Brasil. Rev. bras. cineantropom. desempenho hum., 14*(4) Florianópolis.

Fini, A., Frangi, P., Moria, J., Donzelli, D., & Ferrini, F. (2017). Nature based solutions to mitigate soil sealing in urban areas: Results from a 4-year study comparing permeable, porous, and impermeable pavements. *Environmental Research, 156*, 443–454.

Franco, M.A.R., Osse, V.C., & Minks, V. (2013). Infraestrutura Verde Para As Mudanças Climáticas No C40. Revista LABVERDE n°6—Artigo n°11, junho.

Frota Junior, J. I., Jesuino, I. T., & Martins, M. C. (2018). Análise do modelo de arborização em praças públicas do Município de Fortaleza (Estado do Ceará, Nordeste do Brasil) e sua influência no conforto térmico e estruturas urbanas. *Revista Brasileira de Gestão Ambiental e Sustentabilidade., 5*(11), 883–896. ISSN 2359-1412. https://doi.org/10.21438/rbgas.051107

Gaudereto, G. L., Gallardo, A. L. C. F., Ferreira, M. L., Nascimento, A. P. B., & Mantovani, W. (2019). Avaliação de serviços ecossistêmicos na gestão de áreas verdes urbanas: promovendo cidades saudáveis e sustentáveis. *Ambient. soc., São Paulo, 21*, e01203.

Gehl, J. (2017). La dimensión humana en el espacio público, recomendaciones para el análisis y el diseño. In *Ministerio de Vivienda y Urbanismo Gobierno de Chile y PNUD*. Versión octubre 2017. ISBN: 978-956-9432-70-5.
IPBES. (2016). In S. G. Potts, V. L. Imperatriz-Fonseca, & H. T. Ngo (Eds.), *The assessment report of the intergovernmental science-policy platform on biodiversity and ecosystem services on pollinators, pollination and food production*. Secretariat of the Intergovernmental Science-Policy Platform on Biodiversity and Ecosystem Services. 552 pages.
ISA—International Society of Arboriculture. (2013). *Tree risk assessment manual*. ISA.
Lima, T. H. S. (2006). *O cotidiano nas praças assisenses: uma análise qualiquantitativa. 2006, 108 f. Dissertação de Mestrado (Geografia)*. Universidade Estadual de Maringá. Maringá.
Lima Neto, E. M., & Souza, R. M. (2009). Índices de densidade e sombreamento arbóreo em áreas verdes públicas de Aracajú. *Sergipe. Revsbau, 4*(4), 47–62.
Lima, A. M. L. P., Cavalheiro, F., Nucci, J. C., Sousa, M. A. L. B., Fialho, N. O., & Del Pichia, P. C. D. (1994). Problemas de utilização na conceituação de termos como espaços livres, áreas verdes e correlatos. In *II—Congresso Brasileiro De Arborização Urbana*. (pp. 539–353). Anais.
Lima, G. F., Correa Santos, J. W. M., Albertin, R. M., Martínez-Miranzo, B., Souza, F. L., & Angeoletto, F. (2023). Backyards are a way to promote environmental justice and biodiversity conservation in Brazilian cities. *Diversity, 15*(7), 815.
Lundh, J. (2017). *Indicators for ecosystem services in urban green space management*. Department of Urban and Rural Development, division of landscape architecture—Swedish University of Agriculture Sciences, Ulls väg 27, SE-750 07.
Mace, G. M., Norris, K., & Fitter, A. H. (2012). Biodiversity and ecosystem services: A multilayered relationship. *Trends in Ecology and Evolution, January, 27*(4).
Macedo, S. S. (1999). *Quadro do Paisagismo no Brasil* (p. 144). Silvio Macedo.
Magurran, A. E. (2011). *Medindo a diversidade biológica*. Editora UFPR.
Martini, A., Biondi, D., & Batista, A. C. (2018). *The influence of the different types of urban Forest in the immediate surrounding microclimate*. Ciência Florestal, . 28 (3):997-1007. ISSN 1980-5098 997. https://doi.org/10.5902/1980509833381
MEA—Millennium Ecosystem Assessment. (2005). *Ecosystems and human well-being: Synthesis* (p. 2005). Island Press, Copyright, World Resources Institute.
Melo, M. A., Sanches, P. M., Silva Filho, D. F., & Piratelli, A. J. (2021). Influence of habitat type and distance from source area on bird taxonomic and functional diversity in a Neotropical megacity. *Urban Ecosystems*. https://doi.org/10.1007/s11252-021-01169-5
Moroń, D., Beim, M., Gudowska, A., Angeoletto, F., Celary, W., Cwajna, A., ... & Tryjanowski, P. (2024). Evaluating tramway infrastructure on biodiversity and ecosystem services. *Scientific Reports, 14*(1), 9394.
Moura, G. W., Moura, A. S., & Machado, F. S. (2018). Diversidade de aves em praças de cidades do Triângulo Mineiro: riqueza, similaridade e aspectos biológicos. *Natureza Online, 16*(1), 031–038.
Oliveira, J. D., Scipioni, M. C., Reis, A. R. N., & Ximenes, E. (2019). Diagnóstico da arborização da Praça Centenário, Curitibanos, Santa Catarina, Brasil. *Acta Biológica Catarinense, 6*(3), 23–36.
Pacheco, R. P., & Vasconcelos, H. L. (2007). Invertebrate conservation in urban areas: Ants in the Brazilian Cerrado. *Landscape and Urban Planning, 81*(3), 193–199.
Raupp, M. J., Cumming, A. B., & Raupp, E. C. (2006). Street tree diversity in eastern North America and its potential for tree loss to exotic borers. *Arboriculture & Urban Forestry, 32*(6), 297–304.
Robba, F., & Macedo, S. S. (2002). *Praças Brasileiras*. EDUSP.
Robba, F., & Macedo, S. S. (2010). *Praças Brasileiras*. 3ª ed. São Paulo: EDUSP.
Rocha, S. M., Zulian, G., Maes, J., & Thijssen, M. (2015). *Mapping and assessment of urban ecosystems and their services*. EUR 27706 EN; Joint Research Centre Technical Reports. https://doi.org/10.2788/638737

Rumble, H., Angeoletto, F., Connop, S., Goddard, M. A., & Nash, C. (2019). Understanding and applying ecological principles in cities. In *Planning cities with nature* (pp. 217–234). Springer.

Santamour Júnior, F. S. (1990). *Trees for urban planting: Diversity uniformity, and common sense* (pp. 57–66). Proceedings of the 7th Metria Conference Lisle.

Santangelo, J. S., Ness, R. W., Cohan, B., Fitzpatrick, C. R., Innes, S. G., Koch, S., et al. (2022). Global urban environmental change drives adaptation in white clover. *Science, 375*(6586), 1275–1281.

Santos, A. F., José, A. C., & Sousa, P. A. (2013). Fitossociologia e diversidade de espécies arbóreas das Praças Centrais do Município de Gurupi-TO. REVSBAU. *Piracicaba—SP, 8*(4), 36–46.

Schallenberger, L. S., Araujo, A. J., Araujo, M. N., Deiner, L. J., & Machado, G. O. (2010). Avaliação da condição de árvores urbanas nos principais parques e praças do município de Irati- PR. *REVSBAU., 5*(2), 105–123.

Schjetnan, M., Peniche, M., & Calvillo, J. (2008). *Principios de Diseño Urbano/Ambiental, 2ª*. Edición. Limusa. ISBN: 9786075000015.

Sitte, C. (1992). *A construção de cidades segundo princípios artísticos*. Trad. Ricardo Ferreira Henrique. Editora Ática.

Sousa, R. O., & Oliveira, C. E. (2010). *A praça Como lugar da diversidade cultural* (pp. 1–11). IV Fórum de Educação e Diversidade. Tangará da Serra. Anais.

Souza, A. L., Ferreira, R. A., Mello, A. A., Plácido, D. R., Santos, C. Z. A., Graça, D. A. S., Almeida Júnior, P. P., Barretto, S. S. B., Dantas, J. D. M., De Paula, J. W. A., Silva, T. L., & Gomes, L. P. S. (2011). Diagnóstico Quantitativo E Qualitativo Da Arborização Das Praças De Aracaju, Se. *Revista Árvore, Viçosa-MG, 35*(6), 1253–1263.

Souza, F. L., Fenoglio, M. S., & Angeoletto, F. (2023). To be a Brazilian City dweller, sometimes we must learn to say enough! *Sustainability, 15*(4), 3699.

UN-HABITAT—United Nations Programe for Human Settlements (2020). Nueva Agenda Urbana Ilustrada. Retrived from UN-HABITAT: 12-14-2021. https://publicacionesonuhabitat.org/onu-habitatmexico/Nueva-Agenda-Urbana-Ilustrada.pdf

United Nations. (2015). La Asamblea General adopta la Agenda 2030 para el Desarrollo Sostenible. Objetivos de Desarrollo Sostenible. Retrived from United Nations 12-14-2021: https://www.un.org/sustainabledevelopment/es/2015/09/la-asamblea-general-adopta-la-agenda-2030-para-el-desarrollo-sostenible/

United Nations Climate Change Conference 2021 (02.11.2021). Glasgow Leaders' Declaration on Forests and Land Use. Retrived from United Nations Climate Change: 12-14-2021: https://ukcop26.org/glasgow-leaders-declaration-on-forests-and-land-use/

USDA—United States Department of Agriculture. The large tree argument—the case for large-stature trees vs. small-stature trees. 2014. Available at: https://www.fs.fed.us/psw/topics/urban_forestry/products/cufr_511_large_tree_argument.pdf. Accessed on 27 Apr 2021.

Viezzer, J. (2015). A floresta urbana e o papel das praças. In D. Biondi (Ed.), *Floresta Urbana* (pp. 81–107). O Autor.

Viezzer, J., Biondi, D., Martini, A., & Silva, D. A. (2015). O benefício microclimático proporcionado pela Praça Alfredo Andersen na cidade de Curitiba-PR. *Ciência e Natura, 37*, 138–143. Ed. Especial SIC.

Viezzer, J., Biondi, D., Martini, A., & Grise, M. M. (2018). A vegetação no paisagismo das praças de Curitiba—PR. *Ciência Florestal, Santa Maria, 28*(1), 369–383.

Yang, J., Zhao, L., Mcbride, J., & Gong, P. (2009). Can you see green? Assessing the visibility of urban forests in cities. *Landscape and Urban Planning, 91*(2), 97–104.

Chapter 17
Patch and Landscape Metrics Applied to Urban Green Space Analysis Aiming to Detect Habitat Availability to Birds and Bats in Ponta Grossa, a Medium-Sized City in Brazil

Rosemeri Segecin Moro, Evandro Retamero Rodrigues, and Rodrigo Fernando Moro

Abstract The challenge faced by urban managers is not only to keep many small, isolated green spaces functional but to ensure that species can move freely between them when surrounded by a landscape highly modified by man. To validate that such areas were ecologically meaningful, we compared bat and bird species composition and richness throughout the urban-wildland ecotone to an intra-urban core of a medium-sized Neotropical city in southern Brazil. We applied patch and landscape metrics in the GIS interface to vegetation polygons considering composition parameters related to area, shape, and connectivity. We selected the metrics that best fit in Fragstats 4.2 to assess the landscape's ability to support habitats for regional birds and bats diversity. We evaluated squares, woodlands, riparian forests, and wooded streets that comprise only 13% of the Ponta Grossa urban green spaces. Previous studies registered 61 species of birds that represent circa 38% of the regional avian diversity, and 8 species of bats (50% of bat regional diversity). In the three major woodland areas previously, there are 142 shrub and tree taxa (diversity H′ ranging from 3.68 to 3.87) whereas in 656 wooded streets there are 89 tree species planted, circa 68% of no natives. We considered that squares, riparian forests, and wooded streets hardly could be functional corridors for most species, however, woodlands present significant areas and are next enough to connect structurally as functionally generalist bird and bat species. This framework probably affects the long-term persistence of some bird and bat communities. Therefore, to increase the

R. S. Moro (✉)
Universidade Estadual de Ponta Grossa, Ponta Grossa, PR, Brazil
e-mail: rsmoro@uepg.br

E. R. Rodrigues
Prefeitura Municipal de Ortigueira, Ortigueira, PR, Brazil

R. F. Moro
HCVRN Environment Consultant, Curitiba, PR, Brazil

F. Angeoletto et al. (eds.), *Ecology of Tropical Cities*,
https://doi.org/10.1007/978-3-031-48012-6_17

ability of these urban green spaces to preserve biodiversity, efforts should be made to enhance the landscape functional connectivity by, for instance, increasing the area covered by wooded streets ideally with native species, and restoring the riparian forests.

Keywords Urban ecology · Urban-wildland ecotone · Urban green spaces · Intra-urban · Neotropics

1 Introduction

Urban environmental quality is related to the balance between existing life forms, including people, fauna, and flora species. The anthropization of spaces impacts natural areas to varying degrees depending on the original conditions of ecosystems and the intensity, extent, and timing of human occupation (Grimm et al., 2000). Equally important are the incident abiotic conditions, the degree of local homeostasis, and the resilience of species that live there or visit only frequently. This is especially noticeable in the Neotropical region, an area that has great biodiversity and endemism (Gentry, 1982; MacGregor-Fors, 2010; Boesing et al., 2013; Lozano, 2018; Catenazzi & von May, 2021; Angeoletto et al., 2022).

Urbanization in general results in abrupt changes in the diversity and distribution of various species and affects the quality and quantity of ecosystem services available to the inhabitants of urban areas (Chace et al., 2006; Szlavecz et al., 2011; Seto et al., 2012).

Different methodologies have been applied to analyze the urban setting quality. According to Ahern (2002) and Szlavecz et al. (2011), landscape ecology principles based on the theories of island biogeography and metapopulations are fundamental to evaluating and planning networked green spaces. The space between urban and adjacent wild areas represents an ecological transition that often acts as a semi-permeable biological filter (Puga-Caballero et al., 2014). Out of the five key points developed by Ahern (2002) in designing greenways concepts, we approach two: linearity (comprising streams, roads, streets, and avenues) and connectivity (potential for connecting areas such as squares and parks). In the Matrix/Fragment/Corridor landscape model, the distance between remnants and corridors would be key structural elements in most landscapes to the connectivity possibilities for urban-wild communities (Turner et al., 2001).

2 Urban Green Spaces and the Fragmentation Process

Swanwick et al. (2003) divided urban spaces into built environments (gray spaces) and vegetation environments (green spaces). According to Singh et al. (2010), urban green spaces include all urban parks, remnant forests, and vegetation that, concerning structural diversity, should essentially develop as networks. In Brazilian medium cities, the set of urban green areas includes mainly riparian areas (for permanent preservation purposes), squares, parks, forested lots, and institutional meadows and gardens (MMA, 2021; Lima et al., 2023). The term "urban green spaces" is used in this chapter as a restricted term comprising urban parks, riparian forests, squares, and street trees in the intra-urban area.

The multifunctional character associated with the performance of green areas in urban space covers the preservation of biodiversity, food supply, regulation of the hydrological cycle, promotion of adaptive response to climate change, and the proximity of the population to nature. Thus, urban green spaces are usually categorized and analyzed according to their use by the community and the characteristics of their surroundings (Uy & Nakagoshi, 2007; Park et al., 2014; Lu, 2015); however, a more complete understanding should focus more on habitats than the matrix that contains them.

According to Forman (1995), the process of landscape fragmentation leads to the formation of a mosaic, with a structure composed of patches. In the urban setting, green spaces constitute homogeneous patches within the urban mesh, with a significant role in composing the environment and providing environmental, social, and cultural services (Band et al., 2005). Assessing spatial dynamics at the landscape scale is relevant to the long-term persistence of various animal and plant species in fragmented landscapes and should be considered in planning and management for their preservation (Bastin & Thomas, 1999).

Selecting vegetation cover types and treating them in separate classes is important for understanding the dynamics and provision of ecosystem services at a patch scale (Band et al., 2005; Grafius et al., 2018). According to Rastandeh et al. (2017), the composition and configuration of vegetation patches should be considered a major vehicle for the discipline of landscape architecture to respond to urban biodiversity loss. Spatial patterning of urban vegetation patches is thought to be the key to providing relatively suitable conditions for urban wildlife to remain in cities and support ecosystem services despite the ongoing pressures on wildlife habitats imposed by urban development and climate change.

Numerous studies demonstrate how important large patches of connected vegetation are for increasing native bird diversity in Neotropical cities (Carbó-Ramírez & Zuria, 2011; MacGregor-Fors & Escobar-Ibáñez, 2017).

Among the available tools, landscape metrics that were initially developed for natural areas have been increasingly used to analyze spatial patterns and ecological processes involved in the urban setting (Band et al., 2005). They allow the identification of patterns of green space and their connectivity, the existence of strategic fragments for the preservation of a certain area, and the improvement of biodiversity, among others. Landscape assessment becomes important for the diagnosis of critical areas and the creation of models and strategies for preservation and conservation, seeking to maintain the natural balance between the green, blue, and gray systems of a city.

Urban landscape metrics have been used mainly in countries with temperate climates (Sacco et al., 2013; Szlavecz et al., 2011), in areas with centuries of urbanization and highly modified environments. In Brazil, there are still a few experiences such as Penteado and Alvarez (2007), Castro and Ferreira (2009), Dantas et al. (2017), Barros (2018), Ferreira et al. (2012), Magalhães and Moura (2013), Fonseca et al. (2016), Rocha et al. (2016), and Silva et al. (2019), all located in tropical regions and employing mainly shape and neighborhood indexes without any fauna or flora focus group.

Queiroz and Carvalho (2019), employing qualitative parameters, evaluated the environmental quality of green spaces in the municipality of Ponta Grossa as very low. Our objective in this study is to assess habitat availability in this medium-sized subtropical city in southern Brazil, inserted in the Atlantic Forest hotspot (Mittermeier et al., 2004), with a 200-year history of urban occupation and very intense expansion in recent decades.

3 Methods

3.1 Study Area

Ponta Grossa (25°50′58″S; 50°09′30″W) is located in the Campos Gerais region, Paraná state, southern Brazil (Fig. 17.1). The climate Cfb Köppen's is humid subtropical with mild summers and rainfall evenly distributed throughout the year (1700 mm/yr.; 75–80% of medium atmospheric humidity). The average temperature is 18 °C (22 °C max- 12 °C min), and frosts are frequent in winter (Cruz, 2007). The relief is dissected, with urban occupation on flat tops and slopes of 12–20%, and alluvial surfaces in the valley bottoms. The drainage system includes 12 watersheds comprising over 150 km of streams along the valley bottoms and floodplains, with four main creeks: Olarias, Pilão de Pedra, Ronda, and Madureira.

Ponta Grossa is an important road and rail junction in South Brazil and thus caused an amoeboid growth. The central area and 15 neighborhoods occupy 1172 Km^2, with 355,336 inhabitants (IBGE, 2021). As in other Neotropical cities, several low-income housing projects occupy the suburbs, whereas we could see the progressive verticalization in the center towards the peri-urban areas, surrounded by a mosaic of agricultural land, natural grasslands, and woodlands (Rodrigues et al., 2016; Ortega-Álvarez & MacGregor-Fors, 2011; Puga-Caballero et al., 2014). The street system is mainly orthogonal, with several physical barriers due to the hilly terrain and a complex drainage system. The older streets downtown have narrow

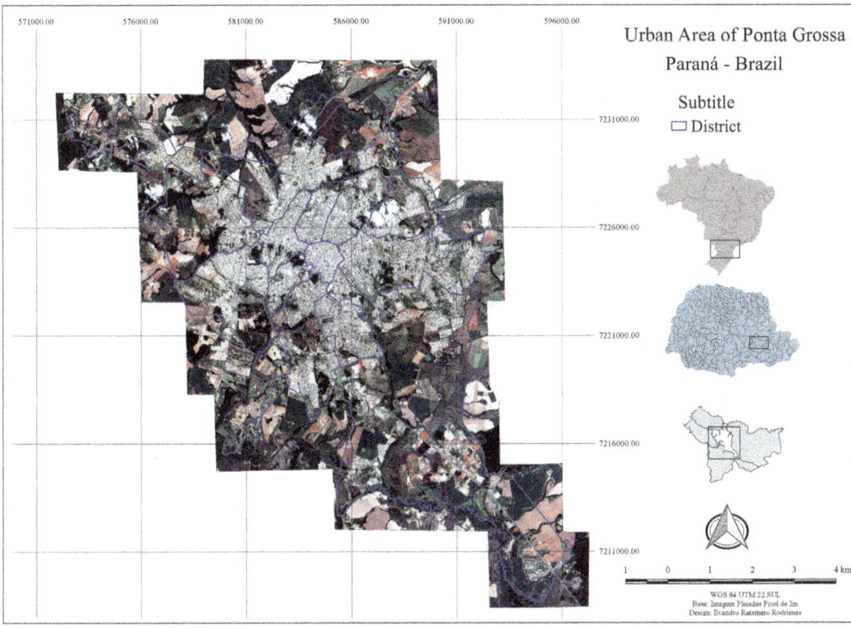

Fig. 17.1 Urban area of Ponta Grossa, Paraná, Brazil

sidewalks not well suited for tree planting and according to IBGE (2021), only 57% of the streets have trees planted.

According to the procedure at MacGregor-Fors (2010) the city was regionalized into intra-urban and peri-urban areas. This one occupies 1.62 Km^2 around the intra-urban area. A third category was created by defining a 1000-meter radius in the intra-urban, a core area, starting from the popular commercial center of the city, from which the intra-urban/peri-urban gradient spread (Fig. 17.2).

The landscape study was developed in four different urban green spaces: woodlands, squares, riverside areas, and wooded streets (Figs. 17.3 and 17.4).

To compose the classes of the Ponta Grossa green system we used photointerpretation in the GIS environment with QGIS and ArcGIS software and subsequent application in Fragstats 4.2 (McGarigal, 2014) to calculate landscape metrics. We used vector datasets of (i) the urban perimeter squares with above 70% of vegetation coverture from existing mapping at a scale of 1:10,000 (Eurich, 2018); (ii) the riparian areas along a 30 m buffer from IPLAN (Ponta Grossa Urban Planning Institute) with raster files from 2013 Pleiades images provided by the Socio-Environmental Studies Laboratory (LAESA) of the State University of Ponta Grossa with 50 cm spatial resolution for the panchromatic band and 2 m for multispectral bands, both fused in Saga Gis 5 to improve spatial resolution. To settlement of the effective tree presence in riparian areas we classified polygons provided by the NDVI analyses and further vectorization; (iii) the same Pleiades images were used by Tadenuma (2019) to settle the street trees upon the total urban grid. To calculate the area

Fig. 17.2 Delimitation of the intra-urban core area (1000 m radius) and part of the intra-urban area with the most significant green spaces (the peri-urban area is not shown). (1) Madureira creek; (2) Getúlio Vargas square; (3) Boca da Ronda Park; (4) Ronda creek; (5) Do Padre creek; (6) Marguerita Masini Park; (7) Barão do Guaraúna Square; (8) Barão do Rio Branco Square; (9) Floriano Peixoto Square (City's ground zero); (10) Ambiental Park; (11) Olarias creek; (12) Universidade creek; (13) Pilão de Pedra creek. (Source: Google Earth Pro 2021)

occupied by street-line trees, we used a canopy projection of a 3 m radius for each individual; and (iv) vector files of urban woodlands mapped by Queiroz and Carvalho (2019) upon 2013 Google Earth images.

3.2 *Vegetation*

Woodlands are represented by two public parks located in the city center that have native tree cover from a mixed ombrophilous forest in a secondary stage of succession. According to Moro et al. (2022), the vegetation of Marguerita Masini Park, with 88 tree species, presented a Shannon index of 3.87. The dominant plant species are *Bauhinia forficata* and *Anadenanthera colubrina* (both Fabaceae), and there was an expressive occurrence of non-native species such as *Hovenia dulcis* (Rhamnaceae),

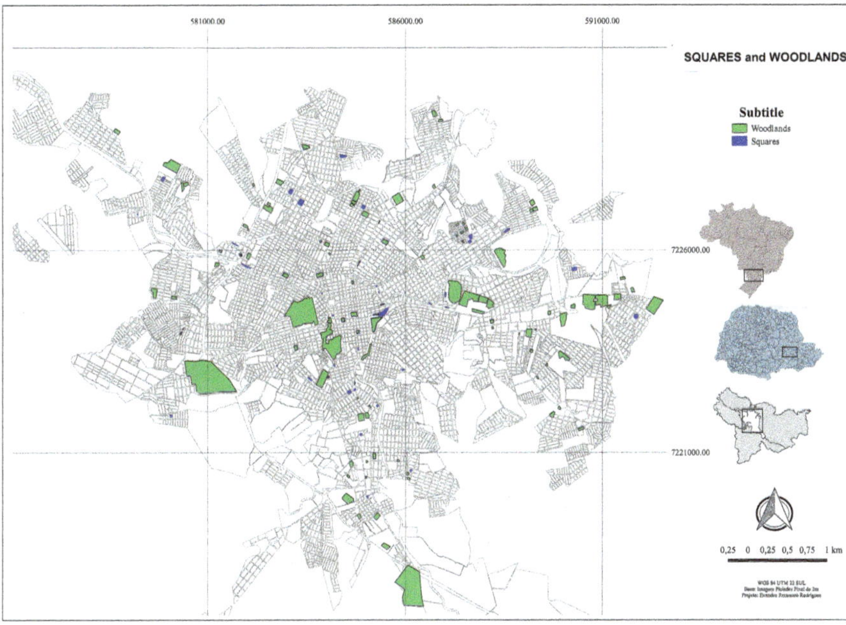

Fig. 17.3 Woodlands and squares in the urban area of Ponta Grossa

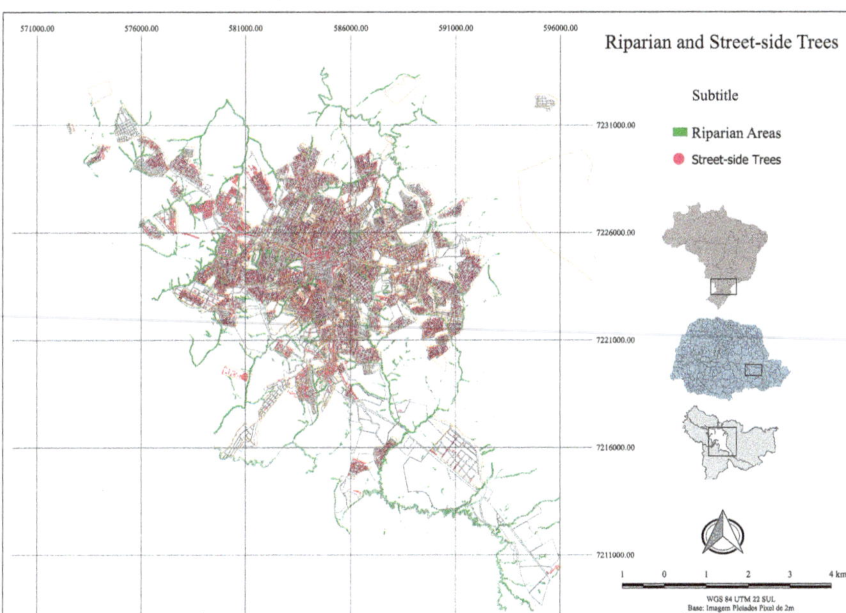

Fig. 17.4 Network of riverside and wooded streets in the urban area of Ponta Grossa

Melia azedarach (Meliaceae), and the ornamental herbaceous *Impatiens walleriana* (Balsaminaceae). To Moro et al. (2022), tree vegetation of Boca da Ronda Park includes 72 species, with a Shannon index of 3.8. The dominant plant species are as follows: *Jacaranda micrantha* (Bignoniaceae), *Myrcia rostrata* (Myrtaceae), *Cinnamomum sellowianum* (Lauraceae), *Casearia sylvestris* (Salicaceae), *Alchornea triplinervia* (Euphorbiaceae), and *Prunus myrtifolia* (Rosaceae). This park also contains the no-native species *Hovenia* and *Melia*, as well as *Pinus taeda*. Honório Park, the third studied urban park, is a private small-forested lot also in the central area. Rocha et al. (2019) pointed to 43 tree species with a Shannon index of 3.7. There are larger-diameter climax species as *Ocotea porosa*, *Ocotea pulchella*, *Ocotea acutifolia*, and *Guarea kunthiana*, in disclimax with small pioneer species such as *Sapium glandulosum*, *Bauhinia forficata*, *Coussaea contracta*, and *Casearia sylvesris*, besides non-native species typical of previously farming-occupied areas such as *Citrus lemon*, *Persea americana*, and *Bixa orellana*.

The public squares, sometimes largely paved, were small green spaces containing a mixture of lawns, ornamental shrubs, and trees, some non-native, located around and between monuments, playgrounds, and other facilities. They were characterized by size between 0.2 and 1.0 ha and a square, sometimes triangular shape. The smaller ones showed little vegetation cover, as they had to compete for space with the existing public facilities (Eurich et al., 2014).

Tree vegetation in riparian areas, when present, showed the same elements seen in woodlands, at a considerably lower Shannon diversity of 2.28 to 2.78 (Diedrichs & Moro, 2000; Moro et al., 2001). The best-preserved areas in the central area were the heads of the creeks Do Padre, Ronda, and Pilão de Pedra.

Finally, Tadenuma (2019) recorded 89 tree species on 656 streets in the urban area of Ponta Grossa, of which only 33.7% were individuals of native species. Of the no native species, *Lagerstroemia indica* predominated, with 21.7%, and *Ligustrum lucidum*, with 14.7%.

3.3 Bats and Birds

To bats, data were obtained previously by Zanon and Reis (2007) in four sampling points, including Boca da Ronda Park, in the intra-urban core. Other points were Olaria Zuk (in the intra-urban area, point 11 at Fig. 17.3), and UTFPR University boundaries (in the peri-urban area, not seen in Fig. 17.3), besides a rural control area 38 Km far from the urban area. The occurrence of nutritional requirements for bats followed Almeida et al. (2005), and Zanon and Reis (2007).

For Birds, data were obtained previously by Moro et al. (2022). The following data were gathered according to Anjos and Graf (1993): occurrence, (a) foraging guild (diet-carnivore, insectivores, frugivore-nectarivore, granivore or omnivore, (b) urban tolerance (urban user species or urban dweller), and (c) native vs. non-native or invasive species.

3.4 Patch and Landscape Metrics

In urban settlements, the impact of spatial configuration on ecological functions may not be very clear due to the complexity and heterogeneity of the fine-scale (Alberti, 2005; Holt et al., 2015). In general, studies tend to address metrics of size, shape, and neighborhood type to determine ecosystem quality (Tratalos et al., 2007; Norton et al., 2016), and for measuring similar, often statistically correlated aspects, some metrics are inherently redundant (McGarigal, 2014).

The selected patch metrics were Number of Patches (NP), Patch area (AREA), LPI (area occupied by the largest fragment), Core Area (CORE), Shape Index (SHAPE), Fractal Dimension (FRAC), Aggregation Index (AI), Interspersion-juxtaposition Index (IJI), and Euclidean Nearest-Neighbor Distance (ENN). The selected landscape metrics were Diversity (SHDI), Uniformity (SHEI), Connectivity (CONNECT), and Proximity (PROX).

CORE corresponded to the measurement of the central area of the patches from the definition of a 50 m buffer, a distance based on classic studies of edges in zoology and botany for the Atlantic Forest (Fontoura et al., 2006). Edge density is defined as the perimeter of a patch of a given land cover type exposed to other land cover types. The larger the value of the core area, the better the quality of the landscape, as it corresponds to larger vegetation blocks that are less exposed to the effects of a highly contrasting edge.

SHAPE measures the complexity of the fragment shape compared to a circle. The further away the fragment values are from 1, the more clipped and complex the fragments are. The fractal dimension (FRAC) measures the complexity of the fragment shapes, so landscapes with simpler, more regular shapes approximate a fractal dimension of (1) while landscapes with more complex shapes show an index closer to (2) Such an index refers to the ability of fragments to occupy space either in an organized and methodical way or distributed in an unsystematic way.

The aggregation metrics AI and IJI represent different ways of assessing the dispersion and interspersion of fragments, that is, the degree of spatial mixing of different types of fragments. AI might vary from 0, i.e., no aggregated fragment, to 100, when there is a single compact class fragment.

The measure of diversity, SHDI, is influenced by two components, richness (the number of different patch types present) and evenness (their distribution and abundance). Thus, rare patch types show a disproportionally large influence on the magnitude of this index. The landscape diversity index, SHDI, does not show a threshold of values—the higher the value, the greater the richness and/or equality in the distribution of areas occupied by the four landscape classes considered.

The uniformity index, SHEI, ranges from 0 to 1, and the upper limit is reached when there is a homogeneous distribution of classes in the landscape, while the aggregation index analyzes contiguous classes in the landscape as a quantitative measure of the levels of aggregation in the spatial pattern of the landscape (He et al., 2000). It ranges from 0 to 1, where zero indicates the lowest level of aggregation, i.e., a landscape formed by larger blocks composed of closer habitats.

After the definition of the target species within the same guild and affinity represented by urban birds and bats (Findley, 1993), the calculated indices were proximity, PROX, nearest neighbor, ENN, and connectivity, CONNECT, with a 1000 meters radius search.

PROX is equivalent to the area of the polygon divided by the sum of the edge-to-edge distances between the polygon in question and all neighbors within a search radius. It considers the size and distance of all patches and evaluates the landscape situated between two opposite conditions: sparse distribution of small patches, or complex aggregations of large patches. Therefore, it considers not only isolation in the sense of minimum distance to the nearest patch (measured by ENN), but also discriminates its spatiality. The value is zero when no neighbors are in the search radius or when the sum of available areas is small and increases as larger areas become closer and more contiguous in the search radius (Lang & Blaschke, 2009). However, this index is not sensitive to any barriers in the matrix, being only a theoretical estimate of the available habitat, and because it is a dimensional index, it only makes sense when compared across areas.

CONNECT measures the ratio of functional junctions between all patches to the total number of possible junctions, depending on the user-defined distance criterion. It starts from zero when the class consists of a few connections and increases as parts of the class are connected.

4 Results and Discussion

4.1 Distribution of Birds

While the urban setting can be stressful for some species due to pollution, habitat fragmentation, etc., others may thrive in the cities because humans create favorable microhabitats or abundant resources for them (Szlavecz et al., 2011). Of the 84 wild species reported by Anjos and Graf (1993) to the Ponta Grossa grasslands and forests (Fig. 17.5), in our study 60 native species were also present in the urban area (71% of the regional pool). Among them, 31 were seen just in the peri-urban area, 18 in the peri-urban as well in the intra-urban area, and just 11 were seen only in the intra-urban core area (18% of the total were seen in the city, and 13% of the regional pool).

Olegário et al. (2020) observed 69 species in a peri-urban forested area in Ponta Grossa, about 5 km away from the intra-urban core. In that 15.5 ha patch at the Federal Technological University of Paraná (UTFPR), researchers found 31 species that were not found elsewhere in the city, which shows the importance of proximity to native areas for urban avifauna diversity. In much more wooded cities like Maringá and Curitiba, Philippsen et al. (2010) recorded 74 species and Anjos (1990) found 88, respectively.

Fig. 17.5 Interior of the urban parks: (**a**) Marguerita Masini; (**b**) Boca da Ronda; (**c**) natural landscape surrounding the municipality of Ponta Grossa, PR, Campos Gerais do Paraná region. (Photos R.S. Moro)

The 18% of species living exclusively in the intra-urban were considered urban dwellers sense Fischer et al. (2015) because they depend on the urban setting to complete their life cycle. This no includes the two observed invasive alien species – sparrow and house pigeon, which agrees with Bastin and Thomas (1999), to whom more than 95% of the birds found in cities are native to the surrounding region.

Among the dwellers in the intra-urban core of Ponta Grossa, however, four are not indicators of altered habitats – *Cacicus haemorrhous* – Red-rumped Cacique, *Zonotrichia capensis* – Rufous-collared Sparrow), *Cyanocorax chrysops*

(Plush-crested Jay), and *Phacellodomus ruber* (Greater Thornbird). Plush-crested Jay and Greater Thornbird are species that never enter the urban grid, seen only inside the Boca da Ronda Park.

The remaining 82% of species living both in the intra-urban and peri-urban areas were considered urban utilizers, i.e. species that could maintain viable populations outside the urban area. To Anjos and Laroca (1989), 58% of the birds observed in Curitiba (South Brazil) were sporadic and Franchin's review (Franchin, 2009) pointed out that about 30% of native birds in Brazil live in both natural and urban areas. Avifauna studies in municipalities that experienced intense urban densification but still maintained large green areas showed that these sites lost fewer species at the local scale (Concepción et al., 2016; Gagné & Fahrig, 2010). Thus, the impact of urban densification on bird communities depends on the scale assessment, whether local (intra-urban) or regional (Haaland & van den Bosch, 2015; Faggi & Perepelizin, 2006).

The urban grid of Ponta Grossa was seen to filter out about 87% of the species that occur in the more preserved areas around the city. These are mainly little birds that, besides their short flight range, probably do not venture more than a few meters into open areas. Others are species that forage along with stream forests but do not approach the intra-urban core through riparian areas due to stream pollution or poor coverture (Fig. 17.6). Therefore, for these species, the urban landscape is disconnected and both stream banks and street-side trees do not constitute functional corridors or stepping stones.

Fig. 17.6 Some urban creeks in Ponta Grossa: (**a**) Pilão de Pedra creek; (**b**) Madureira creek; (**c**) nameless creek on the UTFPR campus; (**d**) Do Padre creek; (**e**) general view of the riparian zone of Olarias creek. (Photos R.S. Moro)

4.2 Distribution of Bats

As for the bats, of the 16 species known in the region, Zanon and Reis (2007) identified 8, from which the generalist frugivorous species *Sturnira lilium* and *Artibeus lituratus* are present both in the intra-urban and peri-urban areas. These opportunistic and/or generalist species show the ability to survive in both natural and disturbed areas, and examination of their diet revealed that they preferentially feed on native species found in the city woods, as well as on eight species of *Solanum* (shrub and small trees), *Piper gaudichaudianum* (forb), and *Ficus enormis* (tall trees). But they also obtain food from courtyards, which include non-native cultivated species such as *Eriobotrya japonica* (Rosaceae) and *Hovenia dulcis* (Rhamnaceae), and even from fruits found on palm trees in the street-side afforestation such as *Syagrus romanzoffianum* (Almeida et al., 2005). According to Szlavecz et al. (2011), the simplistic view that "aliens are always bad" needs revision. Five insectivorous species, *Myotis nigricans, Eptesicus brasiliensise, Eumops auripendulus*, Tadarida *brasiliensis*, and *Histiotus veatus* were present in the intra-urban (Olaria Zuk) but not in the core area. In the peri-urban, it was present the hematophagous *Desmodus rotundus.* Of the 8 avoider bat species, *Myotis ruber* is a rare species whose ecology is not well known, two are hematophagous, two are insectivore, two are nectivourous (rarely frugivorous), while only *Pygoderma bilabiatum* is strictly frugivorous, but rarely found in urban areas.

Due to its urbanization context that began in the mid-1800s, the city's narrow streets are crowded with residential and commercial buildings without an adequate front line and narrow sidewalks. Therefore, the urban matrix does not maintain significant vegetation cover along the streets and avenues (Queiroz & Carvalho, 2019), and the vegetation cover in vacant lots and courtyards progressively decreases when the intra-urban core area is approached. This constitutes a serious obstacle to urban environmental health.

However, some woodlands with representative areas remain, most notably the two large parks in the intra-urban core, which are both structurally connected to the urban creeks of Ronda and Do Padre. The vegetation in these areas is used by several bird and bat species as a complementary habitat and resource source (White et al., 2005). However, trophic barriers exist for non-frugivorous bats, as in the intra-urban core the few insects could not attract nocturnal visitors from the richest peri-urban areas. Similarly, the inexistence of cows and horses in these regions makes them unattractive to hematophagous bats. According to Queiroz and Carvalho (2019), 55% of the street trees in the Ponta Grossa intra-urban zone are not native, and Eurich et al. (2014) pointed out that this rate reaches 74% in the intra-urban core. Considering that the local climate does not present a very marked seasonality, with an abundant supply of nectar throughout most of the year in the surrounding native forests, there is little pressure on these animals to enter urban areas.

For Leveau (2013), the diversity of native vegetation within parks had a strong positive influence on the richness and abundance of avian and insectivorous bird species. Thus, a progressive loss of native vegetation diversity in parks could affect the availability of feeding and breeding habitats for certain urban birds, regardless of the size of these green spaces (van Heezik et al., 2008). Therefore, an increase in the coverage of urban green areas and their connectivity will not necessarily favor the diversity of birds in these spaces unless it is accompanied by a diversification of vegetation in these areas.

4.3 Biological Meaning of the Metrics

The total area of green spaces was 23,320 km^2, corresponding to only 13.53% of the total urban area (172,410 Km^2). This vegetation distributed across the four analysis classes was seen to be fragmented into 4850 patches of different sizes, shapes, and susceptibility levels (Table 17.1, Figs. 17.3 and 17.4).

The urban area of Ponta Grossa showed 0.61 landscape diversity—SHDI, and 0.44 uniformity – SHEI. Unfortunately, these indices are strongly affected by rarer patches, such as the 172 squares and 76 woodlands (below 3% of the total area). The aggregation index 59.2 provided a more accurate answer to point out a landscape where part of the green spaces is quite aggregated, concentrated on the margins of the many streams that cross the city, as well as following the street system.

The fractal dimension and the shape index expressed the systematic reproduction of the same shapes in the classes: riparian areas and linear street-side tree system, with width regulated by law; squares and woods in almost isodiametric polygons, reflecting the distribution of spaces in octagonal blocks.

The interspersion and juxtaposition index observed in the maximum possible overlap for one of the fragment types of the classes was 64.2, demonstrating that most of the woodlands were connected to riparian areas, while the squares and the street-side trees followed the urban planning and did not overlap.

After discounting a core area of 4140 Km^2, the remaining 19,180 Km^2 (82%) are fully exposed to the edge effect. These areas could not provide conditions to harbor a genetically viable population of several species most sensitive to environmental changes (Picket & Thompson, 1978), however, along with Do Padre creek, a

Table 17.1 Metrics of green space classes in the urban landscape of Ponta Grossa

	Woodlands (polygons)	Squares (polygons)	Riparian areas (polygons)	Street-side trees (individuals)
NP	76	172	4602	12,407
TA (Km^2)	4.14 (2.4%)	0.57 (0.33%)	18.21 (10.6%)	0.35 (0.20%)
Average area (m^2)	3978.30	4638.71	3956.97	28.27
SHAPE	1.28 (average)	1.55 (average)	1.02–17.19 (range)	1.51 (mean canopy projection)
Average FRAC	1.46	1.55	1.08	1.51

NP number of patches; *TA* total area; *SHAPE* shape index; *FRAC* fractal dimension synurbic population of *Aramides saracura* was observed. These birds usually leave the riparian area very early in the morning to forage in neighboring backyards. A few years ago, a *Puma concolor*, a medium-sized wild cat, was going up Pilão de Pedra creek and was chased by some dogs and cornered in the basement of a house. It was an extreme case due to large fires in the fields outside the city in a very dry winter, but a good example of what could occur. Other riparian areas, however, due to their level of degradation, cannot even function as corridors.

The largest core urban green spaces are found at the headwaters of the Ronda, Pilão de Pedra, and Olarias creeks, including the Boca da Ronda and Marguerita Masini parks. Within the peri-urban areas, the largest patches were connected to the Cará Cará river, the Fort Sant'Ana Military Base), the campus of three universities, and the Country Club Lagoa. The smallest patches are scattered throughout the territory and quite disconnected from the intra-urban spaces, along with the riparian areas of the various streams that cross the urban area, in the tree-covered streets. The largest fragments (LPI) were two private blocks of wood of 0.693 and 0.4691km^2 respectively, followed by the public parks Boca da Ronda (0.39 Km^2) and Marguerita Másini (0.22 Km^2).

Within a radius of 1000 meters, the landscape proved not to be isolated for bats and generalists birds, with an ENN of 127.53 meters and a slightly larger PROX, 275.28, as these distances could be covered by several flying species in this study. Since the PROX index is sensitive to small areas and patch aggregation, low values were expected. The CONNECT value of 1.9 revealed that in the radius of 1000 meters there is a very small proportion of potentially connectable areas to all sites. By aggregation, the isolation of the patches is greater than the apparent linear proximity of neighboring patches would suggest, as indicated by the indices that also assess their spatial distribution. Thus, only species able to cross moderate distances through the matrix are favored. But considering the average distance covered by generalist bats and birds as well as wind-spread seeds, pollen, and spores, most of the extended patch areas overlap vast regions of the matrix as well as some of the closest patches.

Despite a large number of small fragments, they possibly play a relevant role in the biological function as stepping stones for the avifauna, since they show a low level of complexity in their shape and are close to their neighbors.

The minimum size of green areas to maintain biodiversity in the urban area is still a controversial topic. Large patches of vegetation are likely to contain more diverse communities, thus providing more widespread ecosystem services. Forman and Godron (1986) suggested that the size of patches present in a landscape affects productivity, nutrient and water flow, and species dynamics and thus can be considered an important indicator of biodiversity. Picket and Thompson (1978) proposed the concept of minimal dynamic areas under the natural disturbance regime, which maintains sources of internal recolonization and minimizes extinction. However, a very similar species-area relationship has been found in managed and unmanaged urban spaces (Ferenc et al., 2014; Nielsen et al., 2014), and few studies have been conducted to define them in Neotropical regions (Boesing et al., 2013).

The 76 urban forests totalized 4.13 Km^2, and the two largest areas of native vegetation located in the central area covered 9.4% of the total area – the Park Marguerita Masini is a public space of about 12 ha, and the Park Boca da Ronda, in a massif between public and private areas occupies about 27 ha. Only the Marguerita Masini Park used to be open to visitation and leisure, presenting a basic infrastructure of trails. However, it has been closed for the last fifteen years due to public security problems involving the surrounding community. Boca da Ronda has no infrastructure and has never been open to the community. The remaining parks in the city, namely, Ambiental and Monteiro, were implemented with the repopulation of native species (Ambiental) and non-native species (Monteiro).

The traditional concept that spatial connectivity between vegetation patches can facilitate the colonization process and thus reduce the chance of extinction of plant species does not fully apply to urban settings, since urban barriers could select generalists and even no native species such as those observed in this sample. Still, the concept of urban and regional greenway planning (Ahern, 2002; Angeoletto et al., 2023; Souza et al., 2023) should be pursued to improve local biodiversity.

5 Final Remarks and Conclusions

The few green spaces found in Ponta Grossa, as for most Neotropical cities, are markedly controlled by the norms of use regulation, so the squares and most of the woodlands are constrained within polygons planned for building occupation, with no awareness of the necessary ecological structuring.

The larger patches correspond to woodlands and, even if not many, show significant areas and account for most of the vegetation cover in the urban area since they are close enough to connect significant bird and bat populations. However, the species found are generalist due to the poor quality of the fragments and corridors, which strongly selects species adapted to environments subjected to edge effects and able to pass through ecological steppingstones. The evaluation of the edge effect does not depend only on the calculation of a determined buffer but should take into consideration aspects such as relief and density, size, and quality of the vegetation.

The squares and street-side afforestation follow the grid already implemented but are not continuous nor possibly offer enough or varied food resources to the fauna. So, they are not sufficient in qualitative and quantitative terms to become ecological corridors for most avifauna. Furthermore, areas with scattered trees, such as squares, may not have any functional core area.

The wide hydrological network shows high structural connectivity, however, functional connectivity is only observed for less demanding foraging species. The observed riparian areas are very poorly vegetated in general and there is a tendency to edifications occupy up to the last meter allowed by the law – many riparian strips have less than the prescribed 15 meters. When larger areas are preserved it was usually due to restrictions imposed by slopes. Most urban riparian forests also require an extensive program of river and stream cleaning and enrichment of the tree vegetation.

Yang et al. (2019), based on the abundance and attractiveness to pollinators of several native herbaceous plants, suggest that urban planners should also consider preserving and increasing even the smallest green areas within the city. Thus, their proper planning with a view to better insertion and linear grouping can improve connectivity among fragments and stepping-stones. However, Chollet et al. (2018) advised that these habitats are currently poor refuges for plant and animal communities due to their intensive management. Moreover, we note that preserving or restoring green spaces beyond those strictly protected by applicable law is not a priority of public policies, even if verified a deficit of riparian areas and possibly available space for increased both riverside and street afforestation.

Urban landscape metrics that aim to assess habitat availability must be supported by empirical data from some habitat component to be correctly interpreted and provide results robust enough to guide urban planning to promote a set of ecological services based on the improved landscape structure of the different vegetation cover classes in the urban area. As for increasing habitat availability, in Ponta Grossa, there is no possibility of expanding public green areas with the expropriation of private areas, so it would hardly alter the current situation. Recent authorizations and contradictory court rulings have pointed out the weakness of public management in Brazilian cities regarding the prevention of the suppression of urban forest habitats.

Sites located outside the intra-urban core had more species richness of birds and bats. So, we could conclude that the urban area of Ponta Grossa represents a geographical barrier for wildlife species, once we verified a smaller bird species-richness within the intra-urban area, represented possibly by a subset of regional species pools in the urban-wildland ecotone (peri-urban area). This urban setting represents a semi-permeable filter for birds and bats, with their assemblages representing urban utilizers ranging from making occasional use of urban resources to breeding in developed areas whereas urban dwellers vary from having viable populations in both natural and developed areas to being entirely dependent on developed areas for survival.

References

Ahern, J. F. (2002). *Greenways as strategic landscape planning: Theory and application.* Wageningen University Press.

Alberti, M. (2005). The effects of urban patterns on ecosystem function. *International Regional Science Review, 28*, 168–192. https://doi.org/10.1177/0160017605275160

Almeida, C. G., Moro, R. S., & Zanon, C. M. V. (2005). Feeding habits of two species of frugivorous bats (Chiroptera, Phyllostomidae) from urban disturbed Forest patches in Ponta Grossa, PR, southern Brazil. *Publicatio UEPG, 11*(3/4), 15–21. https://doi.org/10.5212/Publ.Biologicas.v.11i3.0002

Angeoletto, F., Tryjanowski, P., Santos, J., Martinez-Miranzo, B., Leandro, D., Bohrer, J., et al. (2022). Will Brazilian city dwellers actively engage in urban conservation? A case study with the charismatic neotropical Blue-and-yellow macaw (Ara ararauna). *Birds, 3*(2), 234–244.

Angeoletto, F. H. S., Richard, E., Santos, J., Guedes, N., Oliveira, S. L., Albertín, R. M., et al. (2023). When we think of urban afforestation in Brazil, what comes to mind is a mouth that has lost its teeth. *Terr@Plural, 17*, e2321654.

Anjos, L. (1990). Avian distribution in an Araucaria Forest of Curitiba City (South Brazil). *Acta Biologica Paranaense, 19*(1,2,3,4), 51–63.

Anjos, L., & Graf, V. (1993). Riqueza de aves da fazenda Santa Rita, município de Palmeira, Estado do Paraná. *Revista Brasileira de Zoologia, 10*(4), 673–693. https://doi.org/10.1590/S0101-81751993000400013

Anjos, L., & Laroca, S. (1989). Abundância relativa e diversidade específica em duas comunidades urbanas de aves em Curitiba (Sul do Brasil). *Arquivos de Biologia e Tecnologia, 32*(4), 637–643.

Band, L. E., Cadenasso, M. L., Grimmond, C. S., Grove, J. M., & Pickett, S. T. A. (2005). Heterogeneity in urban ecosystems: Patterns and process. In G. M. Lovett, C. G. Jones, M. G. Turner, & K. C. Weathers (Eds.), *Ecosystem function in heterogeneous landscapes* (pp. 257–278). Springer Science + Business.

Barros, M. P. (2018). Métricas da paisagem como ferramenta de avaliação dos efeitos da urbanização na estrutura da paisagem: o caso de Cuiabá, Brasil. *Revista de Gestão & Sustentabilidade Ambiental, 7*(4), 411–432. https://doi.org/10.19177/rgsa.v7e42018411-432

Bastin, L., & Thomas, C. D. (1999). The distribution of plant species in urban vegetation fragments. *Landscape Ecology, 14*, 493–507. https://doi.org/10.1023/A:1008036207944

Boesing, A. L., Araujo, P. C., Santos-Junior, T. Z., & Anjos, L. (2013). Brasil. In I. MacGregor-Fors & R. Ortega-Álvarez (Eds.), *Ecología urbana: experiencias en América Latina* (pp. 26–43). Inecol. http://www1.inecol.edu.mx/libro_ecologia_urbana/ecologia_urbana_experiencias_en_america_latina.pdf. Accessed 10 Jan 2021.

Carbó-Ramírez, P., & Zuria, I. (2011). The value of small urban greenspaces for birds in a Mexican city. *Landscape and Urban Planning, 100*(3), 213–222. https://doi.org/10.1016/j.landurbplan.2010.12.008

Castro, E. C., & Ferreira, N. C. (2009). Diagnóstico do padrão de paisagem com métricas dos remanescentes de vegetação em Goiânia. *Habitus, 7*(1/2), 229–247.

Catenazzi, A., & von May, R. (2021). Systematics and conservation of neotropical amphibians and reptiles. *Diversity, 13*, 45. https://doi.org/10.3390/d13020045

Chace, J., Walsh, J., & J. (2006). Urban effects on native avifauna: A review. *Landscape and Urban Planning, 74*, 46–69. https://doi.org/10.1016/j.landurbplan.2004.08.007

Chollet, S., Brabant, C., Tessier, S., & Jung, V. (2018). From urban lawns to urban meadows: Reduction of mowing frequency increases plant taxonomic, functional, and phylogenetic diversity. *Landscape and Urban Planning, 180*, 121–124. https://doi.org/10.1016/j.landurbplan.2018.08.009

Concepción, E., Obrist, M., Moretti, M., Altermatt, F., Baur, B., & Nobis, M. (2016). Impacts of urban sprawl on species richness of plants, butterflies, gastropods, and birds: Not only built-up area matters. *Urban Ecosystems, 19*(1), 225–242. https://doi.org/10.1007/s11252-015-0474-4

Cruz, G. C. F. (2007). Alguns aspectos do clima dos Campos Gerais. In M. S. Melo, R. S. Moro, & G. B. Guimaraes (Eds.), *Patrimônio natural dos Campos Gerais do Paraná* (pp. 59–72). UEPG Press.

Dantas, M. S., Almeida, N. V., Medeiros, I., & Silva, M. D. (2017). Vegetation diagnosis of remaining Atlantic Forest and associated ecosystems in urban spaces. *Journal of Environmental Analysis and Progress, 2*(1), 87–97. https://doi.org/10.24221/jeap.2.1.2017.1128.87-97

Diedrichs, L. A., & Moro, R. S. (2000). *Zoneamento ecológico da bacia do Arroio Olarias, área urbana de Ponta Grossa, PR* (pp. 273–281). IV Simpósio Nacional de Recuperação de Áreas Degradadas.

Eurich, Z. R. S. (2018). *Indice de qualidade de praças: uma proposta metodológica*. Tesis, Universidade Estadual de Ponta Grossa.

Eurich, Z. R. S., Carneiro, D. C., Maliski, L. F., Gonçalves, N. G. T., & Carvalho, S. M. (2014). Analysis of urban forestry in the downtown area of Ponta Grossa-PR. *Perspectiva Geográfica, 9*(10), 1–16.

Faggi, A., & Perepelizin, P. (2006). Riqueza de aves a lo largo de un gradiente de urbanizaci6n en la ciudad de Buenos Aires Argentina. *Revista del Museo Argentino de Ciencias Naturales, 8*, 289–297.

Ferenc, M., Sedlácek, O., Fuchs, R., Dinetti, M., Fraissinet, M., & Storch, D. (2014). Are cities different? Patterns of species richness and beta diversity of urban bird communities and regional species assemblages in Europe. *Global Ecology and Biogeography, 23*, 479–489. https://doi.org/10.1111/geb.12130

Ferreira, L. V., Parolin, P., Muñoz, S. H., & Chaves, P. P. (2012). O efeito da fragmentação e isolamento florestal das áreas verdes da região metropolitana de Belém. *Documentos-Instituto Anchietano de Pesquisas, 63*, 357–367.

Findley, J. S. (1993). *Bats: A community perspective*. Cambridge University Press.

Fischer, J. D., Schneider, S. C., Ahlers, A. A., & Miller, J. R. (2015). Categorizing wildlife responses to urbanization and conservation implications of terminology. *Conservation Biology, 29*, 1246–1248. https://doi.org/10.1111/cobi.12451

Fonseca, B. M., Ribas, R. P., & Moura, A. C. M. (2016). Applying ecological landscape concepts and metrics in urban landscape management. *Paisagem e ambiente: ensaios, 38*, 7–85. https://doi.org/10.11606/ISSN.2359-5361.V0I38P71-85

Fontoura, S. B., Ganade, G., & Larocca, J. (2006). Changes in plant community diversity and composition across an edge between Araucaria Forest and pasture in South Brazil. *Revista Brasileira de Botânica, 29*(1), 79–91.

Forman, R. T. T. (1995). *Land Mosaics. The ecology of landscapes and regions*. Cambridge University Press.

Forman, R. T. T., & Godron, M. (1986). *Landscape ecology*. John Wiley & Sons.

Franchin, A. G. (2009). *Avifauna em áreas urbanas brasileiras, com ênfase em cidades do Triângulo Mineiro/Alto Paranaíba*. Tesis, Universidade Federal de Uberlândia.

Gagné, S., & Fahrig, L. (2010). The trade-off between housing density and sprawl area: Minimizing impacts to forest breeding birds. *Basic and Applied Ecology, 11*(8), 723–733. https://doi.org/10.1016/j.baae.2010.09.001

Gentry, A. H. (1982). Patterns of Neotropical plant species diversity. In *Evolutionary biology* (pp. 1–84). Springer.

Grafius, D. R., Corstanje, R., & Harris, J. A. (2018). The linking ecosystem services, urban form, and green space configuration using multivariate landscape metric analysis. *Landscape Ecology, 33*, 557–573. https://doi.org/10.1007/s10980-018-0618-z

Grimm, N. B., Grove, J. M., Pickett, S. T. A., & Redman, C. L. (2000). Integrated approaches to long-term studies of urban ecological systems. *Bioscience, 50*, 571–584. https://doi.org/10.1641/0006-3568(2000)050[0571:IATLTO]2.0.CO;2

Haaland, C., & van den Bosch, C. (2015). Challenges and strategies for urban green-space planning in cities undergoing densification: A review. *Urban Forestry and Urban Greening, 14*(4), 760–771. https://doi.org/10.1016/j.ufug.2015.07.009

He, H. S., DeZonia, B. E., & Mladenoff, D. J. (2000). An aggregation index (AI) to quantify spatial patterns of landscapes. *Landscape Ecology, 15*, 591–601. https://doi.org/10.1023/A:1008102521322

Holt, A. R., Mears, M., Maltby, L., & Warren, P. (2015). Understanding spatial patterns in the production of multiple urban ecosystem services. *Ecosystem Services, 16*, 33–46. https://doi.org/10.1016/j.ecoser.2015.08.007

Instituto Brasileiro de Geografia e Estatística (IBGE). (2021). https://cidades.ibge.gov.br/brasil/pr/ponta-grossa/panorama. Accessed 20 May 2021.

Lang, S., & Blaschke, T. (2009). *Análise da Paisagem com SIG*. Oficina de Textos.

Leveau, L. M. (2013). Bird traits in urban–rural gradients: how many functional groups are there?. *Journal of Ornithology, 154*, 655–662.

Lima, G. F., Correa Santos, J. W. M., Albertin, R. M., Martínez-Miranzo, B., Souza, F. L., & Angeoletto, F. (2023). Backyards are a way to promote environmental justice and biodiversity conservation in Brazilian cities. *Diversity, 15*(7), 815.

Lozano, P. (2018). Plant biodiversity of Ecuador: A neotropical megadiverse country: Selected countries in the Americas and Australia. *Global Biodiversity*, 185–211. https://doi.org/10.1201/9780429433634-6

Lu, J. (2015). Landscape ecology, urban morphology, and CBDs: An analysis of the Columbus, Ohio Metropolitan Area. *Applied Geography, 60*, 301–307.

MacGregor-Fors, I. (2010). How to measure the urban-wildland ecotone: Redefining 'peri-urban' areas. *Ecological Research, 25*, 883–887. https://doi.org/10.1007/s11284-010-0717-z

MacGregor-Fors, I. J., & Escobar-Ibáñez, J. F. (2017). *Avian ecology in Latin American cityscapes*. Springer Nature eBook. https://doi.org/10.1007/978-3-319-63475-3

Magalhães, D. M., & Moura, A. C. M. (2013). Landscape morphology metrics for urban areas: Analysis of the role of vegetation in the management of the quality of the urban environment. *Disegnarecon, 6*(11), 81–92. https://doi.org/10.6092/issn.1828-5961/3390

McGarigal, K. (2014). *Fragstats Manual version 4.2.* https://www.umass.edu/landeco/research/fragstats/fragstats.html. Accessed 3 May 2021.

Ministério de Meio Ambiente (MMA). (2021). *Programa Cidades + Verdes*. https://antigo.mma.gov.br/images/agenda_ambiental/areas/ProgramaCidadesVerdes.pdf. Accessed 21 Mar 2021.

Mittermeier, R. A., Gill, P. R., Hoffmann, M., & Pilgrim, J. (2004). *Hotspots revisited: Earth's biologically richest and most endangered terrestrial ecoregions*. CEMEX.

Moro, R. S., Schmitt, J., & Diedrichs, L. A. (2001). Estrutura de um fragmento da mata ciliar do Rio Cará Cará, Ponta Grossa, PR. *Publicatio UEPG, 7*, 19–38.

Moro, R. S., Rodrigues, E. R., Souza, M. K. F., & Moro, R. F. (2022). Urban environmental filters on the movement of birds and bats in a medium-sized city in Southern Brazil. *Terr@Plural, 16*, 1–23. https://doi.org/10.5212/TerraPlural.v.16.2215909.027

Nielsen, A. B., van den Bosch, M., Maruthaveeran, S., & van den Bosch, C. K. (2014). Species richness in urban parks and its drivers: A review of empirical evidence. *Urban Ecosystems, 17*, 305–327. https://doi.org/10.1007/s11252-013-0316-1

Norton, B. A., Evans, K. L., & Warren, H. P. (2016). Urban biodiversity and landscape ecology: Patterns, processes, and planning. *Current Landscape Ecology Report, 1*, 178–192. https://doi.org/10.1007/s40823-016-0018-5

Olegário, A. F., Collaço, S. L., Bertoni, D., Antiqueira, L. M. O., & Dias, J. (2020). Avifauna no campus de Ponta Grossa da Universidade Tecnológica Federal do Paraná: possibilidade de educação em ambiente não formal. *Revista de Estudos Vale do Iguaçu, 35*, 42–53.

Ortega-Álvarez, R., & MacGregor-Fors, I. (2011). Dusting off the file: A review of knowledge on urban ornithology in Latin America. *Landscape and Urban Planning, 101*, 1–10. https://doi.org/10.1016/j.landurbplan.2010.12.020

Park, S., Hepcan, C. C., Hepcan, S., & Cook, E. A. (2014). Influence of urban form on landscape pattern and connectivity in metropolitan regions: A comparative case study of Phoenix, AZ, USA, and Izmir, Turkey. *Environmental Monitoring Assessment, 186*, 6301–6318.

Penteado, H. M., & Alvarez, C. E. (2007). Urban green corridors: A study of the viability of connection of open spaces in Vitória. *Paisagem e Ambiente: ensaios, 24*, 57–68.

Philippsen, J. S., Benedito, E., & Zawadzki, C. H. (2010). Species composition and richness of avifauna in an urban area of southern Brazil. *Acta Scientiarum. Biological Sciences, 32*(1), 55–62. https://doi.org/10.4025/actascibiolsci.v32i1.6118

Picket, S. T., & Thompson, J. N. (1978). Patch dynamics and the design of nature reserves. *Biological Conservation, 13*, 27–37. https://doi.org/10.1016/0006-3207(78)90016-2

Puga-Caballero, A., MacGregor-Fors, I., & Ortega-Álvarez, R. (2014). Birds at the urban fringe: Avian community shifts in different peri-urban ecotones of a megacity. *Ecological Research, 29*, 619–628. https://doi.org/10.1007/s11284-014-1145-2

Queiroz, D. A. H. O., & Carvalho, S. M. (2019). Evaluation of the quality of urban green areas in Ponta Grossa, PR, Brazil. *Terr@Plural, 13*(3), 217–236. https://doi.org/10.5212/TerraPlural.v.13i3.0015

Rastandeh, A., Brown, D. K., & Zari, M. P. (2017). *Biodiversity conservation in urban environments: A review on the importance of spatial patterning of landscapes*. Ecocity World Summit, 12–14 July. Melbourne.

Rocha, N. A., Borges, J. L. C., & Moura, A. C. M. (2016). Conflicts of urban transformation and environmental dynamics according to landscape ecology. *Pesquisa em Arquitetura e Construção, 7*(1), 23–34. https://doi.org/10.20396/parc.v7i1.8644222

Rocha, C. H., Santana, A. C., Linhares, M. A. P. S., Grizafis, V. K. M., Farias, A., & Moro, R. F. (2019). Conservation value of arboreal individuals and management zones in an urban remnant of Template Mixed Forest: A methodological proposal. *Terr@Plural, 13*(2), 513–532. https://doi.org/10.5212/TerraPlural.v.13i2.0014

Rodrigues, A. G., Borges-Martins, M., & Zilio, F. (2016). Bird diversity in an urban ecosystem: The role of local habitats in understanding the effects of urbanization. *Iheringia, Série Zoologia, 108*, e2018017. https://doi.org/10.1590/1678-4766e2018017

Sacco, A. G., Bergmann, F. B., & Rui, A. M. (2013). Bird assemblages in the urban area in the city of Pelotas, Rio Grande do Sul, Brazil. *Biota Neotropica, 13*(2), 153–162.

Seto, K. C., Güneralp, B., & Hutyra, L. (2012). Global forecasts of urban expansion to 2030 and direct impacts on biodiversity and carbon pools. *Proceedings of the National Academy of Sciences of the United States of America, 109*(40), 16083–16088. https://doi.org/10.1073/pnas.1211658109

Silva, A. L., Longo, R. M., Bressane, A., & Carvalho, F. H. M. (2019). Classification of urban forest fragments based on landscape metrics. *Ciencia Florestal, 29*(3), 1254–1269. https://doi.org/10.5902/1980509830201

Singh, V. S., Pandey, D. N., & Chaudhary, P. (2010). Urban forests and open green spaces: Lesson for Jaipur, Rajasthan- India. *Ocasional paper, 1*, 1–23.

Souza, F. L., Fenoglio, M. S., & Angeoletto, F. (2023). To be a Brazilian City dweller, sometimes we must learn to say enough! *Sustainability, 15*(4), 3699.

Swanwick, C., Dunnett, N., & Woolley, H. (2003). The nature, role, and value of green space in towns and cities—An overview. *Built Environment, 29*(2), 94–106. https://doi.org/10.2148/benv.29.2.94.54467

Szlavecz, K., Warren, P., & Pickett, S. T. (2011). *Biodiversity on the urban landscape*, Human Population, pp. 75–101. Springer. https://doi.org/10.1007/978-3-642-16707-2_6

Tadenuma, S. S. K. (2019). *Espacialização da arborização de vias públicas por densidade e níveis de atenção na área urbana de Ponta Grossa (PR)*. MSc Dissertation, Universidade Estadual de Ponta Grossa.

Tratalos, J., Fuller, R. A., Warren, P. H., Davies, R. G., & Gaston, K. J. (2007). Urban form, biodiversity potential, and ecosystem services. *Landscape and Urban Planning, 83*, 308–317. https://doi.org/10.1016/j.landurbplan.2007.05.003

Turner, M. G., Gardner, R. H., & O'Neill, R. V. (2001). *Landscape ecology in theory and practice: Pattern and processes*. Springer.

Uy, P. D., & Nakagoshi, A. N. (2007). Analyzing urban green space pattern and eco-network in Hanoi, Vietnam. *Landscape Ecology Engineering, 3*, 143–157. https://doi.org/10.1007/s11355-007-0030-3

van Heezik, Y., Smyth, A., & Mathieu, R. (2008). Diversity of native and exotic birds across an urban gradient in a New Zealand city. *Landscape and Urban Planning, 87*(3), 223–232.

White, J. G., Antos, M. J., Fitzsimons, J. A., & Palmer, G. C. (2005). Non-uniform bird assemblages in urban environments: The influence of streetscape vegetation. *Landscape and Urban Planning, 71*(2–4), 123–135. https://doi.org/10.1016/j.landurbplan.2004.02.006

Yang, F., Ignatieva, M., Wissman, J., Ahrné, K., Zhang, S., & Zhu, S. (2019). Relationships between multi-scale factors, plant and pollinator diversity, and composition of park lawns and other herbaceous vegetation in a fast-growing megacity of China. *Landscape and Urban Planning, 185*, 117–126. https://doi.org/10.1016/j.landurbplan.2019.02.003

Zanon, C. M. V., & Reis, N. R. (2007). Bats (Mammalia, Chiroptera) in the Ponta Grossa region, Campos Gerais, Paraná, Brazil. *Revista Brasileira de Zoologia, 24*(2), 327–332. https://doi.org/10.1590/S0101-81752007000200010

Chapter 18
Using Landscape Design to Mitigate Urban Climate: Campinas Case

Loyde V. de Abreu-Harbich, Lucila C. Labaki, and Andreas Matzarakis

Abstract Different tree species, and their distribution in urban areas, can help improve thermal microclimate and promote human thermal comfort in these environments. The aim of the current study is to investigate how different tree species and their distribution in urban areas can help improve human thermal comfort in built environments. Firstly, the thermal comfort provided by different tree species, and by their planting design (individuals and clusters), was measured at the free flat area of Campinas County, Brazil. Then, human thermal comfort provided by trees in urban areas (low and high-density construction areas) of Campinas and Santos counties, Brazil, were measured. Benefits provided by the following twelve species were analyzed from 2007 to 2010: *Tabebuia chrysotricha* (*Mart.* ex DC.) Stand., *Jacaranda mimosaefolia* D. Don., *Syzygium cumini* L., *Mangifera indica* L., *Pinus palustris* L. and *Pinus coulteri* L.—individuals; *Lafoensia glyptocarpa* L., *Caesalpinia pluviosa F., Spathodea campanulata* P.Beauv., *Tipuana tipu* F. (individuals and clusters) –, *Delonix indica* F. *and Senna siamea L.* (clusters). *C. pluviosa* F. was capable of reducing the Physiologically Equivalent Temperature (PET) to ranges between 12 and 16° C, for individual trees; and between 12.5 and 14.5° C, for clusters. Streets with shading trees recorded PET decrease by 18° C in low-density areas, as well as by 10° C in high-density areas. Trees planted in high-density areas can also help mitigate the negative effects of heat on facades, in

L. V. de Abreu-Harbich (✉)
Faculty of Architecture and Urban Planning, Mackenzie Presbyterian University, São Paulo, Brazil
e-mail: loyde.harbich@mackenzie.br

L. C. Labaki
Faculty of Civil engineering and Architecture and Urban Planning, State University of Campinas, Campinas, Brazil

A. Matzarakis
Faculty of Environment and Natural Resources, University of Freiburg, Freiburg, Germany

Democritus University of Thrace, Komotini, Greece

F. Angeoletto et al. (eds.), *Ecology of Tropical Cities*,
https://doi.org/10.1007/978-3-031-48012-6_18

addition to increase the aesthetic effect of streets. Trees planted in low-density areas can help improve thermal comfort for pedestrians and cyclists.

Keywords Human thermal comfort · Physiologically equivalent temperature · Mean radiant temperature · Tree planting design · Tropical climate in Brazil (Campinas and Santos counties)

1 Introduction

Urbanization can change different surfaces, building uses, urban configurations, as well as increase the heat stored in buildings and intensify human thermal stress in urban areas (Hwang et al., 2011). Moreover, the energy consumption that has unintentionally increased at building level derives from climatic changes, a fact that leads to remarkable demand for urban energy resources (Akbari & Taha, 1992; Oke, 1984). The development of thermally desirable or neutral outdoor environments has implications that go beyond requirements of both urban and climate design adjusted to buildings. It is important to specify standards capable of determining how urban spaces should be structured and arranged, based on the climate conditions of a given region in order to re-establish thermally acceptable or neutral indoor and outdoor conditions (Alcoforado & Matzarakis, 2010; Givoni, 1989, 1991).

Changes in solar energy flow that reaches the urban surface through physical processes and alterations to some meteorological elements, mainly surface and air temperatures, are one of the causes of heat island formation in different cities. Heating increases throughout the day due to the thermal absorption capacity of different materials used in civil construction, which return part of the received radiation back to the atmosphere (Lombardo, 1985; Oke, 1984). Accordingly, Taha and Meier (1997) has stated that Urban Heat Islands happen in places presenting little urban vegetation, as well as the prevalence of buildings, concrete, asphalt, pollution and vehicles. Materials such as concrete, asphalt and bricks keep releasing the energy derived from solar radiation absorbed throughout the day, even after dusk. Buildings and urban shapes can also modify wind circulation and make it easier for pollutants to accumulate in urban areas and modify outdoor thermal comfort conditions (Li, et al., 2021).

Decreased evapotranspiration caused by a lack of vegetation and water supply, on the one hand, and modification of permeable surfaces due to the presence of buildings in urban areas, on the other hand, are typical changes in urban design capable of promoting heat-island formation and intensity (Lombardo, 1985; Oke, 1984). Using vegetation to reduce solar radiation can help regulate urban microclimate and open opportunities for the possibility of climate change adaptation (Abreu & Labaki, 2010; Green, 1993; Santos Nouri et al., 2018).

The aim of the current chapter is to investigate how different tree species, and their distribution, can help improve human thermal comfort in a given built environment.

The first section (Sect. 18.2) is focused on analyzing the thermal comfort provided by different tree species, as well as by their planting design (individuals or clusters), in free flat areas, whose surroundings did not have buildings or asphalt/concrete pavement, was conducted in Campinas County (Brazil). Second (Sect. 18.3), field studies conducted in Campinas and Santos counties have analyzed the thermal comfort provided by tree clusters in areas with buildings and pavement. This was done to identify the thermal comfort behavior of trees in urban areas with remarkable construction density. Finally (Sect. 18.4), the current chapter presents some planting design recommendations for tropical climates.

2 Benefits Provided by Different Tree Species and by Their Distribution in Urban Areas

Individually planted urban trees can act as shading and wind-shielding elements capable of changing environmental conditions around isolated buildings. On the other hand, tree clusters can help moderating the heat island intensity in urban areas by changing the heat balance in the whole city (Abreu-Harbich et al., 2015; Akbari, 2002; Akbari & Rose, 2008; Angeoletto et al., 2023). Studies have shown that shading and wind speed promotion enable improved thermal comfort in tropical climates (Bueno-Bartholomei, 2003; Lin et al., 2010). Therefore, planting trees is a good mechanism to improve thermal comfort in tropical cities.

Canopy features, such as type, height, age and planting design can directly influence thermal comfort in urban outdoor spaces (Abreu-Harbich et al., 2015; Peixoto et al., 1995). Tree leaves can absorb, reflect and transmit solar radiation, as well as indirectly influence thermal comfort via evapotranspiration (Santamouris, 2001; Scudo, 2002; Emmanuel, 2005; Emmanuel et al., 2007; Abreu & Labaki, 2011).

Trees cast their own shadow, shape and density, which depend on canopy shape and on the solar radiation filtered by it, respectively (Robinette, 1972). The amount of solar radiation filtered by trees depends on the density of their twigs, branches and leaf cover, since these elements influence the overall tree shape and density features (Scudo, 2002; Abreu & Labaki, 2011).

For example, some tree species grown in Brazil can attenuate solar radiation by 76.3–92.8% in summer (Abreu & Labaki, 2010; Bueno-Bartholomei, 2003). A recent study conducted in China has shown that shade tree species *Ficus microcarpa*, *Mangifera indica* and *Bauhinia blakeana* were capable of attenuating solar radiation by 76–93% (Deng et al. 2020). The aforementioned study has shown that solar radiation in shaded areas suddenly increased at certain times, because of gaps in the crown of trees; this process is called the "light spots phenomenon" (LSP).

Scudo (2002) and Vogt et al. (2003) have shown that trees can influence air movement pattern through direction, filtration, obstruction and deflection, as well as showing that this pattern depends on plant features, such as geometry, height, permeability and crown type.

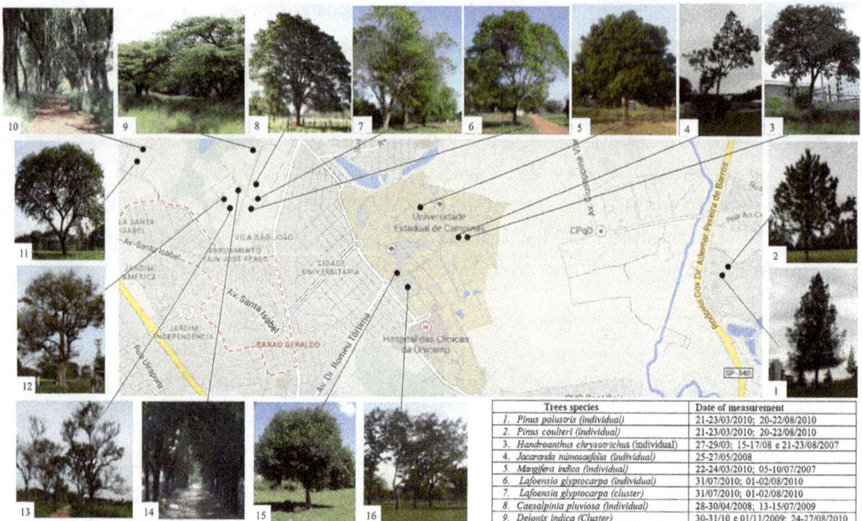

Trees species	Date of measurement
1. *Pinus palustris (individual)*	21-23/03/2010; 20-22/08/2010
2. *Pinus coulteri (individual)*	21-23/03/2010; 20-22/08/2010
3. *Handroanthus chrysotrichus* (individual)	27-29/03; 15-17/08 e 21-23/08/2007
4. *Jacaranda mimosaefolia (individual)*	25-27/05/2008
5. *Mangifera indica (individual)*	22-24/03/2010; 05-10/07/2007
6. *Lafoensia glyptocarpa (individual)*	31/07/2010; 01-02/08/2010
7. *Lafoensia glyptocarpa (cluster)*	31/07/2010; 01-02/08/2010
8. *Caesalpinia pluviosa (individual)*	28-30/04/2008; 13-15/07/2009
9. *Delonix indica (Cluster)*	30-31/10 e 01/11/2009; 24-27/08/2010
10. *Tipuana tipu (cluster)*	21-23/10/2009
11. *Tipuana tipu (individual)*	21-23/10/2009
12. *Spathodea campanulata (individual)*	17-22/07/2009
13. *Spathodea campanulata (cluster)*	17-22/07/2009
14. *Caesalpinia pluviosa (Cluster)*	28-30/04/2008; 13-15/07/2009
15. *Syzygium cumini (individual)*	19-21/03/2010; 18-20/08/2007
16. *Senna siamea (Cluster)*	17-25/07/2007

Fig. 18.1 Investigated sites and species—Individuals and Clusters (Abreu-Harbich et al., 2015)

Individual tree trunk features can also influence shade quality. For instance, tree trunks are often orthotropic/sympodial, orthotropic/monopodial or present plagiotropic branches. Leaf type and shape must also be taken into consideration.

A field study carried out in Campinas County, Brazil, from 2007 to 2010, has investigated the benefits provided by trees at micro scale. Figure 18.1 shows twelve tree species selected in different contexts (both individuals and clusters): *Tabebeuia chrysotricha* (*Mart.* ex DC.) Stand., *Jacaranda mimosaefolia* D. Don., *Syzygium cumini* L., *Mangifera indica* L., *Pinus palustris* L. and *Pinus coulteri* L.—individuals; *Lafoensia glyptocarpa* L., *Caesalpinia pluviosa F., Spathodea campanulata* P. Beauv., *Tipuana tipu* F. —individuals and clusters—, *Delonix indica* F.*and Senna siamea L.*—clusters.

2.1 *Solar Radiation Attenuation*

The Plant Area Index (PAI) of the herein selected tree individuals and clusters was compared between species, both in the shade and in the sun, in order to measure the solar radiation attenuated by their canopies. The current study has used the term "Plant Area Index" (PAI), which refers to all light-blocking elements (such as stems, twigs and leaves), as opposed to Leaf Area Index (LAI), which only concerns leaves (Holst et al., 2004).

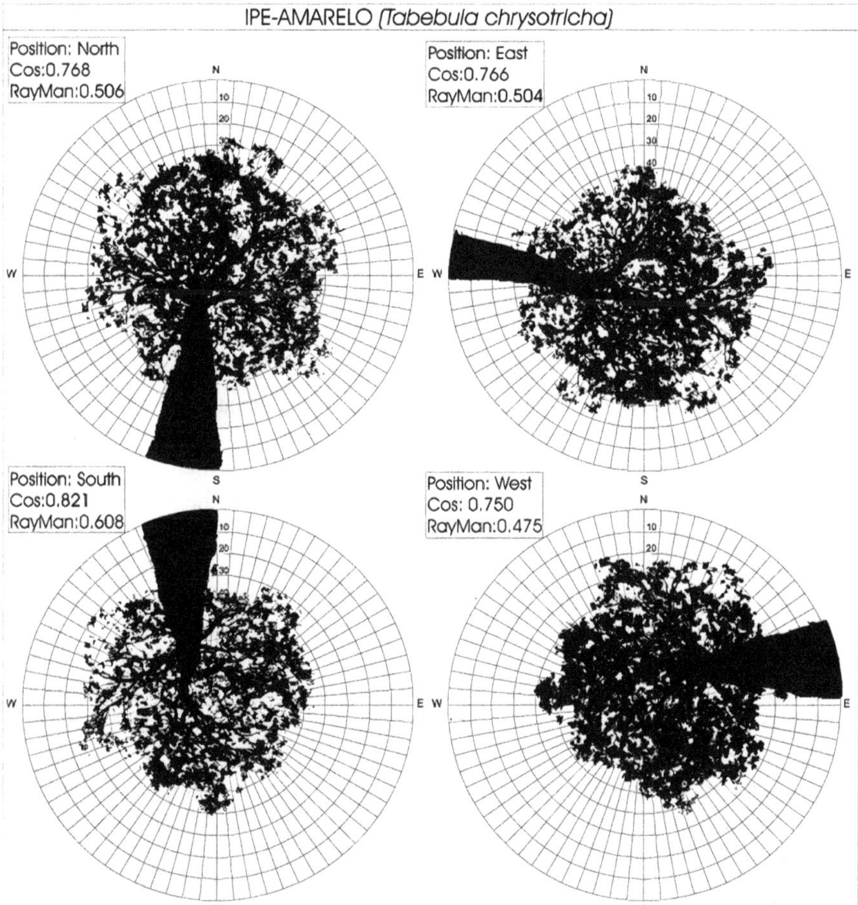

Fig. 18.2 Example of PAI analysis applied to individual trees

Plant Area Index (PAI) was obtained by taking four "fish-eye" photos of each tree, in orthogonal projection, below the canopy and at different orientations (North, South, East and West) for individual trees (Fig. 18.2). The Sky View Factor (SVF) was calculated in RayMan Pro model (Matzarakis et al., 2010), which replaced the sky in the photos by branches and green leaves in order to estimate PAI.

Solar radiation attenuation was measured by two sets of tube solarimeters, TSL type (Delta-T Devices). Sensors in tube solarimeters were connected to a logger, model DL2, also from Delta-T. One set of equipment was installed in the shadow of trees, whereas the second one was placed in the sun. Data were simultaneously collected beneath the crowns of the investigated trees and under sunlight. Measurements started to be taken by 6:00 a.m. and were concluded by 6:00 p.m.; they were recorded every ten minutes. The adopted equipment is capable of measuring mean irradiance (W/m^2) in places such as beneath tree crowns and inside greenhouses, where radiant energy distribution is not uniform. The spectral response corresponds to visible and near infrared radiation (350 nm to 2500 nm).

Solar radiation attenuation depends on tree features such as twig and branch density, as well as leaf cover, which can change from species to species. The rate of radiation attenuated by individual trees or clusters of trees was obtained based on the methodology by Bueno-Bartholomei (2003). This methodology consists in simultaneously measuring solar radiation in the shade and under sunlight, in order to calculate solar radiation attenuation based on the following equation:

$$AT = \frac{S_{\text{sun}} ? S_{sh}}{S_{\text{sun}}} ? 100 \tag{18.1}$$

Wherein: AT is solar radiation attenuation (%); S_{sun} is the area providing the total incident energy (kWh/m^2) collected by the solarimeter under sunlight, within the investigated time interval; and S_{sh} is the area providing the total incident energy (kWh/m^2) collected by the solarimeter in the shade, within the investigated time interval.

Figures 18.3 and 18.4 show results of mean diurnal courses of Mean Solar Radiation recorded for individual trees and clusters of trees, respectively, based on data resulting from field measurements taken within the investigated period-of-time.

There was variation in solar radiation attenuated by individual trees due to specific features of each tree species, with emphasis on crown, trunk, and leaf shape and size. Perennial tree species such as *S. cumini* L. and *M. indica* L. have shown similar behavior in the Summer and in Winter, whereas semi-deciduous and deciduous tree species such as *T. chrysotricha* (*Mart.* ex DC.) Stand.) and *C. pluviosa F.* recorded different results in each season.

Trees' solar radiation attenuation ability is not only associated with their phenology, but also with trunk type, as well as with leaf shape and size. For example, *S. cumini* L., *M. indica* L. and *C. pluviosa F.* have plagiotropic trunk, but different leaf shapes; whereas *S. cumini* L., and *M. indica* L. provide homogeneous shade during the day. On the other hand, *T. chrysotricha* (*Mart.* ex DC.) Stand. and *C. pluviosa F.* present difference in trunk structure—orthotropic sympodial and plagiotropic, respectively—and in leaf shape. The orthotropic monopodial trunk structure found in tree species *P. palustris* L. and *P. coulteri* L. leads to significant variations in solar radiation attenuation during the day, as well as reduces shade quality.

Figure 18.5 shows results observed for plant area index (Cos and RayMan) and the summation of solar radiation attenuated by individual trees and clusters of trees. Plant area index (PAI) can show differences in tree distributions (individuals or clusters), as well as in species' features such as crown and trunk structures, and leaf features. Results have shown that tree species with denser crown—i.e., PAI higher than 80%—can attenuate more solar radiation than others.

The following results can be extracted from this section:

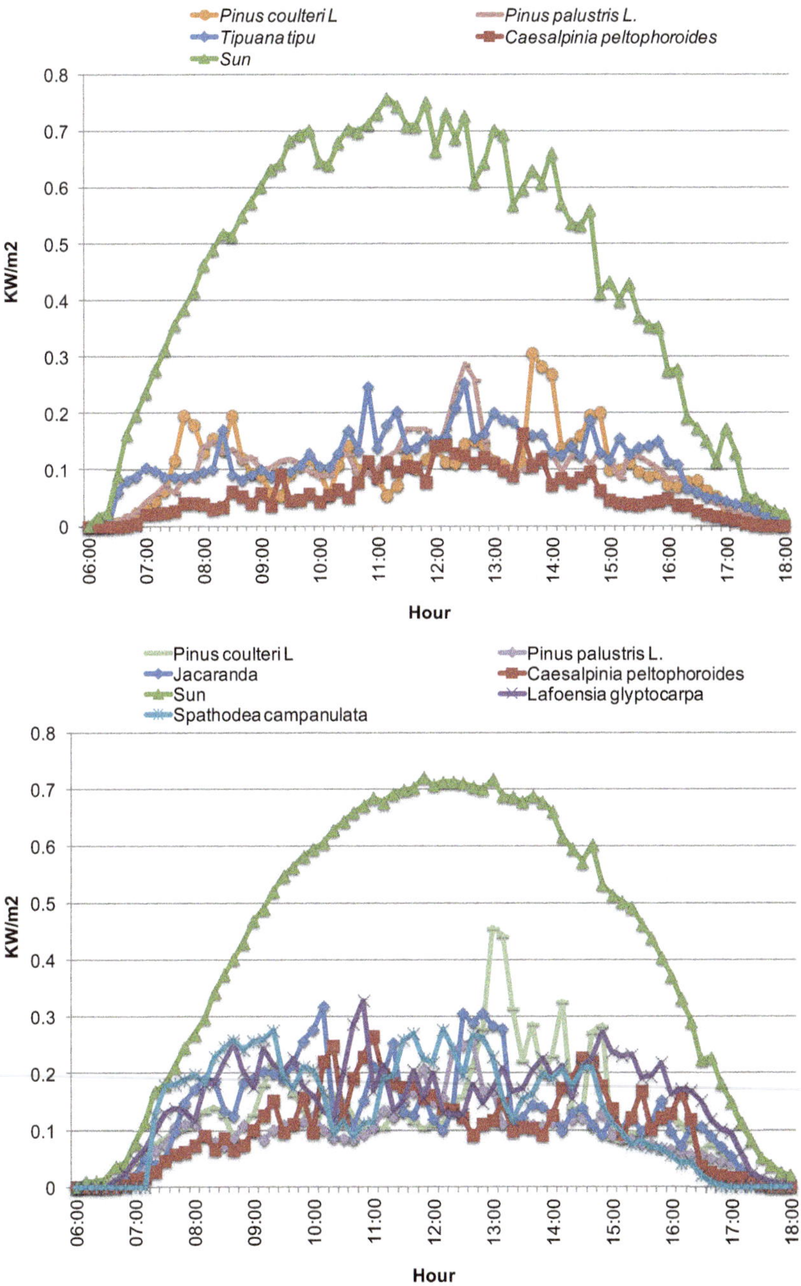

Fig. 18.3 Diurnal courses of Mean Solar Radiation recorded for individual trees based on data derived from field measurements taken from 2007 to 2010

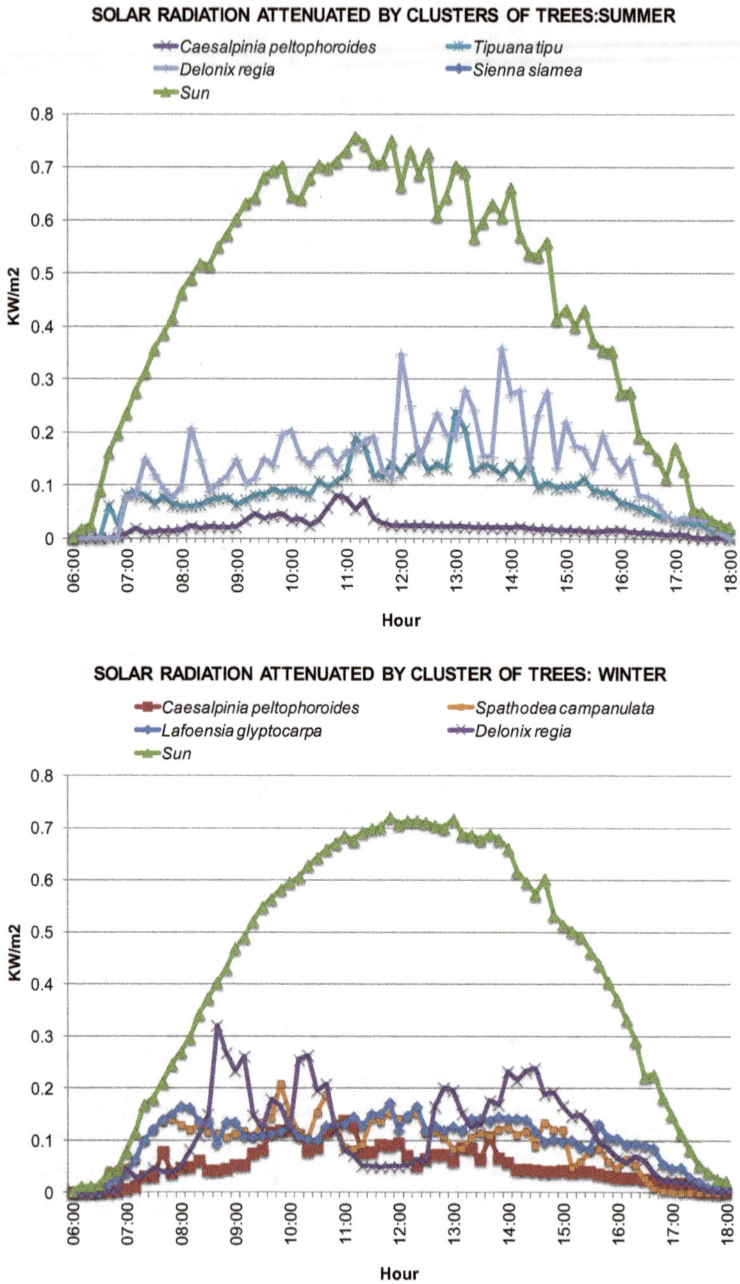

Fig. 18.4 Diurnal courses of Mean Solar Radiation recorded for clusters of trees, based on data derived from field measurements taken from 2007 to 2010

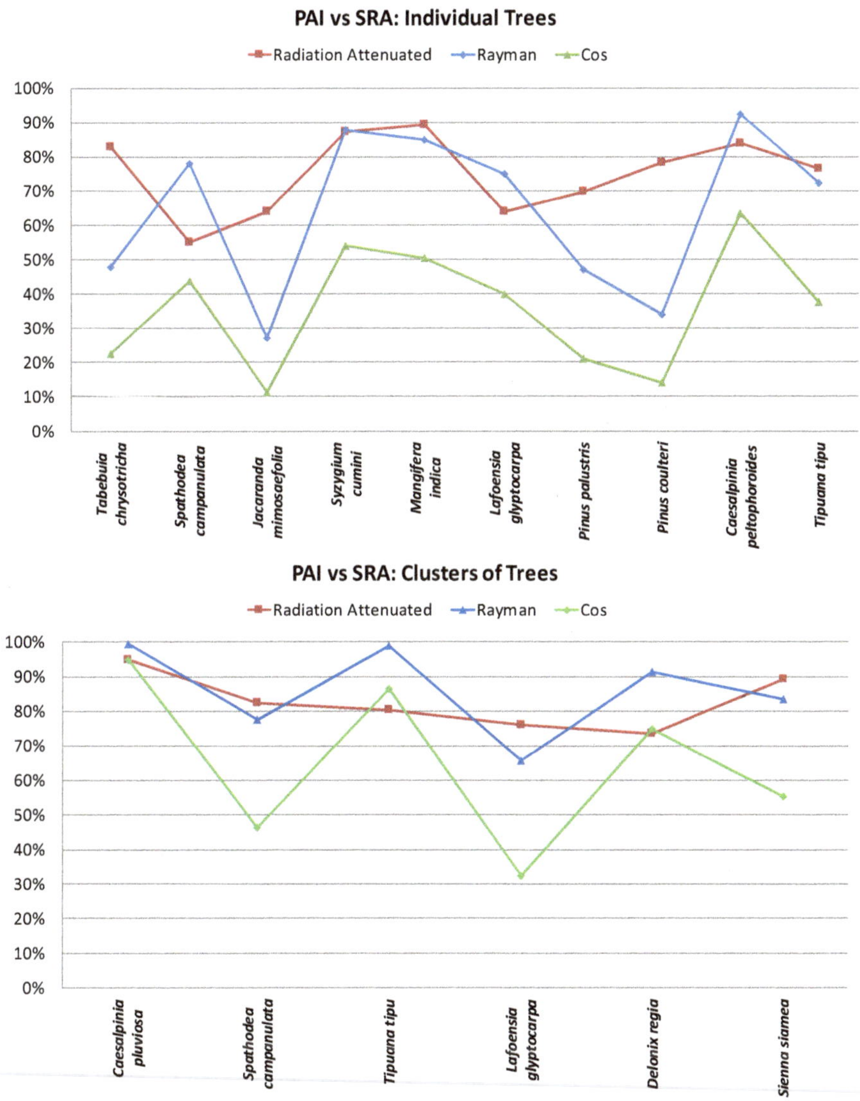

Fig. 18.5 Results of Average Solar Radiation Attenuation and Plant Area Index (PAI) recorded for individual trees and clusters of trees

- Individual species with plagiotropic trunk, such as *S. cumini* L., *M. indica* L. and *C. pluviosa* F., have shown PAI higher than 80% and they were capable of attenuating more solar radiation than the other analyzed species.
- Individual species with orthotropic sympodial trunk, such as *T. chrysotricha* (Mart. ex DC.) Stand., were capable of attenuating solar radiation by more than 80% and they showed PAI lower than 50%. Although the shading area provided

by the crown of these tree species was not large, they showed satisfactory ability to attenuate solar radiation.

- Although species *C. pluviosa* F. and *D. regia* F have plagiotropic trunk types, clusters of *C. pluviosa* F. were capable of attenuating more solar radiation than clusters of *D. regia* due to their leaf features. PAI results recorded for *C. pluviosa* F. (98%) were higher than those observed for *D. regia* (91%).
- The cluster of trees provided solar radiation attenuation approximately 10% higher than that of individual trees, i.e., *C. pluviosa* F and *T. Tipu*.
- It is worth planting trees in clusters to help better attenuating solar radiation in urban areas.

2.2 *Thermal Comfort Improvement*

The thermal comfort improvement provided by trees can be measured by calculating thermal indices, such as Physiologically Equivalent Temperature (PET), Universal Thermal Climate Index (UTCI), among others, in RayMan Pro Software (Matzarakis et al., 2007; Matzarakis et al., 2010). PET and other indices' estimates depend on atmospheric influences, primarily on clouds, as well as on other meteorological variables such as vapor pressure or particles. The primary input parameters for thermal comfort measurement comprised meteorological data—air temperature and humidity, wind speed and global radiation—collected for individual trees and clusters of trees, both in the shade and under sunlight. Data were collected during 3 sunny days, both in the Summer and in Winter. The adopted sampling interval was 10 min, every hour, between 6:00 AM and 6:00 PM.

Air temperature and humidity data were collected with the aid of Testo data logger, model 175-1. Two tripods were fixed in the shadow and in the sun, simultanealy. A Testo anemometer, model 10635-1549, connected to a multifunction recorder, model 445, was used to collect wind speed data at a fixed site. Sensors were positioned at the height of 1.1 m. All recordings were protected from solar radiation by special shelters designed for outdoor measurements.

Air temperature up to 29° C and higher than 35° C was considered comfortable and heat stress, respectively, in the air temperature analysis. With respect to Tmrt, temperature higher than 35° C was considered heat stress, based on Abreu-Harbich et al. (2015). The European thermal comfort range was used (Matzarakis & Mayer, 1996) in association with a tropical model produced for Taiwan by Lin and Matzarakis (2008) and adjusted for São Paulo city by Monteiro and Allucci (2010), in order to classify thermal comfort improvement (Table 18.1). The thermal comfort range set for European climates lies between 18 °C and23 °C, whereas the one set for tropical climates, such as that of São Paulo, lies between 26 °C and 30 °C; temperatures higher than 41 °C and 42 °C in European and tropical climates (Taiwan), respectively, are classified as extreme heat stress. Although the climate in Campinas County is similar to that of São Paulo and Moon Lake (Taiwan), the current study

Table 18.1 Thermal sensations and PET classes applied to Western/Middle European, Taiwan and São Paulo

Thermal Sensation	PET range for Western/Middle European (°C PET) (Matzarakis ; Mayer 1996)	PET ranger for Taiwan (°C PET) (Lin ;Matzarakis 2008)	PET ranger for São Paulo (°C PET) (Monteiro ;Allucci 2009)
Very cold	<4	<14	
Cold	4-8	14-18	<4
Cool	8-13	18-22	4-12
Slightly cool	13-18	22-26	12-18
Comfortable	18-23	26-30	18-26
Slightly warm	23-29	30-34	26-31
Warm	29-35	34-38	31-43
Hot	35-41	38-42	>43
Very hot	>41	>42	

applied PET classes to Taiwan because it presents surrounding conditions similar to the herein analyzed ones.

Figure 18.6 summarizes mean air temperatures, PET and T_{mrt} results observed for shade from 6:00 to 18:00 both in Summer and Winter.

Tree species such as *Mangifera indica, Caesalpinia peltophoroides* and *Tipuana tipu* (individual trees) enabled air temperature ranging from 20 °C to 29 °C for 10 h a day, in Summer, whereas species such as *Pinus paulustri*, *Jacaranda mimosaefolia*, *Tabebuia Chrysotricha*, *Mangifera indica* (individual trees) enabled the very same air temperature range, also for 10 h a day, in Winter.

All analyzed trees (either individuals or clusters) produced shadow capable of reducing PET. Results observed for individual species have shown that *Syzygium cumini*, *Mangifera indica*, *Caesalpinia peltophoroides* and *Tipuana tipu* provided thermal comfort from 7 to 9 h a day, in both seasons. Results observed for clusters of trees have shown that species such as *Caesalpinia peltophoroides* and *Spathodea campanulata* provided thermal comfort from 6 to 9 h a day, in both seasons.

In terms of T_{mrt}, individual trees belonging to species *Syzygium cumini, Mangifera indica, Lafoensia glyptocarpa* and *Caesalpinia peltophoroides* recorded temperatures up to 30 °C throughout the day, in both seasons. On the other hand, clusters of trees belonging to species *Caesalpinia peltophoroides, Spathodea campanulata* and *Lafoensia glyptocarpa* recorded temperatures up to 30 °C throughout the day, in Winter.

The following results can be extracted from this section:

- Trunk structure (plagiotropic or sympodial) and leaf shape can influence thermal comfort sensations;
- Tree Planting design can help improving the thermal comfort provided by trees.

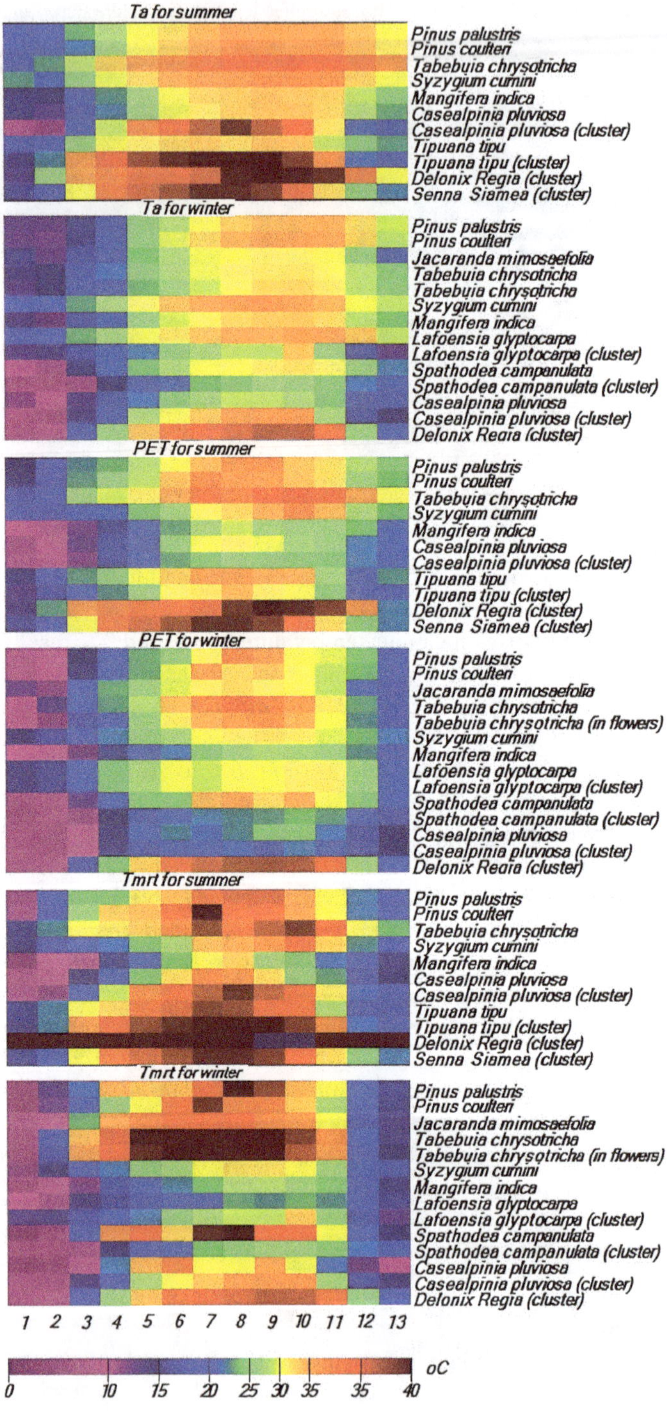

Fig. 18.6 Hourly mean data for: air temperature (**a**), Physiologically Equivalent Temperature (PET) (**b**), and Mean Radiant Temperature (T_{mrt}) (**c**) recorded for different tree species during each investigated in the summer and in the winter

- Thermal comfort analyses applied to vegetation cannot only take into consideration air temperature and humidity. It is necessary to analyze how solar radiation is attenuated by tree crowns, as well as wind permeability in the trees' surroundings.
- Air temperature and T_{mrt} in the shadow of trees cannot be evaluated separately. The time of the thermal comfort index calculation played a key role in the classification of the thermal comfort provided by each species in built environments. This classification indicates the ability of each tree to improve thermal comfort. Results can help developing urban forestry plans focused on urban sustainability.

3 Benefits Provided by Trees for Urban Canyons

The current section has investigated the influence of urban trees and urban configuration on thermal comfort in two tropical cities, namely: Campinas and Santos.

The herein adopted methodological procedures comprised:

- Describing the climate background of selected cities;
- Selecting the study sites by taking into consideration the association between buildings' height and width (H/W) and the presence of vegetation in them;
- Collecting meteorological data about the selected sites in each city;
- Calculating and analyzing thermal comfort indices.

Microclimate data were collected with the aid of a portable weather station used to measure air temperature, air humidity, solar radiation, wind speed and globe temperature. Data were collected for 12 h, in three sunny days—this methodology was based on Correa et al. (2012) and on RUROS Project (Rediscovering the Urban Realm and Open Spaces).

Thermal comfort indices were calculated based on RayMan model application (Matzarakis et al., 2010). Physiologically Equivalent Temperature (PET) was used to classify thermal sensation and heat stress in humans (internal heat production: 80 W, thermal resistance of clothing: 0.9 clo).

3.1 Sites Selected for the Investigated Case

The current study was carried in Campinas and Santos counties, Brazil, Fig. 18.7. These counties were selected to enable investigating trees' ability to improve urban climate in high and low-density zones in 2 different tropical climates. Both counties have similar urban shape, height/width (H/W) ratio of buildings constructions, incidence of trees (species addressed in Sect. 18.2) and use the same materials for paving and facades.

Climate in Campinas County (22° 48′57″ S, 47° 03′33″ W, altitude of 640 m) is classified as Tropical High-Altitude, Cwa (Kottek et al., 2006). Its mean annual air

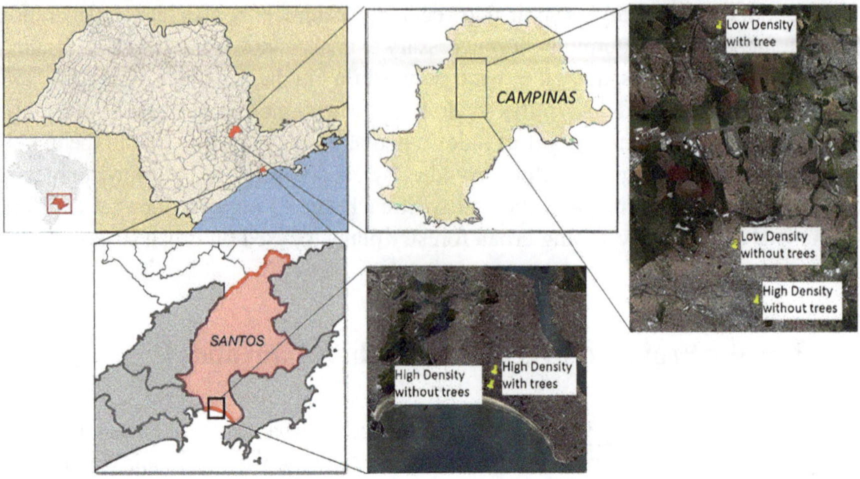

Fig. 18.7 Field Measurement Sites: Campinas and Santos counties, Brazil

temperature is 22.3 ° C, whereas its mean annual rainfall reaches 1411 mm. Southeasterly wind prevails in the region.

Climate in Santos County (23° 56′ 15″S, 46° 19′30″W, altitude of 2 m) is classified as tropical rainforest with no dry season (Kottek et al., 2006). Its mean annual air temperature is 28 °C, whereas its mean annual rainfall reaches 2081 mm. Southeasterly wind with annual mean speed of 1.5 m/s prevails in the region.

Three (3) different sites were selected in Campinas County, namely: A. Resende St. (low-density zone with trees), Gil Vicente St. (low-density zone without trees) and Conceição St. (high-density zone without trees). On the other hand, 2 different sites were selected in Santos County, namely: Washington Luiz Av. (high-density zone with trees) and Galeao Carvalhal St. (high-density zone without trees).

Figure 18.8 shows field measurement sites, Sky View Factor and H/W ratio. It is possible seeing that high-density zones presented similar H/W ratio. The herein selected sites were the ones whose species were analyzed in Sect. 18.2.

C. peltophoroides and *J. mimosifolia* were the prevalent tree species in Campinas County—low-density zone with trees—, whereas *S. cumini* was the prevalent tree species in Santos County—high-density zone with trees. The focus of this analysis lied on sky obstacles such as buildings or trees, as well as on how they influenced the thermal comfort. The presence of water, trees up to 4-meters tall, or vegetation on the pavement of the investigated site were disregarded factors.

3.2 Results of Benefits Provided by Trees to Urban Areas

Three (3) different times—11 p.m., 1 a.m. and 3 a.m.—were used to compare results observed for Campinas and Santos counties.

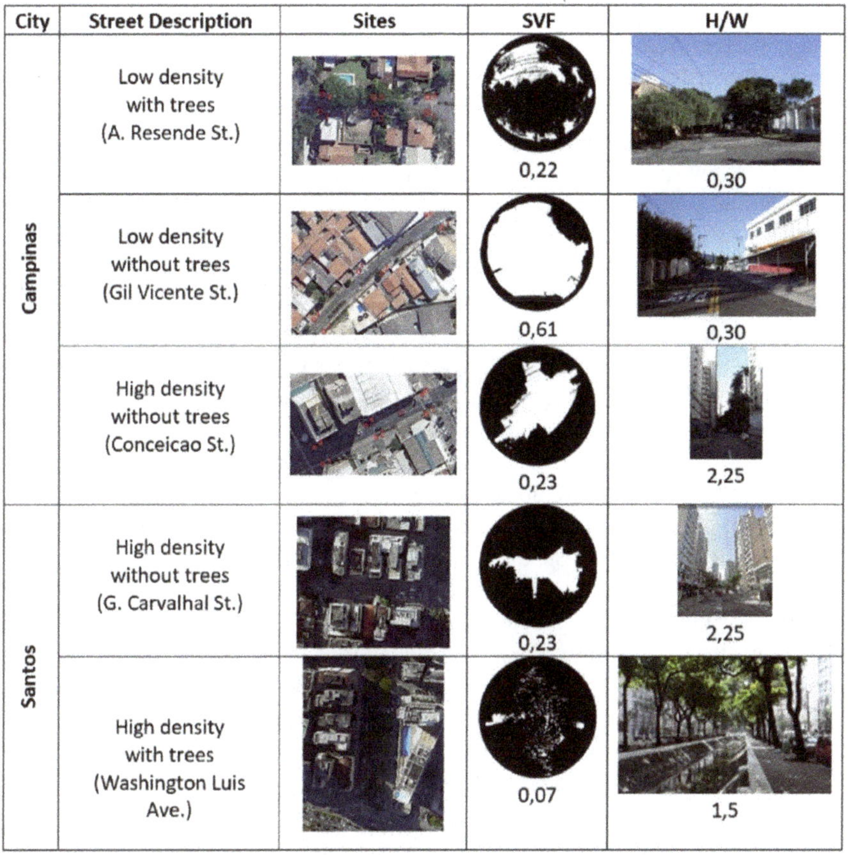

Fig. 18.8 Selected field measurement sites (**a**) Location, (**b**) SVF, (**c**) H/W ratio in Campinas and Santos counties, Brazil

Figure 18.6 shows results recorded for air temperature, PET and wind speed in 3 different selected sites in Campinas, namely: Arthur Resende St. (low-density zone with trees), Gil Vicente St. (low-density zone without trees) and Conceição St. (high-density zone without trees).

The highest air temperature recorded in Campinas County was observed for the low-density zone with trees (34° C). However, in terms of PET, the highest temperatures were 49.5° C (for the low-density zone without trees) and 50° C (for the high-density zone without trees). On the other hand, the highest air temperature recorded in Santos County was observed for the high-density zone with trees (30° C). In terms of PET, the highest temperature was the one recorded for the high-density zone with trees (37° C).

Although the climate in Santos County is warmer than that of Campinas County, air temperature in all analyzed sites remained within the range of 26° C to 34° C. If one compares results observed for high-density zones without trees in both counties, it is possible to state that wind permeability observed in the urban area of Santos County derives from its urban shape and proximity to the sea, and that it can help improve thermal comfort in this county. In addition, the vegetated area and the water-body located close to the site where field measurements were taken can help maintain its air humidity and improve its thermal comfort conditions. Additionally, sites with trees (low- and high-density zones with trees in Campinas County) enabled more thermal comfort than sites without trees.

Based on the comparison between low-density zones without trees and low-density zone with trees in Campinas, trees helped reduce urban temperature by approximately 18 °C. Based on the comparison between these results and that of clusters of species *C. pluviosa* F., in Sect. 18.2, trees planted in streets performed better in reducing PET temperatures than those planted in free flat areas due to materials used both in paving and facades. It is important to highlight that clusters of species *C. pluviosa* F. can reduce urban temperature by 16 °C and promote thermal comfort from 11 p.m. to 15 a.m., in summer, whereas low-density zones with trees only promote thermal comfort at 11 p.m.

Based on the comparison between high-density zones without trees and high-density zones with trees in Santos County, trees helped reducing urban temperature by approximately 10° C. Based on the comparison between these results and the one recorded for individual trees belonging to species *S. Cummini,* in Sect. 18.2, trees planted in streets performed worse in reducing PET temperatures than trees grown in free flat areas due to high-density construction. Although buildings provide shade and reduce PET temperatures, trees planted in streets can provide thermal benefits, as well as improve urban aesthetics.

The following results can be extracted from this section:

- Trees on the streets can reduce PET by 18 °C in low-density sites, as well as by 10 °C in high-density with trees. In addition, they can provide human thermal comfort, both on the streets and in buildings.
- Planting design can influence solar radiation attenuation and help reducing PET temperatures, regardless of urban shape.
- Planting the right tree species on the streets can provide thermal comfort for pedestrians and cyclists, as well as help reestablish thermal conditions in urban environments.

4 Conclusions

The current chapter presented results of field measurements which were taken to investigate the benefits provided by trees in different contexts: individually planted trees, clusters of trees, and trees used as part of urban configuration. The influence of certain tree species on urban microclimate was assessed through field measurements. Individual trees, such as those belonging to species *S. cumini* L., *M. indica* L., and *C. pluviosa F.*, were capable of providing higher thermal comfort throughout the day, in both seasons. These trees presented similar species-related features such as density, trunk structure and crown shape. Other features such as leaf size, shape and color can also influence trees' solar radiation attenuation ability; consequently, they can help improving thermal comfort.

Tropical cities, such as Campinas and Santos, should use trees to control thermal comfort for pedestrians and cyclists. The solar radiation intercepted by tree crowns provides natural protection in outdoor spaces by mitigating temperatures and reducing the energy spent to cool down indoor spaces.

With respect to outdoor spaces, urban forestry is capable of controlling and improving thermal comfort, mitigating air temperature, and controlling relative humidity; consequently, it is capable of providing thermal comfort. As for indoor spaces, shade can help reducing the influence of solar radiation on facades, improving thermal comfort and saving energy spent on cooling processes.

Trees can help improve human thermal comfort in different manners. Individual trees and clusters of trees grown in free flat areas enable more human thermal comfort than the ones grown on sidewalks. However, sidewalks without trees in tropical regions are extremely uncomfortable for pedestrians. Some studies have shown that pedestrians living in tropical cities prefer walking in the shade than in the sun (Lin et al., 2010; Souza et al., 2023); this feature was analyzed in the current study. Results have suggested that trees can mitigate urban heat island effects, as observed in the case of Campinas County. Furthermore, shade trees and low wind speeds can enable more outdoor thermal comfort in tropical cities.

Studies about trees are important tools to assess the contribution of vegetation to mitigate, and to adapt cities to, climate change. Greening cities is a powerful strategy used to reduce the negative impact of urban heat islands on cities. However, urban planning processes must also take into consideration the association between climate, urban configuration and tree features. Measuring the thermal comfort provided by tree species, and by their distribution in urban environments, helps designers' decision-making processes.

These findings provide important information to be used by urban managers, urban planners and architects who aim at improving urban areas and at re-establishing the local population's well-being and urban health as part of their daily life.

Acknowledgments The authors thank the State of São Paulo Research Foundation (FAPESP) and Coordination for the Improvement of Higher Education Personnel (CAPES).

References

Abreu, L. V., & Labaki, L. C. (2010). *Different arboreal species provides specific influence radius in microclimate*. Proceedings of 3rd International Conference on Passive and Low Energy Cooling for the Built Environment, Rhodes. 29 setembro—1 outubro.

Abreu, L. V., & Labaki, L. C. (2011). *Evaluation of direct and indirect effects of different arboreal species on microclimate*. Louvain-la-Neuve.

Abreu-Harbich, L. V., Labaki, L. C., & Matzarakis, A. (2015). Effect of tree planting design and tree species on human thermal comfort in the tropics. *Landscape and Urban Planning, 138*, 99–109. https://doi.org/10.1016/j.landurbplan.2015.02.008

Akbari, H. (2002). Shade trees reduce building energy use and CO2emissions frompower plants. *Environmental Pollution, 116*, 119–126.

Akbari, H., & Rose, L. (2008). Urban surfaces and heat Island mitigation potentials? *Human–Environment System, 11*(2), 85–101.

Akbari, H., & Taha, H. (1992). The impact of trees and white surfaces on residential heating and cooling energy use in four Canadian cities? *Energy, 17*(2), 141–149.

Alcoforado, M. J., & Matzarakis, A. (2010). Urban climate and planning in different climatic zones. *Geographical, 57*, 5–39.

Angeoletto, F. H. S., Richard, E., Santos, J., Guedes, N., Oliveira, S. L., Albertín, R. M., et al. (2023). When we think of urban afforestation in Brazil, what comes to mind is a mouth that has lost its teeth. Terr@. *The Pluralist, 17*, e2321664–e2321654.

Bueno-Bartholomei, C. L. (2003). *Influência da vegetação no conforto térmico urbano e no ambiente construído*. Faculdade de Engenharia Civil, Arquitetura e Urbanismo.

Correa, E., Ruiz, M. A., Canton, A., & Lesino, G. (2012). Thermal comfort in forested urban canyons of low building density. An assessment for the city of Mendoza, Argentina. *Building and Environment, 58*, 219–230.

Deng, J., Pickles, B. J., Smith, S. T., & Shao, L. (2020). Infrared radiative performance of urban trees: spatial distribution and interspecific comparison among ten species in the UK by in-situ spectroscopy. *Building and Environment, 172*. https://doi.org/10.1016/j.buildenv.2020.106682

Emmanuel, M. R. (2005). *An urban approach to climate-sensitive design: Strategies for tropics* (p. 172p). E & FN Spon Press.

Emmanuel, R. H., Rosenlund, H., & Johansson, E. (2007). Urban shading—A design option for the tropics? A study in Colombo, Sri Lanka. *International Journal of Climatology, 27*(14), 1995–2004.

Givoni, B. (1989). Urban Design in Different Climates. *WMO Technical Report, 346*.

Givoni, B. (1991). Impact of planted areas on urban environmental quality. A review. *Atmospheric Environment, 25B*, 289–299.

Green, S. (1993). Radiation balance, transpiration and photosynthesis of an isolated tree. *Agricultural and Forest Meteorology, 64*, 201–222.

Holst, T., et al. (2004). Measuring and modelling plant area index in beech stands. *International Journal of Biometeorology, 48*, 192–201.

Hwang, R.-L., Lin, T.-P., & Matzarakis, A. (2011). Seasonal effects of urban street shading on long-term outdoor thermal comfort. *Building and Environment, 46*(4), 863–870. https://doi.org/10.1016/j.buildenv.2010.10.017

Kottek, M., Grieser, J., Beck, C., Rudolf, B., & Rubel, F. (2006). World map of the Köppen-Geiger climate classification updated. *Meteorologische Zeitschrift, 15*(3), 259–263.

Li, Z., Zhang, H., Juan, Y., Wen, C., & Yang, A. (2021). Effects of building setback on thermal comfort and air quality in the street canyon. *Building and Environment, 208*(15), 108627. https://doi.org/10.1016/j.buildenv.2021.108627

Lin, T. P. (2009). Thermal perception, adaptation and attendance in a public square in hot and humid regions. *Building and Environment, 44*(10), 2017–2026.

Lin, T. P., & Matzarakis, A. (2008). Tourism climate and thermal comfort in Sun Moon Lake, Taiwan. *International Journal of Biometeorology, 52*, 281–290.

Lin, T., Matzarakis, A., & Hwang, R. (2010). Shading effect on long-term outdoor thermal comfort. *Building and Environment. Building and Environment, 45*, 213–221.

Lombardo, M. A. (1985). *Heat Island in Metropolis: Case of Sao Paulo*. Hucited.

Matzarakis, A., & Mayer, H. (1996). Another kind of environmental stress: Thermal stress. WHO collaborating centre for air quality management and air pollution control. *Newsletters, 18*, 7–10.

Matzarakis, A., Rutz, F., & Mayer, H. (2007). Modelling radiation fluxes in simple and complex environments—Application of the RayMan model. *International Journal of Biometeorology, 51*, 323–334.

Matzarakis, A., Rutz, F., & Mayer, H. (2010). Modelling radiation fluxes in simple and complex environments—Basics of the RayMan model. *International Journal of Biometeorology, 54*, 131–139.

Monteiro, L. M., & Alucci, M. P. (2010). *Proposal of an outdoor thermal comfort index for subtropical urban areas*. Proceedings of 3rd International Conference on Passive and Low Energy Cooling for the Built Environment, Rhodes. 29 setembro—1 outubro 2010.

Oke, T. (1984). Methods in urban climatology. *Applied Climatology, 14*, 19–29.

Peixoto, M. C. L. L. C. R. F. (1995). Conforto Térmico Nas Cidades: Avaliação De Efeitos Da Arborização No Controle Da Radiação Solar

Robinette, G. (1972). *Robinette, plants, people, and environmental quality*. USDI, National Park Service.

Santamouris, M. (2001). *Energy and climate in the urban built environment*. James & James.

Santos Nouri, A., Fröhlich, D., Matos Silva, M., & Matzarakis, A. (2018). The impact of *Tipuana tipu* species on local human thermal comfort thresholds in different urban canyon cases in Mediterranean climates: Lisbon, Portugal. *Atmosphere, 9*, 12. https://doi.org/10.3390/atmos9010012

Scudo, G. (2002). Thermal comfort in green spaces Green structures and urban planning Milan. Accessed in: http://www.greenstructureplanning.eu/COSTC11/comfort.htm

Souza, F. L., Fenoglio, M. S., & Angeoletto, F. (2023). To be a Brazilian City dweller, sometimes we must learn to say enough! *Sustainability, 15*(4), 3699.

Taha, H., & Meier, A. (1997). *Mitigation of urban heat islands: Meteorology, energy, and air quality impacts* (pp. 124–163). CREST/JST.

Vogt, J., Lauerbach, H., & Langner, M. (2003). *The influence of urban vegetation on air flow* (pp. 471–474). Lodz, s.n.

Chapter 19
Perpetuating Protection of Brazil's Legal Reserve Area and the Future of Tropical Cities in the Brazilian Amazon

Cláudio Angelo Corrêa Gonzaga, José Guilherme Roquette, Fabio Angeoletto, and Normandes Matos da Silva

Abstract In Brazil, the Legal Reserve is an area in a rural property established by Brazilian law to conserve native vegetation due to its importance in maintaining biodiversity and ecosystem functions for ecological balance. In urban areas, the Law for the Protection of Native Vegetation states that the Legal Reserve is to be converted into green areas. However, it does not make it clear whether this is automatic and mandatory or an option for the public authority, nor what must be the extension of this area or its level of protection. In this chapter, we present an ecological interpretation of Brazilian legislation, arguing that the incorporation of rural areas into the urban perimeter and its subsequent expansion requires the preservation of the Legal Reserve's ecological functions. We demonstrate that the permanent conservation of these vital areas is supported by a robust legal framework, including the Law for the Protection of Native Vegetation (LPNV), the Brazilian Constitution, and the international Kunming-Montreal Global Biodiversity Framework. This interpretation ensures the continuity of essential ecosystem services and biodiversity

C. A. C. Gonzaga (✉)
Programa de Pós-graduação em Ciência Ambiental, Universidade de São Paulo, São Paulo, São Paulo, Brazil
e-mail: claudioangelo@usp.br

J. G. Roquette
CAEx Ambiental, Ministério Público do Estado de Mato Grosso, Cuiabá, Brazil

F. Angeoletto
Programa de Pós-Graduação em Geografia da Universidade Federal de Rondonópolis, Mato Grosso, Brasil

Programa de Pós-Graduação em Gestão e Tecnologia Ambiental, Universidade Federal de Rondonópolis, Rondonópolis, Brasil

PUMAS, Pessoas Unidas pelo Meio Ambiente, Agricultura & Sociedade, Chapada, Brasil
e-mail: angeoletto@ufr.edu.br

N. M. da Silva
Programa de Pós-Graduação em Gestão e Tecnologia Ambiental, Universidade Federal de Rondonópolis, Rondonópolis, Brazil

F. Angeoletto et al. (eds.), *Ecology of Tropical Cities*,
https://doi.org/10.1007/978-3-031-48012-6_19

conservation as urban areas expand. The characteristics and ecosystem functions of these native vegetation fragments provide an ecological equilibrium for present and future generations when incorporated into the urban areas, mainly due to their capacity for mitigating the negative impacts of urbanization on the well-being of the population, conserving urban biodiversity and for adapting to and mitigating climate change.

Keywords Law for the Protection of Native Vegetation · Green areas · Urban environment · Urban biodiversity conservation · Ecosystem functions · Climate change

1 Introduction

The term "urban biodiversity" may sound like an oxymoron since urbanization is associated with land use change and human impacts on the landscape. In its original meaning, urban, from the Latin "Urbanus", refers to "relating to or belonging to the city," as well as "refined" and "sophisticated" as opposed to "the wild," which, according to the dictionary entry, has the meaning of "that which is born or develops in a disorderly way, without control or rules."

However, due to the increasing urbanization of the planet with the resulting increase in urban agglomerations, urban biodiversity has become undeniably important, given that the urban landscape is also the habitat for non-human species (Goddard et al., 2021; Angeoletto et al., 2023; Lima et al. 2023; Santangelo et al. 2022; Souza et al., 2023; Caizergues et al. 2024). Furthermore, the essential ecosystem functions of native vegetation contribute to the ecological balance in the urban landscape by helping to reduce the temperature and relative air humidity, noise, and air pollution. They also aid in flood prevention, improve landscape aesthetics, the relations of humans with nature, and provide other benefits that help to promote well-being (Rumble et al., 2019; Angeoletto et al. 2023; Moroń et al. 2024).

In this context, the Legal Reserve Area (Figs. 19.1 and 19.2) assumes enormous importance since, from the perspective conserving urban biodiversity, it involves conservation of large fragments of native vegetation in the middle of urban landscapes in areas with remarkable biodiversity, such as the Amazon Forest, the Atlantic Forest and the Cerrado (Silva et al., 2017; Vacchiano et al., 2019; Gonzaga et al. 2024).

In Brazilian legislation, the creation of Legal Reserve is found in Art. 3, III of the Law for the Protection of Native Vegetation (LPNV), Federal Law no. 12651/2012, which defines such a Reserve as:

> An area located inside a rural property or landholding, delimited under the terms of Art. 12, with the function of ensuring sustainable economic use of the natural resources of the rural property, assisting the conservation and rehabilitation of ecological processes and promoting biodiversity conservation, as well as shelter and protection for wild life and native flora.

Currently, this area varies from 20% of the area of rural property for properties located outside Brazilian Legal Amazon (BLA), to 80% of the forest areas of rural

Fig. 19.1 Legal Reserve Area near the urban area in Rondonópolis, Mato Grosso, Brazil. (Photo: Roberto Bueno Luiz)

Fig. 19.2 Legal Reserve Area near the urban area in Rondonópolis, Mato Grosso, Brazil. (Photo: Roberto Bueno Luiz)

properties in the BLA (some transition rules establish lower limits of 50%, for these properties). The Legal Reserve is 35% in BLA for properties in the Brazilian savanna (Cerrado), though transition rules for the Cerrado may maintain it at 20%.

In summary, it is a portion of native vegetation (forest or other types of vegetation) that must be conserved on rural properties because of its ecological services. Usage rules are quite restrictive: only some low-impact activities are allowed (collection of fruits, seeds, and small amounts of wood according to a forest management plan approved by environmental agencies).

Although these Reserves are defined for rural properties or landholdings, when a rural property becomes part of the urban area or urban expansion area, the Legal Reserves are transformed into Green Areas, a different classification (under urban law) that will change the characteristics of the protected area. Sometimes, these areas are transformed into public squares, private lots, or institutional spaces donated to the municipality, often resulting in the suppression of native vegetation and not uncommonly its replacement with exotic ornamental vegetation. In this chapter, we argue that this transformation must occur without prejudice to the functions of "assisting the conservation and rehabilitation of ecological processes and promoting biodiversity conservation, as well as providing shelter and protection for wildlife and native flora" which justify the existence of the Legal Reserve.

Acceptance or not of this statement can have dramatic consequences for Brazilian cities, including with regard to adapting to and mitigating the effects of climate change. With that in mind, in this chapter, we investigate this issue: how should one interpret the rule contained in the Law for the Protection of Native Vegetation, which establishes "the transformation of Legal Reserves into Green Areas in urban expansions (…) without the loss of their functions in cases of transmission or modification of the properties"?

2 History of Brazil's Legal Reserve

In Brazil, a country named after a tree (Brazilwood), concern for the conservation of native vegetation is longstanding, dating back to the colonial period, albeit initially for utilitarian reasons. During this era, due to the scarcity of wood used in shipbuilding, the Portuguese crown established its right to exploit hardwood timber from certain tree species. This led to the expression "madeira de lei" (literally "law wood," meaning legally reserved timber) (Dean, 1996).

The definition that an area of native vegetation on private property should be conserved emerged with the advent of the Brazilian Forest Code of 1934 (BFC/1934), which inaugurated the definition that the cut of forests in private properties would be limited to three-quarters of their total area. Thus, a quarter (the "fourth part") of the forests on such rural properties should be conserved. This can be considered the beginning of Brazil's Legal Reserve. This, however, was a flawed solution, since it opened up the possibility that, through successive subdivisions of the property, the area of native vegetation would be successively reduced by three-quarters.

The Brazilian Forest Code of 1965 (Law 4771 of 1965, hereafter BFC/1965) innovated by instituting the Permanent Preservation Area and the Legal Reserve Area with the characteristics currently in force. Furthermore, the BFC/1965 made it clear in its Article 1 that the "forests existing in the national territory" and "other forms of vegetation", are "assets of common interest to all inhabitants of the Country".

Specifically, regarding the Legal Reserve, the BFC/1965 provided (in Art. 16) that:

> In the Southeast, and Center-West regions, in the southern portion of the latter, the felling of native, primary or regenerated forests will only be allowed, provided that, in any circumstance, the minimum limit of 20% of the area of each rural property will respect the forest cover.

In these regions, there was thus a reduction in protection, compared to that provided for in BFC/1934, from ¼ to $\frac{1}{5}$ of the area to be protected on private properties with the advent of BFC/1965.

However, Art. 44 of the BFC/1965 made a distinction between the percentage of Legal Reserve in the Amazon and that in other regions of Brazil. This article established that "in the North region and in the norther part of the Center-West region, suppression of native vegetation would only be allowed "provided that at least 50% of the area of each property remains under tree cover". That means that in these regions, the percentage of protected areas has doubled.

The BFC/1965 would undergo further important changes. In 1986, Federal Law 7511 amended it to increase the extensions of permanent preservation areas (PPAs). PPAs are areas within private properties with even more stringent protection rules and use restrictions; the most common form of PPA being the marginal strips of native vegetation along water courses. The boundaries of the PPAs were again changed by Federal Law 7803/1989, which finally created the expression "Legal Reserve" and amended BFC/1965 to also bring the obligation to maintain the Legal Reserve in the record on the property title, establishing that "the alteration of its destination, in cases of transmission, in any capacity, or of dismemberment of the area is prohibited" eliminating the possibility of successive reduction of the Legal Reserve Areas through subdivision of rural property.

The concept of the Legal Reserve was also mentioned by Federal Law 8171/1991, which stipulated a period of 30 years to fulfill the legal obligation to recompose the Legal Reserve (Brasil, 1991).

In 1996, after the record 29,059 km^2 of deforestation recorded in 1995, Provisional Measure N. 1511/1996 changed the wording of Art. 44 of the BFC/1965 to prohibit clearcutting in 80% of the areas of native vegetation located on properties situated "in the North region and in the North part of the Midwest region". As explained by Cunha (2019):

> The provisional measure was reissued monthly, and new changes were introduced in the BFC/1965. In May 2000, the Forest Code was largely modified by PM n°. 1,956-50/2000, based on an extensive plural debate promoted by the National Council for the Environment (CONAMA), with broad public participation. The text of this standard was reissued without changes in the following months, ending in the PM 2,166-67/2001, which was in effect until May 2012.

Finally, in 2012, the New Brazilian Forest Code, also known as the Law for the Protection of Native Vegetation (its more appropriate name), was approved. The new law maintained the percentages of the Legal Reserve Area provided by BFC/1965 (with its successive amendments); however, it introduced a series of rules to reduce the obligation for its recovery, such as the provision (Art. 67) those small rural properties that on July 22 of 2008 had a Legal Reserve at a percentage lower than the minimum established would not need to recover or compensate the native vegetation deficit.

Additionally, in the rural properties that do not qualify as small rural properties and that, on July 22, 2008, had a Legal Reserve in an extension smaller than that established in the LPNV, the Legal Reserve deficit may be regularized through recovery or offsets with other areas located in the same biome in any states of Brazil, within 20 years (Art. 66).

The new law also included the provision that rural properties where the "suppression of native vegetation occurred in compliance with the Legal Reserve percentages provided by the law in force at the time the suppression occurred" would not be required to recover or compensate the Legal Reserve (Art. 68, LPNV).

3 The Perpetuity of the Protection of Native Vegetation in Brazil's Legal Reserve Area

Reinforcing the rule that prohibits the mischaracterization of the Legal Reserve (which had already been in place since Law 7803/1989), the LPNV provided more explicit rules on the perpetuity of the protection of the Legal Reserve Area by stating, for example, in its Articles 18 and 19 that:

> Art. 18. The Legal Reserve area must be registered with the competent environmental agency through registration in the CAR [Rural Environmental Registry] referred to in Art. 29, and is forbidden to change its destination, in cases of transmission, in any capacity, or dismemberment, except for under the exceptions provided for in this Law. (…)

> Art. 19. The inclusion of rural property in an urban perimeter defined by municipal law does not relieve the owner from maintenance of the Legal Reserve Area, which will only be extinguished concomitantly with the registration of the subdivision of the land for urban purposes approved according to the specific law and in accordance with the guidelines of the master plan referred to in § 1 of Art. 182 of the Federal Constitution.

Further on, in Article 25, item II, the law has a provision for the conversion of the Legal Reserve Area into a Green Area in areas of urban expansion, as an instrument available to the Municipalities for the establishing of Green Areas:

> Art. 25. For establishing urban Green Areas, the municipal government shall have the following instruments: (…)
>
> II - the transformation of Legal Reserves into Green Areas in urban expansions
>
> This understanding is also supported by the Statute for Cities (Law 10,257/2001), which, in setting out the general guidelines of urban policy (Art. 2°, XII) deals with the "protection, preservation and recovery of the natural and constructed environment, cultural, historical, artistic, scenic and archaeological heritage"

The Urban Land Subdivision Law (Law 6766/1979) also protects these areas by establishing that the "land subdivision will not be allowed: (…) V - in areas for ecological preservation (…)."

In the legal definition of the Legal Reserve Area, we have seen that one of its functions is "to assist the conservation and rehabilitation of ecological processes and promote the conservation of biodiversity" i.e., it is an area of ecological interest.

Below we explore some arguments for and against the thesis that in the subdivision of urban land, the Legal Reserve Areas must be converted into Green Areas that preserve the ecological functions of the Legal Reserve Area.

4 Arguments Against the Mandatory Conversion of Legal Reserve Areas into Ecological Green Areas

First, we must clarify that we understand green areas to be *ecological* green areas (such as natural parks) sometimes named in Brazilian urban laws as "contemplative green areas," in opposition to anthropized green areas, such as the construction of squares and gardens.

A first argument against the mandatory conversion of Legal Reserve Areas into Green Areas, regarding the percentage established in Art. 12 of the LPNV, is of a practical nature: although such conversion is feasible in Brazilian states located outside BLA, in the "Amazonian" states, given the much higher percentage of the Legal Reserve Area (which can reach 80% of rural properties located in forest areas and 35% in Brazilian savannah), it is argued that the requirement of Green Areas in the percentages established in the LPNV for rural properties would make urban expansion unfeasible.

A second argument against the thesis concerns the interpretation of the rule provided for in Article 19 of the LPNV, which states that the Legal Reserve Area "will be extinguished concomitantly with the registration of the subdivision of land for urban purposes approved according to the specific legislation and following the guidelines of the master plan referred to in § 1 of the Art. 182 of the Federal Constitution".

Article 182 of the Brazilian Constitution (BC) provides that the municipal master plan, approved by the Municipal Council, is "the basic instrument of the development and urban expansion policy". Some interpret the "specific legislation" mentioned by Art. 182 of the BC as the urban municipal legislation related to the subject.

As a primary instrument of the urban expansion policy, it is argued that the city master plan could provide (or not) for the mandatory conversion of Legal Reserve Areas into Green Areas.

Regarding the first argument (high percentage of the Legal Reserve in properties located in BLA), this is more theoretical than practical. As we explained above, there are so many exceptions to the rule (Articles 66, 67, and 68 of the LPNV) that it is unlikely that the Legal Reserve of a property located in the urban expansion area of a municipality located in BLA has 35% (Brazilian savanna) or 80% (forest) of Legal Reserve Area in practice.

For example: according to the LPNV (Article 67), properties considered "small" rural properties, which in municipalities located in BLA may exceed 800 hectares, will not have to recover the Legal Reserve Area if there was already no native vegetation on the property on July 22, 2008. This will be the situation for most properties in the urban expansion areas of municipalities in BLA.

In addition, properties larger than that and whose Legal Reserve Area on July 22, 2008, did not reach the dimensions described in Art. 12 of the LPNV may offset the deficit with a surplus of native vegetation located in other rural properties or conservation units (Article 66). This means that when a rural property located in the BLA in a forest area does not benefit from the rule of Art. 67 of the LPNV (absence of an obligation to recover the vegetation of the Legal Reserve), it would most likely fall under the rule of Article 66 (possibly offsetting the Legal Reserve with surplus native vegetation on another property).

A third rule that mitigates the obligation of transforming 80% of the Legal Reserve Area in a property into a Green Area is the possibility that the public authorities can reduce the Legal Reserve to up to 50% for recovery purposes when the municipality has more than 50% of its territory occupied by protected areas in the public domain and by ratified indigenous lands (Article 12, paragraph 4, of the LPNV).

Finally, in "extreme" cases of properties with 80% of their Legal Reserve preserved and located in the urban expansion area of a municipality located in BLA we could imagine alternatives more compliant with the Brazilian Constitution than simply disregarding the percentage of a Legal Reserve Area when partitioning urban land.

For example, through a specific law (in analogy to Article 225, paragraph 1, III, of the BC, which requires a law for the alteration or suppression of protected areas) and a declaration of public interest over part of the property, the environmental agency could authorize the compensation of part of the Legal Reserve Area on another rural property in that municipality. This would be similar to the reallocation of Legal Reserves for large infrastructure projects, such as hydropower, highways or railways. A similar rule, although not inspired by ecological purposes, was recently introduced in the Article 94-A of the Environmental Code of Mato Grosso State, located in the BLA, to authorize the relocation of Legal Reserves within rural properties for mineral extraction activities.

In this last hypothesis, urban expansion would not act as a driver for eliminating native vegetation through the mere approval of a land subdivision project for urban purposes. The developer would have to reallocate part of the Legal Reserve to another rural property within the same municipality or metropolitan region and river basin, thus avoiding net loss of biodiversity and ecosystem services in that territory. This interpretation would align with the growing concern for 'No Net Loss' in other countries (Sonter et al. 2018) and with the emerging legal framework in Brazil (as per Article 17 of Brazilian Federal Law 11428/2006), although currently limited to the Atlantic Forest biome.

Regarding the fact that the Green Area is an instrument of urban law different from the Legal Reserve Area, and also the fact that municipal legislation typically provides for specific local norms for the Green Area and its percentage, it must be considered that the competence to legislate on urban rules lies at the Federal and state levels (Art. 24, I, BC). Furthermore, the institutes of environmental law and urban law must be compatible because, in the light of Article 225 of the Brazilian Constitution, there are no possibilities in the Brazilian legal system for an "anti-ecological urban law".

Thus, the fact that Green Areas are an institute of Brazilian urban law is not enough to authorize the suppression of the ecological functions of the area of native vegetation protected as a Legal Reserve.

There is no complete independence between the reach of urban law and environmental law. Urban law cannot be thought of in a Cartesian logic, as if its object were an aseptic three-dimensional space far from ecological concerns, ready to be perfected by concrete constructions. Both in rural and urban areas, human beings depend on air, water, and other forms of life to sustain their fragile existence. Therefore, the urban and ecological aspects of the Green Areas must be considered two sides of the same coin.

5 Arguments in Favor of Mandatory Conversion of Legal Reserve Areas into Ecological Green Areas (Principle of the Perpetuity of Legal Reserve Protection)

The legal definition of the Legal Reserve in the LPNV reflects the connection between this instrument and the conservation of ecological processes and biodiversity. To think that the Legal Reserve Area exercises these functions only when the property is located in rural areas and that the need to conserve ecological processes and biodiversity disappears with urban expansion is a great mistake. Urbanization actually claims more protection for ecological processes and biodiversity conservation because of the well-being of human beings that such protection can entail (Fellowes et al., 2020).

In other words: the city and growing urban expansion make conservation of fragments of native vegetation even more important because these will directly influence the health and well-being of the population since, as demonstrated in several studies of environmental psychology, proximity to green spaces is associated with significantly lower levels of symptoms of depression, anxiety and stress (Angeoletto, 2012; Beyer et al., 2014; Duarte et al., 2017). Promoting Green Areas can thus be considered a strategy for improving the mental health of the urban population.

The preservation of large urban fragments of native vegetation, with the conversion of Legal Reserve areas into ecological green spaces is also an essential tool for mitigating and adapting to climate change. This is because these areas function as carbon sinks (the expansion of carbon sinks is one of the goals of Brazil's National Policy on Climate Change), as previously mentioned, and they can help regulate the local climate, moderating the high temperatures in tropical cities.

The BC establishes (Art. 225, paragraph 1, I) that, in order to promote the right to an ecological equilibrium of the environment, the government must "preserve and restore essential ecological processes and provide the ecological management of species and ecosystems" and, also (Art. 225, § 1, III), define "territorial spaces and their components to be specially protected," establishing that "any use that compromises the integrity of the attributes that justify their protection is prohibited."

If the Legal Reserve is a specially protected area (even if located on private properties), its reduction and subdivision at lower limits than those provided for in Article 12 of the LPNV imply compromising the integrity of the attributes (conservation of biodiversity and ecological processes) that initially justified its creation and existence, which would expressly contradict the constitutional rule provided in Article 225, § 1, III, of the BC.

In this sense, it can be understood that an interpretation consistent with Articles 18 and 25, II, of the LPNV must lead us to conclude that the Legal Reserve Areas must be transformed into ecological Green Areas and that the interventions carried out in these areas should only be those that have low environmental impact (such as the construction of trails and minimal infrastructure for ecotourism, leisure activities, etc.) and public interest that do not compromise their ecological functions and ecosystem services.

The competence to legislate on nature conservation and protection of the environment (Art. 24, VI, BC) and urban law (Art. 24, I, BC) belongs to the Federal level and the states. The city master plan must observe the general norms established by the Union regarding the transformation of the Legal Reserve Area into Green Areas, under penalty of invading the legislative competence of the Union and, thus, being unconstitutional.

6 The Kunming-Montreal Global Biodiversity Framework

Perhaps the strongest argument supporting the mandatory permanent conservation of Legal Reserves, even with the subdivision of rural properties for urban purposes, is based on the Kunming-Montreal Global Biodiversity Framework, adopted at the Convention on Biological Diversity's (CBD) 15th Conference of the Parties (COP). This legal framework explicitly highlights the critical role of biodiversity in cities, particularly in Target 12, and emphasizes several key aspects related to urban biodiversity:

"Significantly increase the area and quality and connectivity of, access to, and benefits from green and blue spaces in urban and densely populated areas sustainably, by mainstreaming the conservation and sustainable use of biodiversity, and ensure biodiversity-inclusive urban planning, enhancing native biodiversity, ecological connectivity and integrity, and improving human health and well-being and connection to nature and contributing to inclusive and sustainable urbanization and the provision of ecosystem functions and services."(Kunming-Montreal Global Biodiversity Framework, 2022).

Consequently, the conversion of Legal Reserves for human use would represent a clear violation of this international environmental agreement, which Brazil has signed and committed to uphold. According to an important 2022 precedent from the Brazilian Supreme Court, the landmark "Brazilian Climate Fund Case", international environmental agreements hold a "supra-legal" normative status, occupying an intermediate position between the Brazilian Constitution and Brazilian laws (Sarlet & Fensterseifer, 2022).

Therefore, any ordinary laws—whether at federal, state, or municipal levels—that violate this protection of urban biodiversity by permitting the land-use change of Legal Reserves for urban purposes may be declared invalid by judges or courts under the "conventionality control".

7 Conclusions

In light of the Kunming-Montreal Global Biodiversity Framework and the constitutional rules that deal with the protection of biodiversity, ecological processes, and protected areas, a given Legal Reserve Area must be converted into an ecological Green Area of the same size, allowing only low-impact interventions (in cases of social interest or public utility) that preserve the ecological functions defined in Article 3, III, of the LPVN, since such functions cannot be suppressed.

This is the only interpretation of Article 25, II, of the LPVN compatible with the right to an ecological equilibrium of the environment provided in Article 225 of the Brazilian Constitution and with the Target 12 of the Kunming-Montreal Global Biodiversity Framework. The practical consequence of this interpretation will be the conservation of the ecological processes of the Legal Reserve, the promotion of quality of life in Brazilian cities, their greater resilience to the climate emergency humanity is experiencing, and the conservation of urban biodiversity.

The arguments invoked against this conclusion are more economic than legal at heart, and, in practice, there are countless exceptions that mitigate the strict observance of the Legal Reserve percentages provided in Art. 12 of the LPNV.

Finally, in those rare cases where cities in Brazilian Amazon must adapt to the maximum reserve percentages provided for in the LPNV, we will be facing the beneficial challenge of imagining the future of tropical cities and their forest neighborhoods. May macaws, toucans and so many species of birds whose wings are tied by the destruction of their biomes paint the skies of our tropical cities with the psychedelic exuberance of their fantastic colors. We urbanites are only waiting for this moment.

References

Angeoletto, F. H. S. (2012). *Planeta ciudad: ecología urbana y planificación de ciudades medias de Brasil*. Tesis Doctoral. Universidad Autónoma de Madrid, España.

Angeoletto, F. H. S., Richard, E., Santos, J., Guedes, N., Oliveira, S. L., Albertín, R. M., et al. (2023). When we think of urban afforestation in Brazil, what comes to mind is a mouth that has lost its teeth. *Terr@Plural, 17*, e2321654.

Beyer, K. M., Kaltenbach, A., Szabo, A., et al. (2014). Exposure to neighborhood green space and mental health: Evidence from the survey of the health of Wisconsin. *International Journal of Environmental Research and Public Health, 11*(3), 3453–3472. https://doi.org/10.3390/ijerph110303453

Brasil. (1991). Lei Federal n° 8.171, de 17 de janeiro de 1991. *Dispõe sobre a política agrícola.* Diário Oficial da União. http://www.planalto.gov.br/ccivil_03/Leis/L8171.htm. Accessed 13 Dec 2021.

Caizergues, A. E., Santangelo, J. S., Ness, R. W., Angeoletto, F., Anstett, D. N., Anstett, J., ... & Johnson, M. T. (2024). Does urbanisation lead to parallel demographic shifts across the world in a cosmopolitan plant?. Molecular Ecology, 33(7), e17311.

Cunha, P. R. (2019). A proteção da vegetação em terras particulares: um breve histórico do Código Florestal Brasileiro. *Revista de Direitos Fundamentais, 1*(1), 61–81. https://doi.org/10.29327/213440.1.1-4

Dean, W. (1996). *A ferro e fogo: a história e a devastação da Mata Atlântica brasileira.* Cia. das Letras.

Duarte, T. E. P. N., Angeoletto, F., Richard, E., et al. (2017). Arborização urbana no Brasil: um reflexo de injustiça ambiental. *Terr@Plural, 11*(2), 291–303.

Fellowes, M. D., Acquaah-Harrison, K., Angeoletto, F., et al. (2020). Map-a-mole: Greenspace area influences the presence and abundance of the European mole Talpa europaea in urban habitats. *Animals, 10*(6), 1097. https://doi.org/10.3390/ani10061097

Gonzaga, C. A. C., Roquette, J. G., Angeoletto, F., Silgueiro, V. D. F., Macedo, L. O. B., Valdiones, A. P., & da Silva, N. M. (2024). The Public Prosecutor's Office's experience using Global Forest Watch to monitor and deter deforestation in the Cerrado. Environmental Conservation, 1–6.

Goddard, M. A., Davies, Z. G., Guenat, S., et al. (2021). A global horizon scan of the future impacts of robotics and autonomous systems on urban ecosystems. *Nature Ecology & Evolution, 5*, 219–230. https://doi.org/10.1038/s41559-020-01358-z

Kunming-Montreal Global Biodiversity Framework (2022). Decision adopted by the conference of the parties to the convention on biological diversity 15/4. *Montreal, Canada.*

Lima, G. F., Correa Santos, J. W. M., Albertin, R. M., Martínez-Miranzo, B., Souza, F. L., & Angeoletto, F. (2023). Backyards are a way to promote environmental justice and biodiversity conservation in Brazilian cities. *Diversity, 15*(7), 815.

Moroń, D., Beim, M., Gudowska, A., Angeoletto, F., Celary, W., Cwajna, A., ... & Tryjanowski, P. (2024). Evaluating tramway infrastructure on biodiversity and ecosystem services. Scientific Reports, 14(1), 9394.

Rumble, H., Angeoletto, F., Connop, S., et al. (2019). Understanding and applying ecological principles in cities. In F. Lemes de Oliveira & I. Mell (Eds.), *Planning cities with nature: Theories, strategies and methods* (pp. 217–234). Springer. https://doi.org/10.1007/978-3-030-01866-5_15

Sarlet, I., & Fensterseifer, T. (2022). Guardian of the Amazon: on the Brazilian Supreme Court's "Climate Fund Case" decision. Verfassungsblog, 6 jul. 2022.

Santangelo, J. S., Ness, R. W., Cohan, B., et al. (2022). Global urban environmental change drives adaptation in white clover. *Science, 375*, 1275–1281. https://doi.org/10.1126/science.abk0989

Silva, N. M., Angeoletto, F., Paranhos Filho, A. C., et al. (2017). The negative influences of the new Brazilian forest code on the conservation of riparian forests. *European Journal of Ecology, 3*(2), 116–122. https://doi.org/10.1515/eje-2017-0019

Sonter, L. J. (2018). Biodiversity Offsets: European Perspectives on No Net Loss of Biodiversity and Ecosystem Services, edited by Wolfgang Wende, Graham Tucker, Fabien Quétier, Matt Rayment, Marianne Darbi 2018. Switzerland: Springer International Publishing. IBSN 978-3-319-72581-9. Biological Conservation, 225, 251–252.

Souza, F. L., Fenoglio, M. S., & Angeoletto, F. (2023). To be a Brazilian city dweller, sometimes we must learn to say enough! *Sustainability, 15*(4), 3699. https://doi.org/10.3390/su15043699

Vacchiano, M. C., Santos, J. W., Angeoletto, F., & Silva, N. M. (2019). Do data support claims that Brazil leads the world in environmental preservation? *Environmental Conservation, 46*(2), 118–120. https://doi.org/10.1017/S0376892918000371

Chapter 20
The Ecological Sensitivity of Cities: Defining a System of Green Corridors

Raquel Weiss, Luis Guilherme Aita Pippi, Letícia de Castro Gabriel, Alice Rodrigues Lautert, Renata Michelon Cocco, and Luan da Silva Klebers

Abstract Population growth and increasing land demand in cities have led to the fragmentation and destruction of the landscape and of green and water-adjacent open spaces. Anthropic actions have systematically caused functional and ecosystemic degradation and loss of the landscape. As an aggravating factor, the master plans of Brazilian cities are, in general, ineffective and lacking in ecological planning, as they neglect the urban landscape and its green and water-adjacent open spaces. The lack of a systemic vision, combined with the scarcity of green and blue infrastructure, has worsened this scenario of negative impacts on landscape connectivity and multifunctionality. Ecological networks formed by green and blue corridors are gaining strength in the planning of urban landscapes, as they directly contribute to the ecosystemic development of human environments in an innovative and resilient way, thus mitigating anthropic impacts. Thus, this study aims to indicate the potential areas for the construction of a system of urban green corridors. It uses GIS and multi-criteria analyses for thematic map spatialization, wherevalues are assigned according to their importance within the study context considering the variables: water resources; land use and land cover; and quantitative landscape metrics. As a result, areas of potential and emerging ecological sensitivity were identified through the hierarchical analysis of green and blue corridors for the implementation of an integrated system of green corridors, interconnecting private and/or public spaces in Santa Maria/RS, Brazil.

Keywords Landscape · Open spaces · Ecological sensitivity · System · Green corridors

---

R. Weiss (✉) · L. G. A. Pippi · L. d. C. Gabriel · A. R. Lautert · R. M. Cocco · L. d. S. Klebers
Universidade Federal de Santa Maria (UFSM), Santa Maria, RS, Brazil
e-mail: raquel.weiss@ufsm.br; leticia.gabriel@ufsm.br

F. Angeoletto et al. (eds.), *Ecology of Tropical Cities*,
https://doi.org/10.1007/978-3-031-48012-6_20

1 Introduction

Most Brazilian cities do not give due importance and consideration to landscape and open space planning. In general, what occurs is the maintenance of leisure and recreation spaces, which belong to or are inserted in the urban mesh during the process of urban growth, and the delimitation of Areas of Permanent Preservation (APP), as a result of environmental legislation. The problem is that the landscape of a city is contemplated and perceived through its open spaces (Pellegrino & Moura, 2017), but these spaces have their own dynamics, as natural ecosystems which are constantly threatened by the land use and occupation-based zoning model if they are not considered from an ecological point of view.

Following the logic observed in large or metropolitan cities, medium-sized cities also encourage densification and verticalization in central areas, while suburban and peripheral areas present dispersion and fragmentation of the urban fabric. Thus, the opportunity to ensure connectivity and different functionalities of the landscape is lost, because anthropic actions are disconnected from environmental variables. Promoting the connection between environment and society is a priority for Brazilian cities. They must be planned to act as conservation spaces for biological diversity (Rumble et al., 2019; Angeoletto et al., 2022, 2023; Lima et al., 2023; Souza et al., 2023).

Another fundamental issue is that medium-sized cities are increasingly playing a key role in regional contexts and in the intermediation between large metropolitan urban centers and small towns and rural areas. They, therefore, present potential for providing natural stores (APP) and sites suitable for the creation of open spaces. However, it is necessary to rethink the urban-rural dichotomy, where urban is defined in opposition to rural, and deal with the multifunctionality and integration of the territory. In this way, landscape ecology presents itself as a groundbreaking field in the perception of landscape as a heterogeneous mosaic composed of interactive units according to a specific scale and observer (Metzger, 2001). Landscape metrics arise as a complement to landscape ecology, providing the ability to quantify and qualify the elements, configurations and patterns of the landscape (Bhatta et al., 2010; Archer, 2013).

The area of the present study is the urban center of the municipality of Santa Maria, in the state of Rio Grande do Sul, Brazil. There are six biomes in the Brazilian territory: Amazon, Caatinga, Cerrado, Atlantic Forest, Pampa, and Pantanal (Brasil, 2020) and the city of Santa Maria, located in the southern region of the country, lies within the transition zone between the Atlantic Forest and Pampa biomes (Follmann, 2018). The Atlantic Forest occupies about 15% of the Brazilian territory, although a considerable part of its original cover has been deforested or altered. It is a biome with substantial plant diversity, composed of native forests and associated ecosystems (Brasil, 2020). The Pampa, on the other hand, is characterized by native fields with herbaceous and bushy species and flat or gently undulating terrain. Forests are present in smaller proportions, on slopes or on the banks of waterways (Brasil, 2018). The Pampa is found in Argentina, Uruguay and southern Brazil, restricted only to the state of Rio Grande do Sul. The structure of its vegetation is simpler, when compared to forests, but no less important from the perspective of biodiversity and environmental services (Brasil, 2020). Santa Maria presents both portions of the Atlantic Forest in the northern region of the city, characterized by hills and forests, as well as the Pampa, with fields in the southern region (Macedo et al., 2018). Thus, Santa Maria is a relevant space with regard to the diversity of its natural and ecological landscape, which incorporates environmental characteristics of two significant biomes of Brazil.

The municipality has an area of 1781.8 km^2 (DEE, 2015), with an urban perimeter (UP) of 13,092 ha and a total population of 283,677 inhabitants (IBGE, 2020a, b), thus being characterized as a medium-sized city. The objective of this study is to present a scenario with ecological perspectives for the urban area of Santa Maria, from an ecological landscape planning perspective, identifying areas of sensitivity and emergent green and blue corridors. The methodology is based on the spatialization of thematic maps using geotechnologies and the application of landscape analysis through the treatment of multicriteria data which, linked to the spatial metrics of the landscape, result in an efficient model of tabulation and data management for ecological landscape planning.

2 Landscape Ecology and Landscape Ecological Planning

Landscape ecology has emerged as an alternative for finding solutions to natural and social conflicts, contributing to the understanding of relationships between heterogeneous landscapes in urban and rural areas. The principles of landscape ecology, when applied to the urban environment, can reorganize rural and urban land use at different scales, be it macro, meso or micro, and, at the same time, transform the landscape, recovering ecosystems and generating ecological functions conducive to sustainability (Queiroga, 2006).

The conception of landscape as an environment resulting from the relationship between society and the biophysical framework presupposes that interactivities occur in territorial and temporal space, leading to adaptations, transformations, readaptations and new transformations. This myriad of possibilities of conformations and configurations, permeated with social interaction with the framework and its resources, is called landscape (Magnoli, 1994; Tardin, 2008; Mcharg, 1969; Lyle, 1999; Macedo et al., 2018).

Landscape ecological planning allows an integrated management of anthropic actions in tune with natural processes. Landscape ecology therefore considers a whole complex of interacting systems and organisms, such as soils, climate, water resources, fauna, flora, built structures, physical environments and community-societies, which differ from each other and are in constant transformation. In this approach, the landscape is configured through an ecological mosaic, represented by the most varied ecosystems, providing multiple understandings and multiscale analyses (Mcharg, 1969; Tardin, 2008; Forman et al., 1996; Forman, 2019).

2.1 Landscape Structure and Methods of Landscape Planning and Ecological Sensitivity

From the perspective of landscape ecology, landscape can be analyzed as a system of spatial patterns, with different possible morphological and ecosystem combinations. As a result of the different spatial patterns, the landscape is characterized by exchanges, flows, gene flows, interactions and impacts from natural and anthropic dynamics inherent to the different configurations.

Landscape ecology proposes a systemic vision which aims at the ecological planning of the landscape. We can mention a few precursors in the development and application of methods of analysis and diagnosis and the proposal of landscape plans and projects in urban, rural, rurban, forest or desert areas.

Forman and Godron (1986) and Forman (1995) stand out as methodologies that define the landscape as an interactive and heterogeneous system. The authors characterize the landscape in accordance with three aspects: structure, function and change. Specifically, the structure of the landscape is composed of a territorial mosaic made up of the elements: matrix, patch or fragment and corridor (Bentrup, 2008; Forman, 1995, 2004, 2008, 2019).

The patch represents a homogeneous area of the landscape characterized by the same type of land use according to a certain scale and system. It is configured as the base component of a given class and is the fundamental unit of measurement and analysis (Bentrup, 2008; Forman, 1995; Weiss, 2016).

The class, as a set of all patches of a given use (Forman, 1995; Lang & Blaschke, 2009), is the main analysis for providing information about the patterns, distributions and variations in landscape development.

To analyze patches and classes, quantitative landscape metrics generate indices to describe landscape configurations and patterns (Bhatta et al., 2010; Archer, 2013). The quantitative data reflect structural characteristics of patches, patch classes and mosaics for geometrical and spatial properties (Herold et al., 2005), such as: shape, size, number, edge, diversity and organization. In this sense, spatial metrics have been increasingly used to quantify, monitor and understand the dynamics of landscape configuration and composition, being incorporated in environmental diagnosis studies to describe changes and compare configurations and interactions of spatial elements (Archer, 2013).

Corridors are elements with the potential to connect landscape fragments, open spaces and parks and, thus, communities and people. Landscape connections denote that all parts are connected to the whole and are necessary for the long-term maintenance of ecological and social functions as well as recreation and tourism. In this way, the constitution of ecological networks depends on several forms of connection: ecological networks, ecological corridors, green, blue and yellow corridors, green belts and contemplation areas (Jongman & Pungetti, 2004; Erickson, 2006; Bentrup, 2008; Pippi, 2014; Forman et al., 1996; Forman, 2019).

Green corridors can be defined based on landscape characteristics and configurations, location and spatial scale and the purpose and strategy of landscape planning.

These landscape connectors are areas that link forest and water remnants. This allows the gene flow between species of fauna and flora and the conservation of biodiversity. They also guarantee the conservation of vegetation, water resources and soil, in addition to contributing to climate and landscape balance. Because of their linearity, they have the ability to penetrate, permeate, and connect different landscapes, city environments, natural habitats and communities. Finally, they have the ability to promote the development of social capital through social integration and cohesion and the democratic promotion of social equity, health and well-being, leisure and recreation, alternative mobility and environmental education (Smith & Hellmund, 1993; Hellmund & Smith, 2006; Pippi, 2014; Flink, 2020).

Blue corridors, on the other hand, are made up of landscape units of the water network and aim at the preservation, conservation and ecological integrity of rivers, streams, lagoons, lakes, waterfalls and springs. Together with the network of green corridors, they can make up an efficient green and blue infrastructure, forging a healthy relationship between urbanization and plant and water resources. Such ecological networks have multiple functions: ecological, recreational, educational, social and tourism (Smith & Hellmund, 1993; Pippi & Trindade, 2013; Pippi, 2014).

McHarg's method (1969) is known for land use suitability analysis and is based on the formulation of an ecological inventory of a given region. In order to obtain data and information on natural processes, several thematic maps are created: vegetation, water resources, altimetry, slope, geomorphology, pedology, geotechnical, drainage, wildlife, urbanization, scenic values, historical values, land use values. Other maps representing different levels of preservation, conservation, and urbanization of specific areas (partial and global) are used to assign weights for ecological sensitivity (high, moderate, and low). By overlapping them, it is possible to identify areas as priority, appropriate, incompatible or vulnerable, compatible to community or ecological land uses (Mcharg, 1969; Mcharg & Steiner, 1998).

For Lyle (1999), ecological sensitivity models are based on the premise that sensitivity is characterized by certain attributes of the landscape and its relationships, which can be intensely or drastically altered by any form of human development. The model identifies areas with high ecological value, due to their productivity and diversity of fauna and flora. Any modification in these sensitive areas can result in losses. The sensitivity model varies in terms of scale and the possibilities for mitigation of impacts. Sensitivities are based on five variables: slope, soil type, geology, hydrography and vegetation. The model utilizes a sequential combination where each attribute is spatialized by determining a range of values: high (10–7), moderate (6–4) and low (3–0). Sensitivity levels are indicated by a color gradient, where the darker the color, the greater the ecological sensitivity, indicating areas that are more sensitive to urban occupation and in need of conservation and protection through Park Systems, Conservation Units and Ecological Corridors. The areas with moderate sensitivity can be occupied with controlled uses and with some limitation and caution. Environmental and ecological parks are among the uses indicated for these areas. The areas with lower value, being less ecologically sensitive sites, can be occupied by anthropic and agricultural uses (Mcharg, 1969; Mcharg & Steiner, 1998; Lyle, 1999).

For Ross (1990, 1994) and Tricart (1977), understanding the potential of natural resources entails the diagnosis of soil, relief, water resources, climate, fauna and flora and components of the geographical and ecological strata that support animal and human life. The analysis of environmental sensitivity requires the spatialization of these natural resources in categories and an integrated assessment, by overlapping, based on the principle of dynamic, physical and biotic functionality of the landscape. This identification of environmental fragilities is important for landscape planning, but the ecodynamic units must still be considered, because the resources are analyzed from a systemic point of view, where exchanges of energy and matter occur, processed through relationships in dynamic and ecosystemic equilibrium. On the other hand, when this dynamism is affected, these relations lose equilibrium and become temporarily or even permanently unstable. Ecodynamic units that are unstable or present emergent instability are intensely transformed by anthropic actions. Stable or potentially stable ecodynamic units are in their natural state, present a dynamic balance and have not been impacted by anthropic actions.

The Steinitz (2012) method, which provides alternative scenarios, both current and future, through analyses and models based on an understanding of the rules that drive the processes of change and testing of variables involved in the characterization and functionality of the landscape, can also be partially incorporated in a propositional simulation. However, the definition of Tricart (1977) and Ross (1994) was improved in our study by defining degrees of valuation for both stability and instability. That is, categories are proposed for each natural resource, ranging from very weak to weak, medium, strong and very strong. By spatializing the landscape and its attributes through its sub-units and elements, it is possible to hierarchize analytical scales to better understand its characteristics and synthesize results into negative and positive aspects. In this panorama, the methods of Saaty (1990) and Lyle (1999) apply hierarchization processes, using weights from 0 to 10. High weights indicate greater importance, while lower weights indicate low importance.

Table 20.1 merges the above-mentioned methodologies, complemented by Bentrup (2008), illustrating the landscape eco-dynamic units adopted and their levels of environmental stability, classified as stable (from 7.01 to 10), transitional (from 3.01 to 7) and unstable (from 0 to 3) environments.

3 Ecological Perspectives: Proposition of Scenarios

This study aims to propose scenarios or ecological perspectives of the urban landscape of Santa Maria from the choice of landscape variables with the potential for implanting green and blue corridors and other important areas in the composition of a system of open spaces to improve connectivity and the environmental, landscape, cultural, social and recreational quality of the city. By combining different methods, we elaborated spatializations using Geographic Information System (GIS) and Multicriteria Decision Analysis through the Analytic Hierarchy Process (AHP),

Table 20.1 Ecodynamic units and levels of environmental stability

Classification	Characteristics	Environmental vulnerability	Environmental sustainability
Stable Environments/ Stable Ecodynamic Units (7.01–10)	Preserved areas with morphodynamics in a natural state; Well-developed vegetation cover and little altered by anthropic actions (maintains natural characteristics or advanced forest regeneration); Well-preserved water resources; Balance between ecological potential and biological exploitation.	**Low Vulnerability: High Sustainability**	High sustainability: good productive capacity of natural resources; High biodiversity: fauna, flora and water habitats; Large core areas (habitats) and existence of ecological connectivities; Protection of sensitive habitats; Increase in natural areas and buffer zones as potential areas for recreation and leisure; Water potential with good surface runoff performance and groundwater accumulation, moderately deep soils with medium to high natural fertility; Dense vegetation with cover that reduces the impact of rainfall and consequently reduces the mechanical processes of morphogenesis; Moderate landform dissecting without violent incision of waterways; Flood control; Presence and quality of water; Quality of the soil: productive.
Transition Environments (3.01–7)	Natural areas moderately modified and degraded by anthropic actions; Ecological potential and biological exploitation tend to suffer easily from changes due to anthropic interventions and may lead to a change in the classification. If no measures are taken to stimulate environmental recovery or mitigate anthropic actions, the area will become unstable. On the other hand, if there are recovery measures and ecological management with environmental control and protection, the area may become stable.	**Moderate vulnerability: moderate sustainability**	Moderate sustainability: reasonable productive capacity of natural resources; Moderate vegetation with partial control of rainfall impacts; Moderate control of flooding.

(continued)

Table 20.1 (continued)

Classification	Characteristics	Environmental vulnerability	Environmental sustainability
Unstable Environments/ Unstable Ecodynamic Units (0–3)	Natural areas intensely modified and degraded by anthropic actions; Morphogenesis is the predominant element of natural dynamics, thus intense erosive processes and depletion of natural resources occur; Natural environments characterized by anthropic degradation, reaching critical situations of imbalance (absence of vegetation, canalization and rectification of water resources, landform dissecting).	**High vulnerability: low sustainability**	Low sustainability: minimal productive capacity with low natural fertility and water deficiency; Inexistence of dense vegetation cover and aggressive bioclimatic conditions; Strong and irregular rainfall and wind variations promoting erosive processes and damage to the quality of surface water; Plains and valley bottoms subject to flooding; Landform with vigorous dissecting; Presence of shallow soils (low nutrient and particle content); Intense bioclimatic aggressions (rainfall and winds).

Adapted from Tricart (1977), Ross (1994) and Bentrup (2008)

which coupled with Landscape Metrics, results in an efficient model of data tabulation and management for ecological landscape planning (Lang & Blaschke, 2009).

Multicriteria Decision Analysis is considered a decision-making support model, being bounded by a base set of criteria that can be measured and evaluated, by assigning weights according to their importance to the research objective. In this case, the variables and their category intervals are given weights according to their magnitude for the construction of ecological sensitivity (Gomes & Gomes, 2014; Lang & Blaschke, 2009; Lyle, 1999; Mcharg, 1969).

The Analytic Hierarchy Process is anchored in three principles: hierarchy construction (criterion levels), priority setting (peer judgment), and logical consistency. In this study, after the criteria were chosen, class categories for land use and land cover and landscape metrics were valuated on a scale of 0–10, where higher weights indicate higher ecological sensitivity and lower weights, lower ecological sensitivity. This step of assigning values for the elaboration of the maps was based on scientific research and applications, the experience of the researchers and field surveys, according to criteria indicated by Weiss and Pippi (2019), Santos et al. (Santos et al., 2010) and Saaty and Vargas (1979).

Among the land use and occupancy classes, spatial metrics of the landscape were applied, which generated quantitative data and revealed qualitative aspects of the integrity of the urban landscape of Santa Maria. As the metrics originated from landscape ecology concepts, the elements were quantified at the fragment-patch and class level, making up the variables in the Analytic Hierarchy Process application.

Three landscape metrics were adopted: shape index (SI)/mean shape index (MSI), core area and area. Based on these scales of analysis, we considered the aspect of spatial configuration, which considers aspects such as patch and edge size, geometry, positioning, and spatial distribution of the fragments. The ArcGIS ArcMap 10.6 software and the Patch Analyst extension were used to generate the data, with a mapping scale of 1/25,000 and a spatial resolution of 5 m per pixel. Five major methodological steps were used for the construction of scenarios, classified into potential and emergent ecological sensitivity scenarios (Fig. 20.1).

In Step A, the following criteria comprised the database: springs (A1), perennial rivers, intermittent rivers and the Vacacaí-Mirim River Dam (artificial reservoir used to supply the population) (A2), land use and coverage (A3) composed of forest-vegetation (A3.1), fields (A3.2), exposed soil and agriculture (A3.3) and urban (A3.4).

In Step B, buffers were established according to the Forest Code (Brasil, 2012), with 50 m for springs (B1). Adopting environmental legislation as criteria, multibuffers were also created with 30 and 50 m for perennial and intermittent rivers (B2) and 100 m for artificial lakes (dam) (B3). This is due to the need for buffer zones that gradually soften the impacts on the immediate APP area, as well as the ecological function of reducing the edge effect on these areas (Bentrup, 2008).

For land use and land cover, the mappings were separated at the class level, i.e., individualized spatialization for each use/cover. For forest-vegetation, the landscape metrics core area (B3.1), shape index (SI) (B3.2) and area (B3.3) were applied. SI (B3.2) generates a numerical index that considers the ratio of each patch's perimeter

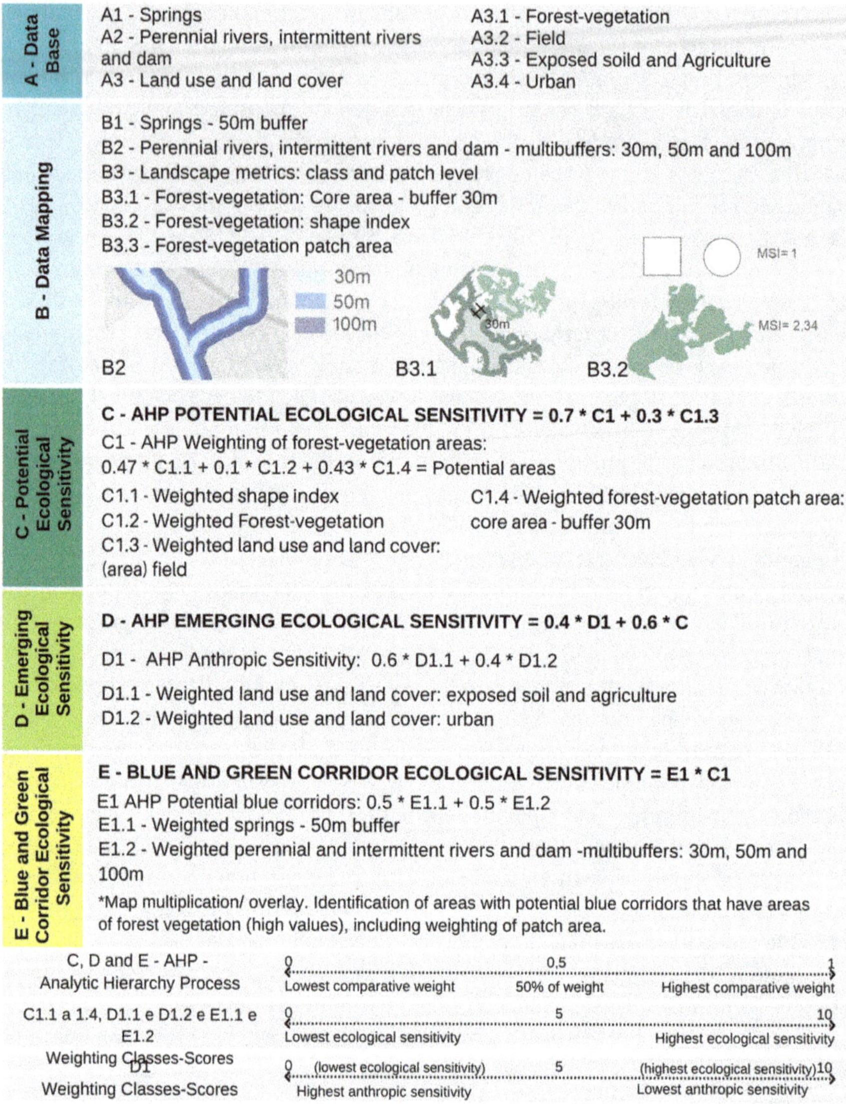

Fig. 20.1 Synthesis of the methodological processes

to the square root of its area, whose ideal value is 1. The ideal ratio between footage and area is found in the shape of a circle or square, where there is the greatest balance between the dimensions of the feature. Patches where this index is close to 1 show environmental quality and ecological relevance, as they suffer a lower edge effect and, consequently, lower degrees of disturbance, configuring a simpler and more solid morphology. On the other hand, patches with values above 1 represent an unbalanced coefficient, i.e., larger perimeter to area ratio, indicating possible edge effects. In this situation, there are geometries marked by indentations or protrusions. This typically occurs in vegetation areas in urban patches or those where the topography imposes on the land, leading to complex geometries and numerous elongated arms, resulting from gradual urban pressure. Therefore, the higher the index, the greater the complexity of the form, denoting patches with large edge effects, major interference and loss of environmental and ecological potential (Fig. 20.1 detail B3.2).

Another metric variable to evaluate the degree of ecological sensitivity is the core area (B3.1), which is configured as a negative buffer. It is one of the most important measures, because it reduces the area of a patch to an effective core area, where its most natural and conserved state is likely to be. The larger the core area, the greater the environmental and ecological potential, since it increases the capacity to shelter diverse flora and fauna, consequently, favoring gene exchange and flow.

Core area is directly related to the shape index. The more jagged and distant from the ideal shape of a circle or square, the smaller the core area will be. In many cases, although a fragment is large, if it has a high SI, it will lose substantial area and even parts of its original geometry.

Finally, the metric area (B3.3) consisted of quantifying the areas of each patch of the forest-vegetation land use and land cover class. The quantitative data allowed us to identify a general panorama of the integrity of the landscape by analyzing the number of patches, average area and minimum and maximum areas. Scenarios with many small patches, for example forest areas, may indicate a landscape that is significantly fragmented, disjointed and losing its ecological, environmental and landscape potential.

3.1 *Definition of Potential Ecological sensitivity and Emergent Ecological Sensitivity*

Methodological Step C consisted of establishing the potential ecological sensitivity of the city of Santa Maria (Fig. 20.2). Based on the studies of Tricart (1977), McHarg (1969) and Lyle (1999), the potential ecological sensitivity was defined as the qualitative level of the landscape considering only natural environmental variables, i.e. without considering land uses arising from anthropic actions. In this step, we sought to identify which areas would be the most ecologically sensitive by evaluating the areas of forest-vegetation using multicriteria decision analysis, followed by AHP applied to the spatial metrics previously mentioned.

In this step, weightings were applied for each metric map (C1.1, C1.2, C1.3 and C1.4) (Fig. 20.2). Five categorization classes were considered from the ArcGIS natural breakdown method, applying a reclassification by assigning weights from 1 to 10 (Table 20.2). Thus, Class 5, the highest class of the shape index, i.e., far above the ideal index 1, received a weight of 1 as it represents the scenario of low potential ecological sensitivity. The fourth group (second highest shape index values) was

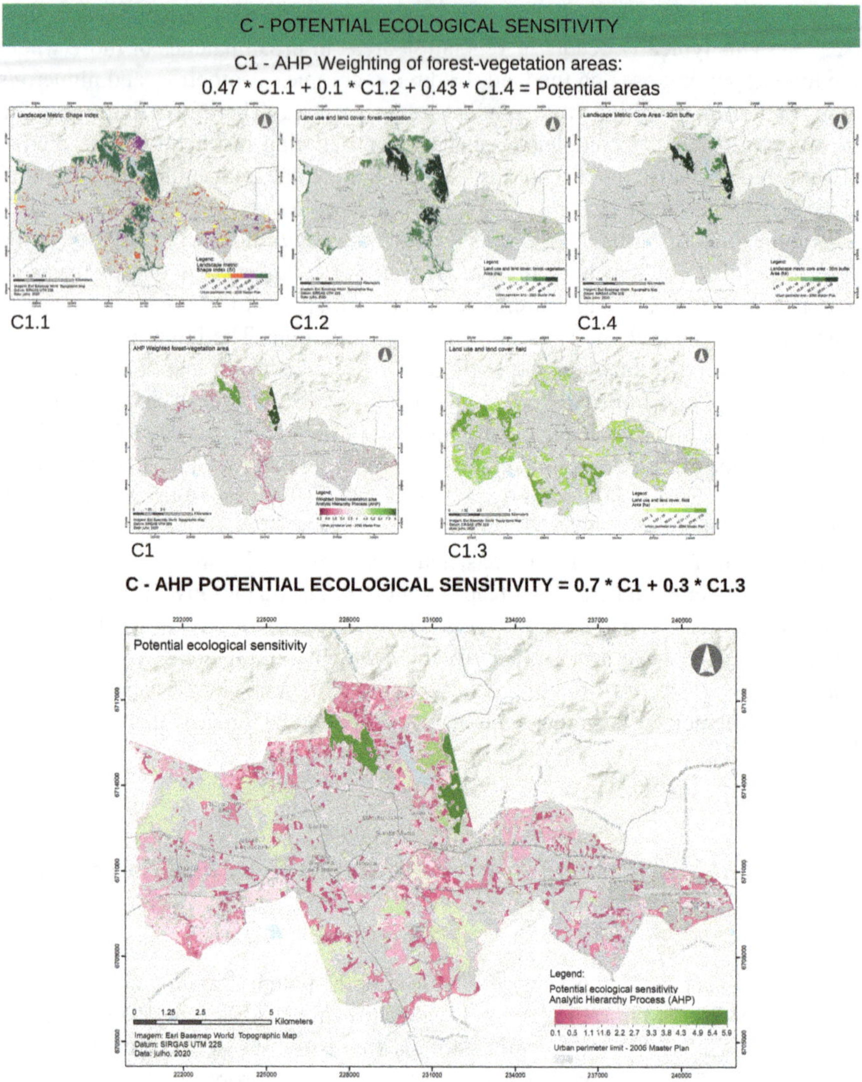

Fig. 20.2 Comparison of maps and levels of potential ecological sensitivity

assigned a weight of 3; the intermediate group was assigned a weight of 5; the second lowest group was assigned a weight of 7, and finally, with the highest weight of 10 indicated the scenario of high potential ecological sensitivity, that is, the class with the lowest shape index values.

The class intervals for Core Area and Area followed the same logic, according to the criteria for potential ecological sensitivity.

After class weights were assigned, Analytic Hierarchy Process (step C1) was carried out by defining weights for forest-vegetation for each metric in order to

Table 20.2 Weights for the spatial metric classes: forest-vegetation and field patches

C1.1. Metrics: Forest-vegetation Patch	**SI**		**C1.2. Metrics: Forest-vegetation Patch**	**Area**	
	Interval	Weight-Reclassification		Interval	Weight-Reclassification
Class 1	1–2	10	**Class 1**	0.0–6 (ha)	1
Class 2	2.01–3	7	**Class 2**	6.01–19 (ha)	3
Class 3	3.01–5	5	**Class 3**	19.01–62 (ha)	5
Class 4	5.01–7	3	**Class 4**	62.01–124 (ha)	7
Class 5	7.01–12	1	**Class 5**	124.01–243 (ha)	10
Areas UP (other uses)	NoData	0	**Areas UP (other uses)**	NoData	0
C1.3. Metrics: Field Patch	**Area**		**C1.4. Metrics: Forest-vegetation Patch**	**Core Area**	
	Interval	Weight-Reclassification		Interval	Weight-Reclassification
Class 1	0.0–7 (ha)	1	**Class 1**	0.0–6 (ha)	1
Class 2	7.01–22 (ha)	3	**Class 2**	6.01–18 (ha)	3
Class 3	22.01–46 (ha)	5	**Class 3**	18.01–42 (ha)	5
Class 4	46.01–76 (ha)	7	**Class 4**	42.01–65 (ha)	7
Class 5	76.01–219 (ha)	10	**Class 5**	65.01–145 (ha)	10
Areas UP (other uses)	NoData	0	**Areas UP (other uses)**	NoData	0

create the ecological sensitivity scenario (Eq. 20.1, Fig. 20.2). Weights were assigned in accordance with Saaty's method (1990). Then, in order verify the coherence of the weights, the consistency ratio (CR) was calculated, which was 0.0047, thus satisfying the CR prerequisite (>0.1). Thus, MSI/SI had a weight of 47% and Core Area presented a weight of 43%, which together made up 90% of the total evaluation. This is due to the fact that both variables provide greater accuracy in defining the levels of ecological sensitivity of a patch and class. With a lower weight, Area was assigned 10%, since there is no direct relationship between the largest fragment/largest area and the best situation of ecological sensitivity, considering that even large patches are subject to edge effects and small core area, in accordance with their geometry.

$$C1_{\text{weighting of forest-vegetation area}} = 0.47^{*}C1.1 + 0.1^{*}C1.2 + 0.43^{*}C1.4 \quad (20.1)$$

After considering the variables related to forest-vegetation, the final potential ecological sensitivity scenario was defined by comparing the result of Map C1 with Field Land Cover C1.3 (natural grass was considered to be Field Land Cover). In this way, it was possible to spatialize the areas in terms of the degree of natural conservation, identifying ecological and environmental relevance. The forest-vegetation areas had the greatest weight in the evaluation, representing 70%, while Field areas were assigned a weight of 30%[1] (Eq. 20.2 and Fig. 20.2).

$$C_{potencial\ ecological\ sensitivity} = 0.7^{*}C1 + 0.3^{*}C1.3 \tag{20.2}$$

To contrast potential ecological sensitivity and evaluate the impact of human actions on the urban perimeter, emergent ecological sensitivity was defined in step D, where variables resulting from anthropic actions were added to the construction of the scenario. These variables are land use and land cover—exposed soil and agriculture (D1.1) and land use and land cover—urban (D1.2). For both, weightings from 1 to 10 were assigned to categorize patch area into 5 classes using the natural breakdown method (Table 20.3). The larger the areas of exposed soil, agriculture and urban area, the lower the weights, as they represent lower levels of ecological sensitivity (Fig. 20.3).

After the processes of weighting and reclassification, the first analytic hierarchy process in step D defined the anthropic sensitivity (step D1). The variable exposed soil and agriculture was assigned 60% of importance in the construction of the scenario, considering these uses to be less impactful to the landscape than urban areas. Urban areas, on the other hand, were attributed an importance of 40%, due to their reduction of environmental and ecological quality (Eq. 20.3 and Fig. 20.3).

Table 20.3 Weights for land use and cover classes: exposed soil/agriculture and urban

D1.1. Exposed soil and agriculture	**Core Area**		**D1.2. Urban**	**Area**	
	Interval	Weight-Reclassification		Interval	Weight-Reclassification
Class 1	0.0–22 (ha)	10	**Class 1**	0.0–2 (ha)	10
Class 2	22.01–62 (ha)	7	**Class 2**	2.01–5 (ha)	7
Class 3	62.01–114 (ha)	5	**Class 3**	5.01–15 (ha)	5
Class 4	114.01–247 (ha)	3	**Class 4**	15.01–31 (ha)	3
Class 5	247.01–393 (ha)	1	**Class 5**	31.01–3749 (ha)	1
Areas UP (other uses)	NoData	0	**Areas UP (other uses)**	NoData	0

[1] The comparison of two variables (n = 2) in AHP give CR = 0.

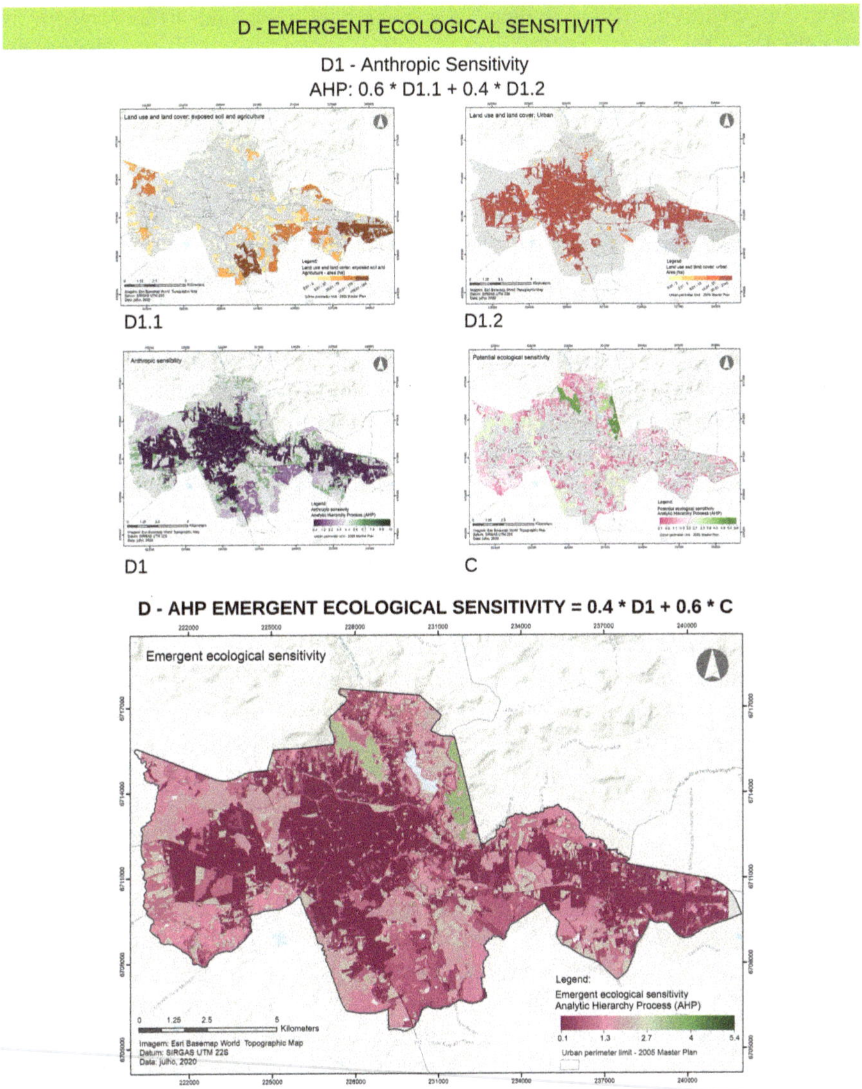

Fig. 20.3 Overlay of maps and levels of emergent ecological sensitivity

$$D1_{anthropic\ sensitivity} = 0.6^{*}D1.1 + 0.4^{*}D1.2 \tag{20.3}$$

Finally, step D entailed the definition of the emergent ecological sensitivity, by comparing anthropic sensitivity and potential ecological sensitivity. Through this process, it was possible to create a scenario to quantitatively and qualitatively demonstrate the degree of ecological sensitivity of the areas of Santa Maria. In this context, the mapping of potential ecological sensitivity (C) is dominant over anthropic sensitivity (D1), assigned percentages of 60% and 40%, respectively (Eq. 20.4 and Fig. 20.3).

$$D_{\textit{Emergent ecological sensitivity}} = 0.4^{*}D1 + 0.6^{*}C \tag{20.4}$$

3.2 Definition of Ecological Sensitivity of Blue and Green Corridors

Step E included environmental and ecological aspects of water resources, given their importance within the urban context of Santa Maria, characterized by numerous rivers and especially by a dam and artificial reservoir. These resources impose a dynamic of distribution and ocupation on the urban perimeter, as well as suffering from environmental vulnerability and anthropic impacts. The ecological sensitivity of blue and green corridors was defined through two phases: first, the identification of potential blue corridors (E1) and, subsequently, emergent blue and green corridors (E) (Fig. 20.4). To guide this process, a scenario was created showing the natural potential of the landscape and of human action.

Potential sensitivity was considered to be an ideal situation, i.e., with buffer areas around springs, rivers and dams, regardless of the presence of vegetation cover. For potential blue corridors (E1), an equal importance was given to each variable, that is, 50% for springs (E1.1) and 50% for perennial and intermittent rivers and dam (E1.2) (Eq. 20.5 and Fig. 20.4).

$$E1_{\textit{potential blue corridors}} = 0.5^{*}E1.1 + 0.5^{*}E1.2 \tag{20.5}$$

For weightings and reclassifications based on the buffers, the classes were also assigned values between 1 and 10, based on their contribution to ecological sensitivity (Table 20.4). Thus, the 50 m buffer for springs was assigned a weight of 10, given its fundamental importance in the hydrographic landscape. The 30 m, 50 m and 100 m multibuffers for rivers and dam were assigned weights of 10, 7 and 4, respectively, considering the role they play as buffers of environmental impacts, as well as legal constraints, vulnerabilities and the degree of anthropization they may suffer.

The second phase of Step E defined emergent blue and green corridors, where weighted forest-vegetation-forest area was included in the analysis. This consisted of an overlay of areas in order to spatialize the forest-vegetation areas within the different buffers, which would indicate the most favorable blue ecological sensitivity scenario. To identify the overlay in these areas, the maps were multiplied (Eq. 20.6 and Fig. 20.4).

$$E_{\textit{emergent blue and green corridors}} = E1^{*}C1 \tag{20.6}$$

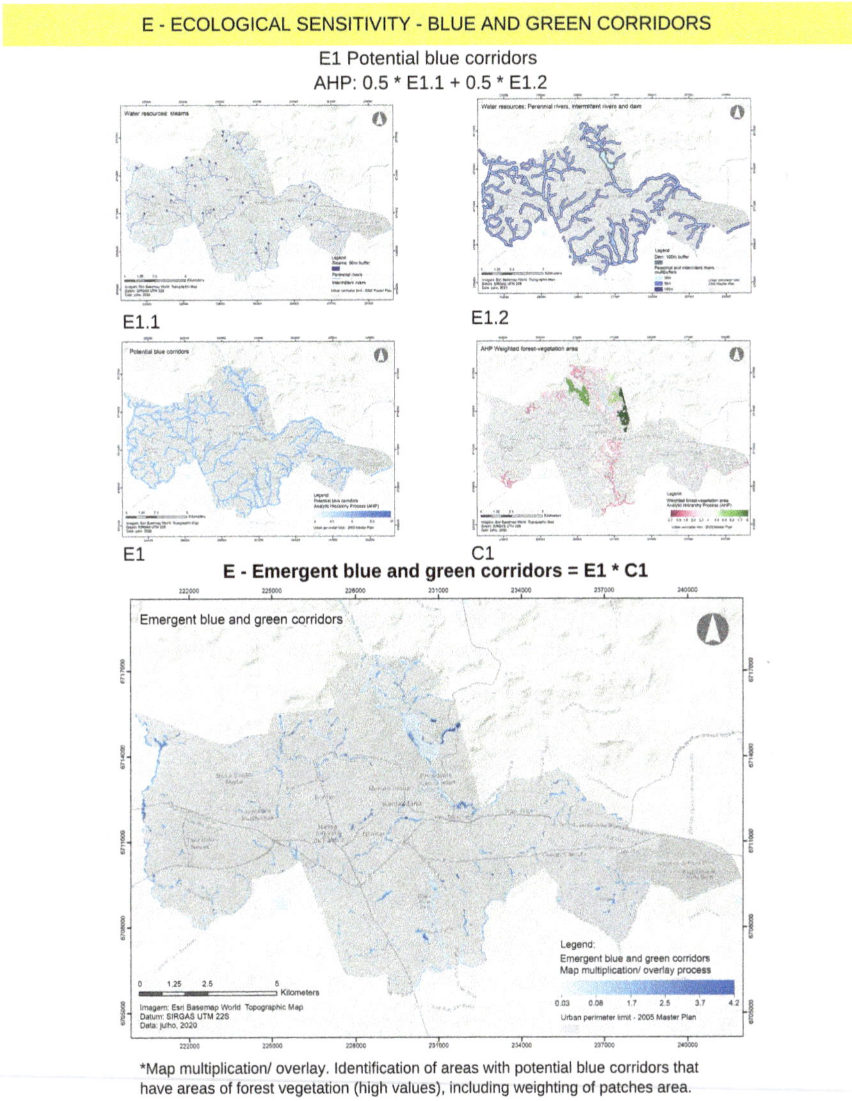

Fig. 20.4 Overlay of maps and levels of emergent blue and green corridor ecological sensitivity

As a final step in the methodological process, we combined the maps of emergent ecological sensitivity and emergent blue and green corridor ecological sensitivity to create a synthesis of the scenario of ecological sensitivity of the urban area of Santa Maria (Fig. 20.5).

Table 20.4 Weights for buffers and multibuffers of springs, perennial and intermittent and rivers and dam

E1.1. Springs: 50m buffer (APP)	**Springs**		**E1.2. Perennial and intermittent rivers and dam: 30m, 50m and 100m buffers**	**Multi Buffers**	
	Interval	Weight-Reclassification		Interval	Weight-Reclassification
Class 1	50 (m)	10	**Class 1**	30 (m)	10
UP Areas (no buffers)	NoData	0	**Class 2**	50 (m)	7
			Class 3	100 (m)	4
			UP Areas (no buffers)	NoData	0

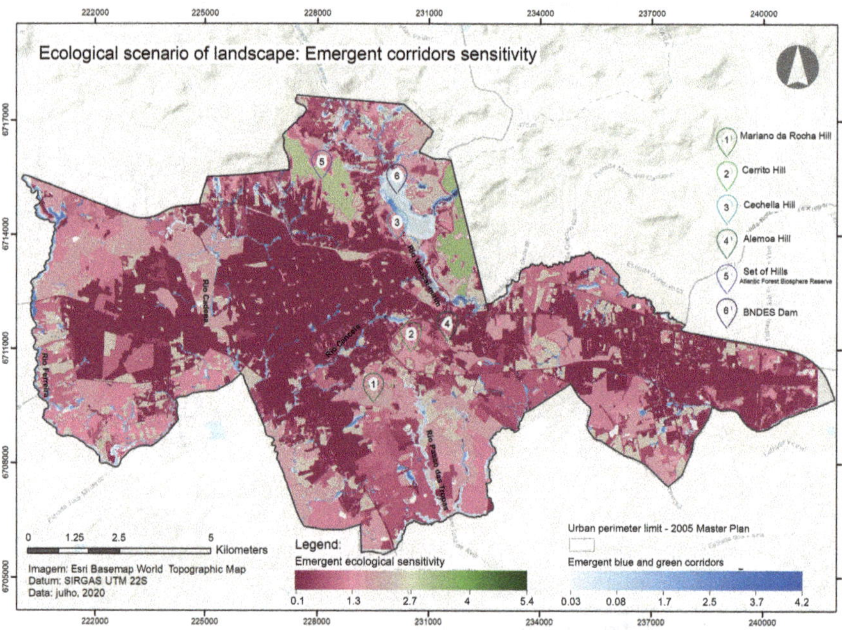

Fig. 20.5 Synthesized map of the ecological scenario of the urban area of Santa Maria

3.3 Discussion

According to the methodological scheme in Fig. 20.1, Map C1, Weighting of Forest-vegetation areas (AHP), is the result of the overlay of landscape metrics applied to the forest-vegetation land cover class. Within the urban perimeter, 2737 forest patches were computed and these present high shape indices, reaching values up to 12, which denotes very complex geometries. According to Map C1.1, most of the patches vary from 3 to 12 (violet and green tone, respectively), especially in patches with larger areas (strong green tone; Map C1.2). This configures a scenario of reduced ecological sensitivity. The low sensitivity is reinforced with application of core area, C1.4, where only 493 larger fragments are retained, but an important area is excluded due to a high SI, altering the original geometry. The largest forest area is located to the north and possesses 243 ha, about 1.9% of the urban perimeter, however with the application of core area, the area is reduced to 145 ha, a loss of about 60%. Much of the reduction of fragments is related to areas of riparian forest, which have elongated and narrow forms, less than the minimum of 30 m defined by the Forest Code. It is worth mentioning that in Map C1.2, there are important areas that present moderate high and high ecological sensitivity, varying from 5.6 to 8, respectively. These areas are mostly either located in a rugged topography (mountain slope), in parks or in the Atlantic Forest Biosphere Reserve, which has made human occupation more difficult, although not fully evaded it.

Definition of potential ecological sensitivity (Fig. 20.2) was concluded with application of Map C1.3 of field areas. The urban area has many fields, totaling 733 patches, with an average area of 4 ha, the largest being 219 ha (1.7% of the urban perimeter). Fields play an important ecological and environmental role, due to their flat or slightly undulated topography, composition of floodplain and wetland areas and because they shelter a good part of the perennial and intermittent rivers. The vast majority of these areas are found on the edges of the urban perimeter and many present medium to large areas (medium to dark green gradient), with weights of 5, 7 and 10 for potential ecological sensitivity, thus representing areas with medium potential ecological sensitivity.

In the definition of anthropic sensitivity (Fig. 20.3), large urban areas and those presenting exposed soil and agriculture were assigned low weights since they negatively affect ecological sensitivity. In maps D1.1 and D1.2, the gradient of warm colors indicates a gradual increase of area. As for spatialization of exposed land and agriculture, 771 patches were recorded, with an average area of 4.1 ha and maximum area of 393.7 ha (3% of the urban perimeter), located mainly in the eastern and southern regions. In the spatialization of urban areas, a patch with an area of 3745 ha (strong red) stands out as it occupies 30% of the urban perimeter. Being connected through the road network, it is morphologically configured as a single spot, although it is characterized by a dispersed distribution towards the east-west axis and along the southern road axis. It is worth noting that there is progressive urban growth in the northern direction, systematically reducing the areas of moderate and high ecological sensitivity identified earlier.

In relation to the blue and green corridors (Fig. 20.4) a hypothetically ideal scenario of sensitivity of the blue corridors was considered, that is, assigning weights to the multibuffers independent of their compliance with environmental legislation and presence of tree cover. Map E1 of potential blue corridors presents the highest weights arising from the sum of the 50 m buffer from springs and the 30 m buffer from APP and subsequently from the sum of the 50 and 100 m buffers. Strong blue colors indicate high propensity for ecological sensitivity and the creation of corridors. This analysis aimed to identify latent connections that can be used to plan an ecological landscape system.

3.4 Discussion of Ecological Sensitivity Maps

The analysis of morphological and functional landscape structure and ecological sensitivity presented in the maps—Potential Ecological Sensitivity, Anthropic Ecological Sensitivity, Emergent Ecological Sensitivity, Emergent Blue and Green Corridors and the Synthesized Map of Ecological Scenario: Sensitivities and Emergent Corridors—constitute instruments that provide important spatial and quantitative information for the ecological planning of the landscape (Figs. 20.2, 20.3, 20.4 and 20.5). In the case of the city of Santa Maria, spatial patterns of landscape variables and elements are presented, including characteristics, potentialities, vulnerabilities, functionalities, flows and scenarios (Forman, 1995, 2004, 2008, 2019; Steinitz, 2012).

After understanding these phenomena, we can promote planning and management with an ecological and systemic perspective, thus ensuring the protection and perpetuation of landscapes and communities. The case of Santa Maria is far from an ideal scenario. None of the maps showed high levels of ecological sensitivity (from 7.01 to 10), i.e. stable environments. The ecological sensitivity levels found were moderate (from 3.01 to 7) or low levels of ecological sensitivity (from 0 to 3), i.e. unstable environments (Fig. 20.5). In Tables 20.2, 20.3, 20.4 and 20.5, the environments and their respective levels of sensitivity were classified using a color gradient (Tables 20.6, 20.7 and 20.8).

Area of agriculture and exposed soil of up to 2 ha was given a weighting of 7, but it should have been assigned a weight of 5 or less, as it is a single incident area and a small size compared to all other areas and should not be recognized as high ecological sensitivity.

In the Synthesized Map of the Ecological Scenario (Fig. 20.5), overlay mapping was used to identify the potential blue corridors and areas of forest-vegetation, which were weighted by core area, area and shape index. The resulting scenario is alarming with regards to the ecological aspects of the landscape and open space system, considering the low sensitivity of the vegetation and water resources of the urban area, constituting unstable environments.

With regard to emergent corridors, low sensitivity areas predominate, while moderate sensitivity areas, when present, are due to the presence of riparian forest near the water resources. In the green and blue corridors, one can clearly see the connected and fragmented areas. The areas with highest sensitivity were adjacent boundary of vegetation areas to the Vacacaí-Mirim River (north), Arroio Ferreira Stream (west), around the dam and in some springs, with a weight of 4.2–3.7.

The green and blue corridors connected with low sensitivity vegetation or no vegetation (riparian forest) presented moderate and low classifications, respectively, with weights ranging from 4.02–2.6 and 2.5–0.03. This is because the water resources have little or no vegetation due to the 30 m buffer for APP, which results in very narrow margins with low ecological value, significantly impacting the functionality of the landscape as habitat and flow of fauna and flora, as well as

Table 20.5 Potential ecological sensitivity

C. Potential Ecological Sensitivity	**Ecological Sensitivity Level (0.1 to 5.9)**	**Low/Unstable Environment**	0.1–2.7 (dark pink to light pink)	**Urban Area**	Dispersed, with the exception of the center of the urban area patch
		Moderate/Transition Environment	2.8–5.9 (light green to dark green)		North: hills near the Vacacaí-Mirim River and dam, the relief is more rugged with vegetation of the Atlantic Forest tree stratum; west and South: flatter relief and vegetation of the southern pampa field.
		High/Stable Environment	Non-existent		Non-existent

Table 20.6 Anthropic ecological sensitivity

D1. Anthropic Ecological Sensitivity	**Ecological Sensitivity Level (0.4–10)**	**Low/Unstable Environment**	0.4–3.4 (dark purple to light purple)	**Urban Area**	Consolidated and densified urban area to the west, south, east and north
		Moderate/Transition Environment	3.5–5.5 (grayish purple to yellow)		Edges to the east, north, south and west of the urban perimeter and in the urban patch fragmentations
		High/Stable Environment	7.8–10 (orange to brown)		Imperceptible*

*Since it is an agricultural area with exposed soil of up to 2 hectares rated 7 for weighting, it should have been assigned a weight of 5 or less for this category. As it is a single small area compared to all the other areas, it is insignificant and should not be recognized as having high ecological sensitivity

Table 20.7 Emergent ecological sensitivity

Emergent Ecological Sensitivity	**Ecological Sensitivity Level (0.1–5.4)**	**Low/Unstable Environment**	0.1–2.7 (light pink to dark pink)	**Urban Area**	Consolidated and densified urban patch in the west, south, east and north and scattered fragmentations throughout the urban area, with emphasis on the Cerrito, Alemoa, and Cechella hills.
		Moderate/Transition Environment	2.8–5.4 (dark green to light green)		North: hills near the Vacacaí-Mirim River and dam, plus small fragments of the Cerrito, Alemoa and Cechella hills.
		High/Stable Environment	Non-existent		Non-existent

Table 20.8 Emergent blue and green corridors

E. Emergent Blue and Green Corridors	**Ecological Sensitivity Level (0.03–4.2)**	**Low/Unstable Environment**	0.03–2.5 (light blue)	**Urban Area Portion**	All major water resource corridors
		Moderate/Transition Environment	2.6–4.2 (dark blue)		Corridors of the Vacacaí-Mirim River (north) and Arroio Ferreira Stream (west). Surroundings of the dam and some springs.
		High/Stable Environment	Non-existent		Non-existent

preventing their future incorporation in linear parks that would ensure more effective use of conservation for leisure and recreation.

The Vacacaí-Mirim River corridor, which incorporates a dam and artificial reservoir with extremely low sensitivity, presents the greatest variation in sensitivity levels among the corridors. In the area north of the urban area, it presents greater integration, while there is considerable fragmentation in the eastern region. The other corridors present, without exception, significant interruptions and discontinuities.

Most of the areas of moderate sensitivity, such as the hills north of the urban area, are close to the Vacacaí-Mirim River corridor. The areas of low sensitivity exert pressure on the emergent corridors, contributing to their classification as moderate sensitivity, in other words, they are in transitional environments and could lose their potential due to anthropic impacts.

As can also be seen in the Ecological Scenario Map, there is a possible configuration of an integrated ecological network of blue and green corridors and other areas of sensitivity in the northern segment of the urban area. Although it presents moderate sensitivity, it has the potential to become a stable environment and reach high ecological sensitivity if environmental control and recovery measures are taken. On the other hand, the map shows how the dam and the Vacacaí-Mirim River are more vulnerable and prone to major environmental impacts and imbalances due to anthropic actions and pressures. The situation becomes even more worrisome when considering that the areas of vegetation cover in the multibuffers of the rivers and dam present low ecological sensitivity. These are small and compact areas, whose shape indices are above 1, resulting from a convoluted geometry, due to losses caused by anthropic actions.

In the central portion of the urban area, there is a continuous corridor along the Cancela Stream and Cerrito hill. Although it also presents low ecological sensitivity, it could acquire increased importance and functionality, if it were integrated with the ecological network located in the northern region of the urban area. This corridor has the potential to strengthen this ecological network and deserves attention because it connects the northern, central and southern segments of the urban area with potential to reach moderate to high ecological sensitivity. Although less evident, the emergent corridor of Arroio Ferreira Stream, located in the west of the urban area, also shows potential as an ecological network, although with less amplitude, since it incorporates only emergent blue and green corridors.

When comparing the maximum and intermediate values of the Ecological Scenario Map of the urban area of Santa Maria with the parameters of theoretical reference (Table 20.1), the highest weights were 5.4 and 4.2. It is worrisome that no areas reached the desired values of high sensitivity and ecological potentiality (7.01–10).

Although the Vacacaí-Mirim River, dam and surrounding hills presented the most positive scenario, the weights are quite serious, reflecting the alarming shape indices and losses of core area. The areas with weights of 2.6 and 2.8 deserve caution, as they may either increase or decrease in ecological sensitivity, requiring emergency attention by public authorities to mitigate this scenario of loss in order to increase or maintain these weights.

4 Final Considerations: Current and Future Challenges for Landscape Planning

Landscape ecology promotes a set of guidelines for landscape design and planning of cities and regions. To apply them, it is necessary to collect and spatialize data of the area to be analyzed and interpreted with geotechnologies. Subsidies and solutions for the planning of these landscapes can then be offered, in order to maintain the integrity of their natural processes and create diverse scenarios: preservation and conservation areas; leisure and recreation areas; anthropic areas with control and/or mitigation of impacts. Most importantly, this process promotes the sustainable management and value of potential ecological areas over time.

Public administrators must pay more attention to the landscape and open spaces of Santa Maria, through strategic landscape planning in order to preserve potential and emergent ecological sensitivities. It is necessary to connect the existing free spaces for leisure and recreation via green and blue corridors through the formation of linear parks, as well as provide new connections and larger park areas, enhancing the functionality, flow dynamics and systemic integration of the landscape.

References

Angeoletto, F., Tryjanowski, P., Santos, J., Martinez-Miranzo, B., Leandro, D., Bohrer, J., et al. (2022). Will Brazilian city dwellers actively engage in urban conservation? A case study with the charismatic neotropical blue-and-yellow macaw (Ara ararauna). *Birds, 3*(2), 234–244.

Angeoletto, F. H. S., Richard, E., Santos, J., Guedes, N., Oliveira, S. L., Albertín, R. M., et al. (2023). When we think of urban afforestation in Brazil, what comes to mind is a mouth that has lost its teeth. *Terr@Plural, 17*, e2321654.

Archer, R. S. (2013). *Measuring and monitoring long term disaster recovery using remote sensing: A case study of post Katrina New Orleans*. Thesis. Geography, University of California, Santa Barbara.

Bentrup, G. (2008). *Zonas de amortiguamiento para conservación: lineamientos para diseño de zonas de amortiguamiento, corredores y vías verdes* (Informe Técnico Gral. SRS-109). Departamento de Agricultura, Servicio Forestal, Estación de Investigación Sur.

Bhatta, B., Saraswati, S., & Bandyopadhyay, D. (2010). Quantifying the degree-of-freedom, degree-of-sprawl, and degree-of-goodness of urban growth from remote sensing data. *Applied Geography, 30*, 96–111.

Brasil. (2012). Lei Federal n° 12.651, de 25 de maio de 2012. *Dispõe sobre a proteção da vegetação nativa; altera as Leis n°s 6.938, de 31 de agosto de 1981, 9.393, de 19 de dezembro de 1996, e 11.428, de 22 de dezembro de 2006; revoga as Leis n°s 4.771, de 15 de setembro de 1965, e 7.754, de 14 de abril de 1989, e a Medida Provisória n° 2.166-67, de 24 de agosto de 2001; e dá outras providências*. http://www.planalto.gov.br/ccivil03/leis/L4771.htm. Accessed 24 Jun 2020.

Brasil. (2018). *Áreas Prioritárias para Conservação, Uso Sustentável e Repartição de Benefícios da Biodiversidade do Bioma Pampa*. Ministério do Meio Ambiente. http://areasprioritarias.mma.gov.br/images/arquivos/Produto01_AP_BiomaPampa_final.pdf. Accessed 18 Mar 2021.

Brasil. (2020). *Biomas*. Ministério do Meio Ambiente. https://antigo.mma.gov.br/biomas.html. Accessed 18 Mar 2021.

DEE. (2015. Departamento de Economia e Estatística. Município de Santa Maria. http://deedados.planejamento.rs.gov.br. Accessed 20 Mar 2021.

Erickson, D. (2006). *MetroGreen: Connecting open space in North American cities*. Island Press.

Flink, C. A. (2020). *The greenway imperative: Connecting communities and landscapes for a sustainable future*. University of Florida Press.

Follmann, F. M. (2018). *Áreas Prioritárias à Conservação Ambiental em Santa Maria/RS: Estratégias para Gestão Ambiental Municipal.* Thesis. Geography, Universidade Federal de Santa Maria, Santa Maria.

Forman, R. T. T. (1995). *Land mosaics: The ecology of landscapes and regions*. Cambridge University Press.

Forman, R. T. T. (2004). *Mosaico Territorial para la Región Metropolitana de Barcelona*. Editorial Gustavo Gili, SA.

Forman, R. T. T. (2008). *Urban regions: Ecology and planning beyond the city*. Cambridge University Press.

Forman, R. T. T. (2019). *Towns, ecology, and the land*. Cambridge University Press.

Forman, R. T. T., & Godron, M. (1986). *Landscape ecology*. Wiley.

Forman, R. T. T., Olson, J. D., & Dramstad, W. (1996). *Landscape ecology principles in landscape architecture and land-use planning*. Harvard University Graduate School of Design. Island press.

Gomes, L. F. A. M., & Gomes, C. F. S. (2014). *Tomada de decisão gerencial: enfoque multicritério*. Atlas.

Hellmund, P. C., & Smith, D. S. (2006). *Designing greenways: Sustainable landscape for nature and people*. Island Press.

Herold, M., Couclelis, H., & Clarke, K. C. (2005). The role of spatial metrics in the analysis and modeling of urban land use change. *Computers, Environment and Urban Systems, 29*, 369–399.

IBGE. (2020a). *Instituto Brasileiro de Geografia e Estatística. Área Territorial de Santa Maria-RS*. https://www.ibge.gov.br/cidades-e-estados/rs/santa-maria.html. Accessed 24 Mar 2021.

IBGE. (2020b). *Instituto Brasileiro de Geografia e Estatística. População estimada de Santa Maria-RS*. https://www.ibge.gov.br/cidades-e-estados/rs/santa-maria.html. Accessed 24 Mar 2021.

Jongman, R., & Pungetti, G. (2004). *Ecological networks and greenways: Concept, design and implementation*. Cambridge University Press.

Lang, S., & Blaschke, T. (2009). *Análise da paisagem com SIG*. Oficina de Textos.

Lima, G. F., Correa Santos, J. W. M., Albertin, R. M., Martínez-Miranzo, B., Souza, F. L., & Angeoletto, F. (2023). Backyards are a way to promote environmental justice and biodiversity conservation in Brazilian cities. *Diversity, 15*(7), 815.

Lyle, J. T. (1999). *Design for human ecosystems. Landscape, land use and natural resources*. Van Nostrand Reinhold.

Macedo, S. S., Queiroga, E. F., Campos, A. C. d. A., Galender, F., & Custódio, V. (2018). *Os Sistemas de Espaços Livres e a Constituição da Esfera Pública Contemporânea no Brasil*. Editora da Universidade de São Paulo.

Magnoli, M. M. (1994). Ambiente, espaço, paisagem. *Paisagem e Ambiente: Ensaios, 1*.

Mcharg, I. L. (1969). *Design with nature*. Doubleday/Natural History Press.

Mcharg, I. L., & Steiner, F. R. (1998). *To heal the earth: Selected writings of Ian L. McHarg*. Island Press.

Metzger, J. P. (2001). O que é ecologia de paisagens? *Biota Neotropica, 1*.

Pellegrino, P., & Moura, N. B. (2017). *Estratégias para uma Infraestrutura Verde*. Manole.

Pippi, L. G. A. (2014). *Social network interaction and behaviors on recreational greenways and their role in enhancing greenway potential*. Thesis. Doctor of Philosophy – Design. North Carolina State University, Raleigh, North Carolina, USA.

Pippi, L. G. A., & Trindade, L. C. (2013). O Papel da Vegetação Arbórea e das Florestas nas Áreas Urbanas. *Paisagem e Ambiente: Ensaios, 31*, 81–96.

Queiroga, E. (2006). Por um Paisagismo Crítico: Uma Leitura sobre a Contribuição de Miranda Magnoli para a Ampliação do Corpus Disciplinar do paisagismo. *Paisagem e Ambiente: Ensaios, 21*, 55–64.

Ross, J. L. S. (1990). *Geomorfologia: Ambiente e Planejamento*. Contexto.

Ross, J. L. S. (1994). *Análise empírica da fragilidade dos ambientes naturais e entropizados* (pp. 63–74). Laboratório de Geomorfologia – Departamento de Geografia, USP.

Rumble, H., Angeoletto, F., Connop, S., Goddard, M. A., & Nash, C. (2019). Understanding and applying ecological principles in cities. In *Planning cities with nature: Theories, strategies and methods* (pp. 217–234).

Saaty, T. L. (1990). How to make a decision: The analytic hierarchy process. *European Journal of Operational Research, 48*, 9–26. http://www.scopus.com. Accessed 12 Apr 2020.

Saaty, T. L., & Vargas, L. G. (1979). Estimating technological coefficients by the analytic hierarchy process. *Socio-Economic Planning Sciences, 6*, 333–336. https://doi.org/10.1016/0038-0121(79)90015-6

Santos, A. R. D., Louzada, F. L. R. D. O., & Eugenio, F. C. (2010). *Arcgis 9.3 Total – Aplicações para Dados Espaciais. 2*. CAUFES.

Smith, D. S., & Hellmund, P. C. (1993). *Ecology of greenways: Design and function of linear conservation areas*. Foreword by Richard T.T. Forman. Copyright by the Regents of the University of Minnesota, USA.

Souza, F. L., Fenoglio, M. S., & Angeoletto, F. (2023). To be a Brazilian City dweller, sometimes we must learn to say enough! *Sustainability, 15*(4), 3699.

Steinitz, C. A. (2012). *Framework for Geodesign: Changing geography by design*. ESRI.

Tardin, R. (2008). *Espaços Livres: Sistema e Projeto Territorial*. 7Letras.

Tricart, J. (1977). *Ecodinâmica*. IBGE/SUPREN.

Weiss, R. A. (2016). *Dinâmica Urbana: da Compreensão à Construção de um Modelo de Identificação dos Padrões de Crescimento Urbano por Meio das Métricas Espaciais Da Paisagem: O Caso do Distrito da Lagoa da Conceição*. Thesis. Architecture. Universidade Federal de Santa Catarina.

Weiss, R., & Pippi, L. G. A. (2019). Análise Multicritério na definição de vulnerabilidade ambiental. *Terra Plural, 13*, 272–295.

Chapter 21
Public Policies and Biodiversity Conservation in Conflict in a Brazilian Agribusiness City

Jeater W. M. C. Santos, Marcelo C. Vacchiano, Marco Antonio M. de Souza, Cathiane M. de Oliveira, Sergio S. Negri, Normandes Matos da Silva, Nestor Alexandre Perehouskei, and Mark D. E. Fellowes

Abstract This chapter examines the importance of planning instruments and environmental management regulations resulting from Brazilian public policies in making urban growth sustainable. To this end, we present the case of updating the Municipal Master Plan (MMP) of Rondonópolis, a medium-sized city located in the southeast of the state of Mato Grosso and in the transition zone of two important Brazilian biomes (Cerrado and Pantanal), the latter of which is already on the list of global biodiversity hotspots. Rondonópolis is a city that since the 1970s has experienced vigorous economic growth, which has boosted the expansion and subdivision of urban land, as well as accelerated the degradation of its ecosystems. In 2013, the largest railroad terminal in Latin America was installed in its urban space. Since the installation of this terminal would imply the occurrence of many positive and negative impacts for the municipality, and especially for the city of Rondonópolis, the

In memoriam of the author Nestor Alexandre Perehouskei.

J. W. M. C. Santos (✉) · M. A. M. de Souza · C. M. de Oliveira · S. S. Negri
Graduate Program in Geography, Federal University of Rondonópolis, Rondonópolis, Brazil
e-mail: jeater@ufr.edu.br

M. C. Vacchiano
Ministério Público do Estado de Mato Grosso, Cuiabá, Brazil

N. M. da Silva
Programa de Pós-Graduação em Gestão e Tecnologia Ambiental, Universidade Federal de Rondonópolis, Rondonópolis, Brazil

N. A. Perehouskei
In Memoriam, Maringá, Brazil

M. D. E. Fellowes
Department of Biological Sciences, Royal Holloway University of London, Egham, Surrey, UK

F. Angeoletto et al. (eds.), *Ecology of Tropical Cities*,
https://doi.org/10.1007/978-3-031-48012-6_21

federal government established several conditions for granting the environmental license for its construction, including carrying out socio-environmental studies to support the review of its Municipal Master Plan (MMP). By law, the prerogative of carrying out work to update an MMP, the main instrument of Brazilian urban policy, belongs to municipal public managers. However, in the case of Rondonópolis, they found themselves forced to observe the technical guidelines indicated by a researcher from the Federal University of Rondonópolis. This was due to a decision handed down by the judiciary after intervention by the Public Ministry of the State of Mato Grosso, as managers had been found to not respect the conditions of the environmental licensing of this terminal. The result was interesting as it makes it possible to consider the potential of the planning and management instruments of Brazilian public policies, particularly in relation to the mitigation of the impacts of the main vectors of economic growth related to local biodiversity. This is central to meeting desired sustainability goals of this city, as well as being an unprecedented experience in terms of management, with the work of updating a MMP being carried out by a scientist, and not by those ruled by politicians.

Keywords Cerrado · Pantanal · Commercial agriculture · Commodities · Intermodal terminal · Environmental degradation · Urban planning · Environmental management · Conservation units

1 Introduction

Brazilian territory covers an area of 8,510,417.7 km^2, which is politically distributed and managed by a population of about 208,000,000 people in its 5576 municipalities (IBGE, 2022). Ultimately, municipalities are responsible for implementing policies for the conservation of natural resources and also the enormous existing biodiversity in the six biomes that occur in Brazil, especially since some of them are already on the list of global biodiversity hotspots (Angeoletto et al., 2022).

While the Brazilian State has established policies for the conservation of nature for a many tens of decades [for example in 1861 Dom Pedro II issued a decree transforming a portion of the mountains of the Tijuca massif into Floresta da Tijuca in the city of Rio de Janeiro (Fernandez, 2016)], their implementation is still very much linked to political forces, where historically there is a hegemony of large economic groups, weakening democratic public management and constituting a constant obstacle for Brazilian society.

Overcoming this challenge has become an urgent socio-environmental need (Souza et al. 2023). Since the 2000s, the process of urbanization has accelerated globally, and cities are now the place where the majority of the population lives (UN, 2019). But that growth and increasing density of people, without concurrent guarantees of the conservation of its biodiversity, is a problem that is even more serious in the cities of developing countries. In Brazil, even if the urbanization process has concentrated more people in its medium and large cities, the small ones have also grown significantly and, therefore, must also be considered in this challenge since they occupy most of its territory (Angeoletto, 2012).

More than ever, the planning and environmental management regulations made available by state public policies must be effective in guaranteeing territorial planning and sustainable urban development in all cities, seeking to reconcile economic growth with biodiversity conservation, maintaining quality of life for their populations (Santangelo et al., 2022; Lima et al., 2023). It is the obligation and responsibility of the Brazilian State to institute guidelines for urban development that includes the right to housing, basic sanitation and urban transport. It must also, through its municipal entities, promote adequate territorial ordering with planning for the use, subdivision and occupation of urban land, and in order to fulfill this desideratum, the master plan is raised to the level of a constitutional requirement (Brazil, 1988).

In this sense, and realistically, in understanding the economic and political realities that characterize Brazil, the present chapter aims to discuss the strategies and potential of Brazilian public policies to promote the environmentally sustainable development of cities, and as a result, guarantee the conservation of biodiversity with environmental quality for its population. To this end, the case of updating the Municipal Master Plan of Rondonópolis, a municipality located in the southeast of the state of Mato Grosso in the Center-West Region of Brazil is analyzed. This city, both for its urban functions (especially the logistics and industrialization of the grain harvest in the state of Mato Grosso), and for its status as a medium-sized city (about 240,000 inhabitants in 2021), is an interesting municipality to consider the efficiency of urban planning and environmental management regulations, the central theme of the present chapter.

According to the classification of the Brazilian Institute of Geography and Statistics (IBGE), Rondonópolis is considered a Regional Capital,[1] that is, an urban center with a high concentration of management activities, trade in goods and industries, but with an even lower level compared to that of the Brazilian metropolises. Such centers tend to have a smaller reach in terms of regional influence and attractiveness for trade in goods and services compared to the pull exerted by the larger metropolises.

Rondonópolis also stands out in economic terms as an important commercial, industrial and logistical center for commercial agriculture in the Cerrado in the state of Mato Grosso. In 2022 it was the 15th largest exporting municipality in Brazil.[2] Therefore, it can be considered one of the most important agribusiness cities in Brazil and has experienced a significant expansion of its regional influence (centrality), due to the implementation of a railroad terminal in the city in 2013.

Considering this case of the updating of the Municipal Master Plan of Rondonópolis (the main instrument of urban planning and environmental management of Brazilian municipalities) is of considerable interest as the territory of Rondonópolis is located in the transition zone of two important Brazilian biomes

[1] https://www.ibge.gov.br/cidades-e-estados/mt/rondonopolis.html

[2] https://balanca.economia.gov.br/balanca/outras/panoramamunicipaisbraslileiros.comex.pdf

(the Cerrado and Pantanal). On the other hand, the unusual way in which the review of the Municipal Master Plan of Rondonópolis approved in 2006 (hereinafter referred to in this text only as PDM/ROO/2006), was carried out, and is still in force, also deserves a more detailed analysis. This is because, in an unprecedented way in the history of Brazil, the coordination of these works was removed from the municipal executive power by court decision and was transferred to a professor at the Federal University of Rondonópolis.

This process of updating the PDM/ROO/2006 was initiated in 2014, that is, two years before the 10-year period determined by Federal Law 10.251/2001 (Brazil, 2001). This advance was due to a condition for the environmental licensing process of the project "Intermodal Railroad Terminal for the extension of the Malha Norte Railway – ALL" (Fig. 21.1), built in Rondonópolis by the company América Latina Logística (ALL), and is one of the largest Brazilian companies in this sector, which was acquired by Grupo Consan, currently integrating "Rumo Logística S/A". The installation of what is the largest intermodal terminal (railroad), in Latin America in the city of Rondonópolis in 2013 promoted the connection of the municipal territory via the Ferronorte railroad with the largest ports in the south and southeast of Brazil. This resulted in Rondonópolis acting as a significant regional hub, attracting a very high number of trucks transporting the soybean crop from the municipalities of Mato Grosso.

In practice, the terminal promoted the opening of a "rail corridor" for the flow of grain production from the state of Mato Grosso, which is currently the largest

Fig. 21.1 Intermodal Terminal of the América Latina Logística Company opened in 2013 in the city of Rondonópolis—Mato Grosso—Brazil. (Source: Zig Koch, 2018)

producer of soy in Brazil. In numbers, this terminal allowed the annual outflow of agricultural commodities from Rondonópolis to be doubled, which before its installation, already amounted to around 10 million tons (Macedo et al., 2015; Urbaniza, 2017).

Considering that the implementation of this huge enterprise in the municipality of Rondonópolis would certainly promote, at a regional level, many positive and negative impacts from an economic, social and environmental point of view, at a regional level, it was imperative that its master plan be updated and, therefore, such an update was put in place as one of the constraints of its environmental licensing process.

Thus, the Municipality of Rondonópolis, in partnership with the company ALL, jointly prepared a work proposal to update the PDM/ROO/2006, which would be fully funded by the company. However, since this proposal did not meet the environmental licensing requirements of the enterprise determined by experts from the Brazilian Institute of the Environment (IBAMA), the implementation of the enterprise ran the risk of being suspended in early 2016 due to an injunction granted by the judiciary in a public civil action filed by the Public Ministry of the State of Mato Grosso. The state judiciary also determined that their coordination would be the responsibility of a researcher from the Graduate Program in Geography at the Federal University of Rondonópolis who prepared a new proposal in order to contemplate all the conditions established by the study of environmental impact (EIR/RIMA), of the project's environmental licensing process. This proposal was effectively implemented, thus generating an unprecedented experience in terms of public management in Brazil, which will be presented in detail throughout the present chapter.

The arrival of the railroad and the installation of the intermodal terminal in the municipality of Rondonópolis significantly boosted agricultural production in the region and the urban development of the cities in that region. More specifically for the economy of the city of Rondonópolis, this has greatly and quickly boosted its urban functions of logistics, commerce, industry and even the dynamics of its real estate market. But on the other hand, it also brought several negative impacts such as the disorderly expansion of the urban perimeter, lack of a structured road network to absorb the huge amount of heavy vehicles attracted by the terminal and even the significant increase in atmospheric pollution in the city as a result of this additional fleet of heavy vehicles and the intensification of the activities of its industrial plants (Angeoletto et al., 2019).

In the case of the southeastern region of the state of Mato Grosso, where the city of Rondonópolis is located, most income is generated in the countryside as a result of the implementation in the 1970s of a productive system that can be called commercial agriculture/capital intensive industry (Peixinho, 1998; Negri, 2008, 2010). Since then, this production system has been intensely degrading the natural landscapes of the Cerrado biome, even in specially protected urban and rural territorial spaces in the municipalities of this region (Santos, 2010; Silva & Anjos, 2010; Vacchiano et al., 2019; Souza, 2017; Oliveira, 2020).

Therefore, either because of the regional characteristics of the natural environment where it is located, that is, because of its expressive urban growth, or even because of the unprecedented way in which the revision of its PDM/ROO/2006 was carried out, Rondonópolis can be considered a true "laboratory" to evaluate the efficiency of urban planning and environmental management instruments created in the context of Brazilian urban and environmental public policies.

2 The Ideal of Sustainable Cities in Brazilian Public Policies

After more than 10 years of processing, Federal Law 10.257 (known as the "Statute of Cities"), was finally approved in 2001 (Brazil, 2001), with the mission of regulating Brazilian urban policy provided for in Articles 182 and 183 of the Federal Constitution, approved in turn in 1988 (Brazil, 1988). However, the Statute of Cities did not consider the entire scope of the Brazilian urban issue since the Participatory Municipal Master Plan (PMMP), the main planning instrument proposed by it, is not mandatory for the planning of small cities (considered as being those with a population of less than 20,000 inhabitants), even though they represent about 70% of Brazilian municipalities (Fernandez, 2016), and occupy about 50% of the 8,510,417km^2 of the surface of the national territory.

The enactment of this law significantly altered the way in which the Brazilian State began to regulate the use of municipal territories. Since then, the PMMP ruled that urban planning covers the entire municipal territory, that is, it must include both urban spaces and rural areas in the municipality. In addition, it also made the participation of different segments of society mandatory in the planning process, which thus ceased to be a purely technical act of the executive branch, but rather constitutes a political act of the entire municipal population (Lopes & Henrique, 2010).

However, it should be recognized that Brazilian initiatives to standardize the process of ordering the use of municipal territories started well before the Statute of Cities and the Federal Constitution itself. The first initiatives in this sense can be verified through technical instruments aimed at identifying and mapping homogeneous regions (areas that have similar physical or socioeconomic characteristics), of a given territory. Such instruments were foreseen both in Law 4.504 of November 30, 1964 called "Land Statute" (Brasil, 1964) as in Law 6766/1979, which deals with the subdivision of urban land and Law 6938/1981, the "National Environmental Policy Law" (Brazil, 1981).

In the Land Statute, zoning was used as an instrument to identify homogeneous regions from the perspective of the national rural land structure in order to subsidize the planning of a national agrarian policy. In the context of Law 6.938/1981, zoning came to be seen as an instrument with a wider scope of application in terms of planning. In the context of this law, its use aimed at maintaining the ecological balance of the landscapes of municipal territories through the control of areas for the implementation of economic enterprises with potentially or effectively polluting activities. Thus, the environment and biodiversity of municipal territories came to be seen as public property to be necessarily protected. The subdivision of urban land, through subdivisions and divisions of areas, determined lots for buildings and openings of new circulation routes, public places or any alteration in urban roads was limited by Law n° 6.766/79, which also dealt with the minimum infrastructure necessary for urban environments.

Recognizing the effective potential of zoning as a planning tool, in 2002 the federal government regulated item II of Art. 9 of Law no. 6938/1981 through Decree no. 4297 (Brazil, 2002), thus establishing the technical criteria for the Ecological-Economic Zoning of Brazil (EEZB), which has since become one of the most important Brazilian planning instruments of the integrated territory.

Even though the initiatives to standardize the uses of municipal territories predate the institution of Law 10.257/2001, it was only after that when the Brazilian urban policy provided for in Articles 182 and 183 of the Federal Constitution effectively received a set of instruments for territorial planning and environmental management with real potential for effectiveness:

> For all intents and purposes, this law, called the City Statute, establishes norms of public order and social interest that regulate the use of urban property in favor of the collective good, the safety and well-being of citizens, as well as environmental balance. (Brazil, 2001)

The Brazilian State, when defining the foundations of urban policy through the Statute of Cities, also makes this law an important instrument of environmental management and ordering of the urbanization process (Carvalho & Braga, 2001).

However, in order for the guidelines and instruments for environmental management provided for both in Laws 6766/79 and 6938/1981 and in the Statute of Cities to be effective in their purposes, it is necessary for municipalities to have an up-to-date and high-quality mapping of their territories. Without the existence of maps (topographic and thematic), of the uses of the municipal territory, updated and produced in adequate scales, it is practically impossible for the planning instruments to deliver the objectives established by the public policies of territorial organization.

3 Planning and Management Instruments to Achieve Sustainable Cities

As explained above, Brazilian environmental and urban policies provided public managers with various planning and management instruments so that they could carry out strategic and participatory planning for the sustainable urban development of municipalities. However, the vast majority of these are underutilized when not totally ignored by them (Villaça, 1999). According to this author, only the "Soil Use and Occupation Zoning" and the "Environmental Licensing of Effectively or Potentially Polluting Activities/Enterprises" planning instruments have been used by Brazilian public managers. The reason for the underutilization of these instruments may be the small budget that most municipalities (especially the small ones), have to keep their official cartography up to date (since this demands high and frequent investment), as well as the low level of qualifications and diversity of academic training of municipal managers (Angeoletto et al., 2016, 2023; Rumble et al., 2019; Souza et al., 2023).

Environmental licensing is an important tool for preventing and mitigating environmental damage. In the case of this study, it was in this instrument that IBAMA identified the need for improvements in the urban planning of Rondonópolis due to the impacts that the installation of the railroad terminal would have on the dynamics of the city, which is why it concluded that there was a need to update the master plan in force at the time, which, according to the records of the public civil action filed, would no longer take advantage of the new demands created and strengthened by the railway project in the municipality and in the region.

3.1 The Environmental Zoning

Villaça (1999) argues that urban planning in Brazil has been limited to the elaboration and application of the zoning of the use and occupation of the soil and control of subdivisions. That is, in practice the master plan does not exist, which makes zoning the real expression of what has been produced under the label of urban planning in Brazil. This distortion of territorial planning is not an exclusive problem of Brazilian public managers, because according to Souza (2010: 251), this has also been observed for a long time in the United States. In that country, zoning emerged as a simple technical instrument aimed at segregating unhealthy areas of cities with a high potential for the spread of diseases. But over time, it became the very expression of American city planning. In the words of Souza “the part became the whole.”

Neves (2006) suggests that zoning is the action of regionalizing a given territory from the definition of rules and criteria that unfortunately are almost always guided by the interests of groups that are able to influence the election of politicians (public managers). That is, zoning is carried out above all in order to order the use and occupation of urban spaces with a view to the economic exploitation of their spaces (real estate projects), and natural resources.

The first step to guarantee the appropriate implementation of zoning of a municipal territory is an understood set of definitions of the spatial categories to be used in the zoning of the use and occupation of the urban and rural land of this territory, which ideally should start from those defined in the zoning of the territory of the state where the municipality is located. This is because, after the federal government established the Ecological-Economic Zoning (EEZ), as the main territorial planning instrument in the country (Brazil, 1981), several Brazilian states carried out this zoning, especially those that make up the Legal Amazon region as in the case of Mato Grosso. It is also used to define public environmental policies, as in the case of the allocation of portions of areas located inside private rural properties for environmental protection purposes (Brazil, 2012). In addition, zoning is also one of the instruments provided for in the scope of Brazilian urban policy via the Statute of Cities, therefore, mandatory in the preparation of Municipal Master Plans. Thus, with zoning at the state level, it is essential that zoning at the level of municipalities within a state are harmonized with land use categories adopted at the state level.

The concept of the EEZ, defined in Art. 2 of Decree 4.297/2002, which regulated item II of Art. 9 of Law 6938/81, is quite broad and objective in its function of promoting territorial order in order to guarantee the sustainable development of Brazilian municipalities:

> Art. 2, The EEZ, an instrument for organizing the territory to be compulsorily followed in the implementation of plans, works and public and private activities, establishes measures and standards of environmental protection aimed at ensuring the quality of the environment, water resources and soil and the conservation of biodiversity, guaranteeing sustainable development and improving the living conditions of the population.

Normally, zoning is prepared on two geographic scales: the first, more general, is where the main categories of land use in the municipal territory called "macrozones" are established. The second, more detailed, is where the subcategories of land use and occupation that make up the area of each macrozone are defined.

In the structuring of Master Plans for the territories of Brazilian municipalities, at least two qualitative categories of land use and occupation macrozones are usually defined: (a) Urban Macrozone; (b) Rural Macrozone. In a second moment, an attempt is made to define the areas "with vocation" or specific demands for land use and occupation within each of these two main groups. As a result, land uses for residential, industrial, commercial, leisure, etc. normally comprise subcategories of the Urban Macrozone. Areas already with land use for agricultural production, large forest cultivation projects, mining, installation of highly polluting or uncomfortable industries, hydroelectric power plants, wind power plants, etc., normally make up the subcategories of the Rural Macrozone.

However, it should be noted that the area of the Urban Macrozone may contain areas with effectively rural land use and occupation. This is because, in Brazil, defining the boundaries of areas considered urban within the municipal territory is more a political decision than a technical one. And this explains the great weight that the particular interests of some politicians, aligned with the interests of economic groups, can exert on the disorderly expansion process of Brazilian cities, despite the large number of planning instruments made available by Brazilian public policies.

The area of the Urban Macrozone is defined by means of a municipal law project, called "law of the limit of the urban perimeter." The urban perimeter is actually subdivided into two categories of spaces: (a) the "consolidated urban area" of a municipal or district headquarters, that is, the one that already has the minimum urban equipment/infrastructure; (b) a strip of area surrounding the first one, called the "urban expansion area." This second, in turn, is subdivided into urban areas that are not yet effectively urbanized (the suburbs in the English sense - "subcities"), and areas with effectively rural land use. This portion of the rural area that integrates the area of the urban perimeter works as a kind of "reserve space" for the expansion of the city, because once it is defined, it is previously approved for new subdivisions of the land.

From the foregoing, it can be seen that there is a conceptual, technical and legal difference in Brazilian urban policy between what is considered an "urban area" and an "effectively urbanized area," the latter being also called a "consolidated urban area." And for an urban area to be considered effectively urbanized it must meet the requirements of Law n° 5.172/1966, popularly known as the National Tax Code (NTC). It is based on this law that the municipal government collects taxes on urban territorial property. Thus, in paragraph 1 of Art. 32 the NTC establishes that:

> For the purposes of this tax, an urban area is defined by municipal law; observing the minimum requirement of the existence of improvements indicated in at least 2 (two) of the following items, built or maintained by the public power: I - curbs or sidewalks with rainwater pipes; II - water supply; III - sanitary sewer system; IV - public lighting network, with or without poles for home distribution; V - primary school or health center within a maximum distance of 3 (three) kilometers from the considered property.

Although it is a tax law, it is used to define the requirements for a portion of an area to be considered urban or rural, since, depending on the case, there is an incidence of Urban Territorial Tax or Rural Territorial Tax, respectively.

As urban expansion results in land values substantially increasing, sometimes there are exaggerations and distortions in the definition of their size in the zoning of the municipal territory. In some cases, this distortion is so high that it is even greater than the consolidated urbanization area of the city. Such distortion is normally due to pressure, exerted by representatives of business groups on representatives of the municipal legislative and executive powers so that they prioritize their speculative commercial interests in their decisions rather than technical issues of urban planning.

Due to their strategic importance for achieving the goals of sustainable cities established in Brazilian urban policy, the areas protected by environmental legislation should be previously identified and mapped both within the limits of the Urban Macrozone and the Rural Macrozone for the effectiveness of planning and territorial organization. As a result, both environmentally protected areas and those that do not have this protection, but present some vulnerability capable of generating negative effects on human health and safety, or even for the conservation of biodiversity, should constitute their own category (restricted or prohibited social use), in the zoning of urban land use.

Unfortunately this is not observed in Brazilian master plans, which are restricted to zoning aimed at meeting the interests of the real estate market (Nazareth, 2018). Most of the time, such zoning is not concerned with the conservation of biodiversity in the urban space, considering that the only function of the city's "green areas" (which are mandatory reserves in allotment projects), would be to serve as leisure spaces for the population. As a result, such "green areas" are almost always just large lawns practically devoid of shrub and tree species (mainly fruit trees), which reduces their effectiveness for the conservation of biodiversity in urban areas.

Such "green areas" are often referred to in urban allotment projects as "amenities," that is, a factor that adds value, increasing income from urban land for a small number of owners. Thus, even urban biodiversity is marketed as a commodity of real estate capital, without being recognized for its true value as fundamental for the conservation of nature in urban ecosystems that helps the survival and quality of life of city dwellers.

This situation does not change even when significant portions of the municipal territory constitute areas protected by environmental legislation (whether federal, state or municipal), such as Environmental Protection Areas (EPAs), Permanent Preservation Areas (PPAs), Green Areas and Institutional and Ecological Corridors (ECs) areas. This situation is usually the source of intense conflicts between the environmental protection guidelines of such PPAs, and the urban development guidelines established in the zoning of urban land use in the context of their Municipal Master Plans (Poubel & Campos, 2019; Fonseca & Fonseca, 2013; Pizella, 2015).

3.2 *The Environmental Licensing of Effective or Potentially Polluting Activities*

Environmental licensing is also one of the most important instruments of Brazilian environmental policy provided by Law No. 6938/81 with the purpose of guaranteeing the conservation of natural resources and the environmental quality of urban and rural spaces in municipalities. It must be applied when approving undertakings and activities considered as having actual or potential capacity to promote significant degradation of the environment and natural resources. Due to its importance, the responsibility for its application is shared between the union, states and municipalities, considering the potential damage to the environment that a given undertaking/activity may promote. Thus, those with greater potential for damage, that is, with greater capacity to promote environmental impacts at the national or regional level, fall under the competence of the union via the action of its executive body, the Brazilian Institute of the Environment and Renewable Natural Resources (IBAMA). On the other hand, undertakings capable of promoting environmental impacts only at the subregional and local levels are under the responsibility of state environmental agencies and licensing municipalities.

It should be clarified that not all municipalities have the prerogative of being a licensing entity. To be a licensing municipality, it must set up a Municipal Council for the Environment since there must be effective participation of society in these licensing processes, and an environmental agency with a technical staff qualified for this function. In addition, the municipality must also have an environmental policy expressed either by means of a specific law (for example a "Municipal Environmental Code"), or that such matter is adequately covered in the items that constitute the Municipal Participatory Master Plan of the municipality (Brazil, 2001).

Data from the survey by the Profile of Brazilian Municipalities (MUNIC), for the year 2021, shows that only 21% of them have specific legislation for a prior study of environmental impact (IBGE, 2022), and are therefore those that possibly represent the totality of licensing municipalities in Brazil. According to Resolutions number 01 of 1986 and 237 of 1997 of the National Council for the Environment (CONAMA 1986, 1997), for undertakings considered with effective or potential capacity to promote environmental impact and significant degradation to the environment at national or regional level for licensing purposes, IBAMA will require entrepreneurs (public or private persons), to carry out an Environmental Impact Study (EIS). The present study will assess the effects of the proposed development on the physical, biotic and socioeconomic environments of the area that will be directly and indirectly affected by it.

In addition to the EIS, it is also mandatory to produce its respective Environmental Impact Report (EIR), which must contain a synthetic version of the assessments/considerations and recommendations produced in the EIS and, be written in simpler language so that its content is accessible to all populations in general and not just for specialists in environmental issues. The results of an EIR must be appreciated in sessions of public hearings and with special participation of the population that inhabits the area that will be affected by the activity/proposed undertaking. For medium-sized especially small-sized undertakings (with sub-regional and local environmental impacts), the states and municipalities may require, for the purpose of environmental licensing, a more simplified environmental study than the EIS/EIR. And it is exactly at this point that the effectiveness of the application of this environmental instrument begins to be compromised since it is not uncommon for state agencies, and especially municipal ones, to require very simplified environmental studies to release this environmental license.

According to Agra Filho (2021), this flexibility, instituted mainly from Complementary Law n° 140/2011, has even resulted in the option taken by some licensing bodies, to adopt the simplified procedure of Environmental Licensing by Adhesion and Commitment (ELAC), *consisting of a single phase and electronic, through only the commitment declared by the entrepreneur to observe the current environmental norms.* Thus, in this case, the entrepreneur does not need to carry out any prior study in the area of the enterprise to obtain an environmental license, just their personal judgment that the activity will not cause significant environmental impact. Thus, the conservation of nature in the urban environment becomes totally dependent on the socio-environmental conscience of the developer.

This problem is more evident in the case of activities/undertakings with potential impacts only on a local scale that are licensed by the environmental agency of the licensing municipality. This is because, in this case, the required environmental studies are normally carried out by only one professional based on insufficient data or coming only from secondary sources and often with the use of inadequate methodologies that lead to superficial conclusions and weak recommendations of support or rejection of the request (MPF, 2004; Motta & Pêgo Filho, 2013). The low quality of these environmental studies is usually due to the fact that entrepreneurs allocate little resource to them, even if the cost of these studies is practically irrelevant when compared to the other costs of the project. As a result, many times such studies are carried out only with the intention of complying with a formality with the municipal environmental agency for the release of the licensing of the enterprise, and not to effectively reach the goals established in the Brazilian environmental policy.

Even such bodies normally requiring very simplified environmental studies, developers resort to strategies to escape these requirements, as is the case of closed vertical or horizontal residential condominiums, which are understood by Queiroga (2002), as a form of "clandestine land subdivision." This is because, although such condominiums are in fact a form of urban land subdivision, they do not need to comply with all the rules determined by the federal land subdivision law regarding free spaces and areas reserved for community equipment in subdivisions (Brazil, 1979).

Vertical residential condominiums are true "machines for the artificial production of urban land" and, consequently, one of the main instruments for multiplying land rent in cities, since they pile up dozens of urban properties (and people), on the land of a single building that can have more than 30 floors. As a result, they considerably densify the space of the city's neighborhoods with people and cars, but without contributing a single square meter of new green areas to it. The horizontal residential condominiums, are usually walled (a modern version of the feuds of the Middle Ages), always claiming security reasons, and thus, do not need to transfer to the municipality their streets, squares, green areas, space for community facilities and, much less, integrate into the urban and environmental policies of the municipality. This is a serious issue that needs to be tackled in the legal system of urban and environmental policies in Brazil since it effectively contributes to making the achievement of the goal of sustainable cities advocated by them inaccessible.

3.3 Specially Protected Territorial Spaces (SPTSs)

Another important issue that is directly related to the conservation of biodiversity in Brazilian municipalities is the preservation of the different categories of fragments of the native vegetation cover of their biomes, which are found in both urban and rural spaces and which are generically called "Specially Protected Territorial Spaces." (SPTSs).

The SPTS is one of several environmental management instruments created by Law 6.938/1981 to make Brazilian environmental policy viable (Brazil, 1981). Art. 9 of this law provides for the administrative possibility of the federal, state and municipal public authorities to establish SPTSs in municipal territories with the purpose of protecting their attributes considered special, whether in environmental, cultural, archaeological, esthetic, etc. terms.

Initially, item VI of Art. 9 of this law defined as SPTS categories only ecological reserves and stations, environmental protection areas and areas of relevant ecological interest. However, subsequently, the importance of adopting such an environmental management instrument to achieve the objectives of the national environmental policy was reinforced by the wording given by item III of Art. 225 of the Federal Constitution, which determined the mandatory to:

> Define, in all units of the federation, territorial spaces and their components to be specially protected, with alteration and deletion permitted only by law, any use that compromises the integrity of the attributes that justify their protection being prohibited. (Brazil, 1988: 131)

Subsequently, in order to put the provisions of Law 9.638/81 in line with the rules made by the Federal Constitution of 1988, the wording of its item VI was changed through the approval of Law no. 7.804 of 1989, then proceeding to textually quote the SPTS: "VI - the creation of territorial spaces specially protected by the federal, state and municipal public authorities, such as areas of environmental protection, of relevant ecological interest and extractive reserves" (Brazil, 1989).

However, the definition of what are areas of environmental protection and areas of relevant ecological interest remained quite imprecise in the writing of these laws, giving rise to different interpretations even by jurists (Aráujo et al., 2013) and, consequently, also for their use by public managers. Therefore, debate exists around what should be the understanding of the scope of a SPTS, more specifically, whether it should be restricted to the concept of conservation units (CUs), as established under Law 9.985/2000 (Brazil, 2000), or should it also include protected areas due to non-ecological attributes. In the view of many authors cited by Araújo and Gannem (2006) SPTSs are not restricted to Conservation Units (CUs), since they are a type of territorial spaces that demand special protection due to attributes of different natures, while the second would be just one of their species (special areas due to their ecological attributes).

As a result, a SPTS can refer to special areas depending on other natures of attributes, as is the case, for example, of indigenous lands, lands of traditional communities, extractive reserves, natural cavities, archaeological sites, etc. The problem is that these and other types of SPTSs do not find legal provision within the scope of Law 9.985/2000, therefore, it is necessary for legislators from the three spheres of public management to act in order to provide through the creation of specific laws, environmental protection that they need due to their special attributes. However, since municipal public managers have preferred to be guided exclusively by the determinations of Law 9.985/2000 (Law of the National System of Conservation Units – NSCU), to define the areas that should or should not receive environmental protection in the municipal territory, many areas that should be protected because they are effectively SPTs do not receive such protection.

The exclusive use of the NSCU by public managers to define areas that should or should not be environmentally protected in urban spaces has also generated other distortions that have undermined the efficiency of Brazilian public policies that seek to conserve the country's biodiversity. This is because the NSCU law also does not recognize the Legal Reserve Areas (LRAs), created by the Brazilian Forest Code as environmentally protected areas, despite their unique importance for the conservation of the biodiversity of Brazilian biomes (Brazil, 2012).

Therefore, as long as the rural property is located in the rural area of the municipality, its owner will preserve the LRA by force of the forest code law. But when the limit of the urban perimeter of the city is extended, and the rural property becomes located within the urban zone, the vegetation of its LRA will be suppressed so that the owner of the rural property can use all its area for lucrative urban land subdivision projects. Thus, despite the eminently ecological nature of the LRA and its relevance for the conservation of Brazilian biodiversity, once the rural property where it is found becomes located in the urban space of a city, they will be fatally suppressed for not having their own ecological importance recognized neither by the NSCU law nor by municipal managers, except if local legislation requires the creation of green spaces, when these areas (or part of them), will have priority for preservation purposes. Thus, most of the time that rural properties are located within the limits of the urban perimeter of cities, municipal managers prefer to ignore the eminently ecological nature of their LRA and not declare them as SPTSs in the zoning of land use of the master plan.

To avoid this damage to the conservation of biodiversity in Brazilian cities, public managers should recognize the LRA as a SPTS in the zoning of urban space in the master plans, regardless of the position of the NSCU pertaining to them. This simple change in the destination of land use of these rural properties does not imply in automatic removal of the legal protection of their LRA as explicit in Art. 19 of Law 12.651/2012:

> The insertion of the rural property in an urban perimeter defined by municipal law <u>does not relieve the owner or squatter of the maintenance of the legal reserve area</u>, which will only be extinguished concomitantly with the registration of the subdivision of the land for urban purposes approved according to the specific legislation and in accordance with the guidelines of the master plan referred to in § 1 of Art. 182 of the Federal Constitution. (Brazil, 2012) (Our underlining)

However, this is not what has been verified in practice, since the owners/possessors of properties/rural land, wishing to maximize the profits of their real estate investments, seek in any case to find some way to suppress their LRA and thus expand the areas of its subdivision projects.

Art. 12 of Law 12.651/2012 defines the percentage of the rural property area that must be maintained with native vegetation cover as a LRA, which is always significant even though it varies significantly according to the political-administrative region of Brazil and the type of vegetation cover of the biome that occurs in the area where the property is located:

> I - Located in the Legal Amazon:
> a) 80% (eighty percent), in the property located in a forested area;
> b) 35% (thirty-five percent), in the property located in the Cerrado area;
> c) 20% (twenty percent), on the property located in the Campos Gerais area;
> II - Located in other regions of the country: 20% (twenty percent). (Brazil, 2012)

Based on the percentages expressed in Art.12 of Law 12.651/2012, it is possible to estimate the impact that the suppression of LRA vegetation can have for the sustainability of Brazilian cities. The destruction of native vegetation fragments that are or should be protected in SPTSs, and especially those resulting from PPAs and LRAs, has certainly contributed significantly to the production of cities with low rates of vegetation cover per inhabitant in Brazil. This is the reality observed in large (Fernandes, 2021; Machado et al., 2010; Mello-Théry, 2011), medium (Costa & Ferreira, 2009; Pina et al., 2011) and even small (Harder et al., 2006; Lima, 2014;

Louzada, 2022) cities, which implies a significant reduction in biodiversity and in the ecosystem services of these environments, and consequently also in the quality of life for the population.

From the above, one can understand the great importance of SPTSs as a planning/management instrument for nature conservation in urban environments and, therefore, to achieve the goal of sustainable cities in Brazil.

4 The Environmental Sustainability of Brazilian Agribusiness Cities

With the change that occurred in the 1970s in the productive base of the countryside in the Center-West of Brazil due to the introduction of a system based on intensive agriculture, this region experienced strong economic growth, but also intense and accelerated degradation of the natural landscapes of the Cerrado biome (Gonzaga et al. 2024). Due to the strong connection of the urban functions of the cities in this region with this new productive system (including Rondonópolis), they are now called by some authors as "agribusiness cities", and have as their main characteristics socio-spatial inequalities and intense degradation of the natural environment (Elias & Pequeno, 2007; Frederico, 2011).

This industrial agricultural production system was scientifically developed for this region and is highly centered on soybean cultivation. In this system, soybeans are cultivated with a high level of mechanization of agricultural operations (on average above 70%), and a great appeal for the use of chemical inputs (Santos, 2002). Thus, it dispenses with labor in the field, but increases its demand in urban activities (trade of goods, specialized services, agroindustries, etc.), that support this production system.

Immigration to the cities and rural villages of the region of this labor released from the field by the new productive system made them grow rapidly, resulting in many of these rural villages also becoming new cities in the regional urban network. As a result, a great demand for housing, basic sanitation and transport systems, among other urban equipment and services, quickly emerged in these new urban centers (Elias & Pequeno, 2007; Negri, 2010; Frederico, 2011).

However, these cities were not prepared to offer such infrastructure and basic urban services to receive this new population contingent. This is because, at that time, the Brazilian State itself still did not have urban and environmental planning instruments to manage the process of city growth (Antonello, 2013). As a result, the growth of the urban space of these cities occurred without territorial and environmental planning, both in terms of providing basic and social urban services, as well as the conservation of their biodiversity.

Although happening in a disorderly way, the process of urban expansion of the cities materializes in the form of real estate developments aimed both at supporting the agricultural production system (industrial and commercial subdivisions), and to house the growing population of these cities (residential subdivisions). However, subdivisions for residential purposes have very different construction standards according to the economic class for which they are intended in Rondonópolis. Thus, for the population of the less economically favored classes, subdivisions are built with houses of low construction standard, with a built area of about 55 m^2, delivered on very small plots (area of 200 m^2). These make up the popular residential housing that are locally called "housing groups," and can be both horizontal and vertical, but always exhibiting a high population density and buildings in the allotted area (Fig. 21.2).

For the more economically favored classes, subdivisions are built for buildings of medium and high construction standards, and even in horizontal and vertical condominiums with huge and super luxurious houses or apartments. In the case of luxury horizontal condominiums, they are usually built far from the city center, surrounded by very high walls to guarantee the safety of its residents, and housing extremely large houses with a built-in area that can exceed 1000 m^2 and, offering a wide range of specialized services to its residents (including ecological services) (Fig. 21.3). Thus, the distribution of wealth generated from the commercialization of commercial/industrial agricultural commodities in the Brazilian Cerrado gives rise to the socio-spatial segregation of the rich and poor population in the process of producing the urban space of agribusiness cities (Negri, 2008).

In essence, the city is seen and used by the politically and economically dominant classes in Brazil (which are almost always the same actors), as a space of opportunities for carrying out big business. Businesses that guarantee the expansion of their own capital, to the point that the city itself can now be understood as a business (Carlos et al., 2015), and not as the space that has become par excellence the "home of man" and, therefore, needs to be environmentally sound and ecologically balanced.

Fig. 21.2 Panoramic view of Dona Neuma Residency in the city of Rondonópolis-Mato Grosso-Brazil in May 2020. (Source: Rondonopolis City Hall 2020)

Fig. 21.3 Aerial photograph showing the panoramic view of the Condomínio Vilagge do Cerrado area in Rondonópolis-MT-Brazil—September 2022. (Source: Oliveira, 2020)

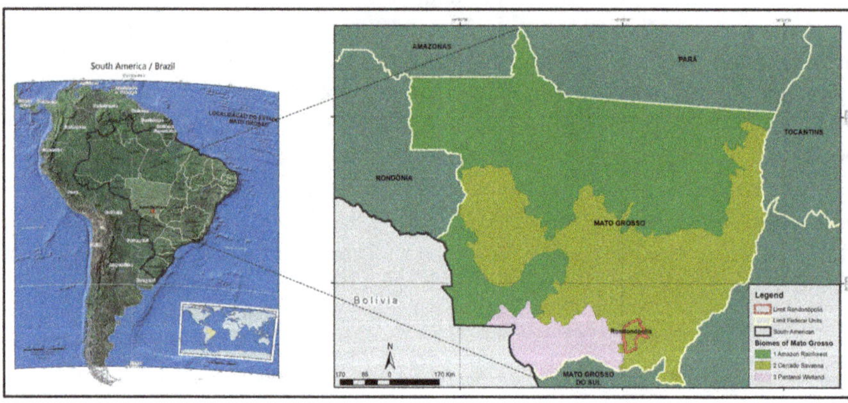

Fig. 21.4 Location of Rondonópolis in the Context of the Transition Zone of the Cerrado and Pantanal Biomes in the Southeast of the State of Mato Grosso. (Source: IBGE, 2021; MMA, 2020; ANA, 2021)

5 Urban Planning and Sustainable Development in Rondonópolis

Rondonópolis is a municipality with close to 250,000 inhabitants (IBGE, 2022), located in the southeastern region of the state of Mato Grosso in the center of Latin America and Brazil and in the transition zone between the Cerrado and Pantanal biomes (Fig. 21.4).

Its strategic location in the transition zone of these two biomes and the fact that it is a city that has shown intense economic development since the 1980s makes it a place of special interest to discuss issues related to nature conservation policies and especially in urban spaces.

Despite being a relatively old municipality by Brazilian standards (its official foundation dates back to the 1920s), Rondonópolis only experienced strong economic growth after the 1970s, which was mainly driven by capital-intensive commercial agriculture and the trade in production goods and agroindustries linked to it (Peixinho, 1998; Negri, 2010). This model of agricultural production system was scientifically developed to be implemented in the "Cerrado Lands," which naturally have acidic and nutrient-poor soils and, therefore, to become economically viable they demand high doses of chemical inputs (Santos, 2002). As a result, the efforts of the federal government to promote the capitalization of agriculture in the Center-West region of Brazil were decisive, especially through the Cerrados Development Program (POLOCENTRO), and the Japanese-Brazilian Cooperation Program for the Development of the Cerrados (PRODECER) (Peixinho, 1998).

5.1 The Challenges of Planning to Harmonize the Relationship Between Society and Nature in the Formation of the Initial Urban Nucleus of Rondonópolis

The history of the formation of the city of Rondonópolis clearly demonstrates the challenges that arise in the process of formation of a human settlement in an area where there is an ecosystem that was still quite preserved and unknown to the pioneers. Therefore, this reality demands that territorial planning (a technical and cultural product of human societies), takes into account environmental issues and the conservation of its biodiversity for the success of this settlement.

Around the year 1900, as a result of a cluster of immigrant residences associated with garimpeiro activities (gold and diamond mining), that then developed along the banks of the courses of the Vermelho, Ponte de Pedra and Poxoréu rivers (Demamann, 2011), begins the formation of the "village" that gives rise to the urban area of the city of Rondonópolis. The urban site of this settlement was located on the right bank of the Vermelho River and close to its confluence with the mouth of the Ribeirão Arareau, that is, swampy environments vulnerable to the frequent flooding of these rivers (Tesoro & Mesgravis, 1993). Thus, it can be said that the process of occupation and settlement of the population in this still practically wild and hydrologically dynamic and challenging space was carried out based on an urban planning that took environmental issues into account. What actually happened can be seen in the definition of the zoning of use of their lands, proposed in the first urban plan of the Heritage of Rondonópolis, which was prepared in 1918 by Lieutenant Otávio Pitaluga (Fig. 21.5).

According to this plan, the land subdivision process should prioritize the environmental protection of the bottoms of floodable valleys along the Vermelho River

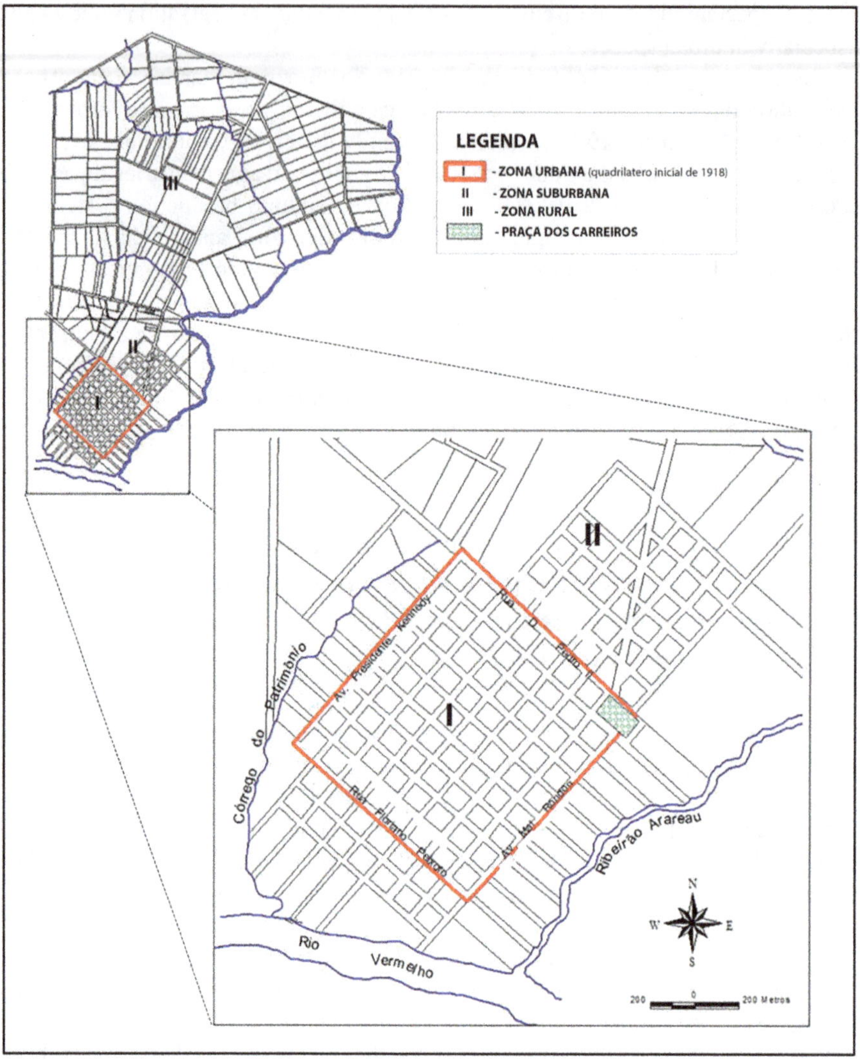

Fig. 21.5 Urban Plan for the Initial Urban Nucleus of Rondonópolis Heritage
Legend: I – Urban Zone (delimited by the red line); II – Suburban Zone; III – Rural Area; Area found in green its "Praça dos Carreiros"
Source: SEPLAN—Secretary of Planning, Coordination and General Controllership of the Municipality of Rondonopolis. Adapted from Demamann (2011: 45)

and its tributaries at the limits of the urban fabric (Suzuki, 1997). However, if on the one hand nature imposed limitations on territorial occupation and on the location of buildings for urban purposes in this place, on the other hand it was decisive in providing food to the pioneer population that sought to survive in this "new world". This is because the pioneers still did not know the fruits of the Cerrado biome and also because the waiting time for the crops to be harvested from their gardens was long. Thus, it was from the rivers that framed the area of the settlement, especially the Red River that was the main source of food in the early years of occupation. These rivers were full of fish of different species and sizes [Curi cited by Demamann (2011)].

The urban plan prepared by Lieutenant Otávio Pitaluga (Fig. 21.3), determined the territorial ordering of the Heritage of Rondonópolis in three zones: the urban zone with 100 ha, the suburban zone with practically the same area, and the rural (with about 1800 ha), and even changed the name of the village from "Rio Vermelho" to "Rondonópolis" (Negri, 2008: 37). However, by 1947, just 29 years after the approval of this first urban norm, Domingos de Lima, second elected mayor of Rondonópolis decided to carry out the expansion of the quadrilateral of the initial urban zone of the city, increasing its area from 100 to 223 ha, even before the approval of its first two residential subdivisions, which only occurred in 1952.

In 1955, Daniel Martins Moura, mayor of Rondonópolis at the time, approved the third residential subdivision in the city, but placing it approximately 5 km from the area defined as an urban area in its urban plan (Negri, 2008). That is, this residential neighborhood was deliberately allocated in the rural area far from the initial urban core of the city, leaving its population totally unattended by the few services and urban equipment that the initial urban core had. Therefore, expansion of the limit of the legal urban zone far above the real demand for growth and the creation of urban areas outside of the limits of the legal urban zone, were political actions that compromised the conservation of biodiversity and the quality of life of the population, hallmarks of the disorderly urbanization of the city of Rondonópolis since its founding.

5.2 *Public Policies and Urban and Environmental Planning in Rondonópolis*

The role of urban planning in the conservation of biodiversity in cities has not yet attracted much interest from Brazilian researchers (Angeoletto et al. 2019b). Early discussion sought to understand how such planning has managed to preserve the vegetation cover of cities, often by evaluating vegetation coverage per inhabitant, and comparing these values with indexes recommended by national and international institutions specializing in the issue. This, however, is not enough.

An aspect more directly related to the conservation of biodiversity that can be explored in this question is to also evaluate how much of the city's vegetation cover is composed of native vegetation of the biomes where the cities are located. However, studies with this approach are even rarer in Brazil, despite the fact that Brazilian urban policy seeks to provide managers with various instruments to achieve the figure of sustainable cities, one of its most important goals.

As explained above, since the beginning of the process of formation of the urban space in the city of Rondonópolis, territorial planning had to take into account the environmental challenges imposed by the landscapes of the Cerrado biome in the

South of the State of Mato Grosso. This biome and its ecosystems were little known by the pioneers of this city, who came from other regions of the country where the environmental realities were very different. Thus, environmental concerns have always been present in the territorial and urban planning of Rondonópolis, and as a result, they have also been present in the municipal legislation that structures its urban and environmental policy. As can be seen, in the first master plan approved in the municipality in 1975 through Complementary Law 447/75 (PDM/ROO/1975), which was prepared 26 years before this type of regulation was required for all Brazilian municipalities with more than 20,000 inhabitants through Federal Law 10,251 and through the enactment of its Organic Law on May 5, 1990, which was certainly elaborated already reflecting the effects of Law 6.938 /81 that established the National Environmental Policy and also the Federal Constitution of Brazil approved in 1988.

This matter is dealt with in detail in this organic law through its Articles 276 and 278, the first of which reproduces entire excerpts from the Brazilian Federal Constitution of 1988, which mentions that everyone has the right to an ecologically balanced environment, essential to a healthy quality of life, and imposing on the municipality and the community the duty and right to preserve and defend it. Article 278 mentions that the municipality must formulate a "Municipal Code of the Environment" and through its eight items, this article also establishes the guidelines and technical, legal and institutional instruments for implementing the municipality's environmental policy (Table 21.1).

In 1994, the second Municipal Master Plan of Rondonópolis was approved (PDM/ROO/1994), which presented significant progress in municipal democratic management with the creation of the Municipal Planning Council of Rondonópolis (MPCR). This is a collegiate body of an advisory and supervisory nature, whose

Table 21.1 Main guidelines of the environmental policy of the Municipality of Rondonópolis defined in the organic law

Rational use of natural resources in order to ensure their perpetuation;
Mandatory carrying out of a prior environmental impact study to be approved in a public hearing for the installation in the municipality of work with the potential to promote a major environmental impact;
Promotion of environmental education as a strategy for the preservation of the environment;
Stimulate and promote the recovery of native vegetation cover in public or private degraded areas, aiming at maintaining the ecological balance;
Control and regulate the production, sale and use of substances that pose risks to human life and the environment;
Protect natural cavities, archaeological sites, remarkable natural landscapes and other assets of tourist, scientific and cultural value in the municipal territory under the law;
Define territorial spaces to be specially protected (SPTS), through the creation of environmental conservation units and the preservation of assets of cultural value;
Promote the anthropic-environmental zoning of the municipal territory by establishing consistent and differentiated policies for the preservation of its natural environments of relevant ecological, hydric and biological interest.

Source: Rondonópolis (1990). Organization Santos, J.W.M.C

objective was to issue opinions on any plans, programs, global or specific projects of an urban and environmental nature, which were forwarded to the Municipal Executive. Its importance lay in the opening it provided for popular participation in decisions on municipal planning through a composition that included members of public power (executive and legislative), and representatives of neighborhood associations and various other institutions representing society civil society in Rondonópolis. But unfortunately this advice was never actually implemented. PDM/ROO/1994 failed to produce the Municipal Environmental Code in the set of laws emanating from it.

However, Law 2.120/94 dealing with the Urban Land Subdivision of Rondonópolis (Rondonópolis, 1994), one of the legal diplomas issued by PDM/ROO/1994, harmonized with some of the guidelines of the municipal environmental policy established in the Organic Law of Rondonópolis. As a result, it instituted norms and institutional instruments aimed at preventing the city's vegetation cover from becoming increasingly reduced as it grew.

Thus, through § 1 of item IV of its Art. 5, determined that "the springs or forests existing in the areas to be parceled out must be preserved" and it is up to the City Hall to demarcate them and, furthermore, they cannot be included in the percentages destined to the reserves destined to the public areas of the subdivision. Through item III of its Art. 6, it was determined that 35% of the area of the city's new subdivision projects should be reserved for public use, with 20% for the implementation of circulation routes and 15% for recreation areas and institutional use. That is, at least 10% of the last 15% would be for areas free of buildings, and with ecological, recreational and cultural purposes, as can be inferred from the concept of recreation areas that was given by item X of Art. 2 of Law 2.120/94:

> X - Recreation area, consisting of squares, parks, woods and gardens, is the one reserved for cultural, civic, sports and contemplative activities of the population. (Rondonopolis, 1994)

Although the Organic Law and the Urban Land Subdivision Law have represented significant advances in the implementation of the environmental policy of Rondonópolis, it was only fully established with the elaboration of the Municipal Environmental Code, which was approved only in 2002, that is, 12 years after its provision in the Organic Law, finally aligning itself with the guidelines of the National Environmental Policy.

This is because this Environmental Code made available several technical, legal and institutional instruments that came to complement those already created by the municipal government in order to adjust the national environmental policy (Table 21.2).

5.3 Problems in the Conservation of Biodiversity in the City of Rondonopolis

Despite all the technical, legal and institutional instruments that the National Environmental Policy has made available in Brazil, the country still suffers from a serious crisis of urban environmental problems. This makes it clear that it is not

Table 21.2 Main environmental management instruments made available to the environmental policy of the Municipality of Rondonópolis through the municipal environmental code

Municipal Development and Environment Fund—fundamental to finance the actions of this policy;
Environmental Planning—main instrument of this policy in the sense of establishing guidelines capable of reconciling economic growth and conservation of the municipality's natural resources in order to move towards its sustainable development;
Environmental Inspection—which began to be systematically carried out by inspectors from the Municipal Environment Secretariat
Environmental Audit—instrument that, among other objectives, is intended to verify the effective or potential levels of pollution and environmental degradation caused by activities or works carried out in the municipality;
Environmental Information and Registration System (EI/RS)—responsible for collecting and systematizing all information of environmental interest to the municipality in order to support all actions of the municipal environmental policy.

Source: Rondonópolis (2002). Organization Santos, J.W.M.C

enough to have such regulations, political will, commitment from public managers and mainly broad democratic participation of the population in public management is also necessary, especially at the municipal level.

The city of Rondonópolis is a good example of this, because despite having managed to implement a well-structured environmental policy, it presents several problems to the conservation of its local biodiversity. Problems that tend to get worse due to the pressure to dismantle this policy that has been taking place on the part of agents of local economic and political power and which effectively began to have effects already during the approval of the PDM/ROO/2006 and intensified during the work to update it that took place between 2016 and 2021.

Many of these environmental problems in the city of Rondonópolis have already been the subject of research produced by researchers from the Federal University of Rondonópolis, especially those who work in the Graduate Program in Geography, and served as a basis for carrying out several studies (Geotechnical Charts of Urban and Rural Zones, Thematic Mappings at scales of 1:10,000, 1:50,000 and 1:250,000 of Urban and Rural Zones; Urban and Rural Mobility Plan), of the first phase of the work to update the PDM/ROO/2006, which is called "Reading the Local Reality," materials that are available for access and download at: http://www.rondonopolis.mt.gov.br/plano-diretor/materiais/.

5.3.1 Urban Expansion and Vegetation Coverage Reduction

Despite the environmental policy of Rondonópolis having made available the environmental licensing and the obligatory reserve of 10% of the areas of the projects of new subdivisions to be designated as "green areas", they have not been able to fulfill this objective. Studies of Rondonópolis' vegetation cover shows that it has been significantly reduced due to the disorderly subdivision of urban land, especially those approved from the 1980s onwards (Pereira, 2009; Duarte et al., 2017; Souza,

2017; Oliveira, 2020). Such failures are almost always due to the lack of an approval protocol for projects that would require joint approvals from both the Municipal Housing and Urban Planning Department and the Municipal Environment Secretariat. As this does not happen, and communication between these two secretariats is insufficient, the projects are approved with serious flaws.

Pereira (2009) evaluated how much the urban fabric (effectively urbanized area) of Rondonópolis had vegetation cover in the years 1986 and 2008, and found that it was about 40% larger in 1986. In other words, the process of growth of the urban fabric of the city in this period took place at the expense of the loss of a large part of its vegetation cover.

Duarte et al. (2017), continuing the analyses initiated by Pereira (2009), evaluated the effect of the urban expansion of the city of Rondonópolis between the years 2006 and 2015 on its vegetation cover. The author found that in this period the urban vegetation cover was also significantly reduced with the expansion of the city. The author also mentions that another policy that could contribute significantly in this regard would be the adoption of a master plan for afforestation of roads in the city. Indeed, Art. 130 of the PDM/ROO/2006 now requires the presentation of a tree planting project for roads and landscaping in the projects for new subdivisions in the city. In fact, Art. 134 also stipulated a period of 10 years for the municipal public power to regulate and execute, in the form of a law, a Master Plan for Urban Afforestation in Rondonópolis. However, this is yet another important environmental planning and management instrument that, despite being provided for in the municipal environmental policy, was never implemented.

Souza (2017) found that less than 25% of "green areas" in Rondonópolis were actually implemented by the municipal government. The author also noted that the government prioritized the implementation of these green areas in neighborhoods where the population has a better income level and which are located closer to the central region of the city. On the other hand, his study verified that the municipal government of Rondonópolis also authorized the suppression of native vegetation cover of the Cerrado in areas with environmental protection in the city in flagrant disregard of the municipal environmental legislation and the 2006 Municipal Master Plan. For example, he portrayed the case of the Conservation Unit Zone (CUZ 55), which had its vegetation suppressed for the implementation of the Terra Nova Residential Condominium subdivision (Fig. 21.6).

In 2013 and 2014, intense summer rains, characteristic of the tropical climate of the Cerrado biome, caused floods and severe damage to the properties of residents of the Terra Nova Residential Condominium (Fig. 21.7). This was strongly associated with the suppression of vegetation at CUZ n° 55 due to the failure of the municipal executive power to guarantee the preservation of this environmental unit.

Oliveira (2020), verified that the way in which the municipal environmental agency carries out the environmental licensing of residential or industrial land subdivisions considered to have a reduced power of environmental impact (that is, with an area of less than 100 ha), has also favored this disorderly and even criminal suppression of vegetation cover of the urban area of Rondonópolis. This is because for the environmental licensing of subdivisions of this size, developers are not obliged

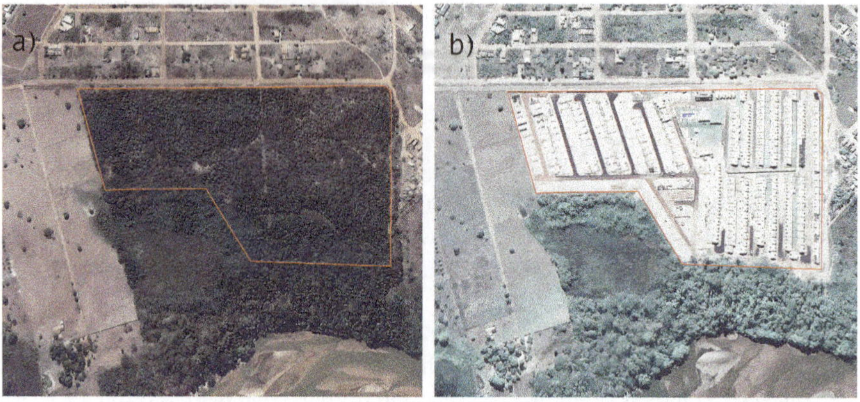

Fig. 21.6 Image showing the vegetation cover of Conservation Unit Zone 55 fully conserved on 08/02/2005 (left), and almost completely suppressed on 08/03/2009 (right), due to the irregular implantation of the Terra Nova Residential Condominium allotment regarding the same

Fig. 21.7 Record of the effects of water runoff from a heavy rain (about 100 mm in 1 h), that occurred on 12/29/2013 in the southeastern sector of the city of Rondonópolis: (**a**) massive runoff flowing into the Terra Nova Residential Condominium area; (**b**) height that the water reached inside one of the houses in this condominium. (Source: Trassi et al., 2020: 62,64)

to carry out the EIS/EIR. In this case, the required environmental study is the "simplified environmental report" (SER) that does not adequately characterize the physical and biotic environments of the area where the project will be implemented, much less the direct and indirect impact that it may generate.

According to Oliveira (op cit), the SER is solely based on a bibliographical review of studies that carry out this characterization only at the regional level, thus being able to ignore important characteristics of the physical, biotic and socioeconomic environments that exist in the area where the allotment will be implemented, which greatly weakens the effectiveness of the environmental licensing in question. One of the most deleterious effects on biodiversity resulting from this is the one observed when the SER does not demonstrate, for example, the existence of important fragments of native Cerrado vegetation still preserved in the area where the project will be implemented and therefore removed. This type of failure in the environmental licensing process of urban land subdivision projects makes it possible, for the environmental agency to provide a retrospective license for developers who remove all vegetation (primary or secondary), as part of "land clearing". In other words, the developer removes important fragments of native vegetation that were preserved in that area so that later the government can implement vegetation cover in its "green area", which, if by luck it has a minimum of tree species, will represent a high cost to the public purse.

In our view, this distortion in the environmental licensing process for subdivision of urban land could be mitigated if the municipal environmental agency required a less simplified environmental study from entrepreneurs. A study that adequately characterizes the existing vegetation cover on the plot of land to be parceled out. And if there is vegetation cover with tree species on this plot, it was obligatorily destined to constitute the green area of this subdivision. In the same sense, this reflects the position of the municipal environmental policy regarding whether or not to preserve the vegetation in the Legal Reserve Areas (LRAs) of rural properties. When the position of the municipality, expressed through its Municipal Master Plan and other municipal legislation, is for the conservation of the vegetation of the LRA, the chances of the city growing while maintaining an adequate index of vegetation cover per inhabitant and preserving its biodiversity will be greater. In contrast, if the municipal environmental policy is to not conserve the fragments of native vegetation existing in the LRA, the chances of this environmentally sustainable growth will be smaller.

That is why it is very important that the municipal environmental policy has a clear position in relation to this issue, and even establishes what is the minimum rate of vegetation cover that the city must have (per inhabitant or km^2). It should be understood that the establishment of this index, in addition to being an important action to maintain the ecological balance of the urban environment, also has a profound impact on the quality of life and health of people living in cities. Unfortunately, the conservation of LRAs has not been the predominant option in the environmental policy of Brazilian municipalities since the pressure of real estate market agents on the members of the municipal legislative and executive power to remove it has been intense. Together, they actively contribute to the process of urban expansion resulting in "desert cities".

Evidently, the best option in terms of the environment and conservation of local biodiversity would be to preserve the fragments of vegetation in the LRA as an additional strategy so that the city's vegetation cover rates do not become increasingly reduced as it expands. Fortunately, this was the option adopted in the environmental policy of Rondonópolis since the approval of its Organic Law in 1990 until the approval of the PDM/ROO/2006. Even in this PDM, around 50 fragments of native cerrado vegetation were identified and mapped as LRAs of rural properties remaining in the urban space of the city. Then these LRAs were transformed into areas of legal environmental protection called Conservation Unit Zones (CUZs), and started to appear in the Urbanistic and Environmental Macrozoning of the municipality. Furthermore, these CUZs were elevated to the category of SPTSs in the urban area of the city of Rondonópolis as a strategy to provide them with the maximum possible legal environmental protection.

However, during the work to update the PDM/ROO/2006, these CUZs became the target of heavy attacks by the owners of these properties, including the explicit support of local politicians and even city hall officials. They wanted at all costs that these CUZs were deconstructed, so that these vegetation fragments could be suppressed allowing the areas to be subdivided.

This is because most of these LRAs are currently located in highly valued regions of the city of Rondonópolis, which increases the desire of their owners to be able to use them to generate profit, by suppressing the vegetation preserved in these areas in order to be able to divide them into new real estate developments.

5.3.2 Irregular Occupation of Permanent Preservation Areas (PPAs)

Another problem that has contributed significantly to reducing the effectiveness of the municipality's environmental policy is the irregular occupation of permanent preservation areas (PPAs), on the banks of rivers in the urban space. Studies carried out on a stretch of about 16 km in which the Arareau river runs through the urban fabric of the city of Rondonópolis to its mouth on the Vermelho river (Santos et al., 2016, 2019; Santos & Rubim, 2019; Santos & Oliveira, 2019), found that both its waters and the vegetation of its PPAs have suffered various types of environmental degradation in recent years.

The degradation of the water quality of the Arareau River is directly associated with the suppression of its riparian vegetation and irregular occupation of the PPA strips along its banks (Santos et al., 2019). Other impacts associated with these first two were also identified, such as landfilling with construction debris at various points along its course, several points of its river canal already heavily silted up, processes of erosion going up its banks which in some cases were already even reaching houses and circulation routes in subdivisions located nearby, among other environmental impacts.

Santos et al. (2016), observed about 20 clandestine pipes on its banks, which are used for the irregular release of effluents of domestic origin and also from commercial and industrial activities directly into its river bed. In relation to irregular land occupations within the limit of its permanent preservation area (PPA), it was observed that these occur for various land uses, such as agricultural cultivation, small animal breeding, fish farming, nurseries for exotic plant species, residences and even for the implementation of commercial/industrial establishments. That is, these various forms of irregular occupation of the land in its PPA clearly demonstrate that the structure of the environmental policy (in whatever sphere of government it may be), is not managing to promote the environmental protection of the biodiversity of the ecosystems of the water bodies in the urban area of Rondonópolis.

More specifically in relation to the irregular occupations of its PPA range for residential use, Santos and Rubim (2019), found that among houses built in subdivision projects and others outside these and without any type of project, 185 families were living within the subdivision strip of the Arareau River PPA in 2018.

This irregular occupation results from two different factors. The first is that part of the area of several residential subdivisions in the city was parceled out within the limits of the PPA of that river by the municipal government. The second is related to the irregular and clandestine occupation of its banks by poor families who did not have the resources to buy a property in the city.

Regarding the former, this probably occurred due to the difficulty that the municipal government had in determining what was the real width of the river's fluvial canal in its urban section at the time when such subdivisions were implanted in the city(1970s/1980s). Thus, believing that such width was less than 10 m and guided by the rules for determining the PPA range of rivers stipulated by the Forestry Code of 1965 (Law No.4.771/1965), meters in its urban stretch. However, the Complementary Law of the Municipality n. 043/2006 defines since that time, for Ribeirão Arareau, the Environmental Protection Zone in the urban perimeter is 50 meters. Notwithstanding the federal and municipal legislation in force, the municipal executive branch, following its tradition of considering the banks of urban water courses as parcelable areas, approved such subdivisions with the limits of their lots extending to the banks of the Arareau River (Fig. 21.8). This generated problems both for the owners of these lands and also for the conservation of this water body and the biodiversity associated with it.

The second cause of this irregular occupation of the Arareau river PPA strips for residential purposes proved to be more serious in socio-environmental terms, but fortunately represented less than 20% of the verified cases (less than 30 houses), and was also more spatially restricted to the region from the mouth of the river. These houses do not have environmental sanitation for their solid waste and effluents. As a result, they represent a high risk for these residents since they are built practically on the bank of that river, especially for children who play on that bank all the time and are therefore very exposed to the risk of falling into that river. As for the conservation of the body of water and its biodiversity, it also configured a highly impactful situation since both solid waste and effluents from these houses are released directly into its bed.

Moreover, not even the houses that were built within the boundaries of the Arareu River PPA had an adequate solution for sewage disposal since the solution adopted for them was the septic tank. This is because the proximity of the houses to the river canal makes any attempt to collect sewage through the public sewage network unfeasible. Sinkholes represent a high risk of environmental contamination for the waters of the Arareau River, mainly due to the variation in the level of its water table during the episodes of intense rains that are characteristic of the tropical climate in the Cerrado region.

The extending of the development limits of the land to the banks of the rivers also allowed several companies to build (partially or even completely) their buildings within the range of the PPA on the Arareau River. Santos and Oliveira (2019), found that 28 companies ranging in size from micro to medium-sized companies had buildings within the limits of the Arareau river PPA range. And 15 of them developed activities classified as medium to high environmental risk, such as, for example, storage of pesticides and other chemical products, which could generate a

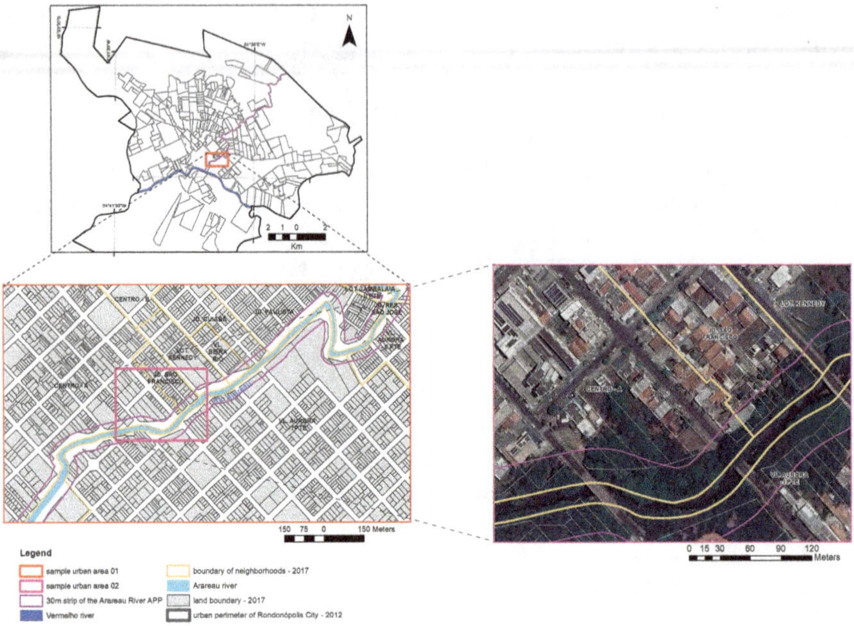

Fig. 21.8 Plots of subdivisions in the center of the city of Rondonópolis encroaching on the 30-m strip of the Permanent Preservation Area of the Arareau River. (Source: City Hall of Rondonópolis, 2017—Organization: Santos, J.W.M.C.)

serious environmental problem in the event that such deposits were affected by the waters of the river in one of its full episodes.

These issues became even more complex after the amendment of Law n. 4771/65 by Law 7803/89, which increased to 50 m the minimum width of the PPA strip for rivers with a canal width between 10 and 50 m. As a result, the number of properties and buildings that are within the limits of the permanent preservation area of this river has significantly increased, requiring that the municipal government seek alternatives to overcome these environmental management challenges in the municipality.

5.3.3 Ecological Simplification of the Urban Forest

From the approval of PDM/ROO/2006, the municipal environmental policy of Rondonópolis started to demand that new subdivisions also considered the afforestation of roads and landscaping. And for that, it demanded that such street tree planting projects follow the following technical criteria: (a) Exact and maximum distance of 60 cm from the edge of the curb or its projection; (b) That the trees should be planted following the direction of the axis of the lots as a parameter and should keep a minimum distance of eight meters from the intersection of streets or avenues; and

(c) The trees should be of suitable plant species under the electricity distribution networks.

Leite (2018), evaluating in quantitative and qualitative terms the urban forest of the city of Rondonópolis concluded that afforestation has good quantitative indices (using the indicator of trees per linear km). However, in qualitative terms, afforestation falls far short of what would be necessary to meet the socio-environmental demands of the population of Rondonópolis. This is because between 60% and 80% of the trees that make up this urban forest are of a single species, *Licania tomentosa* (popular name Oiti), which is not native to the Cerrado biome. In ecological terms, this is not recommended as the presence of native tree species can promote native insect and bird diversity, as well as the ecological integrity of urban ecosystems. Their results support Duarte (2017), who pointed out that it was urgent for the public power to draw up a master plan for urban afforestation for Rondonópolis as determined by Art. 134 of the PDM/ROO/2006, which even stipulated a period of 10 years for the municipal executive power to carry out such a master plan.

A similar study of the environmental conditions of public squares in the city of Rondonópolis (Rondonopolis, 2017a) found that these squares are not able to meet the demands of the population nor adequately perform ecological functions for the city. In quantitative terms, it was found that the number of trees present in the vegetation of the 31 squares in Rondonópolis is far below its capacity. In 23 of them this situation was classified as "very bad," in six as "bad" and two did not even have any tree species. In qualitative terms, 54 tree species were verified in the vegetation of the squares of Rondonópolis. Most of them, 43 species (79.6%), are of exotic origin, many invasive, and two endangered species: the exotic *Paubrasilia echinata* (origin: Atlantic Forest Biome) and *Dypsis decaryi* (African origin).

6 Adjustments to the Urban and Environmental Policies of Rondonópolis Proposed in the PDM/ROO/2006 Update

The municipality of Rondonópolis has many technical, legal and institutional instruments, such that public managers can, through planning and environmental management, support the conservation of biodiversity and the quality of life of the population. However, despite this, there are many environmental problems both in urban and rural spaces, which demand that adjustments be suggested in the instruments and processes that structure its environmental policy.

As previously mentioned, although a team from the municipal government of Rondonópolis started work to update the PDM/ROO/2006 in 2014, it was removed from this task in 2016 by a court decision, as the team did not observe the various environmental conditions that were required by IBAMA in the release of the preliminary license for the installation of an enterprise that had the potential to cause environmental impacts at a regional level, the Intermodal Railroad Terminal from the company América Latina Logística—ALL.

As one of the conditions for its environmental licensing, the updating of the master plan of the municipalities that were within the area of its direct impacts, there was a determination that the works of updating the PDM of Rondonópolis occur a little before the date that obligatorily should occur. Normally the Municipal Master Plan must be reviewed every 10 years, but the development of the Intermodal Railroad Terminal meant that this needed to occur earlier, and, as determined by the Public Prosecutor's Office in the records of Public Civil Action n. 0009937-23.2015.8.11.0003, which was processed at the 4th Civil Court of Rondonópolis/MT, it would be incumbent upon the company América Latina Logística—ALL to bear all the expenses necessary to carry out the new PDM.

After more than a year of various legal battles, with lawsuits proposed, injunctions and precautionary measures granted, as well as appeals decided by the Court of Justice of the State of Mato Grosso, the Public Ministry, the company América Latina Logística—ALL and the Municipality of Rondonópolis signed, on 09/22/2016, an agreement in which inputs would be purchased and teams hired to collect all the data and information necessary for a master plan that would prove to be effective and participatory. The studies and acquisition of inputs would be financed by the ALL; the municipality should set up and maintain teams to carry out administrative activities, as well as actively participate in the construction of the new PDM. The differential, however, was that it was agreed that the general technical coordination of all the works would be under the responsibility of a professor/researcher from the Federal University of Rondonópolis (FUR), who assumed the task of planning and coordinating the execution of the works of update of the PDM/ROO/2006 and, above all, in order to fully meet all the conditions of its environmental licensing. Thus, Professor Jeater Santos, at the time was coordinator of the Graduate Program in Geography at FUR, was entrusted with the mission.

With this innovative work methodology, the aim was to meet some objectives, among which the following stand out: (a) Understand on what basis the municipality's growth and development process is taking place; (b) Delimit actions that seek, on the one hand, to take advantage of the potential and, on the other, to remedy/mitigate the deficiencies of the socioeconomic and environmental reality of the municipality; (c) Prioritize the right of all citizens to decent housing in a healthy environment and with equal access to urban equipment and services; (d) Create conditions for the harmonious and orderly growth of urban/rural spaces.

To this end, a work plan was presented, composed of three stages: (1) Acquisition of updated remote sensing products with adequate spatial resolution for the multiscale representation/analysis of urban and rural spaces in the territory of the municipality of Rondonopolis; (2) Elaboration of previous technical studies, updating the outdated official municipal cartography; (3) Elaboration of the proposal to update the new master plan in accordance with the requirements established by Law 10.257/2001.

This work plan was presented and approved at a public hearing held in the second half of 2016. Representatives of public authorities (Judiciary, Legislative, Executive and Public Ministry), of organized civil society and also members of the Monitoring Management Nucleus (MMN), participated in this hearing. The MMN is the committee officially constituted by the municipal executive power, composed of representatives of all stakeholders, composed of 60% civil society representatives and 40% of the public power. Its role is to monitor the work to revise the Municipal Master Plan, as well as to approve the new proposal for carrying out the work. It was also agreed at this hearing that all materials that were prepared in the works for the revision of the PDM/ROO/2006 should be approved in public hearings and made available on the internet through the PDM page on the website of the municipal government of Rondonópolis. Simple rules, but decisive for a transparent and democratic process, ensuing the PDM reflects the interests of all citizens.

Such remote sensing products referred to a mosaic of aerial orthophotos with a spatial resolution of 15 cm and also a digital elevation model produced with LIDAR technology (light detection and ranging), and with metric spatial resolution. For the rural areas of the municipality, free products were obtained, which refer to satellite images with spatial resolution ranging from 2 to 30 m, as well as images of a digital elevation model produced and made available by NASA with a spatial resolution of 30 m.

The resources made available by ALL were also essential to pay for stages 02 (previous technical studies), and 03 (preparation of the new proposal for the Municipal Master Plan), foreseen in the work plan. Thus, it became clear that without this money, the update of the PDM/ROO/2006 would not be able to be carried out in order to meet conditions for the environmental licensing of the multimodal terminal installed in the city. Almost 2.5 million reais (which at the time corresponded to just over US$725000), were invested.

6.1 Technical Products Designed to Support the PDM/ROO/2006 Update

Three technical products were prepared to support the update of the 2006 municipal master plan for Rondonópolis: (a) A set of thematic mappings on environmental and urban aspects of the urban and rural areas of the municipality; (b) Geotechnical charts to verify the existence of areas with environmental risk to land subdivisions or economic uses of other types in the municipal territory; (c) A Master Plan for Urban and Rural Mobility to analyze the risks of conflict between internal traffic flows in the city of Rondonópolis and the flow of heavy vehicle traffic from other municipalities that would be attracted to the intermodal terminal.[3]

6.1.1 Thematic Mappings of the Municipal Geographic Reality

These thematic mappings aimed at detailing the urban and environmental characteristics of urban and rural spaces in the municipal territory. This mapping effort generated a set of 76 cartographic products (cartograms, image-cards, maps, etc.),

[3] Available for consulting and download at: http://rondonopolis.mt.gov.br/plano-director/materiais/

elaborated in three cartographic scales. Products at 1:10,000 to represent the elements of urban spaces of the municipality; in 1:50,000 to represent the elements of the municipality's rural space and, products in 1:250,000 to represent connections at the regional and national level of physical and socioeconomic aspects of the municipality. As an example, Fig. 21.9 shows one of these products that supported the analyses of phase 1 of the preparation of the proposal for the new Municipal Master Plan.

6.1.2 The Geotechnical Charts

The geotechnical charts of the municipal territory were prepared seeking to identify and map the existence of areas that had restrictions on processes of subdivision of land in urban space, or for agricultural, industrial or any other use in rural areas.

The geotechnical map of the urban space was prepared starting with the mapping at a scale of 1:10,000, and in the sectors of the urban perimeter identified as having a greater risk to urbanization due to natural physical processes (erosion, soil erosion or landslides, floods etc.), its mapping was produced on a scale of 1:5000, considering areas not yet parceled out, and on a scale of 1:2500 for those that were already parceled out. In the areas that had already been parceled out and that presented environmental risks, the mapping was produced with a higher level of detail, in order to guide the public authorities on how they should be monitored and not allow new parceling over them.

As for the rural spaces of the municipality, the geotechnical map was based on analyses produced on mapping carried out at a scale of 1:50,000 and, in the same way as in the case of the maps of the urban area, the sectors that were identified with greater risk to natural physical processes were mapped in greater detail at 1:25,000 scale. This methodology proved to be quite efficient both in terms of identifying the urban and rural sectors of the municipality with environmental risks and their causes, as well as reducing efforts and costs for the production of mappings.

As an example, Fig. 21.10 shows one of the products of these geotechnical maps, which is a map of levels of susceptibility for the occurrence of natural processes in the physical environment in the area of the urban perimeter of the city of Rondonópolis.

6.1.3 The Urban and Rural Mobility Plan of Rondonopolis

The municipality's Urban and Rural Mobility Plan sought to answer two questions. The first was related to the quality of the urban and rural mobility service that the Rondonopolis road system provided to its population, including people with special needs. The second, no less important, aimed at estimating how much traffic in the city would be affected by the additional traffic it would receive during the harvest period of the regional crop as a result of the huge flow of trucks from other municipalities that would be attracted by the Multimodal Terminal of the company

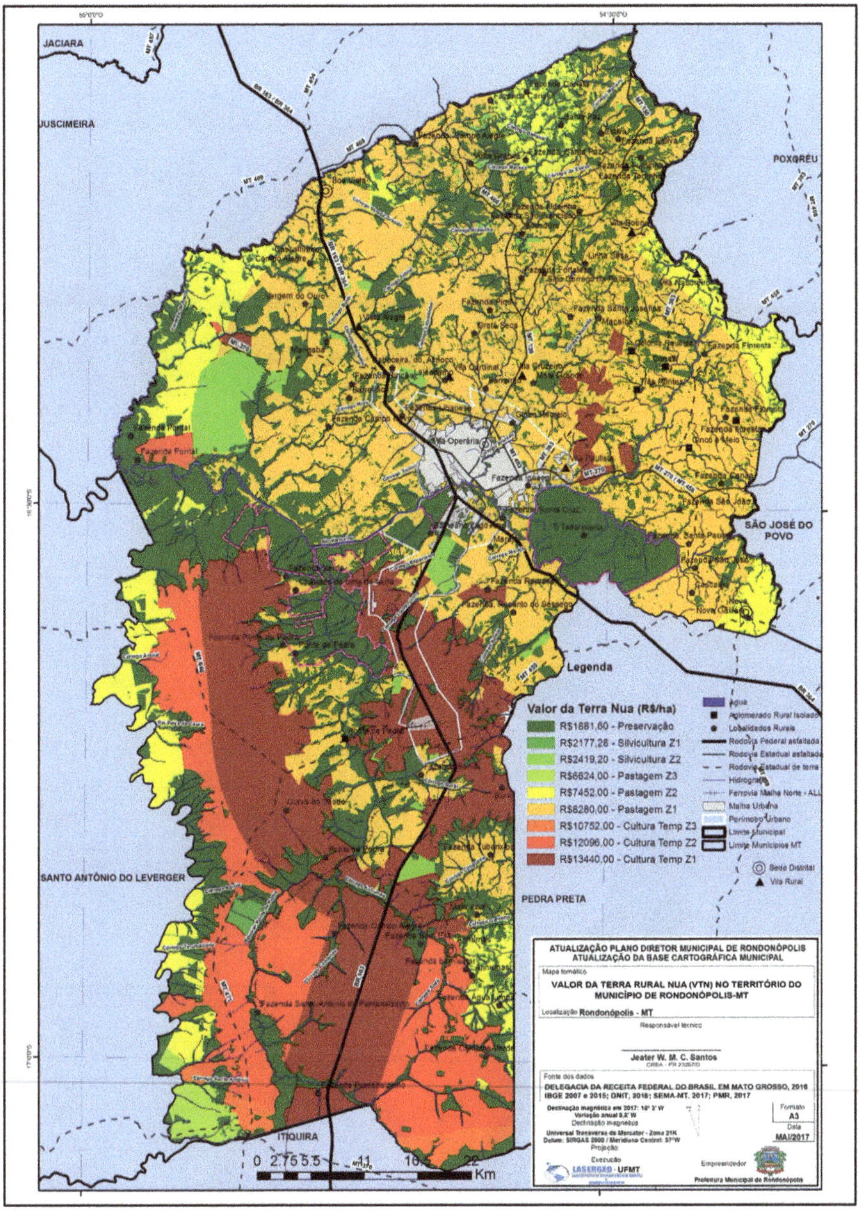

Fig. 21.9 Map with the variation in the value of rural land in the territory of the municipality of Rondonópolis prepared to support the update of the 2006 Municipal Master Plan. The caption indicates a lower value in areas with preserved native vegetation or forestry (shades of green), and a higher value in areas with agricultural use (tones of yellow and red). (Source: Rondonopolis, 2017b)

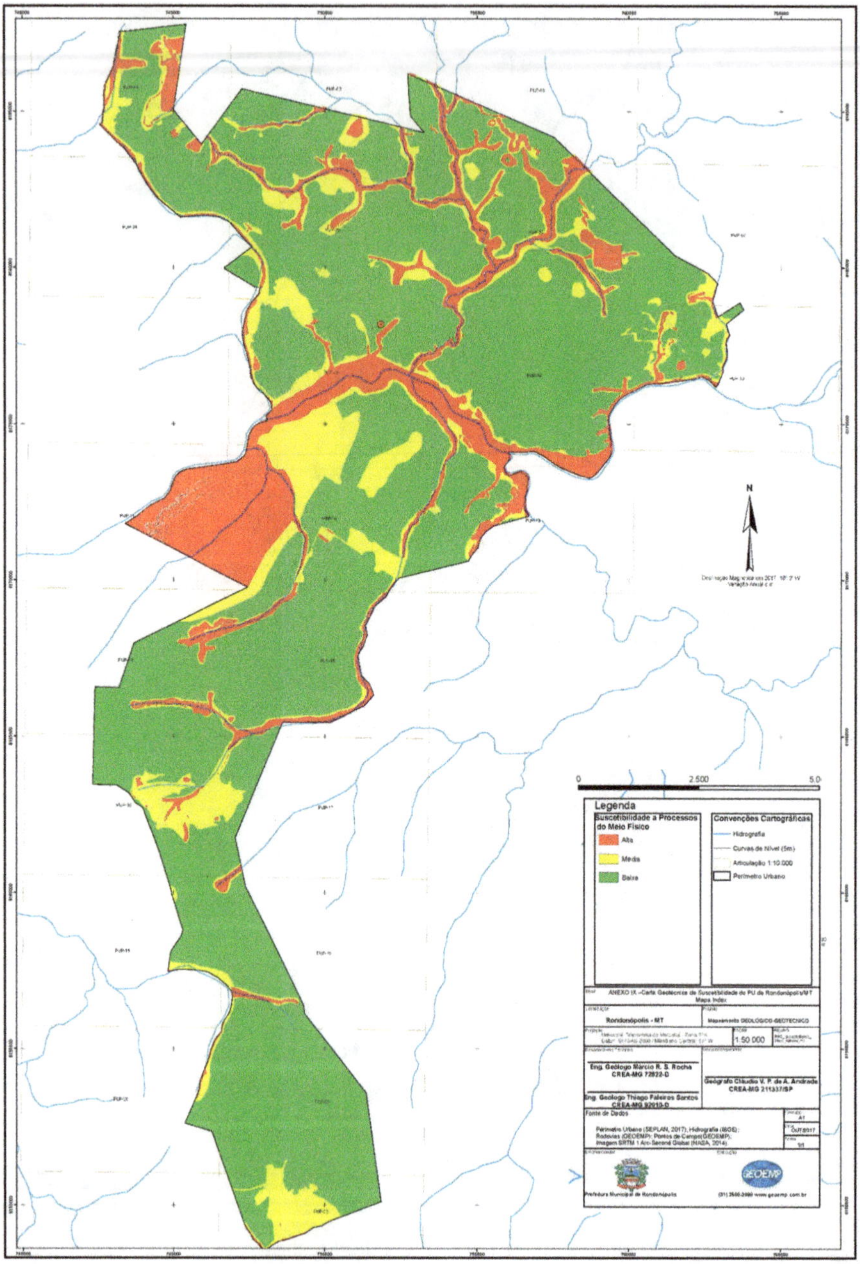

Fig. 21.10 Geotechnical map representing the levels of susceptibility of occurrence of natural processes in the physical environment in the area of the urban perimeter of the city of Rondonópolis. The legend indicates high susceptibility in the red areas, medium in the yellow areas and low in the green areas. (Source: Rondonopolis, 2017b)

ALL. The company took advantage of several thematic mappings that constituted the first technical product of the PDM/ROO/2006 update work plan (Sect. 6.1.1 of this text). However, in order to answer the two questions initially proposed, it was necessary to produce a huge set of data, which until then the Municipality of Rondonópolis still did not have.

To answer the first question, it was necessary to assess in detail the entire infrastructure of the city's road system. In the urban area, the situation of bridges, sidewalks, horizontal and vertical signaling on the roads, the physical situation of the roads, the traffic load on the roads by different modes, etc., were evaluated. In the rural area, the quality of the mobility service offered to the population residing there was evaluated, that is, from the situation of the roads through which intercity transport bus lines pass and, above all, the school transport bus lines.

The analyses and materials produced to answer the first question made it possible to make several important proposals for adjustments in the urban mobility service in the city of Rondonópolis. These adjustments also aimed to make this mobility less restricted to motorized modes, and thus encourage walking (including for people with special needs), and bicycles (including the possibility of integration on public transport buses). As a result, some of the plan's proposals referred to the expansion/implementation of bus lines, bike lanes and lanes, adjustments to the physical structure of sidewalks and bridges, and also the reordering of the hierarchy of urban roads in terms of their actual traffic flows and position in the urban fabric. In terms of population mobility in rural Rondonópolis, the study found that there were many deficiencies due to the large number of unpaved municipal roads and many bridges still made of wood and in poor condition. And that this was the situation of many roads where several bus lines of school transport in the municipality passed. It also verified a very low offer of intercity bus lines for the transport of the rural population that lives in farms and settlements. This was a very serious situation, especially for residents who did not own a car, who therefore had no possibility of public or private transport/mobility.

In order to answer the second question, that is, how the flow of trucks that would be attracted by the ALL terminal would reflect on the normal traffic in the city, it was necessary to obtain data that would allow for a reliable estimate of such flow. To this end, logistics data were collected from the company itself, which revealed that the flow of heavy vehicles that would be attracted to the intermodal terminal was very large, on the order of 10,000 trucks a day, and that such flow could even triple within just 20 years (Macedo et al., 2015; Urbaniza, 2017). Therefore, the flow of trucks on federal and state highways in the municipality should be highly impacted during the regional harvest outflow periods, fatally affecting traffic in the city of Rondonópolis.

In possession of these data and also those that were produced to answer the first question of the present study, it was verified the need to carry out major infrastructure works (bridges, viaducts, etc.), in order to separate the flow of heavy vehicles coming from other municipalities for the intermodal terminal of the ALL company and the transit of vehicles in the city itself. As a result, the company proposed that the complement of the "ring road" be established, so that the flow of vehicles coming from BR 163 and 364 and from MT 130 and 270 would pass as little as possible through the urban network of the city of Rondonópolis. This proposal was integrated with another one from the government of the state of Mato Grosso, which provided for the construction of another bridge over the Vermelho River. Thus, the most expensive part of the works to make this ring road viable would be the responsibility of the state government.

6.2 The Proposal for Updating the 2006 Municipal Master Plan of Rondonopolis

In the third stage of the work plan agreed with the Follow-up Management Nucleus (FMN), the contracted consulting company started to carry out the phases that constitute the formal process of revising the PDM.

Thus, the phases carried out were as follows: 1 – Reading the Municipal Reality (a diagnosis of the main problems and also potentialities that the municipality presents to be considered in its urban development plan); 2 – Development of Guidelines, Proposals, Strategies and Management Instruments to face such problems and take advantage of such potentialities; 3 – Update of the Law that establishes the new Municipal Master Plan (which summarizes the new planning proposal of the municipality), as well as of the others that establish the instruments that allow to implement this new planning and management proposal; 4 – Development of the Action Plan; 5 – Delivery of the proposal and approval during a Municipal Conference in which it is presented and debated with representatives of the entire community of the municipality.

These five phases of the third stage of the work plan were summarized in a final report that actually materializes the proposed update of the PDM/ROO/2006, which was presented and discussed with the FMN in several meetings during the end of 2019. Unfortunately, due to the COVID19 pandemic at the beginning of 2020, it was not possible to hold all the public hearings that were foreseen in the work plan, for the presentation of the drafts of bills that make up the proposal for the new PDM in Rondonópolis. Nor was it possible to hold the Municipal Conference, which is the act provided for by the Statute of Cities for the "delivery" and final discussion of the proposal with the entire society of Rondonópolis. Due to this unusual situation, the PDM/ROO/2021 proposal was only approved by the FMN at a meeting held in February 2021, thus ending the work plan that was designed to update this PDM.

Once the new proposal has been approved at the level of the municipal executive power, it should be sent by them already in the form of a draft bill to the Municipal Legislative Chamber of Rondonópolis, which, as determined by law 10.251/2001, is the body responsible for ultimately approving it as complementary municipal laws that would then come into force. However, the municipal government of Rondonópolis, in an act of total disrespect and insubordination to the Statute of Cities - federal law 10.251/2001) (Brasil, 2001), did not send the proposal approved by the FMN to the municipal legislative chamber, putting at risk all the efforts that the society of Rondonópolis did to elaborate this new proposal that effectively contains advances to guarantee the sustainable development of the municipality.

In November 2021, the Public Ministry of the State of Mato Grosso filed another action to execute the Municipality of Rondonópolis to trigger the executive process, and this action is still ongoing in the 4th Civil Court of Rondonópolis (ACP no. 1027780-71.2021.8.11.0003). On 06/28/2022, the draft of the New Master Plan was drawn up, which received the number PL 34 to trigger the legislative procedure, with no news emerging about the submission of the various other bills or the scientific technical studies carried out.

6.3 Adjustments to the Environmental Planning and Management Instruments Proposed in PDM/ROO/2021

With regard to environmental problems, proposals for adjustments to these instruments were presented with the aim of:

- **Mitigating City Land Cover Loss as It Grows**

Proposals in this regard were presented in the text of the draft laws of the Municipal Master Plan, the Municipal Environmental Code and the Subdivision of Urban Land. They are:

(a) Obligation to also include the following information as mandatory in the environmental study reports (simplified or not), of the environmental licensing request process:

  - Georeferenced location of the proposed area for the development and boundaries of conservation units and indigenous lands on an image-map produced on a large scale (at least 1:10,000), from a mosaic of aerial orthophotos with centimetric spatial resolution;
  - Spatialization also on that image-card of information about the flora species existing in the area proposed for the development, produced from *in loco* surveys with the aid of GPS receiver and camera;

(b) Obligation of the environmental license as a prerequisite for approval by any other Secretariat of the City Hall of land subdivision projects.
(c) Obligation of new proposals for expanding the urban perimeter of the city to fully comply with the determinations of Art. 42B of Law 10.251/2001 which includes a set of criteria that curb its disorderly expansion.
(d) Maintain the policy of incorporating the LRA of rural properties that become part of the urban perimeter of the city, as conservation unit zones (CUZs) and, with total restriction of any attempt to subdivide the land over them;

(e) Mandatory destination of tree fragments (of native vegetation or not), existing in the proposed areas for new allotment projects to compose the “green areas” of their public reserve.
(f) Elaboration and immediate approval of the “Municipal Master Plan for Afforestation of Roads,” following the recommendations of species diversity and use of native Cerrado species in the composition of this urban forest;
(g) Obligation for tree species to represent at least 50% of the area designated as “green areas” in the public reserves of subdivisions, and in 70% of areas approved as new public squares in the city, implanted in land belonging to the municipal executive power itself.

– **Protect the Biodiversity Contained in the Conservation Units and River Valleys of the Urban Spaces of the Municipal Territory**

It was proposed to establish buffer zones around Conservation Units located in urban spaces in strips of at least 250 m to a maximum of 500 m, depending on the size of their areas. These strips, in practice, impose restrictions on the subdivision of land around such CUs, favoring the conservation of the biodiversity they shelter.

– **Prevent the subdivision of areas of the Municipal Territory that have Environmental Vulnerability**

In order to prevent projects for new subdivisions of urban or rural land from being approved in areas that have environmental vulnerability, it made mandatory to include in the environmental study report, of the process of requesting the environmental licensing, a map depicting the location georeferenced image of the project’s direct influence area on the geotechnical charts that are now also part of the official cartography of the municipality of Rondonópolis.

– **Guarantee the protection of Water Bodies in the Municipal Territory**

In order to guarantee the protection of water bodies in the municipality, the following were proposed:

(a) Maintenance in the draft update of the Municipal Environmental Code law of the footage originally proposed for the same in 2002 in relation to the tracks of the Environmental Protection Zones (EPZs), around the rivers, springs and lakes existing in the municipal territory. Keeping the area of the EPZ strips in the draft of the new Municipal Environmental Code represents an enormous contribution to the conservation of biodiversity.
(b) The policy of removing irregular occupations (by buildings for residential or commercial/industrial purposes), from the environmental protection zones (EPZs), on the banks of urban rivers was maintained. In the case of occupations by residential dwellings, families must be relocated to houses in housing complexes built in partnership with the state and federal governments, in which part of the properties were destined to receive families found living in such a situation in the city. In the case of occupations by companies, these are obliged to demolish their irregular buildings and carry out the environmental recovery of

the EPZ under the guidance of technicians from the Municipal Environment Secretariat (SEMMA). In both cases, when an EPZ area is recovered from illegal occupation, it is fenced off and systematically monitored by environmental inspections to prevent it from being illegally occupied again.

(c) The establishment of the Rural Controlled Occupation Zone (RCOZ), in its Rural Macrozone (Fig. 21.11), defined as the area of such RCOZ, the area of catchment basins located in the northeast sector of the municipal territory, and upstream of the point where water is collected from the Vermelho River to supply the city. The Vermelho River is the current main source of surface water for the population of the city of Rondonópolis, but since its volume has been decreasing in recent years, it is necessary to protect other basins in the municipal territory with the potential to complement this supply.

As a result, the area proposed for RCOZ also includes the catchment basins of the Tadarimana and Arareau rivers in the municipal territory since they also have great potential to be surface sources for supplying the city. Effectively, in terms of protection of water resources, the establishment of this zone, which corresponds to about 1/3 of the area of the municipal territory (Fig. 21.11), represents greater planning power and environmental management to the public power in Rondonópolis, which as a result passes to have one more management instrument to prevent disorderly land occupations and uses from occurring in this area of high strategic environmental interest for the municipality.

- **Reduce city air pollution**
  (a) The bottom of the river valleys of the Vermelho and Arareau rivers are restricted zones for the construction of buildings with more than three floors in order to favor the process of natural ventilation of the air and the dispersion of atmospheric pollution in the city.
  (b) The Municipal Environment Secretariat (SEMMA) should install air quality monitoring devices in at least two points of the urban network of the city of Rondonópolis. As a result, this mini monitoring network will make it possible to monitor the temporal variation of air pollutant indices in the city, depending on the pace of industrial activities and urban traffic, in interaction with the pace of climate at the local level.

7 Final Considerations

Brazil has several technical, legal and institutional instruments developed in the context of its urban and environmental policies that have a high potential to contribute decisively to the planning and environmental management of urban and rural spaces in our cities (Angeoletto et al., 2017; Silva et al., 2017; Angeoletto et al., 2018). Here, we present the case of updating the Master Plan 2006 of the municipality of Rondonópolis-MT, which is an unprecedented experience in terms of urban

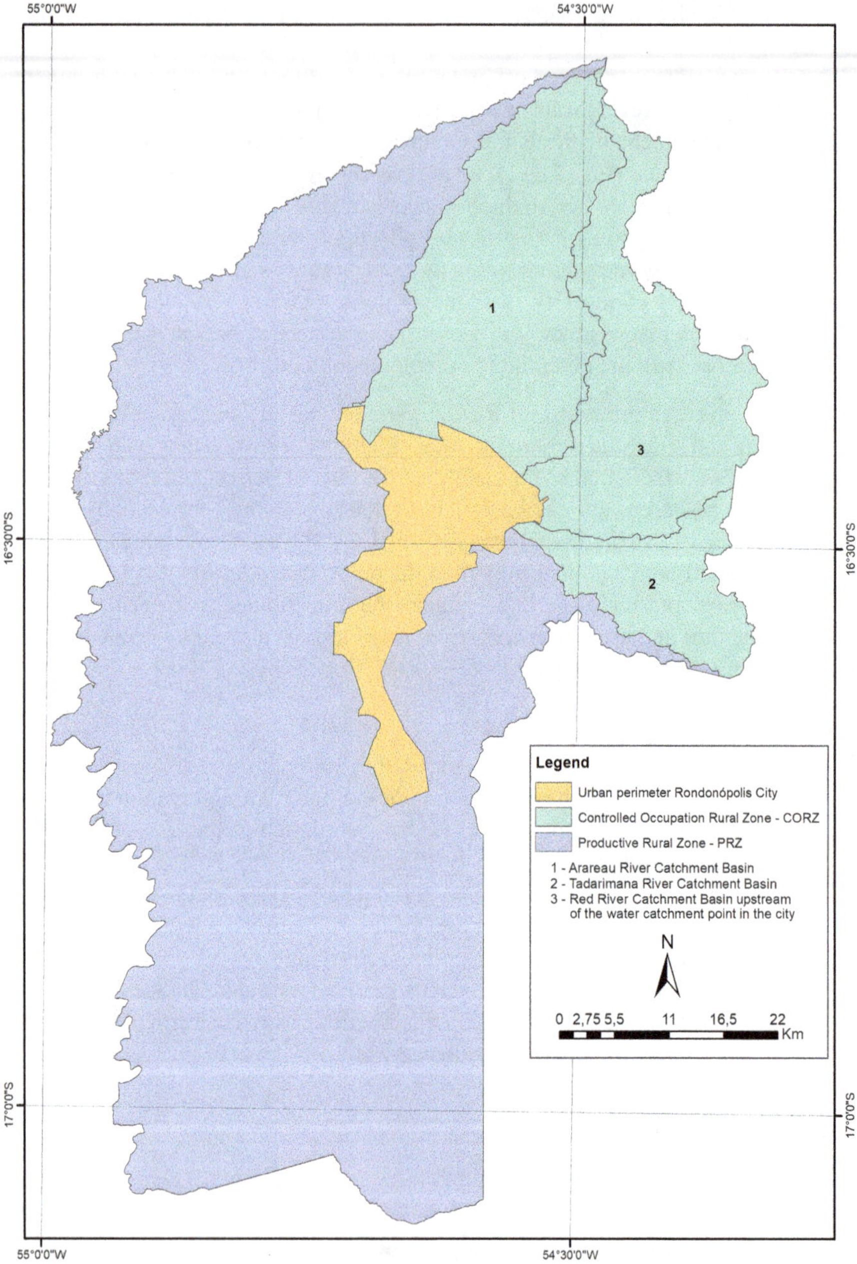

Fig. 21.11 Area of the Watersheds that make up the Rural Controlled Occupation Zone (ZOCR) in the Territory of the Municipality of Rondonópolis defined in the final proposal for the revision of its 2006 Municipal Master Plan. (Source: Rondonopolis City Hall 2021)

planning in Brazil. In addition, the municipality has several unusual characteristics of its geographic reality (logistic, economic, urban, environmental, etc.), which make it an interesting laboratory to assess how much the instruments of Brazilian public policies can really contribute to the achievement of the ideal of sustainable cities.

Regarding the ability of Brazilian urban and environmental policy to conserve biodiversity

Due to the powers made available by urban and environmental policies (at the federal, state and municipal levels), there is no doubt that there is a true arsenal of resources available for public managers in the urban and environmental planning of cities. They can guarantee positive results for nature conservation in urban spaces and, consequently, ensuring that our cities offer healthy environments both for the population and for the biodiversity they harbor. In other words, it is not due to the lack of technical, legal and institutional regulations that Brazilian cities are still the scene of intense socio-environmental problem. Instead, this is due to the lack of political commitment of managers to use them properly and also to the poor governance of the population in general, which has no real opportunity to participate in decision-making in relation to urban and environmental planning of the counties.

Regarding the efficiency of planning instruments and environmental management of urban and environmental policies in the municipality of Rondonópolis

Commenting more specifically on the reality of the municipality of Rondonópolis, the planning and management instruments made available in public policies at the federal level are also adequately incorporated into the instruments of urban and environmental policy in the municipality. Since the beginning of its history, Rondonópolis has had planning instruments that aim to guarantee the orderly growth of its urban space, as well as the conservation of its biodiversity. However, also in Rondonópolis, historically the action of capitalist groups in partnership with politicians without social commitment, resulted in urban and rural spaces with numerous socio-environmental problems that have generated incalculable losses in terms of biodiversity and quality of life for its population. The population's lack of democratic participation in decisions and actions taken in the context of municipal public policies significantly contributes to their failure to achieve their goals. This situation is so serious that not even full transparency of public power decisions/acts is guaranteed to the municipal population, a fact that significantly reduces the possibility of guaranteeing a minimum level of governance to the population and, therefore, planning is carried out more to serve the particular interests of a few rather than the collective social interests.

In Relation to the Unprecedented Experience in the Way of Conducting the Updating Works of the PDM 2006 of Rondonópolis (PDM/ROO/2006)

The experience of having a scientist at the head of the coordination of the work for preparing the proposal for updating the PDM/ROO/2006 with technical competence and job stability obtained through a public tender in an important public institution of education and research, proved to be very positive for several reasons. And one of the most important was that this condition significantly hindered lobbying from representatives of economic power on the elaboration of these public planning and management instruments. The coordination of the work to update the PDM/ROO/2006 had as one of its differentials the establishment of the obligation to make the whole process of this work, as well as the products that were generated with it, totally transparent for the population of Rondonópolis and, also, for any other Brazilian citizen who was interested in following its development. For the first time, all decisions in the updating process were taken collectively with the Public Prosecutor's Office and with the members of the follow-up management core, including the approval of the executive work schedule itself, and published on the master plan page on the City Hall website from Rondonópolis. This situation illustrates why the existing model of public management needs to be rethought in Brazil.

The proposal made it possible to update some of the instruments that the municipality already had, such as the environmental code, the urban and rural land subdivision laws of the municipality, the urban and environmental zoning of the municipal territory, among others, which had outdated norms and were insufficient to guarantee the planning and management of the municipal territory. The Municipal Environmental Code was updated so that it harmonized with some definitions established in the law of the National System of Nature Conservation Units (Law 9.985/2000) and, above all, make suggestions for adjustments for the creation of green areas in the new subdivisions.

To finalize these considerations, the following questions are asked by way of reflection on the outcome of the case of updating the PDM of Rondonópolis:

1. Why did the 2020–2024 administration of the Municipality of Rondonópolis only send the PDM/ROO/2021 proposal to the City Council (actually only a small part of it), after a new action was filed by the Public Ministry and determined by the Judiciary?
2. Could it be that this municipal management and its advisory team (CODEUR), do not want this proposal to be implemented in the municipality because they consider that it is impossible to reconcile economic growth with biodiversity conservation?
3. Or is it due to the fact that even though it was an alternative proposal elaborated by academics and not by politicians which is the usual procedure, it presents several ways to properly use the instruments made available by Brazilian public policies in the proposition of concrete strategies of urban planning and environmental management in the territory of the municipality of Rondonópolis?

- **Final Recommendations**

In light of the above considerations and, based on the experience acquired during the years of carrying out studies and inter-institutional articulations, we believe it is important to make some recommendations so that public policies can achieve more consistent results in the sense of producing truly sustainable urban spaces in socio-environmental and ecological terms:

1. **Establishment of Partnerships for Updating Territorial Planning Instruments**

The first recommendation is that the municipal public authorities establish as a rule, whenever possible, the establishment of agreements between municipal governments and research institutions (preferably public), for carrying out processes for updating their Municipal Master Plans. This recommendation is even more emphatic in the case of small municipalities, which in Brazil represent around 70% of the 5570 municipalities in the country according to estimates of the Brazilian population in 2021 prepared by the IBGE.

2. **Obligation to update the Official Cartography of the Municipality and Preparation of Previous Technical Studies**

The second recommendation is that regardless of the size of the municipality, it must have an updated official cartographic base and 100% public access to its entire population. This cartographic base must be composed of a broad set of thematic mappings that are able to provide up-to-date information, produced in adequate cartographic scales, on social, economic and environmental aspects of the urban and rural spaces of the municipal territory. These maps are fundamental in describing the reality of the municipal territory and, thus, allow scientists, public managers, and the common citizen themselves to effectively know what the real problems of the municipality are and what instruments of planning and environmental management Brazilian public policies have to face them.

3. **Adoption of More Effective Rules in terms of Environmental Conservation for the Approval of Urban Allotment Projects**

The third recommendation concerns how to use the instrument of environmental licensing in the context of municipal environmental policy. As evidenced in this text, the "loosening" of the use of this instrument in the case of environmental licensing of projects considered to have "low environmental impact" may have a deleterious result on the conservation of local biodiversity. Thus, it is recommended that the size of the enterprise not be the only criterion to be considered to determine the level of potential environmental impact of urban land subdivision projects. But also consider: (a) The natural socio-environmental vulnerability of the area where it is intended to install them; (b) The use of the soil in the area immediately adjacent to it, as this could also determine, for various reasons, the impediment of its operation in the location indicated in its application for environmental licensing; (c) That the environmental characterization of the development area be carried out with the support of mappings produced on a multi-scale basis from updated images (satellite and aerial photos), as well as on-site surveys by multidisciplinary teams capable of identifying and recording the occurrence of species of fauna and flora of the biome in question, especially native species and those considered threatened.

It is believed that the discussion of the case of updating the Municipal Master Plan of Rondonópolis provides a novel example of why it is so difficult to reconcile economic growth and conservation of biodiversity in cities, despite the various planning instruments and environmental management made available by Brazilian public policies. If we are to finally achieve the ideal of sustainable cities, we must ensure that planning and environmental management regulations are appropriately and independently implemented. Brazilian cities could be a home for biodiversity in the face of increased habitat destruction, for the benefit of nature and of people, but only if existing rules are followed.

Acknowledgments We would like to thank Professor Fabio Angeoletto, a pioneer of urban ecology in Brazil. His seminal articles and ideas have been an inspiration to us, scientists committed to conserving biodiversity in Brazilian cities.

References

Agra Filho, S. S. (2021). *Licenciamento ambiental no Brasil/Severino Soares Agra Filho*. EdUFBA.

ANA (2021) - https://metadados.snirh.gov.br/geonetwork/srv/api/records/7cfd53c4-b4e1-4aba-a79b-857a19649df6 (limites paises Am do Sul).

Angeoletto, F. (2012). Planeta ciudad: ecología urbana y planificación de ciudades medias de Brasil. Tesis Doctoral. Universidad Autónoma de Madrid, España.

Angeoletto, F., Santos, J. W. M. C., Sanz, J. P. R., Silva, F. F. d., & Albertin, R. M. (2016). Tipología socio-ambiental de las ciudades medias de Brasil: aportes para un desarrollo urbano sostenible. *Urbe. Revista Brasileira de Gestão Urbana, 8*, 272–287.

Angeoletto, F., Sanz, J. P. R., Albertin, R. M., & Silva, F. F. D. (2017). The grass is always greener on the other side of the fence: the flora in urban backyards of different social classes. *Ambiente & Sociedade, 20*(01), 01–20.

Angeoletto, F., Fellowes, M. D., & Santos, J. W. (2018). Counting Brazil's urban trees will help make Brazil's urban trees count. *Journal of Forestry, 116*(5), 489–490.

Angeoletto, F., Leandro, D., & da Silva; Fellowes, Mark de. (2019). The consequences of Brazil's lack of transport planning is written in the blood of sparrows. *Urban Geography, 40*(8), 1191–1197.

Angeoletto, F., Fellowes, M. D., Essi, L., Santos, J. W., Johann, J. M., da Silva Leandro, D., & Mendonca, N. M. (2019b). Ecología urbana y planificación: una convergéncia ineludible. Revista Electrônica em Gestão, Educação e Tecnologia Ambiental, 23(1), 1–8.

Angeoletto, F., Tryjanowski, P., Santos, J., Martinez-Miranzo, B., Leandro, D., Bohrer, J., et al. (2022). Will Brazilian city dwellers actively engage in urban conservation? A case study with the charismatic neotropical Blue-and-yellow macaw (Ara ararauna). *Birds, 3*(2), 234–244.

Angeoletto, F. H. S., Richard, E., Santos, J., Guedes, N., Oliveira, S. L., Albertín, R. M., et al. (2023). When we think of urban afforestation in Brazil, what comes to mind is a mouth that has lost its teeth. *Terr@Plural, 17*, e2321654.

Antonello, I. T. (2013). Potencialidade do planejamento participativo no Brasil. *Sociedade & Natureza, 25*, 239–254.

Araújo, S. M. V. G., & Gannem, R. S. (2006). *Conflitos em relação ao conceito de espaços territoriais especialmente protegidos*. Disponível em: https://www.terrabrasilis.org.br/ecotecadigital/images/conflitos.pdf. Acesso em: 03 Junho 2022.

Araújo, S. M. V. G. d., Guimarães, P. C. V., & Fazzolari-Corrêa, S. (2013). Licenciamento Ambiental: base normativa e perspectivas. In D. M. d Motta, & B. Pêgo Filho (Orgs). *Licenciamento ambiental para o desenvolvimento urbano: avaliação de instrumentos e procedimentos* (pp. 61–87).

Brasil. Lei 4.504 de 30 de novembro de 1964. *Dispõe sobre o Estatuto da Terra, e dá outras providências*. Disponível em: https://www.planalto.gov.br/ccivil_03/leis/l4504.htm. Acesso em: 30 de dezembro de 2020.

Brasil. Lei n° 6.766, de 19 de dezembro de 1979. *Dispõe sobre o Parcelamento do Solo Urbano e dá outras Providências*. Disponível em: https://www.planalto.gov.br/ccivil_03/Leis/L6766.htm. Acesso em: 10 de outubro de 2021.

Brasil. Lei n° 6.938, de 31 de agosto de 1981. *Dispõe sobre a Política Nacional do Meio Ambiente, seus fins e mecanismos de formulação e aplicação, e dá outras providências*. Disponível em: http://www.planalto.gov.br/ccivil_03/Leis/L6938.htm. Acesso em: 05 março 2022.

Brasil. Constituição (1988). *texto constitucional promulgado em 5 de outubro de 1988, com as alterações determinadas pelas Emendas Constitucionais de Revisão nos 1 a 6/94, pelas Emendas Constitucionais nos 1/92 a 91/2016 e pelo Decreto Legislativo no 186/2008*. – Brasília: Senado Federal, Coordenação de Edições Técnicas, 2016. Disponível em: https://www2.senado.leg.br/bdsf/bitstream/handle/id/518231/CF88_Livro_EC91_2016.pdf. Acesso em: 05 março 2022.

Brasil. Lei 7.804 de 17 de julho de 1989. *Altera a Lei n° 6.938, de 31 de agosto de 1981, que dispõe sobre a Política Nacional do Meio Ambiente, seus fins e mecanismos de formulação e aplicação, a Lei n° 7.735, de 22 de fevereiro de 1989, a Lei n° 6.803, de 2 de julho de 1980, e dá outras providências*. http://www.planalto.gov.br/ccivil_03/leis/l7804.htm. Acesso em: 05 março 2022.

Brasil. Lei n° 9.985, de 18 de julho de 2000. *Regulamenta o art. 225, § 1o, incisos I, II, III e VII da Constituição Federal, institui o Sistema Nacional de Unidades de Conservação da Natureza e dá outras providências*. https://www.planalto.gov.br/ccivil_03/leis/l9985.htm. Acesso em: 27 junho 2022.

Brasil. Lei n°. 10.257, de 10 de julho de 2001 *(Estatuto da Cidade), que "regulamenta os arts. 182 e 183 da Constituição Federal, estabelece as diretrizes gerais da política urbana e dá outras providências"*. Disponivel em: http://www.planalto.gov.br/ccivil_03/leis/leis_2001/l10257.htm. Acesso em: 05 março 2022.

Brasil. Decreto 4.297 de 10 de julho de 2002. *Regulamenta o art. 9°, inciso II, da Lei no 6.938, de 31 de agosto de 1981, estabelecendo critérios para o Zoneamento Ecológico-Econômico do Brasil – ZEE, e dá outras providências*. Disponível em: http://www.planalto.gov.br/ccivil_03/decreto/2002/d4297. htm Acesso em: 10 abril 2021.

Brasil. Lei Complementar n° 140, de 25 de maio de 2012. *Fixa normas, nos termos dos incisos III, VI e VII do caput e do parágrafo único do art. 23 da Constituição Federal, para a cooperação entre a União, os Estados, o Distrito Federal e os Municípios nas ações administrativas decorrentes do exercício da competência comum relativas à proteção das paisagens naturais notáveis, à proteção do meio ambiente, ao combate à poluição em qualquer de suas formas e à preservação das florestas, da fauna e da flora; e altera a Lei no 6.938, de 31 de agosto de 1981*. Disponível em https://www.planalto.gov.br/ccivil_03/leis/lcp/lcp140.htm. Acesso em: 18 maio 2023.

Brasil. Lei n° 12.651, de 25 de maio de 2012. *Dispõe sobre a proteção da vegetação nativa; altera as Leis n°s 6.938, de 31 de agosto de 1981, 9.393, de 19 de dezembro de 1996, e 11.428, de 22 de dezembro de 2006; revoga as Leis n°s 4.771, de 15 de setembro de 1965, e 7.754, de 14 de abril de 1989, e a Medida Provisória n° 2.166–67, de 24 de agosto de 2001; e dá outras providências*. https://www.planalto.gov.br/ccivil_03/_ato2011-2014/2012/lei/l12651.htm. Acesso em: 11 abril 2022.

Carvalho, P. F. d., & Braga, R. (orgs.) (2001). Perspectivas de Gestão Ambiental em Cidades Médias. LPM-UNESP pp. 111 a 119.

Carlos, A. F. A. et al. (2015). A crise urbana e os caminhos de construção de uma teoria crítica. Anais ENANPUR, v. 16, n. 1.

Conama, R. (1986). n. 01/86, de 23 de janeiro de 1986-Licenciamento ambiental–Normas e procedimentos. Publicada no DOU, de, v. 17, pp. 2548–2549.

Conama, R. (1997). n. 237/97. Regula o procedimento de Licenciamento Ambiental.

Costa, R. G. S. D., & Ferreira, C. C. M. (2009). Análise do índice de áreas verdes (IAV) na área central da cidade de Juiz de Fora. *Revista Brasileira de Arborização Urbana, Piracicaba, 4*(1), 39–57.

Demamann, M. (2011). Terezinha Mundt. Rondonópolis-MT: campo, cidade e centralidades. Tese de Doutorado. Universidade de São Paulo. Disponível em: https://www.teses.usp.br/teses/disponiveis/8/8136/tde-23042012-134329/en.php. Acesso em 20/05/2021.

Duarte, T. E. P., Angeoletto, F. H. S., Santos, J. W. M. C., da Silva Leandro, D., Bohrer, J. F. C., Vacchiano, M. C., & Leite, L. B. (2017). O papel da cobertura vegetal nos ambientes urbanos e sua influência na qualidade de vida nas cidades. *Desenvolvimento em questão, 15*(40), 175–203.

Elias, D., & Pequeno, R. (2007). Desigualdades socioespaciais nas cidades do agronegócio. *Revista Brasileira de Estudos Urbanos e Regionais (RBEUR), 9*(1), 25–39. Disponível em: https://www.redalyc.org/pdf/5139/513951695003.pdf. Acesso em 10/10/2022.

Federal, S. (2001). *Constituição da República Federativa do Brasil de 5 de outubro de 1988*. Senado Federal.

Fernandes, M. E. B. V. (2021). *Espacialização de áreas verdes públicas e sua relação socioeconômica em Recife – PE*. 37 f. Trabalho de Conclusão de Curso (Bacharelado em Engenharia Florestal) – Departamento de Ciência Florestal, Universidade Federal Rural de Pernambuco, Recife.

Fernandez, A. C. F. (2016). O sertão formal da política brasileira de conservação da natureza. Revista Brasileira de Ciência. *Política*, 165–204.

Fonseca, M. F., & Fonseca, M. L. (2013). Planos diretores municipais e o zoneamento urbano: uma análise da (des) articulação entre os instrumentos de planejamento territorial no entorno do reservatório de Salto Grande (SP). Sociedade & Natureza, v. 25, 61–74.

Frederico, S. (2011). As cidades do agronegócio na fronteira agrícola moderna brasileira. *Caderno Prudentino de Geografia, 1*(33), 5–23. Disponível em: https://revista.fct.unesp.br/index.php/cpg/article/view/1933. Acesso em 02/07/2022.

Gonzaga, C. A. C., Roquette, J. G., Angeoletto, F., Silgueiro, V. D. F., Macedo, L. O. B., Valdiones, A. P., & da Silva, N. M. (2024). The Public Prosecutor's Office's experience using Global Forest Watch to monitor and deter deforestation in the Cerrado. *Environmental Conservation*, 1–6.

Harder, I. C. F., Ribeiro, R. d. C. S., & Tavares, A. R. (2006). Índices de área verde e cobertura vegetal para as praças do município de Vinhedo, SP. *Revista Árvore, 30*, 277–282.

IBGE (2021) - https://www.ibge.gov.br/geociencias/organizacao-do-territorio/malhas-territoriais/15774-malhas.html?=&t=downloads (limites UFs do Brasil).

IBGE. (2022). *Pesquisa de Informações Básicas Municipais—Perfil dos Municípios Brasileiros 2021*. Instituto Brasileiro de Geografia e Estatística.

Leite, L. B. (2018). Flora Arbórea de Rondonópolis: Um Estudo Ecológico Urbano. Rondonópolis. Dissertação (Mestrado) Universidade Federal de Mato Grosso, Programa de Pós-Graduação em Geografia. Rondonópolis.

Lima, V. (2014). Análise da qualidade ambiental urbana: o exemplo de Osvaldo Cruz/SP. *Geografia em Questão, 7*(2).

Lima, G. F., Correa Santos, J. W. M., Albertin, R. M., Martínez-Miranzo, B., Souza, F. L., & Angeoletto, F. (2023). Backyards are a way to promote environmental justice and biodiversity conservation in Brazilian cities. *Diversity, 15*(7), 815.

Lopes, D. M. F., & Henrique, W. (Eds.). (2010). *Cidades médias e pequenas: teorias, conceitos e estudos de caso*. Publicações SEI.

Louzada, R. M. (2022). *Análise da qualidade ambiental urbana em Álvares Machado-SP*. Tese (Doutorado em Geografia)—Programa de Pós-Graduação em Geografia—UNESP, Presidente Prudente.

Macedo, L. O. B., et al. (2015). Uma Análise Prospectiva dos Efeitos da Implantação do Complexo Intermodal da Ferronorte ao Município de Rondonópolis-MT. *Revista de Estudos Sociais, 17*(33), 51–71.

Machado, R. R. B., Pereira, E. C. G., & Andrade, L. d. H. C. (2010). Evolução temporal (2000–2006) da cobertura vegetal na zona urbana do município de Teresina–Piauí–Brasil. *Revista da Sociedade Brasileira de Arborização Urbana, 5*(3), 97–112.

Mello-Théry, N. A. d. (2011). Conservação de áreas naturais em São Paulo. *Estudos avançados, 25*, 175–188.

MMA (2020) - http://mapas.mma.gov.br/i3geo/datadownload.htm (Limite de Biomas do Brasil).

Motta, D. M. da, & Pêgo Filho, B. (Orgs). (2013). *Licenciamento ambiental para o desenvolvimento urbano: avaliação de instrumentos e procedimentos.*

MPF – Ministério Público Federal. (2004). *Deficiências em Estudos de Impacto Ambiental: síntese de uma experiência.* Ministério Público Federal – 4ª Câmara de Coordenação e Revisão.

Nazareth, P. A. (2018). Planos diretores e instrumentos de gestão urbana e ambiental no Estado do Rio de Janeiro. *Revista Do Serviço Público Brasília, 69*(1), 211–240.

Negri, S. M. (2008). *O processo de segregação sócio-espacial no contexto do desenvolvimento econômico da cidade de Rondonópolis-MT.* 180f. Tese (doutorado) – Universidade Estadual Paulista, Instituto de Geociências e Ciências Exatas.

Negri, S. M. (2010). *Uso desigual do território em Rondonópolis no processo de expansão do agronegócio da soja em Mato Grosso.* 176 f. Tese (doutorado) – Universidade Estadual Paulista, Instituto de Geociências e Ciências Exatas. Disponível em: http://hdl.handle.net/11449/104315

Neves, E. M. S. C. (2006). A política ambiental e os municípios brasileiros. Tese (Doutorado) Universidade Federal Rural do Rio de Janeiro, Instituto de Ciências Humanas e Sociais.

Oliveira, C. M. d. (2020). *O Potencial das manchas de habitat preservadas no espaço urbano, periurbano e rural de Rondonópolis-MT para conservação da biodiversidade.* 142f. Dissertação (Mestrado em Geografia) PPGEO – Universidade Federal de Mato Grosso.

Peixinho, D. M. (1998). *Onças vermelhas e amarelas: a ocupação dos cerrados e a dinâmica sócio-espacial em Rondonópolis-MT.* Dissertação (Mestrado) Universidade de São Paulo, Faculdade de Filosofia, Letras e Ciências Humanas, São Paulo.

Pereira, E. S. (2009). *Expansão urbana e impactos ambientais no perímetro urbano de Rondonópolis—MT: uma comparação da situação da cobertura vegetal e das áreas de proteção ambiental nos anos de 1986 e 2008.* Monografia (Especialização em Educação Ambiental)—Universidade Federal de Mato Grosso.

Pina, José Hermano Almeida et al. A influência das áreas verdes urbanas na qualidade de vida: o caso dos Parques do Sabiá e Victório Siquierolli em Uberlândia-MG. 2011.

Pizella, D. (2015). A relação entre Planos Diretores Municipais e Planos de Bacias Hidrográficas na gestão hídrica. Revista Ambiente & Água, v. 10, n. 3, 635–645.

Poubel, I., Campos, F. (2019). Conflitos territoriais entre o zoneamento urbano municipal e as diretrizes ambientais: O caso do Corredor Ecológico Duas Bocas–Mestre Álvaro, Grande Vitória–ES, Brasil. GOT: Revista de Geografia e Ordenamento do Território, n. 18, p. 57–61.

Queiroga, A. E. d. (2002). *Os parcelamentos ilegais do solo e a desapropriação como sanção: o caso dos "condomínios irregulares" no Distrito Federal.* Sergio Antonio Fabris Editor.

Rondonópolis. (1990). *Lei Orgânica do Município.* Promulgada em 05 de maio de 1990.

Rondonópolis. Lei 2.120 de 14 de março de 1994. Institui o Parcelamento Urbano do Município de Rondonópolis, e dá outras providências.

Rondonópolis. Lei Complementar n°12 de 30 de dezembro de 2002. Institui o Código Ambiental de Rondonópolis e dá outras providências.

Rondonópolis – Prefeitura Municipal de Rondonópolis/Programa de Pós-Graduação em Geografia/UFMT. (2017a). *Produção de Mapeamentos Temáticos para a Fase de Diagnóstico do Processo de Atualização do PDM/ROO/2006 - Relatório 03: Mapeamentos Produzidos em 1:10.000* (157 p).

Rondonópolis – Prefeitura Municipal de Rondonópolis/Programa de Pós-Graduação em Geografia/UFMT. (2017b). *Produção de Mapeamentos Temáticos para a Fase de Diagnóstico do Processo de Atualização do PDM/ROO/2006 - Relatório 02: Mapeamentos Produzidos em 1:50.000* (108 p). UFMT.

Rumble, H., Angeoletto, F., Connop, S., Goddard, M. A., & Nash, C. (2019). Understanding and applying ecological principles in cities. Planning cities with nature: Theories, strategies and methods, 217–234.

Santangelo, J. S., Ness, R. W., Cohan, B., Fitzpatrick, C. R., Innes, S. G., Koch, S., et al. (2022). Global urban environmental change drives adaptation in white clover. *Science, 375*(6586), 1275–1281.

Santos, J. W. M. C. (2002). *Clima e produtividade da soja nas terras de Cerrado do sudeste de Mato Grosso* (388 p). Tese (doutorado)—Universidade de São Paulo, Faculdade de Filosofia, Letras e Ciências Humanas.

Santos, M. (2005). *Da Totalidade ao Lugar*. EDUSP.

Santos, J. W. M. C. (Ed.). (2010). *Produção do espaço e transformações socioambientais das paisagens do Mato Grosso* (203 p). EdUFMT.

Santos, J. W. M. C., & Oliveira, H. O. d. (2019). Perfil Socioeconômico das Empresas Situadas dentro da Área da APP do Rio Arareau na Área Urbana de Rondonópolis-MT. In *ANAIS DO XIX SIMPÓSIO BRASILEIRO DE SENSORIAMENTO REMOTO, 2019*. Anais eletrônicos… São José dos Campos, INPE. Disponível em: https://proceedings.science/sbsr-2019/papers/perfil-socioeconomico-das-empresas-situadas-dentro-da-area-da-app-do-rio-arareau-na-area-urbana-de-rondonopolis-mt?lang=pt-br. Acesso em: 07 abr. 2023.

Santos, J. W. M. C., & Rubim, I. G. (2019). Perfil Socioeconômico das Famílias que Residem Dentro da Área da APP do Rio Arareau em Rondonópolis-MT. In *ANAIS DO XIX SIMPÓSIO BRASILEIRO DE SENSORIAMENTO REMOTO, 2019*. Anais eletrônicos Santos:Instituto Nacional de Pesquisas Espaciais. Disponível em: https://proceedings.science/sbsr-2019/papers/perfil-socioeconomico-das-familias-que-residem-dentro-da-area-da-app-do-rio-arareau-em-rondonopolis-mt?lang=pt-br. Acesso em: 07 abr. 2023.

Santos, J. W. M. C., Loverde-Oliveira, S., & Angeoletto, F. (2016). Diagnóstico socio-ambiental de las áreas de preservación permanente del río Arareau en la zona urbana de Rondonópolis, MT. *Terr@Plural, 10*(1), 35–45.

Santos, J. W. M. C., et al. (2019). Diagnóstico de impactos na Área de Preservação Permanente do arroio urbano Arareau, Rondonópolis, Brasil. *Revista Geográfica Venezolana, 60*(2), 300–312.

da Silva, E. B., Anjos, A. F dos. (2010). O monitoramento do desmatamento e as ações de conservação do bioma Cerrado na primeira década do século XXI. In: PELÁ, M.; CASTILHO, D. (Org.). Cerrados: perspectivas e olhares. Goiânia: Vieira (pp. 71–91).

Silva, N. M., Angeoletto, F., Paranhos Filho, A. C., Vacchiano, M. C., Bohrer, J. F., & Candido, A. K. A. (2017). The negative influences of the new Brazilian forest code on the conservation of riparian forests. *European Journal of Ecology, 3*(2), 116–122.

Souza, M. L. (2010). *Mudar a Cidade: uma introdução crítica ao planejamento e à gestão urbanos* (6th ed.). Bertrand Brasil.

Souza, M. A. M. d. (2017). *Análise da política de criação, implementação e conservação dos espaços territoriais especialmente protegidos em Rondonópolis/MT*. Dissertação (Mestrado em Geografia) UFMT, Rondonópolis, MT.

Souza, F. L., Fenoglio, M. S., & Angeoletto, F. (2023). To be a Brazilian city dweller, sometimes we must learn to say enough! *Sustainability, 15*(4), 3699.

Suzuki, J. C. (1997). *De povoado a cidade: a transição do rural ao urbano em Rondonópolis*. Dissertação (Mestrado) – Universidade de São Paulo, São Paulo.

Tesoro, L. L. L. M., & Mesgravis, L. (1993). *Rondonópolis-MT: um entroncamento de mão única; o processo de povoamento e de crescimento de Rondonópolis na visão dos pioneiros, 1902–1980*. Tese (Doutorado) - Universidade de São Paulo, São Paulo.

Trassi, J. d. S., Santos, J. W. M. C., & Negri, S. S. (2020). Inundações E Repercussões Socioambientais na Cidade de Rondonópolis, MT, Brasil. *Terr@Plural, 14*, 1–21.

United Nations, Department of Economic and Social Affairs, Population Division. (2019). *World urbanization prospects: The 2018 revision (ST/ESA/SER.A/420)*. United Nations.

Urbaniza Engenharia Consultiva. (2017). *Relatório Técnico – Estudos de Mobilidade e Circulação Urbana e Rural: Modelagens e Propostas para a Mobilidade Urbana e Rural.* Rondonópolis. 93p.

Vacchiano, M. C., Santos, J. W. M. C., Angeoletto, F., & Silva, N. M. (2019). Do data support claims that Brazil leads the world in environmental preservation? *Environmental Conservation, 46*(2), 118–120.

Villaça, F. et. al. (1999). Dilemas do plano diretor. CEPAM. O município no século XXI: cenários e perspectivas. São Paulo: Fundação Prefeito Faria Lima–CEPAM, p. 237–247.

Index

F. Angeoletto et al. (eds.), *Ecology of Tropical Cities*,
https://doi.org/10.1007/978-3-031-48012-6

GPSR Compliance
The European Union's (EU) General Product Safety Regulation (GPSR) is a set of rules that requires consumer products to be safe and our obligations to ensure this.

If you have any concerns about our products, you can contact us on

ProductSafety@springernature.com

In case Publisher is established outside the EU, the EU authorized representative is:

Springer Nature Customer Service Center GmbH
Europaplatz 3
69115 Heidelberg, Germany

www.ingramcontent.com/pod-product-compliance
Lightning Source LLC
Chambersburg PA
CBHW071632020725
29038CB00032B/30
9783031480119